The Handbook of Reliability, Maintenance, and System Safety through Mathematical Modeling

The Handbook of Reliability, Maintenance, and System Safety through Mathematical Modeling

Edited by

Amit Kumar
Department of Mathematics, Lovely Professional University, Phagwara, Punjab, India

Mangey Ram
Department of Mathematics; Computer Science & Engineering, Graphic Era Deemed to be University, Dehradun, Uttarakhand, India

ELSEVIER

ACADEMIC PRESS
An imprint of Elsevier

Academic Press is an imprint of Elsevier
125 London Wall, London EC2Y 5AS, United Kingdom
525 B Street, Suite 1650, San Diego, CA 92101, United States
50 Hampshire Street, 5th Floor, Cambridge, MA 02139, United States
The Boulevard, Langford Lane, Kidlington, Oxford OX5 1GB, United Kingdom

Notices

Knowledge and best practice in this field are constantly changing. As new research and experience broaden our understanding, changes in research methods, professional practices, or medical treatment may become necessary.

Practitioners and researchers must always rely on their own experience and knowledge in evaluating and using any information, methods, compounds, or experiments described herein. In using such information or methods they should be mindful of their own safety and the safety of others, including parties for whom they have a professional responsibility.

To the fullest extent of the law, neither the Publisher nor the authors, contributors, or editors, assume any liability for any injury and/or damage to persons or property as a matter of products liability, negligence or otherwise, or from any use or operation of any methods, products, instructions, or ideas contained in the material herein.

Library of Congress Cataloging-in-Publication Data
A catalog record for this book is available from the Library of Congress

British Library Cataloguing-in-Publication Data
A catalogue record for this book is available from the British Library

ISBN: 978-0-12-819582-6

For information on all Academic Press publications visit our website at https://www.elsevier.com/books-and-journals

Publisher: Candice Janco
Editorial Project Manager: Sara Valentino
Production Project Manager: Kumar Anbazhagan
Cover Designer: Miles Hitchen

Typeset by TNQ Technologies

Contents

Contributors

Adarsh Anand
Department of Operational Research, University of Delhi, Delhi, India

Igor Bolvashenkov
Institute of Energy Conversion Technology, Technical University of Munich, Munich, Germany

Dejan Lj. Branković
Department of Hydro and Thermal Engineering, University of Banja Luka, Faculty of Mechanical Engineering, Banja Luka, Republic of Srpska, Bosnia and Herzegovina

Svetlana R. Dumonjić-Milovanović
Partner Engineering Ltd., Banja Luka, Republic of Srpska, Bosnia and Herzegovina

Jerzy K. Filus
Department of Mathematics and Computer Science, Oakton Community College, Des Plaines, IL, United States

Lidia Z. Filus
Department of Mathematics, Northeastern Illinois University, Chicago, IL, United States

Ilia Frenkel
Center for Reliability and Risk Management, SCE-Shamoon College of Engineering, Beer Sheva, Israel

Shen Guixiang
School of Mechanical and Aerospace Engineering, Jilin University, Changchun, China

Vandana Gupta
Department of Operational Research, University of Delhi, Delhi, India

Hans-Georg Herzog
Institute of Energy Conversion Technology, Technical University of Munich, Munich, Germany

Valentina Z. Janičić Milovanović
Routing Ltd., Banja Luka, Republic of Srpska, Bosnia and Herzegovina

Jörg Kammermann
Institute of Energy Conversion Technology, Technical University of Munich, Munich, Germany

Amisha Khati
Department of Mathematics, Statistics and Computer Science, G.B. Pant University of Agriculture and Technology, Pantnagar, Uttarakhand, India

Pardeep Kumar
Department of Mathematics, Lovely Professional University, Phagwara, Punjab, India

Akshay Kumar
Department of Mathematics, Graphic Era Hill University, Dehradun, Uttarakhand, India

Amit Kumar
Department of Mathematics, Lovely Professional University, Phagwara, Punjab, India

Amit Kumar
Department of Mathematics, Birla Institute of Technology and Science Pilani, Pilani Campus, Pilani, Rajasthan, India

Dinesh Kumar Kushwaha
Department of Industrial and Production Engineering, Dr. B. R. Ambedkar National Institute of Technology, Jalandhar, Punjab, India

K.C. Lalropuia
Department of Operational Research, University of Delhi, Delhi, India

Zdravko N. Milovanović
Department of Hydro and Thermal Engineering, University of Banja Luka, Faculty of Mechanical Engineering, Banja Luka, Republic of Srpska, Bosnia and Herzegovina

Snježana Z. Milovanović
Department of Materials and Structures, University of Banja Luka, Faculty of Architecture, Civil Engineering and Geodesy, Banja Luka, Republic of Srpska, Bosnia and Herzegovina

Kuldeep Nagiya
Department of Mathematics, S.K.I.C. (U.P. Secondary Education Board, Prayagraj), Aligarh, Uttar Pradesh, India

Dilbagh Panchal
Department of Industrial and Production Engineering, Dr. B. R. Ambedkar National Institute of Technology, Jalandhar, Punjab, India

Ljubiša R. Papić
DQM Research Center, Čačak, Serbia

Mangey Ram
Department of Mathematics; Computer Science & Engineering, Graphic Era Deemed to be University, Dehradun, Uttarakhand, India

Anish Sachdeva
Department of Industrial and Production Engineering, Dr. B. R. Ambedkar National Institute of Technology, Jalandhar, Punjab, India

Chandra Shekhar
Department of Mathematics, Birla Institute of Technology and Science Pilani, Pilani Campus, Pilani, Rajasthan, India

S.B. Singh
Department of Mathematics, Statistics and Computer Science, G.B. Pant University of Agriculture and Technology, Pantnagar, Uttarakhand, India

Panagiotis H. Tsarouhas
International Hellenic University, Department of Supply Chain Management (Logistics), Katerini, Greece

Shreekant Varshney
Department of Mathematics, ICFAITech, Faculty of Science and Technology, ICFAI Foundation for Higher Education (IFHE), Hyderabad, India

Zeng Wenbin
Institute of Energy Conversion Technology, Technical University of Munich, Munich, Germany; School of Mechanical and Aerospace Engineering, Jilin University, Changchun, China

Editors' Biographies

Dr. Amit Kumar works as an assistant professor in the Department of Mathematics at Lovely Professional University, Punjab, India. He has taught several core courses in pure and applied mathematics at the undergraduate and postgraduate levels. He earned his bachelor's and master's degrees from Chaudhary Charan Singh University, Meerut, India, in 2006 and 2009, respectively. In 2016, he completed his doctorate in applied mathematics from Graphic Era (deemed to be a university), Dehradun, Uttarakhand, India, in the field of reliability theory. He published several research papers and book chapters in various esteemed international journals and books from publishers including Taylor & Francis, Springer, Emerald, World Scientific, and Inderscience, and in many national and international journals of repute. He also presented his works at national and international conferences. He is a reviewer of many international journals including those published by Elsevier, Springer, Emerald, John Wiley, and Taylor & Francis. His fields of research are operations research, reliability theory, fuzzy reliability, and system engineering. He is a lifetime member of the Indian Science Congress. He received the ***Research Appreciation Award*** for his research contribution for 2017 from Lovely Professional University, Punjab, India.

Dr. Mangey Ram received a doctoral degree with a major in mathematics and a minor in computer science from G. B. Pant University of Agriculture and Technology, Pantnagar, India. He has been a faculty member for approximately 12 years and has taught several core courses in pure and applied mathematics at the undergraduate, postgraduate, and doctorate levels. He is currently a research professor at Graphic Era (deemed to be a university), Dehradun, India. Before joining Graphic Era, he was briefly a deputy manager (probationary officer) with Syndicate Bank. He is editor in chief of the *International Journal of Mathematical, Engineering, and Management Sciences* and the guest editor and member of the editorial board of various journals. He is a regular reviewer for international journals, including those published by IEEE, Elsevier, Springer, Emerald, John Wiley, and Taylor & Francis. He has published more than 200 research publications by IEEE, Taylor & Francis, Springer, Elsevier, Emerald, World Scientific, and many other national and international journals of repute, and also presented his works at national and international conferences. His fields of research are reliability theory and applied mathematics. Dr. Ram is a senior member of the IEEE; a life member of the Operational Research Society of India, Society for Reliability Engineering, Quality and Operations Management in India, and Indian Society of Industrial and Applied Mathematics; and a member of the International Association of Engineers in Hong Kong and the Emerald Literati Network in the United Kingdom. He has been a member of the organizing committees of a number of international and national conferences, seminars, and workshops. He was bestowed the Young Scientist Award by the Uttarakhand State Council for Science and Technology, Dehradun, in 2009. He was awarded the Best Faculty Award in 2011, the Research Excellence Award in 2015, and the Outstanding Researcher Award in 2018 for his significant contributions to academics and research at Graphic Era (deemed to be a university), Dehradun, India.

Preface

In the current era of industrialization, almost every system is facing dynamic challenges to maintain itself in the competitive scenario with other, similar systems and industries. These challenges include maintaining the system's various performance measures, such as reliability, availability, maintainability, cost, and failure factors, and many more parameters associated with them. The past few decades have witnessed rapidly changing technology and high-tech industrial processes and their advancements in various systems. As a result, one has to pay more attention to these reliability measures as well as the system's maintenance strategy to optimize overall performance. These analyses for a system can be performed through different methods and techniques including Markov modeling, stochastic modeling, simulation, the Lz-transform approach, the fuzzy Lambda-Tau approach, the universal generating function, game theoretic modeling, the six-sigma approach, and mathematical modeling. **The Handbook of Reliability, Maintenance, and System Safety Through Mathematical Modeling,** focuses on these aspects and provides sound, thorough, and concrete knowledge associated with reliability parameters of different structures that will be useful for researchers, academicians, and the related industry.

Amit Kumar,
Punjab, India

Mangey Ram,
Uttarakhand, India

Acknowledgments

The editors acknowledge Elsevier for this opportunity and professional support. Also, we would like to thank all the chapter authors and reviewers for their availability for this work.

1 Reliability, availability, and maintainability analysis of an industrial plant based on Six Sigma approach: a case study in plastic industry

Panagiotis H. Tsarouhas
International Hellenic University, Department of Supply Chain Management (Logistics), Katerini, Greece

1. Introduction

Many quality management techniques in the industry sector have been applied over the past decade. Continuous improvement was most common because of its unique ability to provide businesses with competitive advantages [7,14]. Six Sigma (SS) has been a popular method of continuous improvement to enhance the operational efficiency of an organization, raise its profitability, and decrease its costs [35,37]. Therefore, quality improvement has been the most sought-after term for many years, addressed not only in large companies, but also in small industries [27]. The implementation of this powerful tool has not only decreased the defect level, but also in product development, customer retention evaluation, cycle time management, improved productivity, and market share [17,25].

SS not only focuses on decreasing system variations and defects, but also helps companies to build goal-oriented business planning [22]. The SS methodology offers guidelines to help workers learn how to do their job and overcome potential problems. It also increases production line productivity and production capacity, through eliminating operational waste such as extracting redundant parts and unnecessary activity and reducing maintenance cycle times ([38]). Harry and Schroeder [21] stated that SS is a powerful business development technique that allows businesses to use simple and powerful quantitative methods to achieve and maintain operational excellence. SS not only helps to maintain the quality standards for execution, which are of paramount importance on the business process outsourcing industry, but also allows reaching the overall effect of organizations [11]. Hakimi et al. [20] concentrated on improving the quality of the yogurt production process in Company A by changing the factors affecting yogurt acidity and deciding the optimum level of these factors. Antony [4] reported that SS is a well-established methodology aimed at detecting and removing faults, failures, or anomalies in business processes or structures by concentrating on process quality features of critical importance to customers.

https://doi.org/10.1016/B978-0-12-819582-6.00001-0

Following the success of Motorola's SS, DMAIC's approach has become common in companies focused on improving quality. DMAIC is a systemic technique that is an acronym for define, measure, analyze, improve, and control, for tracking when it is very difficult to identify a defect or source of an issue by routine inspection [1]. DMAIC also correlates to some prior problem-solving production, such as plan-do-check-act and the Juran and Gryna seven-step process [12]. The DMAIC method, along with associated statistical and nonstatistical evaluation methods, is fundamental framework to attain progress in SS and project goals [47]. Tong et al. [39] used the DMAIC method to improve the screening process's sigma rate in the printed circuit board industry. Al-Mishari and Suliman [3] addressed the shortcomings of current system reliability approaches by incorporating them into the SS DMAIC approach.

Several studies have been conducted using the statistical analysis approach to analyze the reliability, availability, and maintainability (RAM) of industrial systems. RAM is a key performance factor and a good starting point for system upgrades in process analysis [34]. De Sanctis et al. [13] proposed a method for enhancing industry performance and suggested some maintenance strategies for managing high cost, safety, and environmental issues. For this purpose, Reliability Availability Maintainability Durability (RAMD) analysis was carried out by considering the equipment of the oil gas sector as an object of case study. Rahimdel et al.[30] carried out the reliability-based maintenance schedule of the hydraulic system of a rotary drilling machine and estimated the desired level of reliability for the interval of preventive maintenance. In reciprocating compressors, Corvaro et al. [10] examined the component's actions and efficiency, then defined the critical components to improve system's reliability by computing the system's RAM. In another study, Eti et al.[15] presented an approach to Reliability Availability Maintainability Sustainability (RAMS) implementation and risk analysis as a guide to maintenance policies to reduce failure frequency and maintenance costs. Sanctis et al. [13] integrated RAM and reliability-centered maintenance analysis to increase the performance and availability of operating systems. Niwas and Garg [26] suggested a model for analyzing an industrial system's organizational actions using the product guarantee theory. The Markovian approach is used to assume constant system failure and repair rates. Reliability, MTBF, availability, and profit numerical results were derived. The RAM study helps to model maintenance schedules and improve any system's availability [24,33]. Tsarouhas [40,41,43,44] examined the statistically measured RAM of various industrial systems such as cheese production line, wine packaging, etc. For failure, the best-fit distribution has been established and repair and maintenance policies are introduced to increase plant performance. In the form of the RAM index, [31] introduced a model of reliability, availability and maintenance to measure system performance. Rajpal et al [9] proposed a methodology to improve the reliability in the cement industry. System's MTBF and MTTR estimated over 2 years and analysis of RAMD indices were computed. Aggarwal et al. [2] determined RAM indices for measuring and improving the efficiency of a dairy system's skim milk powder process. Effective reliability and maintenance programs require careful data collection and analysis, as well as developing reliability models to assist in the decision-making processes [18].

2. Theoretical background

2.1 Six Sigma

SS was a company-wide theory of quality improvement initially developed and adopted by Motorola in 1987. Sigma (σ) is a letter from Greek alphabet used to define the standard deviation of a random variable in mathematical terms. It is a statistical measurement unit that represents the likelihood of an error occurring. The main goal in SS approach is to reduce the number of defects per million opportunities (i.e., 3.4 defects). Defects in the system create cost increases due to cleaning, rework, maintenance, repair, etc. SS relies on rigorous statistical techniques and implements mechanisms for monitoring quality, cost, process, workers, and responsibility [16]. It has been used by many other companies to improve the quality performance of the organization [29]. Among the most powerful processes of SS is DMAIC that improve systems from their current level to new higher efficiency [5]. The SS DMAIC process is defined as follows [12,36,46]:

1. Define: Identify the problem and its requirements.
2. Measure: Collect data to confirm and quantify the issue and assess the current situation.
3. Analyze: Discover root causes and deepen understanding of the process and problem through thorough analysis of influencing factors.
4. Improve: Design and implement the system performance adjustments.
5. Control: To make sustainable improvements, rearrange the control and process management framework.

2.2 RAM analysis

RAM's purpose is to define critical points within the plant or procedure where it is possible to implement changes to improve them. The efficiency of the plant is thus affected by RAM, which consists of three main parameters, i.e., reliability, availability, and maintainability.

Reliability is the likelihood that a machine (or equipment) can perform a prescribed operation for a specified period of time under the defined operating conditions. If T is the continuous random variable representing the system's time between failure (TBF) with $T \geq 0$, the reliability can be expressed as [45],

$$R(t) = P(T \geq t) \tag{1.1}$$

If $F(t)$ is the system's unreliability,

$$R(t) + F(t) = 1 \tag{1.2}$$

The unreliability is therefore the likelihood of a failure occurring before time t,

$$F(t) = 1 - R(t) = P(T < t) \tag{1.3}$$

$F(t)$ is also called the cumulative distribution function of the failure distribution.

The hazard or failure rate function is defined in the reliability theory as

$$\lambda(t) = f(t)/R(t) \tag{1.4}$$

where $f(t)$ is the fault distribution's probability density function and is defined by

$$f(t) = \frac{dF(t)}{dt} = -\frac{dR(t)}{dt} \tag{1.5}$$

The expected value of $f(t)$ is the mean time between failure ($MTBF$),

$$\text{MTBF} = \int_0^\infty t \cdot f(t)dt = \int_0^\infty R(t)dt \tag{1.6}$$

where the MTBF can be calculated from failure analysis.

The availability (A) of a machine can be determined by the ratio of the lifetime to the total time between failures [8]:

$$A = \frac{\text{Life time}}{\text{Total time}} = \frac{\text{Life time}}{\text{Life time} + \text{Repair time}} \tag{1.7}$$

where Life time represents MTBF, and the Total time corresponds to MTBF and the mean time to repair (MTTR). The MTTR can be calculated through the maintenance analysis and is the repair time. Thus, Eq. (1.7) becomes:

$$A = \frac{\text{MTBF}}{\text{MTBF} + \text{MTTR}} \tag{1.8}$$

Maintenance is the likelihood that a defective machine will be returned to operational efficiency within a specified period of time when the repair work is carried out in compliance with the approved procedures. If T_r is the continuous random variable representing the system's time to repair (TTR) with a probability density function of $r(t)$, then the maintainability is [6]:

$$M(t) = P(T_r \leq t) \tag{1.9}$$

The function of repair rate is,

$$\lambda_r(t) = r(t)/(1 - M(t)) \tag{1.10}$$

The mean time to repair (MTTR) is the average amount of time it takes to restore a machine to operational status after it fails to operate, and thus:

$$\text{MTTR} = \int_0^{\propto} (1 - M(t))dt \tag{1.11}$$

3. Project methodology

The research methodology for conducting the SS study on performance efficiency and quality improvement at a plastic plant is based on the DMAIC process (define, measure, analyze, improve, and control), and is as follows.

3.1 Define phase

The plastic plant is an automatic transmission line composed of various machines in sequence. As a consequence, when a failure occurs, it not only affects the particular machine that has the failure, but also all the machines located downstream, creating an output gap down the line. In fact, during the stoppages, there are quality issues. The actual production rate of the line can therefore be significantly lower than the nominal production rate.

3.2 Measure phase

In this phase, the failure and repair data collections need to be established. The production line operates in an 8-h shift per working day and usually stops during the weekends. Records have the exact time the machine fails, and the exact time between failures. Both the TBF and TTR failures for the entire line itself were recorded in seconds. The equipment's TBF is defined as the time elapsed since the equipment is switched on and starts operating after a failure, until it is rebooted and stopped by a new failure. On the other hand, the failed TTR equipment is described as the time from the moment the equipment falls and stops until it rises and reoperates.

3.3 Analyze phase

The main steps of this phase performed with the statistical tools for RAM analysis are as follows [19,23,28,32,42]:

(a) The first step is the collection, sorting, and classification of the failure and repair data.
(b) Then, we identify the frequency of failure data for the system which has been shown with histograms.
(c) In the third step, the Anderson and Darling tests are used of failure data sets with MINITAB professional software using goodness-of-fit test. The parameters are calculated by the least squares test for the best suited statistical distributions.
(d) RAM analysis for the system, for different time intervals, was computed.
(e) Finally, the goal is to enhance performance and to devise a better maintenance plan using knowledge of critical equipment and faults.

3.4 Improve phase

The research proposed solutions to established factors (i.e., MTBF, MTTR, and so on) by drawing findings from the RAM study. This stage is aimed at reducing the failures and their effects, removing the root causes and standardizing new operations management (i.e., training programs for operators/technicians, Total Productive Maintenance (TPM) technique, maintenance planning schedule, etc.), and improving the performance, quality, and efficiency of the production system.

3.5 Control phase

At this stage, the goal was to maintain the momentum and sustain the long-term positive change. The new maintenance plan and training programs have been implemented. SS requires continuous monitoring and regulation of the output by means of some kind of statistical process control (SPC). Therefore, the collection and analysis of numerous data from the maintenance department in this phase provided useful information about the level of control and efficiency of the maintenance process of the entire system.

4. Case study

4.1 Production process

The plastic industry is an automated production system with advanced technology and speed. Blow molding is the most common process of creating on-scale hollow plastic objects. Typical uses include containers, toys, components for vehicles, manufacturing equipment, and packaging. It is a production process used by inflating the hot plastic tube inside the mold to create hollow plastic pieces until it bends into the desired shape. Blow molding, including injection molding and extrusion, is a continuous process that can be fully automated, resulting in high rates of production and low unit costs. It works at much lower pressures than injection molding, resulting in lower tooling costs. The manufacturing process consisted of three phases: (i) *mold setup*, small plastic pellets are melted and molded into a hollow tube (depending on the blow molding subtype) called preform; (ii) *molding*, the preform is molded and filled with pressure air until it assumes the shape of the inside of the mold; and (iii) *cooling and release*, the mold element cools until it is stable enough to be expelled.

When a failure occurs in the plastic plant, most of the line upstream of the failure continues to operate unproduced, resulting in a downstream output deficit due to failure. Moreover, we will wait for the temperature of system to drop in order to fix the failure, to clean the molds of the residues, and then restart again to warm up the equipment. Thus, an additional production void from the interruption will be created, since all in-process material upstream and downstream of the interruption may need to be scrapped due to performance degradation during the interruption. As a result, the actual production rate of the plant can be significantly lower than its nominal production rate.

4.2 Operations management

Failure and repair data for the plastic plant are obtained from the archives of the maintenance staff by the end of each shift reported in print by the responsible technicians (mechanical and electrical). Those records contained a total of 183 days, which is nearly 10 months. During this period, the plant operated in an 8-h shift per day excluding weekends and holidays. On the basis of the data reported, the registrations-failures amounted to 364 for the entire production plant.

Therefore, the plant operated for a total of 87,360 min. Out of the 87,360 continuous working minutes, the plant operated without failure for 83,845 min and was under repair for the remaining 3515 min. Thus, 95.97% $\left(\frac{83845}{87360}\times 100\right)$ of the total operating period the plant was functioning properly, while the remaining 4.03% $\left(\frac{3515}{87360}\times 100\right)$ the plant was under repair. Moreover, the failure rate for the system is about two failures $\left(\frac{364}{183}\right)$ per shift.

The maintenance policy for the plastic system is preventive and corrective maintenance. The line's preventive maintenance takes place mainly during the weekends in order to keep the production line operational and to avoid potential failures. Prevention maintenance shall be focused on the implementation of particular preventive measures such as cleaning (air and water filters, tubes, etc.), checking, and replacement of air valves, relays, machinery oil, etc. The aim of this maintenance is to predict the likelihood of product loss in order to take "preventive action" on time to ensure the reliability of the production line. On the other hand, corrective maintenance is planned and conducted whenever a failure occurs; the process allows the maintenance staff to take immediate action to return the system to operating condition.

In Fig. 1.1, the summaries of TBFs and TTRs for the entire plastic plant were shown. The graphical summaries include three graphs, and the following observations were made: (a) both TBFs and TTRs

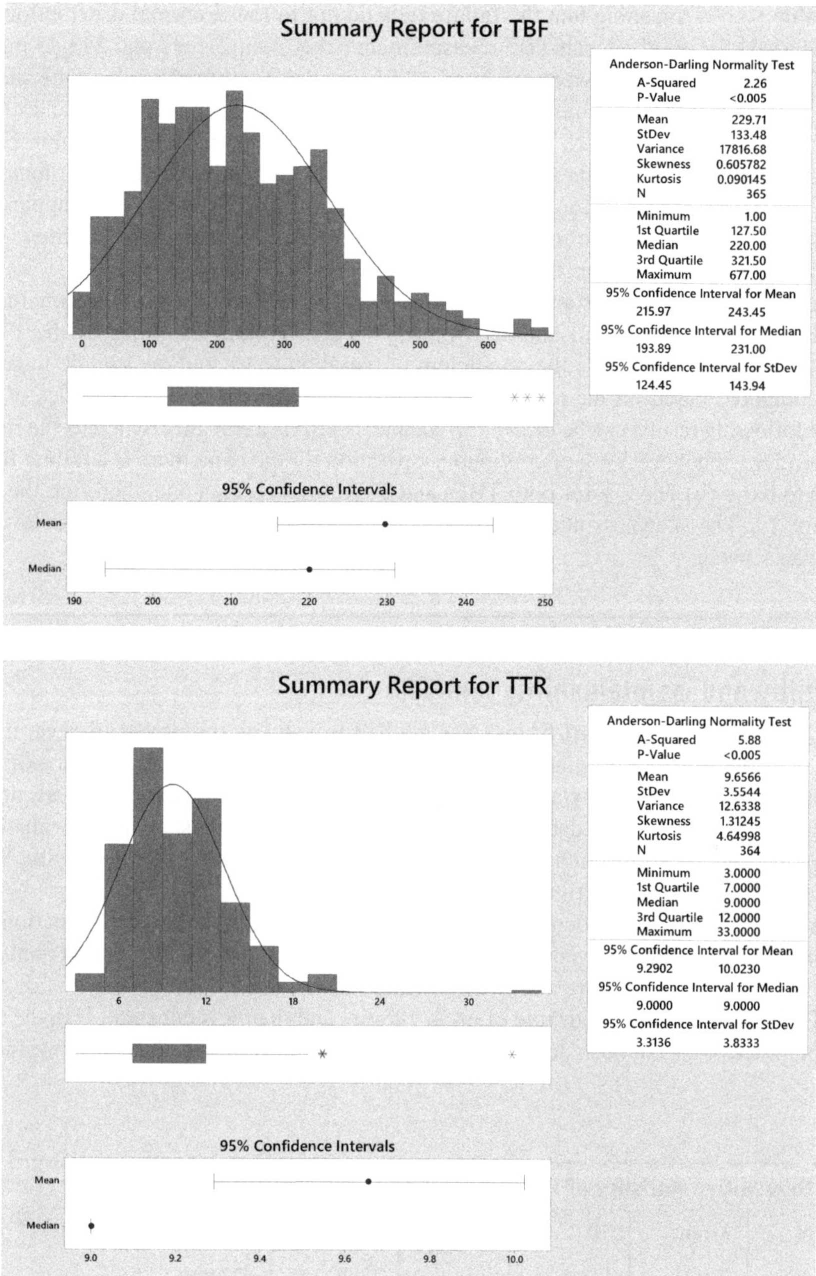

FIGURE 1.1

Summary reports for the TBF and TTR of the plastic plant.

have the P-value $\leq .005$, meaning that the failure data do not follow a normal distribution. (b)With a confidence interval of 95%, the mean TBF measurement is between 215.97 and 243.45 min, whereas the mean TTR measurement is between 9.29 and 10.02 min. (c) The boxplots show the shape, central tendency, and variability of the failure data. For the TBFs for the system, the Q1 is the 25th percentile and indicates that 25% of the data are less than or equal to 127.5 min, the median (Q2) is 220 min, and the Q3 indicates that 75% of the data are less than or equal to 321.5 min. On the other hand, the Q1 of the TTRs is 7 min, the Q2 stand for 9 min, and the Q3 is 12 min. It is therefore apparent that the majority of stoppages in the plastic industry are linked to micro downtimes which interrupt the production line for a couple of minutes but with high frequency.

The application of descriptive statistics to fault data is very helpful in drawing conclusions about identifying the most important faults and determining the distributions identified by the TTFs and the TTRs. The standard deviation *(SD)*, the coefficient of variation (*CV*), and the sample mean at system levels were computed based on the records. Table 1.1 shows the descriptive statistics of the plastic plant, and the following results can be drawn: (a) the mean TBF is 229.7 min, whereas the mean TTR is 9.65 min. Therefore about every 4 h of continuous operation for the plant there is a failure that requires about 10 min to fix it. (b) The CV for both TBFs and TTRs are less than one, therefore the system has low variability. (c) The skewness and the kurtosis are positives meaning that the failure data have mode < median <mean.

5. Results and discussion

5.1 Reliability and maintainability analysis

The Anderson–Darling goodness-of-fit test was applied to validate the fitness of several theoretical distributions (i.e., Weibull, lognormal, loglogistics distribution, etc.) by using the maximum probability estimation method. The test statistics of several theoretical distributions of TBF and TTR are listed in Table 1.2, focused on fault data from the whole plant. The smallest statistical value implies the most suitable distribution for the data. It is observed that the TBFs and TTRs follow the Weibull and the lognormal distributions, respectively.

The plastic plant's probability density function, survival function, and hazard function are shown graphically in Fig. 1.2 for time between failure (TBF) and time to repair (TTR). It was pointed out that: (a) The TBF has an increasing failure rate, which means that the line is highly likely to fail as time goes by. (b) The TTR has a growing repair rate of up to 18 min, and then it is constant. If in the first 18 min the restoration process has not been done, then the probability of fixing the fault is constant over time.

Table 1.1 Descriptive statistics of the failure data for the plastic plant.

Variable	N	Mean	SD	CV	Minimum	Maximum	Skewness	Kurtosis
TBF	365	229.71	133.48	0.5811	1.00	677.00	0.61	0.09
TTR	364	9.657	3.554	0.3681	3.000	33.000	1.31	4.65

Table 1.2 The Anderson–Darling statistics for TBF and TTR of the plastic plant.

Distribution	TBF	TTR
Weibull	0.634	5.692
Lognormal	9.36	2.584
Exponential	25.01	72.915
Loglogistic	4.28	3.235
3-Parameter Weibull	0.719	3.096
3-Parameter lognormal	0.644	2.59
2-Parameter exponential	24.853	40.318
3-Parameter loglogistic	1.184	3.186
Smallest extreme value	10.623	27.15
Normal	2.314	5.949
Logistic	2.215	5.377

The following findings are taken on the basis of Weibull distribution for the plastic production line TBF (see Table 1.3): (a) The period within which 25% of the faults are expected to occur (Q1) corresponds to 124.551 min of operation, the time within which 75% of the faults (Q3) are expected to occur is 309.933 min of operation, while the time within which half of the failures (IQR) is expected to occur is equal to 185.382 min of operation. (b) It is evident that from the percentiles (95% CI) the time within which the 5% of the failures are anticipated to occur amounts up to 45.8371 min. (c) From the probabilities of survival (95% CI), it was found that after an hour (60 min) of operation the plant is 92.16% likely to function properly. After 3 h (180 min) of operation the system's probability of working properly is 58.10%, and after a shift (480 min) the plant's likelihood of proper function decreases to 5.24%.

As for Table 1.4, the findings of the TTR model are as follows, focused on lognormal distribution: (a) 25% of the failures (Q1) shall be repaired within the first 7.12909 min, 75% of the failures (Q3) shall be repaired within 11.5261 min, while half of the failures (IQR) shall be repaired within 4.39696 min. (b) It can be perceived from the percentiles (95% CI) that 30% of the failures will be repaired within 7.52053 min, and (c) from the probability of survival (95% CI), the probability of repairing the plant in less than 5 min is 1−0.95260 = 0.0474 or 4.74%, and the probability of repairing the system in less than 15 min is 1−0.07865 = 0.9213 or 92.13%.

5.2 Improve and control

One of the best ways to improve the system's continuous operation in the plastic plant project is to keep the entire system at a high level of reliability. For this reason, the reliability-based scheduling maintenance may be the adequate strategy. Thus, the preventive maintenance intervals that were computed by the reliability models are used to get desired efficiency, quality, and safety of operation of the plant. Table 1.5 indicates the maintenance intervals that would reach different levels of operating reliability, i.e., 0.50, 0.60, 0.70, 0.80, and 0.90%. Therefore, maintenance must be done until 180 min (or 3 h) to reach 60% reliability for the plastic factory because after the device has operated for 3 h

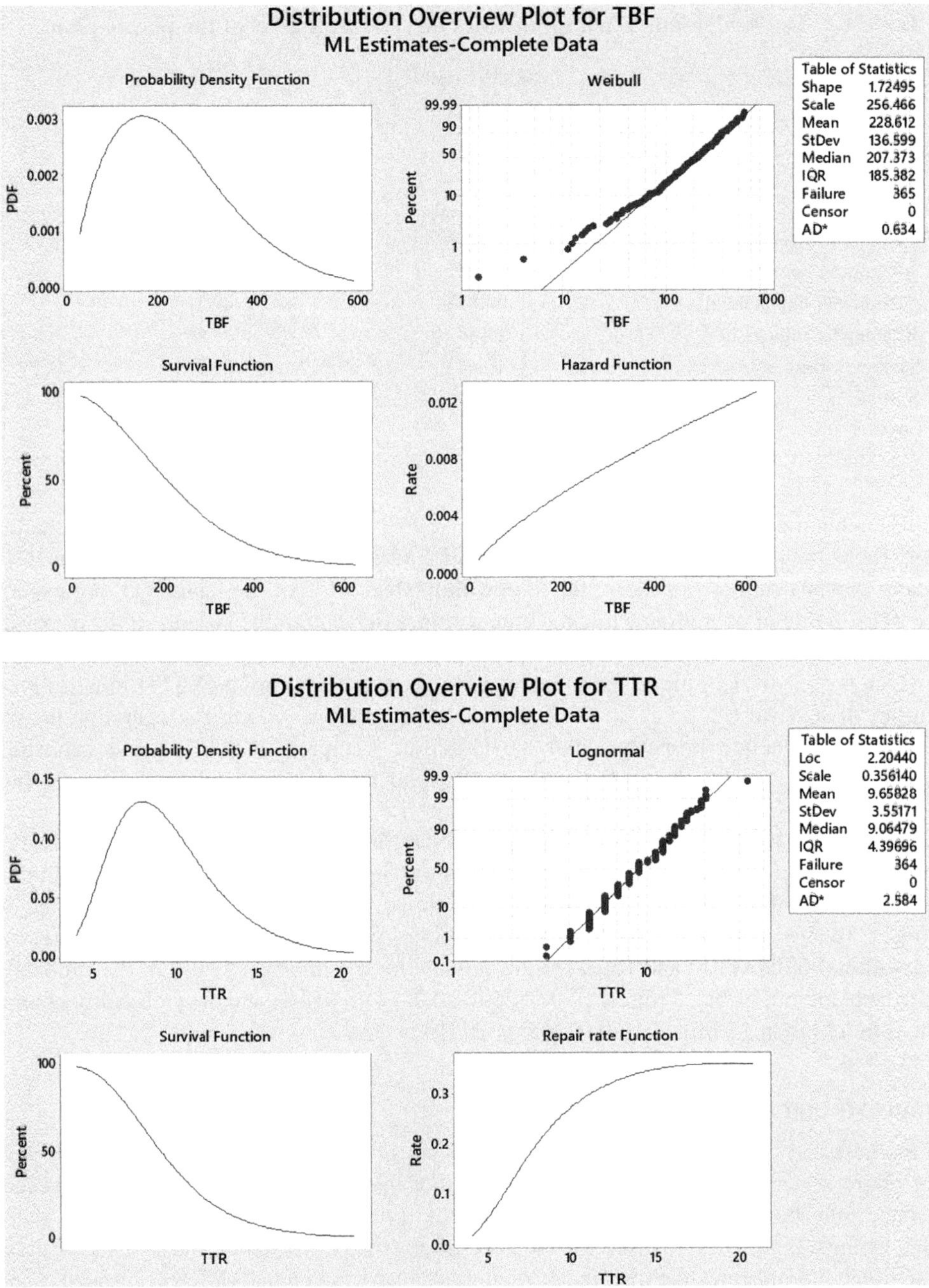

FIGURE 1.2

Survival function, probability density function, hazard function, and repair function for TBF and TTR of the plastic plant are shown graphically in accordance with the necessary distributions and their statistics.

Table 1.3 Distribution analysis of TBF, applying the Weibull distribution for the plastic plant, with 95% CI.

Distribution analysis: Variable TBF	
Uncensored value	365

Estimation method: maximum likelihood				
Distribution: Weibull				
Parameter estimates (95.0% normal CI)				
Parameter	**Estimate**	**Std. Error**	**Lower**	**Upper**
Shape	1.72495	0.0721631	1.58916	1.87235
Scale	256.466	8.16047	240.960	272.969
Loglikelihood = −2283.498				
Characteristics of distribution (95.0% normal CI)				
	Estimate	**Std. Error**	**Lower**	**Upper**
Mean (MTTF)	228.612	7.11634	215.081	242.994
Standard deviation	136.599	5.93184	125.454	148.734
Median	207.373	7.36589	193.427	222.324
First Quartile (Q1)	124.551	6.23264	112.915	137.386
Third Quartile (Q3)	309.933	9.41862	292.011	328.954
Interquartile Range (IQR)	185.382	7.16877	171.850	199.978
Table of percentiles (95.0% normal CI)				
Percent	**Percentile**	**Std. Error**	**Lower**	**Upper**
1	17.8172	2.22508	13.9489	22.7584
5	45.8371	3.99098	38.6459	54.3663
10	69.5744	4.93747	60.5400	79.9570
20	107.495	5.91940	96.4969	119.746
30	141.081	6.49031	128.917	154.393
40	173.743	6.92755	160.682	187.865
50	207.373	7.36589	193.427	222.324
60	243.792	7.92807	228.738	259.837
70	285.605	8.78672	268.892	303.356
80	337.944	10.2762	318.391	358.697
90	415.926	13.3762	390.518	442.987
99	621.629	25.4416	573.712	673.548

Table of survival probabilities (95.0% normal CI)			
Time	**Probability**	**Lower**	**Upper**
1	0.999930	0.999841	0.999969
5	0.998878	0.997960	0.999383
10	0.996296	0.993889	0.997756
20	0.987807	0.981751	0.991862
30	0.975613	0.965531	0.982773
60	0.921627	0.899655	0.938952
120	0.763539	0.725516	0.797047
150	0.672701	0.631150	0.710682
180	0.581020	0.538295	0.621256
240	0.409899	0.369241	0.450076
300	0.269671	0.233897	0.306612
360	0.166146	0.136538	0.198299
420	0.096171	0.073402	0.122529
480	0.052436	0.036366	0.072614
720	0.002650	0.000993	0.006151
960	0.000059	0.000008	0.000304

Table 1.4 Distribution analysis of TTR, using the lognormal distribution for the plastic plant, with 95% CI.

Distribution analysis: variable TTR				
Uncensored value				364
Estimation method: maximum likelihood				
Distribution: Lognormal				
Parameter estimates (95.0% normal CI)				
Parameter	**Estimate**	**Std. Error**	**Lower**	**Upper**
Location	2.20440	0.0186668	2.16781	2.24098
Scale	0.356140	0.0131994	0.331187	0.382974
Log-likelihood = −943.090				
Characteristics of distribution (95.0% normal CI)				
	Estimate	**Std. Error**	**Lower**	**Upper**
Mean (MTTF)	9.65828	0.185918	9.30067	10.0296
Standard deviation	3.55171	0.170292	3.23315	3.90166

Table 1.4 Distribution analysis of TTR, using the lognormal distribution for the plastic plant, with 95% CI.—cont'd

Log-likelihood = −943.090

Characteristics of distribution (95.0% normal CI)

	Estimate	Std. Error	Lower	Upper
Median	9.06479	0.169211	8.73913	9.40257
First Quartile (Q1)	7.12909	0.147438	6.84590	7.42400
Third Quartile (Q3)	11.5261	0.238373	11.0682	12.0029
Interquartile Range (IQR)	4.39696	0.185259	4.04845	4.77548

Table of percentiles (95.0% normal CI)

Percent	Percentile	Std. Error	Lower	Upper
1	3.95860	0.142253	3.68938	4.24746
5	5.04601	0.144480	4.77063	5.33728
10	5.74302	0.144673	5.46635	6.03369
20	6.71714	0.145912	6.43716	7.00929
30	7.52053	0.149725	7.23273	7.81979
40	8.28271	0.157073	7.98050	8.59636
50	9.06479	0.169211	8.73913	9.40257
60	9.92071	0.188136	9.55874	10.2964
70	10.9261	0.217527	10.5080	11.3609
80	12.2329	0.265728	11.7231	12.7650
90	14.3079	0.360432	13.6186	15.0320
99	20.7574	0.745923	19.3457	22.2721

Table of survival probabilities (95.0% normal CI)

Time	Probability	Lower	Upper
1	1.00000	1.00000	1.00000
5	0.95260	0.93468	0.96634
10	0.39139	0.35184	0.43210
15	0.07865	0.05944	0.10224
20	0.01314	0.00790	0.02115
25	0.00220	0.00104	0.00443
30	0.00039	0.00014	0.00098
35	0.00007	0.00002	0.00023
40	0.00002	0.00000	0.00006
45	0.00000	0.00000	0.00002
50	0.00000	0.00000	0.00000
55	0.00000	0.00000	0.00000
60	0.00000	0.00000	0.00000

Table 1.5 Reliability-based preventive maintenance (PM) time intervals at line level.

Reliability level %	50	60	70	80	90
Time intervals for PM	210	180	140	110	70

without failure, there is only a 0.60 possibility that it will not function properly. Consequently, the initial preventive maintenance period proposed should be a low level of reliability (i.e., 50%) and may be followed to increase performance and reduce the number of repairs for the production plant. Furthermore, on the basis of the benefits obtained in terms of cost, safety, and operational efficiency, a higher value of the level of reliability can be gradually adjusted.

In order to control the system, the generation of SPC charts for the production line is updated daily and displayed in an open area to be accessible to manufacturing managers, technicians, and maintenance personnel. Therefore, they can be notified directly about the system's performance, quality, and reliability. Thus, it is possible to act quickly on any failure that appears on the system before more serious issues arise.

6. Conclusions

In this chapter, SS technique in a plastic plant under real working conditions was computed. RAM analysis was applied to make inferences during the project using several statistical tools and techniques. The key findings of the study can be summarized as follows:

(a) During 95.97% of the total operational period the line had been working properly, while in the remaining 4.03% of the total operating time the line was under maintenance, with a failure rate of about two failures per shift for the system.
(b) The mean TBF is 229.7 min, and the mean TTR is 9.65 min. Therefore, about every 4 h of continuous operation for the plastic line there is a failure that requires about 10 min to get fixed.
(c) The TBF faces a rising failure rate, which implies that as time passes the system is likely to fail. The TTR has a rising repair rate of up to 18 min, and then it's steady. Thus, if the restoration process has not been done in the first 18 min, the chance of fixing the fault is constant over time.
(d) The maintenance intervals that would maintain different levels of reliability in the operation of the production line have been determined.
(e) We outline the actions required for enhancing and managing the production system.

SS and RAM analysis are two powerful tools that guarantee the continuous improvement of the plastic production system. The advantage of these techniques is that unplanned failures are reduced, thus ensuring that the line operates with a high level of quality and efficiency. In addition, significant savings can be accomplished by minimizing downtime-related costs, reducing maintenance costs, and ensuring the continued safety of workers at the factory.

References

[1] C.R. Abhilash, J.J. Thakkar, Application of Six Sigma DMAIC methodology to reduce the defects in a telecommunication cabinet door manufacturing process: a case study, Int. J. Qual. Reliab. Manag. 36 (9) (2019) 1540–1555.

[2] A. Aggarwal, S. Kumar, V. Singh, Performance modeling of the skim milk powder production system of a dairy plant using RAMD analysis, Int. J. Qual. Reliab. Manag. 32 (2) (2015) 167–181.

[3] S. Al-Mishari, S. Suliman, Integrating Six-Sigma with other reliability improvement methods in equipment reliability and maintenance applications, J. Qual. Mainten. Eng. 14 (1) (2008) 59–70.

[4] J. Antony, Can Six Sigma be effectively implemented in SMEs? Int. J. Prod. Perform. Manag. 57 (2008) 420–423.

[5] V. Arumugam, J. Antony, M. Kumar, Linking learning and knowledge creation to project success in Six Sigma project: an empirical investigation, Int. J. Prod. Econ. 141 (2013) 388–402.

[6] M. Ben-Daya, O.S. Duffua, A. Raouf, J. Knezevic, D. Ait-Kadi, Handbook of Maintenance Management and Engineering, Springer-Verlag London Limited, NY, USA, 2009.

[7] M. Butler, M. Szwejczewski, M. Sweeney, A model of continuous improvement programme management, Prod. Plann. Contr. 29 (5) (2018) 386–402.

[8] H.F. Castro, K.L. Cavalca, Availability optimization with genetic algorithm, Int. J. Qual. Reliab. Manag. 20 (7) (2003) 847–863.

[9] D. Choudhary, M. Tripathi, R. Shankar, Reliability, availability and maintainability analysis of a cement plant: a case study, Int. J. Qual. Reliab. Manag. 36 (3) (2019) 298–313.

[10] F. Corvaro, G. Giacchetta, B. Marchetti, M. Recanati, Reliability, availability, maintainability (RAM) study on reciprocating compressors, Petroleum 3 (2) (2017) 266–272.

[11] H. De Koning, R.J. Does, S. Bisgaard, Lean six sigma in financial services, Int. J. Six Sigma Compet. Advant. 4 (1) (2008) 1–17.

[12] J. De Mast, J. Lokkerbol, An analysis of the Six Sigma DMAIC method from the perspective of problem solving, Int. J. Prod. Econ. 139 (2) (2012) 604–614.

[13] I. De Sanctis, C. Paciarotti, O. Di Giovine, Integration between RCM and RAM: a case study, Int. J. Qual. Reliab. Manag. 33 (6) (2016) 852–880.

[14] T.J. Douglas, W.Q. Judge, Total quality management implementation and competitive advantage: the role of structural control and exploration, Acad. Manag. J. 44 (1) (2001) 158–169.

[15] M.C. Eti, S.O.T. Ogaji, S.D. Probert, Reducing the cost of preventive maintenance (PM) through adopting a proactive reliability-focused culture, Appl. Energy 83 (2006) 1235–1248.

[16] J.R. Evans, W.M. Lindsay, An introduction to Six Sigma and process improvement, Stamfort Cengage Learn. (2014) 74–83.

[17] J.E. Ferreira, I.S. Lopes, Improvement of scrap request process with Six Sigma methodology, in: Proceedings of the World Congress on Engineering 2010, Vol. III, London, 2010, p. 30. June-2 July.

[18] A.H.S. Garmabaki, A. Ahmadi, J. Block, H. Pham, U. Kumar, A reliability decision framework for multiple repairable units, Reliab. Eng. Syst. Saf. 150 (2016) 78–88.

[19] A. Gustafson, H. Schunnesson, U. Kumar, Reliability analysis and comparison between automatic and manual load Haul Dump machines, Qual. Reliab. Eng. Int. 31 (3) (2015) 523–531.

[20] S. Hakimi, S.M. Zahraee, J. Mohd Rohani, Application of Six Sigma DMAIC methodology in plain yogurt production process, Int. J. Lean Six Sigma 9 (4) (2018) 562–578.

[21] M.J. Harry, R. Schroeder, Six Sigma: The Breakthrough Management Strategy Revolutionizing the World's Top Corporations, Random House. Inc., New York, NY, 2000.

[22] K. Linderman, R.G. Schroeder, S. Zaheer, A.S. Choo, Six sigma: a goal-theoretic perspective, J. Oper. Manag. 21 (2) (2003) 193–203.

[23] D.M. Louit, R. Pascual, A.K. Jardine, A practical procedure for the selection of time-to-failure models based on the assessment of trends in maintenance data, Reliab. Eng. Syst. Saf. 94 (2009) 1618–1628.

[24] S. Malik, P.C. Tewari, Performance modeling and maintenance priorities decision for water flow system of a coal based thermal power plant, Int. J. Qual. Reliab. Manag. 35 (4) (2018) 996−1010.
[25] P. Muraleedharan, S. Balamurugan, R. Prakash, Six Sigma DMAIC in manufacturing industry: a literature review, Int. J. Innov. Res. Sci. Eng. Technol. 6 (9) (2017) 18288−18293.
[26] R. Niwas, H. Garg, An approach for analyzing the reliability and profit of an industrial system based on the cost free warranty policy, J. Braz. Soc. Mech. Sci. Eng. 40 (265) (2018) 1−9.
[27] R. Panat, V. Dimitrova, T.S. Selvamuniandy, K.S. Ishiko, D. Sun, The application of Lean Six Sigma to the configuration control in Intel's manufacturing R&D environment, Int. J. Lean Six Sigma 5 (4) (2014) 444−459.
[28] R.B. Patil, S.B. Kothavale, L.Y. Waghmode, G.S. Joshi, Reliability analysis of CNC turning center based on the assessment of trends in maintenance data - a case study, Int. J. Qual. Reliab. Manag. 34 (9) (2017) 1616−1638.
[29] E. Psomas, The underlying factorial structure and significance of the six sigma difficulties and critical success factors: the Greek case, TQM J. 28 (4) (2016) 530−546.
[30] M.J. Rahimdel, M. Ataei, R. Khalokakaei, S.H. Hoseinie, Reliability-based maintenance scheduling of hydraulic system of rotary drilling machines, Int. J. Min. Sci. Technol. 23 (2013) 771−775.
[31] P.S. Rajpal, K.S. Shishodia, G.S. Sekhon, An artificial neural network for modeling reliability, availability and maintainability of a repairable system, Reliab. Eng. Syst. Saf. 91 (2006) 809−819.
[32] S.K. Roy, M.M. Bhattacharyya, V.N.A. Naikan, Maintainability and reliability analysis of a fleet of shovels, Min. Technol. 110 (3) (2001) 163−169.
[33] S. Saraswat, G.S. Yadava, An overview on reliability, availability, maintainability and supportability (RAMS) engineering, Int. J. Qual. Reliab. Manag. 25 (3) (2007) 330−344.
[34] S. Saraswat, G.S. Yadava, An overview on reliability, availability, maintainability and supportability (RAMS) engineering, Int. J. Qual. Reliab. Manag. 25 (3) (2008) 330−344.
[35] R.G. Schroeder, K. Linderman, C. Liedtke, A.S. Choo, Six sigma: definition and underlying theory, J. Oper. Manag. 26 (4) (2008) 536−554.
[36] A.B. Sin, S. Zailani, M. Iranmanesh, T. Ramayah, Structural equation modelling on knowledge creation in Six Sigma DMAIC project and its impact on organizational performance, Int. J. Prod. Econ. 168 (2015) 105−117.
[37] M.V. Sunder, Six sigma-A strategy for increasing employee engagement, J. Qual. Participation 36 (2) (2013) 34−38.
[38] B. Tjahjono, P. Ball, V.I. Vitanov, C. Scorzafave, J. Nogueira, J. Calleja, M. Minguet, L. Narasimha, A. Rivas, A. Srivastava, S. Srivastava, A. Yadav, Six sigma: a literature review, Int. J. Lean Six Sigma 1 (3) (2010) 216−233.
[39] J. Tong, F. Tsung, B. Yen, A DMAIC approach to printed circuit board quality improvement, Int. J. Adv. Manuf. Technol. 23 (7/8) (2004) 523−531.
[40] P. Tsarouhas, A practical approach for reliability and maintainability analysis of repairable systems: a case study of polypropylene production for food packaging applications, Int. J. Reliab. Saf. 9 (4) (2015) 255−271.
[41] P. Tsarouhas, I. Arvanitoyannis, T. Varzakas, Reliability and maintainability analysis of cheese (feta) production line in a Greek medium-size company: a case study, J. Food Eng. 94 (3−4) (2009) 233−240.
[42] P. Tsarouhas, Application of statistical approaches for analyzing the reliability and maintainability of food production lines: a case study of mozzarella cheese, in: D. Granato, G. Ares (Eds.), Chapter in Mathematical and Statistical Methods in Food Science and Technology, John Wiley & Sons, Inc., 2014, pp. 491−510.
[43] P. Tsarouhas, Reliability, availability and maintainability (RAM) analysis for wine packaging production line, Int. J. Qual. Reliab. Manag. 35 (3) (2018) 821−842.

[44] P. Tsarouhas, Statistical analysis of failure data for estimating reliability, availability and maintainability of an automated croissant production line, J. Qual. Mainten. Eng. 25 (3) (2019) 452–475.

[45] H. Wang, H. Pham, Reliability and Optimal Maintenance, Springer-Verlag London Limited, NJ, USA, 2006.

[46] D.Y. Yeh, H. Cheng, M.L. Chi, A modified two tuple FLC model for evaluating the per-formance of SCM: by the Six Sigma DMAIC pro-cess, Appl. Soft Comput. 7 (2007) 1027–1034.

[47] X. Zu, L.D. Fredendall, T.J. Douglas, The evolving theory of quality management: the role of Six Sigma, J. Oper. Manag. 26 (5) (2008) 630–650.

CHAPTER

2 Impact of the degree of hybridization on the reliability and system safety of a Helicopter's traction drive based on *Lz*-transform approach

Igor Bolvashenkov[1], Jörg Kammermann[1], Zeng Wenbin[1,2], Ilia Frenkel[3], Shen Guixiang[2], Hans-Georg Herzog[1]

[1]*Institute of Energy Conversion Technology, Technical University of Munich, Munich, Germany;* [2]*School of Mechanical and Aerospace Engineering, Jilin University, Changchun, China;* [3]*Center for Reliability and Risk Management, SCE-Shamoon College of Engineering, Beer Sheva, Israel*

1. Introduction

The electrification of different types of vehicles is very topical today because of well-known benefits. A similar challenge emerges also in the domain of air transportation, in particular, for electrification of helicopters, to provide more reliable and sustainable flying with a minimum of energy consumption and without further environmental damage. That is why the attempts of application of the electrical traction drive for the helicopter have caused great interest now. However, preliminary studies have shown [1,2] that at the current stage of technological development, the implementation of a purely electric version of the helicopter traction drive is practically impossible. Therefore, it is crucial to determine an appropriate hybrid traction drive topology that meets various specified project requirements and to quantitatively demonstrate the impact of the degree of hybridization (DoH), which describes the ratio of electrical and mechanical energy [3,4], on the performance of different topology at an early design stage. Due to this fact, an evaluation method based on the *Lz*-transform approach then proposed for exploring the impact of DoH on the reliability and system safety of all possible helicopter traction drive topologies (conventional, hybrid-electric, and full-electric) in this chapter.

As conventional option, the Airbus helicopter EC135 (H135) with two gas turbine engines and a gearbox, described in Ref. [5] has been selected. Then, three hybrid-electric traction drives with different topologies are presented for DoH impact analysis on availability and system fault tolerance. Although the application of purely electric traction drive is unrealistic now [6,7], three different full-electric traction drive topologies are still proposed for the comparative analysis, which can certify the advantages and significance of electrification. The comparative availability analysis of all these

The Handbook of Reliability, Maintenance, and System Safety through Mathematical Modeling.
https://doi.org/10.1016/B978-0-12-819582-6.00002-2

kinds of traction drive is evaluated considering the failure and repair rates of components, which can assist to determine the optimal propulsion system through its lifetime. Moreover, in order to help a researcher or an engineer to make appropriate decisions for safety and sustainability improvement of the helicopter at an early design stage, the sensitivity analysis of components of different traction drive topologies is calculated and identified.

The availability and DoH impact analyses are carried out by means of the *Lz*-transform approach. A brief description about the DoH and the *Lz*-transform approach is given in Section 2. Section 3 explains the construction and operation of the conventional and hybrid-electric traction drive topologies. The graphs of the availability and the DoH influence analyses for a hybrid helicopter traction drive are illustrated in Section 4. Then the topologies of full-electric traction drive and the corresponding availability analyses are depicted in Section 5. Consequently, Section 6 presents the sensitivity analysis results for the system's elements. In summary, the decisions about how to determine an appropriate traction drive topology for the helicopter and how to improve the reliability and system safety can be made at an early design stage.

2. Description of the degree of hybridization and the *Lz*-transform method

2.1 Degree of hybridization

It is shown in Refs. [3,4] that the DoH can be used as a universal tool for preliminary judging different aspects relevant for hybrid vehicles efficiency. The quantitative dependency of reliability for the hybrid-electric helicopter propulsion system from the DoH is considered in a previous work [8] of the authors. Hence, the dependencies of availability and fault tolerance for all proposed helicopter's traction drive topologies from the DoH are investigated in this section.

There are different mathematical versions of the DoH but they all have the same goal—to estimate the ratio of the installed power of electric traction motors and of gas turbine engines:

$$\text{DoH} = \frac{P_{EM}}{P_{EM} + P_{GTE}} \tag{2.1}$$

with P_{EM} being the electric machine power and P_{GTE}, the gas turbine engine power available for the propulsion system. The DoH in Eq. (2.1) can change from zero (for the conventional vehicle) up to one (for a pure electric vehicle).

In this chapter, the DoH equation is modified from the reliability perspective—to evaluate the ratio of energy required for the flight charged by the electric energy storage and by the gas turbine engine:

$$\text{DoH} = \frac{E_{El}}{E_{El} + E_{Mech}} \tag{2.2}$$

with E_{El} being the electric energy flight and E_{Mech}, the mechanical energy flight available for the traction drive. Therefore, the dependencies of reliability and system safety for the helicopter's traction drive from the DoH can be demonstrated.

2.2 *Lz*-transform method

The *Lz*-transform approach has been used for many systems for their performance evaluation throughout the lifetime [9–11]. Lisnianski [12] introduced the method by where one can find the detailed content and mathematical proofs. A brief introduction comes in the following paragraphs.

A discrete-state continuous-time (DSCT) Markov process $X(t) \in \{x_1, x_2, \dots, x_K\}$ which has K possible states and the performance level associated with any state i is x_i, $i = 1, 2, \dots, K$. The Markov process is completely defined by the set of possible states $\mathbf{x} = \{x_1, x_2, \dots, x_K\}$, the transition intensities matrix $= [a_{ij}]$, $i, j = 1, 2, \dots, K$ and by the initial state probability distribution, given by $p_0 = \{p_{10} = Pr\{X(0) = x_1\}, \dots, p_{K0} = Pr\{X(0) = x_K\}\}$. Therefore, according to Ref. [13], the *Lz*-transform of a DSCT Markov process $X(t)$ is defined by the following expression:

$$L_z\{X(t)\} = \sum_{i=1}^{K} p_i(t) z^{g_i} \tag{2.3}$$

where $p_i(t)$ is the probability that the process is in state i at time instant $t \geq 0$ for a given initial state probability distribution p_0, g_i is the performance level of state i, and z is a complex variable.

Without losing generality, we consider that any element e in MSS can have k_e different states corresponding to different performance, represented by the set $g_e = \{g_{e1}, \dots, g_{ek_e}\}$, where g_{ei} is the performance rate of element e in the state i, $i \in \{1, 2, \dots, k_e\}$, and $e \in \{1, 2, \dots, n\}$, where n is the number of elements in the MSS.

The first step of *Lz*-transform is to build a Markov model of stochastic process for each multistate element in MSS. Based on this model, the state probabilities $p_{ei} = Pr\{G_e(t) = g_{ei}\}$, $i \in \{1, 2, \dots, k_e\}$ for every MSS's element can be obtained as a solution of the corresponding system of differential equations under the given initial conditions. These probabilities define the output stochastic process $G_e(t)$ for each element e in the MSS. Then, the individual *Lz*-transform for each element e can be found as

$$L_z\{G_e(t)\} = \sum_{i=1}^{k_e} p_{ei}(t) z^{g_{ei}}, \ e = \{1, 2, \dots, n\} \tag{2.4}$$

The next step is based on the determined *Lz*-transform of each element e and system structure function f, given by $G(t) = f(G_1(t), G_2(t), \dots, G_n(t))$, where the *Lz*-transform of the output stochastic process for the entire MSS should be defined. Using the Ushakov's operator Ω_f [14] over all *Lz*-transforms of individual elements, one can obtain the resulting *Lz*-transform $L_z\{G(t)\}$ associated with output performance of stochastic process $G(t)$ of the entire MSS:

$$L_z\{G(t)\} = \Omega_f\{L_z[G_1(t)], L_z[G_2(t)], \dots, L_z[G_n(t)]\} \tag{2.5}$$

The resulting *Lz*-transform is associated with output performance of stochastic process for the entire MSS:

$$L_z\{G(t)\} = \sum_{k=1}^{K} p_k(t) z^{g_k} \tag{2.6}$$

Therefore, the availability of helicopter's traction drive can be easily derived from the resulting *Lz*-transform in the following form:

$$A(t) = \sum_{g_k \geq 0} p_k(t) \tag{2.7}$$

3. Topologies of the conventional and hybrid-electric traction drives

In Ref. [15], the authors argue that no conceptual model exists which is applicable to any powertrain layout, while being sensitive to the different powertrain architecture. Therefore, this chapter analyzes the reliability of a set of predefined traction drive topologies. The classification principles are presented in Ref. [16]. The configurations include the conventional powertrain of the Airbus helicopter EC135 as a reference and hybrid-electric powertrains, as well as full-electric powertrains for comparison.

3.1 Conventional traction drive topology

The traction drive of the Airbus helicopter EC135 consists of a fuel tank (FT), gas turbine subsystem (SGTE), consisting from two similar gas turbine engines (designated as GTE_C), a gearbox (GB), and a propeller (P). Note that the two gas turbine engines coprovide the energy needed for normal flight and even if either fails, the remaining gas turbine can continue to work until emergency landing. The structure and corresponding reliability block diagram (RBD) are shown in Fig. 2.1.

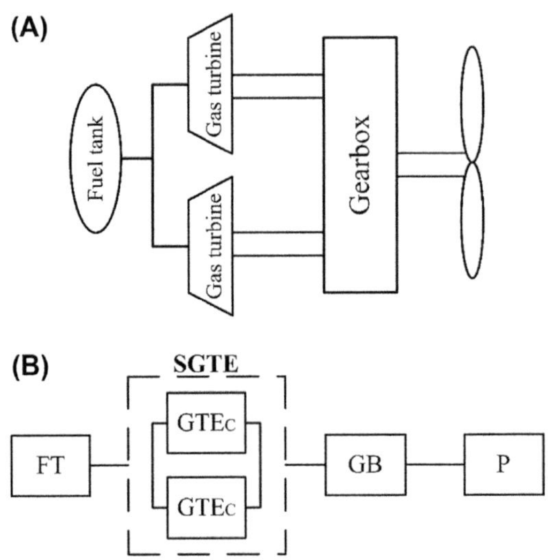

FIGURE 2.1

(A) Structure of conventional traction drive, (B) reliability block diagram of conventional traction drive.

3.2 Hybrid-electric traction drive topologies

The outstanding feature of hybrid-electric traction drive topologies is two energy sources, the fuel tank and the battery storage supplying power for the operation of the traction drive, which is only done by the fuel tank in the conventional EC135. The configurations of hybrid-electric powertrains are usually serial hybrid, parallel hybrid, and combined hybrid. Three different hybrid-electric traction drives, composed of different components, are presented to show their different reliability characteristics. Note that the switchboard should be included in all hybrid-electric powertrains while considering reliability because the switchboard is used to switch between operation scenarios of the helicopter's propulsion system. Moreover, designated the battery storage of hybrid-electric traction drive topologies as SBS (small battery storage) while that designated as LBS (large battery storage) in full-electric topologies because it is possible to install an additional amount of electric battery cells due to more free space in full-electric embodiments.

3.2.1 Serial hybrid traction drive topology

The serial hybrid traction drive consists of a fuel tank (FT), a gas turbine engine (GTE), a gearbox (GB), an electric generator (EG), an electric energy converter (EEC), an electric motor (EM), a small battery storage (SBS), and a propeller (P). The structure of this powertrain is shown in Fig. 2.2A. The gas turbine engine electric generator feeds the traction multiphase electric motor through the multilevel inverter in normal failure-free operation. In failure case of the fuel tank, gas turbine engine, gearbox, or electric generator, it is possible to achieve the continuation of the flight for completing the task based on the value of battery storage capacity. Thus, the RBD is shown in Fig. 2.2B, where "SB" represents the switchboard.

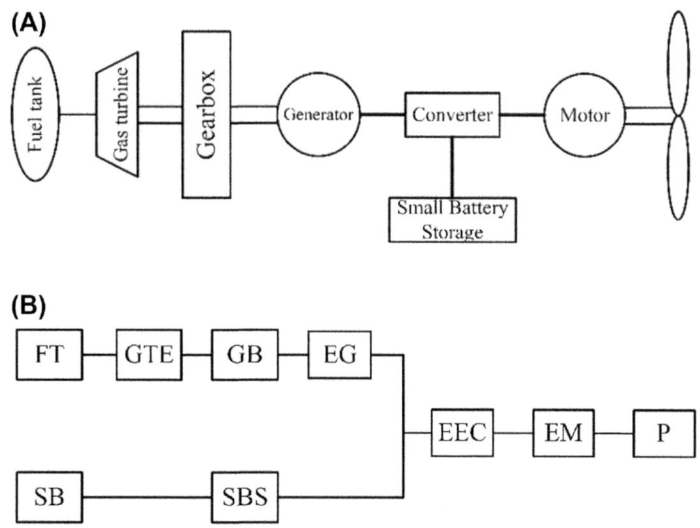

FIGURE 2.2

(A) Structure of serial hybrid traction drive, (B) reliability block diagram of serial hybrid traction drive.

3.2.2 Parallel hybrid traction drive topology

The parallel hybrid traction drive consists of a fuel tank (FT), a gas turbine engine (GTE), a gearbox (GB), an electric energy converter (EEC), an electric motor (EM), a small battery storage (SBS), and a propeller (P). The structure of this powertrain is shown in Fig. 2.3A. In the normal operation scenario, the gas turbine engine and the electric motor rotate the propeller through the gearbox simultaneously. In the failure case of either the fuel tank, the gas turbine engines, or the small battery storage, the converter, the motor, it is possible that another energy source can power the propeller to complete the task performance or an emergency landing. Thus, the corresponding RBD is presented in Fig. 2.3B.

3.2.3 Combined hybrid traction drive topology

A fuel tank (FT), a gas turbine engine (GTE), a gearbox (GB), a shaft (S), an electric generator (EG), an electric energy converter (EEC), an electric motor (EM), a small battery storage (SBS), and a propeller (P) compose the combined hybrid traction drive; its structure is shown in Fig. 2.4A. In the normal failure-free operation of this powertrain, there are two possible embodiments of the flight. In the first "electric" version of flight, the gas turbine electric generator feeds the multiphase electric traction motor through multilevel converter. In the second, the "mechanical" version of the flight, the gas turbine engine directly rotates the propeller of the helicopter through the gearbox and the shaft.

Therefore, in the failure case of the fuel tank, the gas turbine, or the gearbox, a further short-time flight to the safe landing of helicopter can be carried out by the electric energy of the battery storage. Thus, the corresponding RBD is shown in Fig. 2.4B, where "S" represents the shaft and "SB" is the switchboard.

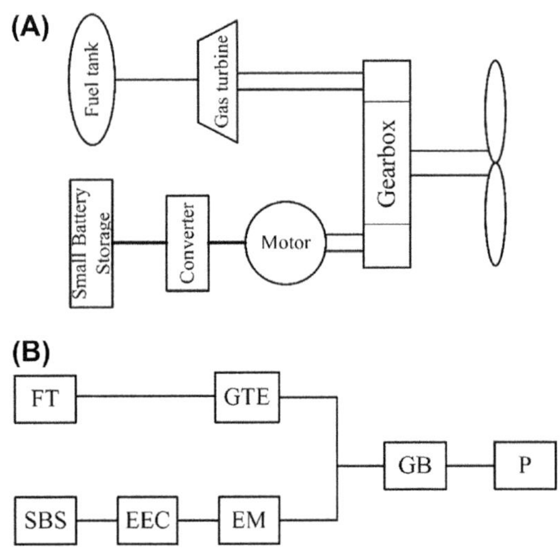

FIGURE 2.3

(A) Structure of parallel hybrid traction drive, (B) reliability block diagram of parallel hybrid traction drive.

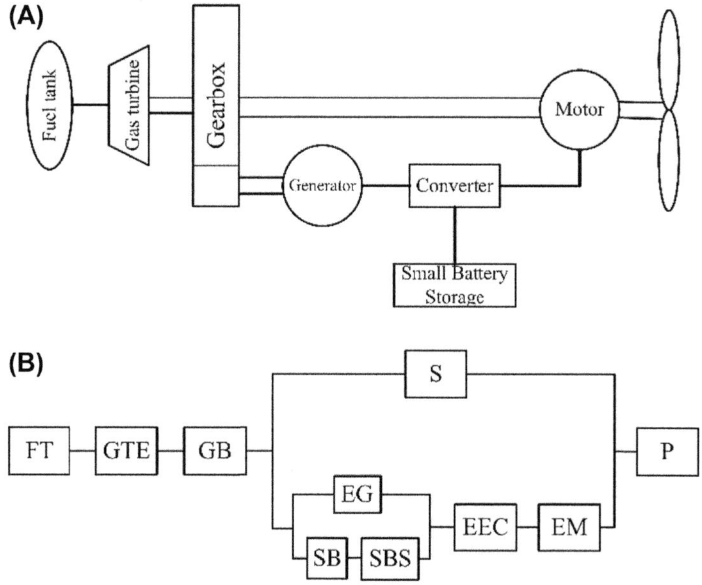

FIGURE 2.4

(A) Structure of combined hybrid traction drive, (B) reliability block diagram of combined hybrid traction drive.

4. The impact of DoH on availability and fault tolerance of hybrid-electric traction drive

After systematically describing the characteristics and operating scenarios of conventional and hybrid-electric helicopter propulsion systems, the availability and fault tolerance of each traction drive can be determined through *Lz*-transform, which has been introduced for DSCT Markov processes. Then the procedure for analyzing the influence of DoH on availability and fault tolerance of hybrid-electric traction drive is shown in this section. The following sections represent a step of the program.

4.1 Elements description

The first step of availability analyses is to build the Markov models of system elements. The helicopter's traction drive can be considered as a repairable system to calculate the availability, consisting from a repairable element having two states (fully working and fully failed). The state space diagram of the repairable system's elements is presented in Fig. 2.5.

The following system of differential equations is built to calculate probabilities of each state:

$$\begin{cases} \dfrac{dp_1^{(i)}(t)}{dt} = -\lambda_i p_1^{(i)}(t) + \mu_i p_2^{(i)}(t) \\ \dfrac{dp_2^{(i)}(t)}{dt} = \lambda_i p_1^{(i)}(t) - \mu_i p_2^{(i)}(t) \end{cases} \tag{2.8}$$

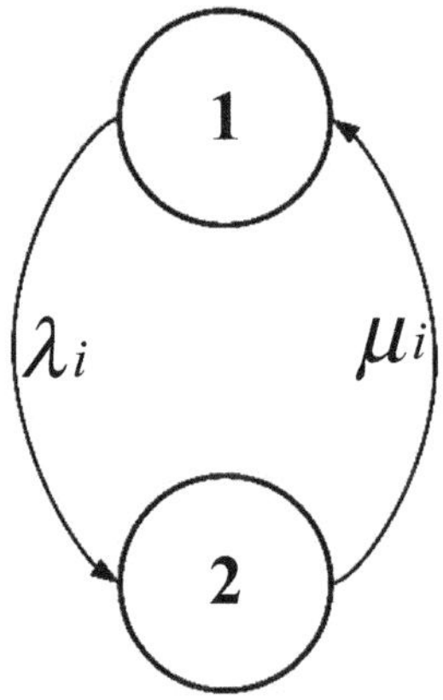

FIGURE 2.5

State space diagram of the repairable system's elements.

where $i = FT, GTE_C, GTE, EG, SB, GB, P, S, EEC, EM, SBS$, and the initial conditions are: $p_1^{(i)}(0) = 1; p_2^{(i)}(0) = 0$.

For the numerical solution of this system, MATLAB was used to obtain $p_1^{(i)}(t), p_2^{(i)}(t)$. Simultaneously, fully working and fully failed indicate the relative power or performance of an element is 100% (for simplification—1) or 50% (for simplification—0.5) and 0, respectively. Therefore, for the system elements the output performance of stochastic processes can be obtained as follows:

$i = FT, GTE, EG, SB, GB, P, S, EEC, EM, SBS$	
$\begin{cases} g^{(i)} = \left\{g_1^{(i)}, g_2^{(i)}\right\} = \{1, 0\} \\ p^{(i)}(t) = \left\{p_1^{(i)}(t), p_2^{(i)}(t)\right\} \end{cases}$	$\begin{cases} g^{GTE_C} = \left\{g_1^{GTE_C}, g_2^{GTE_C}\right\} = \{0.5, 0\} \\ p^{GTE_C}(t) = \left\{p_1^{GTE_C}(t), p_2^{GTE_C}(t)\right\} \end{cases}$

The sets $g^{(i)}$, $p^{(i)}(t)$ $i = FT, GTE_C, GTE, EG, SB, GB, P, S, EEC, EM, SBS$ define the Lz-transforms for each element as follows:

Fuel tank:

$$Lz\{g^{FT}(t)\} = p_1^{FT}(t)z^1 + p_2^{FT}(t)z^0$$

Gas turbine engine for serial hybrid, parallel hybrid, and combined hybrid propulsion systems:

$$Lz\{g^{GTE}(t)\} = p_1^{GTE}(t)z^1 + p_2^{GTE}(t)z^0$$

Gas turbine engine for conventional traction drive:

$$Lz\{g^{GTE_C}(t)\} = p_1^{GTE_C}(t)z^{0.5} + p_2^{GTE_C}(t)z^0$$

Electric generator:

$$Lz\{g^{EG}(t)\} = p_1^{EG}(t)z^1 + p_2^{EG}(t)z^0$$

Switchboard:

$$Lz\{g^{SB}(t)\} = p_1^{SB}(t)z^1 + p_2^{SB}(t)z^0$$

Gearbox:

$$Lz\{g^{GB}(t)\} = p_1^{GB}(t)z^1 + p_2^{GB}(t)z^0$$

Propeller:

$$Lz\{g^{P}(t)\} = p_1^{P}(t)z^1 + p_2^{P}(t)z^0$$

Shaft:

$$Lz\{g^{S}(t)\} = p_1^{S}(t)z^1 + p_2^{S}(t)z^0$$

Electric energy converter:

$$Lz\{g^{EEC}(t)\} = p_1^{EEC}(t)z^1 + p_2^{EEC}(t)z^0$$

Electric motor:

$$Lz\{g^{EM}(t)\} = p_1^{EM}(t)z^1 + p_2^{EM}(t)z^0$$

Small battery storage for hybrid applications:

$$Lz\{g^{SBS}(t)\} = p_1^{SBS}(t)z^1 + p_2^{SBS}(t)z^0$$

Moreover, for the subsystem of gas turbine engines used in conventional traction drive, where two similar elements connected in parallel, *Lz* transform may be presented as follows:

$$\begin{aligned} Lz\{G^{SGTE}(t)\} &= \Omega_{f_{par}}\left(Lz\{g^{GTEc}(t)\}, Lz\{g^{GTEc}(t)\}\right) = \Omega_{f_{par}}\left(p_1^{GTEc}(t)z^{0.5} + p_2^{GTEc}(t)z^0\right)^2 \\ &= \left\{p_1^{GTEc}(t)\right\}^2 z^{2\cdot 0.5} + 2\cdot p_1^{GTEc}(t)p_2^{GTEc}(t)z^{0.5} + \left\{p_2^{GTEc}(t)\right\}^2 z^0 \\ &= p_1^{SGTE}(t)z^1 + p_2^{SGTE}(t)z^{0.5} + p_3^{SGTE}(t)z^0 \end{aligned}$$

where

$$\begin{cases} p_1^{SGTE}(t) = \left\{p_1^{GTEc}(t)\right\}^2 \\ p_2^{SGTE}(t) = 2\cdot p_1^{GTEc}(t)p_2^{GTEc}(t) \\ p_3^{SGTE}(t) = \left\{p_2^{GTEc}(t)\right\}^2 \end{cases}$$

4.2 Multi-state models for conventional and hybrid-electric traction drive topologies

4.2.1 Topology of conventional traction drive

Components of a conventional (C) traction drive connected in series, so the composition operator $\Omega_{f_{ser}}$ is used to obtain the whole system's Lz-transform, where the powers of z are found as minimum of the powers of the corresponding terms, shown as follows:

$$Lz\{G^C(t)\} = \Omega_{f_{ser}}\left(Lz\{g^{FT}(t)\},\ Lz\{g^{SGTE}(t)\}, Lz\{g^{GB}(t)\}, Lz\{g^P(t)\}\right)$$
$$= \Omega_{f_{ser}}\left(\begin{array}{c} p_1^{FT}(t)z^1 + p_2^{FT}(t)z^0, p_1^{SGTE}(t)z^1 + p_2^{SGTE}(t)z^{0.5} + p_3^{SGTE}(t)z^0, \\ p_1^{GB}(t)z^1 + p_2^{GB}(t)z^0, p_1^P(t)z^1 + p_2^P(t)z^0 \end{array}\right)$$

Using the following notations

$$P_1^C(t) = p_1^{FT}(t)p_1^{SGTE}(t)p_1^{GB}(t)p_1^P(t)$$
$$P_2^C(t) = p_1^{FT}(t)p_2^{SGTE}(t)p_1^{GB}(t)p_1^P(t)$$
$$P_3^C(t) = p_1^{FT}(t)p_3^{SGTE}(t)p_1^{GB}(t)p_1^P(t) + p_2^{FT}(t)p_1^{GB}(t)p_1^P(t) + p_1^{GB}(t)p_2^P(t) + p_2^{GB}(t),$$

the whole system's Lz-transform is defined in the following form:

$$Lz\{G^C(t)\} = P_1^C(t)z^1 + P_2^C(t)(t)z^{0.5} + P_3^C(t)z^0 \tag{2.9}$$

In line with Eq. (2.7), the availability for the conventional helicopter traction drive is:

$$A^C(t) = P_1^C(t) + P_2^C(t) = p_1^{FT}(t)\{p_1^{SGTE}(t) + p_2^{SGTE}(t)\}p_1^{GB}(t)p_1^P(t) \tag{2.10}$$

4.2.2 Topology of combined hybrid traction drive

The combined hybrid (CH) traction drive topology is selected to demonstrate the availability modeling process of hybrid-electric traction drives through Lz-transform. According to the normal operation scenarios and the RBD, the composition operators $\Omega_{f_{ser}}$ and $\Omega_{f_{par}}$ are used to calculate the whole system's Lz-transform.

First, using the composition operators $\Omega_{f_{ser}}$ to obtain the Lz-transform $Lz\{g^{EEE}(t)\}$ for the electric generator, electric energy converter, and electric motor (EEE), connected in series, the powers of z are found as the minimum of the powers of the corresponding terms:

$$Lz\{g^{EEE}(t)\} = \Omega_{f_{ser}}\left(Lz\{g^{EG}(t)\}, Lz\{g^{EEC}(t)\}, Lz\{g^{EM}(t)\}\right)$$
$$= \Omega_{f_{ser}}\left(p_1^{EG}(t)z^1 + p_2^{EG}(t)z^0, p_1^{EEC}(t)z^1 + p_2^{EEC}(t)z^0, p_1^{EM}(t)z^1 + p_2^{EM}(t)z^0\right)$$
$$= P_1^{EEE}(t)z^1 + P_2^{EEE}(t)z^0$$

where

$$\begin{cases} P_1^{EEE}(t) = p_1^{EG}(t)p_1^{EEC}(t)p_1^{EM}(t) \\ P_2^{EEE}(t) = 1 - P_1^{EEE}(t) \end{cases}$$

Then, using the composition operators $\Omega_{f_{par}}$ to obtain the Lz-transform $Lz\{g^{EEES}(t)\}$ for the parallel connection between EEE and the shaft (EEES), the powers of z are calculated as the sum of powers of the corresponding terms:

$$\begin{aligned} Lz\{g^{EEES}(t)\} &= \Omega_{f_{par}}\left(Lz\{g^{EEE}(t)\}, Lz\{g^{S}(t)\}\right) = \Omega_{f_{par}}\left(P_1^{EEE}(t)z^1 + P_2^{EEE}(t)z^0, p_1^S(t)z^1 + p_2^S(t)z^0\right) \\ &= P_1^{EEES}(t)z^2 + P_2^{EEES}(t)z^1 + P_3^{EEES}(t)z^0 \end{aligned}$$

where

$$\begin{cases} P_1^{EEES}(t) = P_1^{EEE}(t)p_1^S(t) \\ P_2^{EEES}(t) = P_1^{EEE}(t)p_2^S(t) + P_2^{EEE}(t)p_1^S(t) \\ P_3^{EEES}(t) = P_2^{EEE}(t)p_2^S(t) \end{cases}$$

Finally, the composition operator $\Omega_{f_{ser}}$ is used in order to obtain the system's normal operation Lz-transform $Lz\{G_1^{CH}(t)\}$ for the EEES and other elements, connected in series, where the powers of z are found as the minimum of the powers of the corresponding terms:

$$\begin{aligned} Lz\{G_1^{CH}(t)\} &= \Omega_{f_{ser}}\left(Lz\{g^{FT}(t)\}, Lz\{g^{GTE}(t)\}, Lz\{g^{GB}(t)\}, Lz\{g^{EEES}(t)\}, Lz\{g^{P}(t)\}\right) \\ &= \Omega_{f_{ser}}\begin{pmatrix} p_1^{FT}(t)z^1 + p_2^{FT}(t)z^0, p_1^{GTE}(t)z^1 + p_2^{GTE}(t)z^0, p_1^{GB}(t)z^1 + p_2^{GB}(t)z^0, \\ P_1^{EEES}(t)z^2 + P_2^{EEES}(t)z^1 + P_3^{EEES}(t)z^0, p_1^P(t)z^1 + p_2^P(t)z^0 \end{pmatrix} \\ &= P_{11}^{CH}(t)z^1 + P_{12}^{CH}(t)z^0 \end{aligned}$$

where

$$\begin{cases} P_{11}^{CH}(t) = p_1^{FT}(t)p_1^{GTE}(t)p_1^{GB}(t)p_1^P(t)\left(P_1^{EEES}(t) + P_2^{EEES}(t)\right) \\ P_{12}^{CH}(t) = 1 - P_{11}^{CH}(t) \end{cases}$$

Therefore, using the whole system's Lz-transform, it is possible to obtain the expression of the availability of the combined hybrid helicopter propulsion system under normal operation scenarios in the following form:

$$A_1^{CH}(t) = P_{11}^{CH}(t) = p_1^{FT}(t)p_1^{GTE}(t)p_1^{GB}(t)p_1^P(t)\left(P_1^{EEES}(t) + P_2^{EEES}(t)\right) \tag{2.11}$$

Moreover, in the failure case of the fuel tank, the gas turbine, or the gearbox, a further short-time flight to enable a safe landing of the helicopter can be determined by the remaining electric energy of the battery storage. Then, the composition operator Ω_{fser} leads to the system's Lz-transform $Lz\{G_2^{CH}(t)\}$ for the switchboard, small battery storage, electric energy converter, electric motor, and the propeller, connected in series, where the powers of z are found as the minimum of the powers of the corresponding terms:

$$\begin{aligned} Lz\{G_2^{CH}(t)\} &= \Omega_{fser}\left(Lz\{g^{EG}(t)\}, Lz\{g^{EEG}(t)\}, Lz\{g^{EM}(t)\}\right) \\ &= \Omega_{fser}\left(p_1^{EG}(t)z^1 + p_2^{EG}(t)z^0, p_1^{EEC}(t)z^1 + p_2^{EEC}(t)z^0, p_1^{EM}(t)z^1, p_2^{EM}(t)z^0\right) \\ &= p_1^{EEE}(t)z^1 + p_2^{EEE}(t)z^0 \end{aligned}$$

where

$$\begin{cases} P_{21}^{CH}(t) = p_1^{SB}(t)p_1^{SBS}(t)p_1^{EEC}(t)p_1^{EM}(t)p_1^{P}(t) \\ P_{22}^{CH}(t) = 1 - P_{21}^{CH}(t) \end{cases}$$

For the combined hybrid traction drive, the availability of the partial failure operation scenarios has the following form:

$$A_2^{CH}(t) = P_{21}^{CH}(t) = p_1^{SB}(t)p_1^{SBS}(t)p_1^{EEC}(t)p_1^{EM}(t)p_1^{P}(t) \tag{2.12}$$

Therefore, the availability of the combined hybrid traction drive has the following form:

$$\begin{aligned} A^{CH}(t) &= A_1^{CH}(t) + A_2^{CH}(t)\cdot\left(p_2^{FT}(t) + p_2^{GTE}(t) + p_2^{GB}(t)\right) \\ &= p_1^{FT}(t)p_1^{GTE}(t)p_1^{GB}(t)p_1^{P}(t)\left(P_1^{EEES}(t) + P_2^{EEES}(t)\right) \\ &\quad + p_1^{SB}(t)p_1^{SBS}(t)p_1^{EEC}(t)p_1^{EM}(t)p_1^{P}(t)\cdot\left(p_2^{FT}(t) + p_2^{GTE}(t) + p_2^{GB}(t)\right) \end{aligned} \tag{2.13}$$

Similarly, the availability of serial hybrid (SH) and parallel hybrid (PH) traction drives can be easily obtained through Markov models and *Lz*-transform approach, which are shown as follows:

For serial hybrid:

$$\begin{aligned} A^{SH}(t) = {} & p_1^{EEC}(t)p_1^{EM}(t)p_1^{P}(t)\{p_1^{FT}(t)p_1^{GTE}(t)p_1^{GB}(t)p_1^{EG}(t) + p_1^{SB}(t)p_1^{SBS}(t)\left(p_2^{FT}(t) + p_2^{GB}(t)\right. \\ & \left.+ p_2^{GTE}(t) + p_2^{EG}(t)\right)\} \end{aligned} \tag{2.14}$$

For parallel hybrid:

$$A^{PH}(t) = p_1^{P}(t)\{p_1^{FT}(t)p_1^{GTE}(t)p_1^{GB}(t) + p_1^{SBS}(t)p_1^{EEC}(t)p_1^{EM}(t)\left(1 - p_1^{FT}(t)p_1^{GTE}(t)p_1^{GB}(t)\right)\} \tag{2.15}$$

Consequently, the plots of availability of conventional and hybrid-electric traction drive topologies are depicted in Fig. 2.6 based on the failure rates and repair rates of each element presented in Table 2.1.

4.3 Impact of DoH on availability and fault tolerance of hybrid-electric traction drive

In accordance with the definition of DoH and Eq. (2.2), the impact of DoH on the availability of the hybrid-electric traction drives is shown as follows.

It is clearly depicted in Fig. 2.7 that the DoH has a positive impact on the availability of hybrid-electric traction drives. Moreover, the quantitative analyses of the impact can be obtained by measuring the converging values of availability under various DoH, where the results are shown in Fig. 2.8:

In summary, the DoH has a positive impact on the availability of hybrid-electric propulsion system, which indicates that the DoH is an essential factor in the design process of helicopter's traction drive. In addition, the proposed method can provide a quantitative value of DoH to meet specific project's requirements on availability and fault tolerance, which is ultimately beneficial to appropriate helicopter's traction drive decision-making.

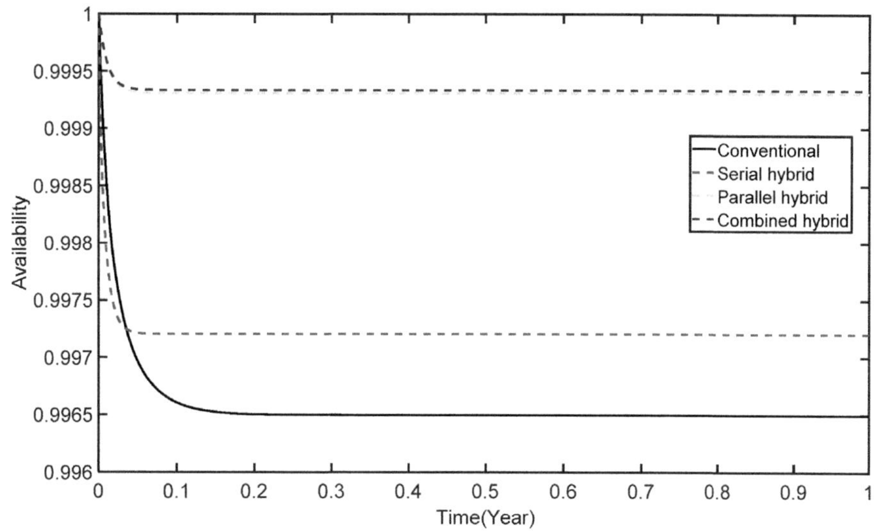

FIGURE 2.6

Availability comparison for conventional and hybrid-electric traction drives.

Table 2.1 Failure and repair rates of each system's element.

System element	Failure rates (year^{-1})	Repair rates (year^{-1})
Fuel tank	0.06	48
Gas turbine engine	1.6	55
Electric generator	0.1	175
Switchboard	0.05	750
Gear box	0.09	110
Propeller	0.06	95
Shaft	0.02	138
Electric energy converter	0.3	584
Electric motor	0.2	120
Small battery storage	0.4	250
Large battery storage	0.08	438

5. Full-electric traction drive topologies and corresponding availability analyses

The remarks in Section 4 show that the DoH has a positive impact on availability and fault tolerance of helicopter's traction drive. Therefore, three full-electric traction drive topologies are analyzed in this

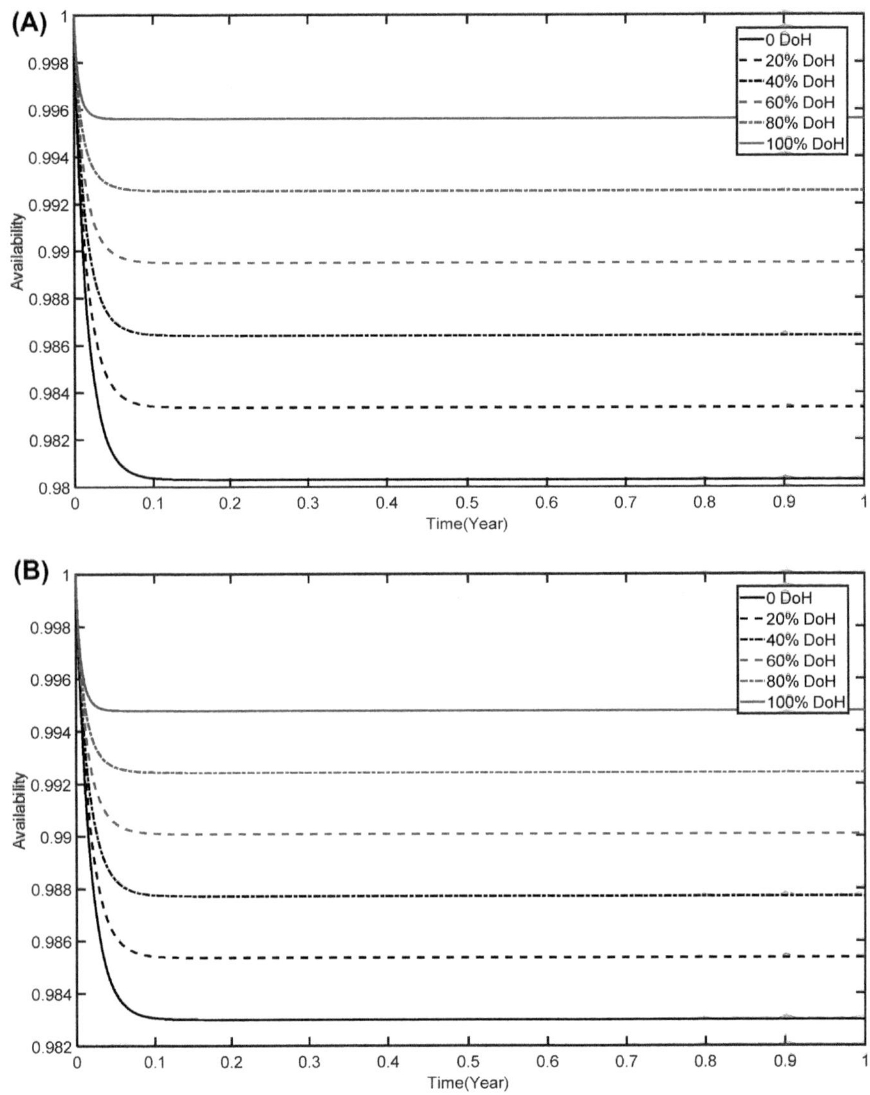

FIGURE 2.7

Impact of DoH on the availability of hybrid-electric traction drives: (A) serial, (B) parallel, and (C) combined.

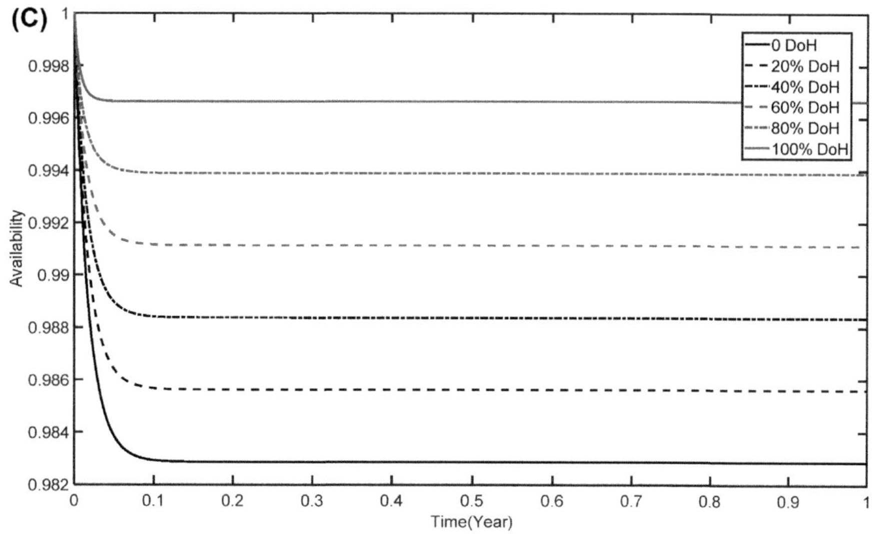

FIGURE 2.7 Cont'd

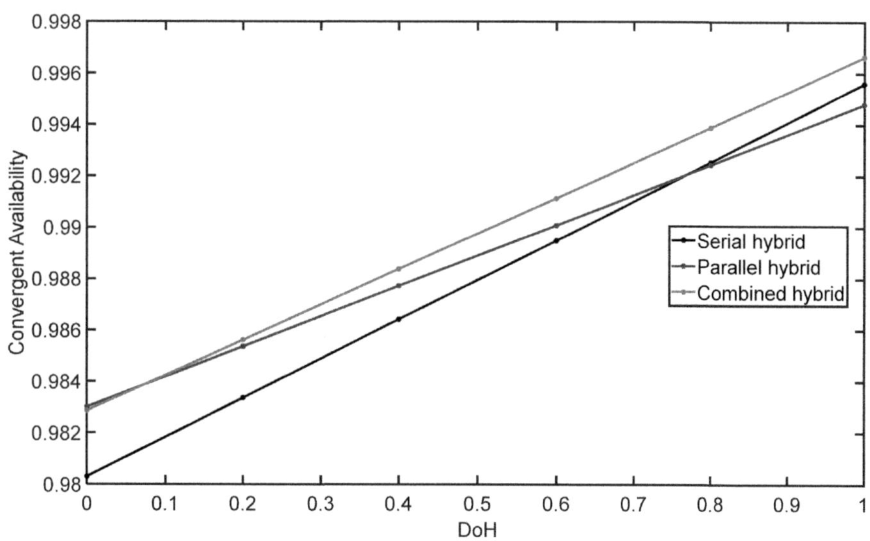

FIGURE 2.8

Quantitative Impact of DoH on the increment of the availability of hybrid-electric traction drives.

section that the energy of helicopter flight is only provided by the large battery storage (designated as LBS), which can function as the reference for designing a pure electric traction drive in the future. The details of each topology and analyses results are presented in the following sections.

5.1 Full-electric traction drive topologies

5.1.1 Single-line electric traction drive topology

The single-line electric propulsion system consists of a large battery storage (LBS), an electric energy converter (EEC), an electric motor (EM), and a propeller (P). The components are connected in series and the structure, as well as the corresponding RBD, is shown in Fig. 2.9.

5.1.2 Dual-line 1 electric traction drive topology

The topology of the dual-electric 1 traction drive is shown in Fig. 2.10A. The system consists of two subsystems, each consisting of an electric energy converter (EEC) and an electric motor (EM), a large battery storage (LBS), a gearbox (GB), and a propeller (P). The two subsystems work in parallel to rotate the propeller, and the RBD is shown in Fig. 2.10B.

5.1.3 Dual-line 2 electric traction drive topology

The topology of the dual-electric 2 traction drive is shown in Fig. 2.11A. The system consists of two subsystems, each consisting of an electric energy converter (EEC), an electric motor (EM), a propeller (P), and one large battery storage (LBS) power for both subsystems. The two subsystems work in parallel powered by the battery storage. The entire RBD is shown in Fig. 2.11B.

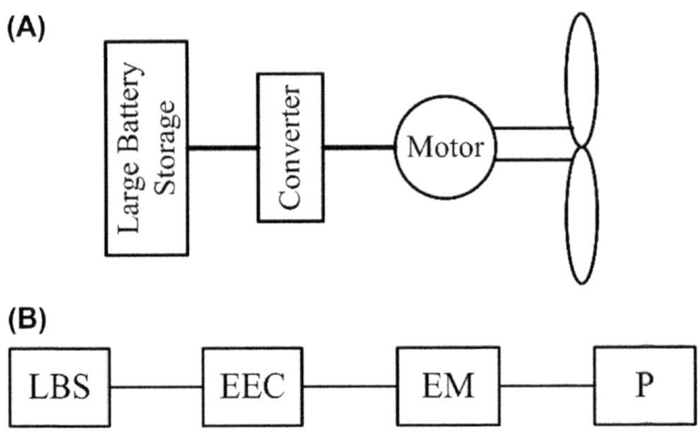

FIGURE 2.9

(A) Structure of single-line electric traction drive, (B) reliability block diagram of single-line electric traction drive.

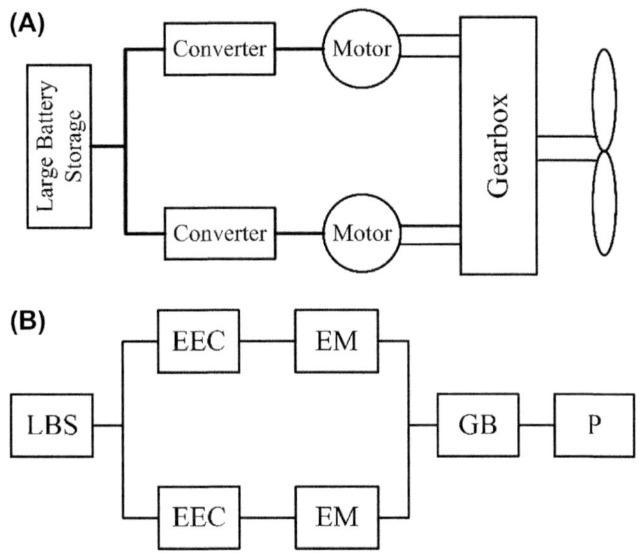

FIGURE 2.10

(A) Structure of dual-electric 1 traction drive, (B) reliability block diagram of dual-electric 1 traction drive.

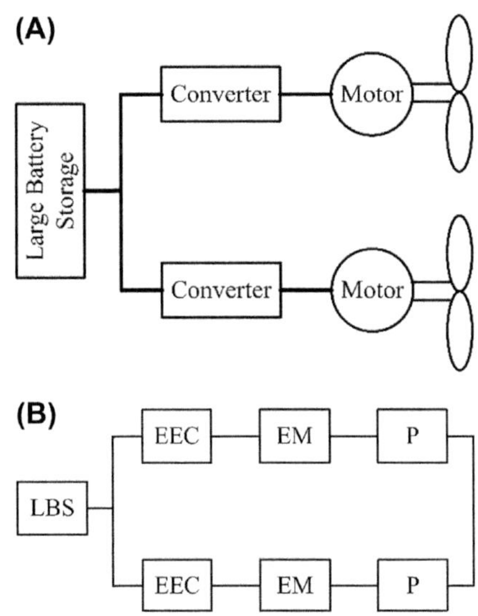

FIGURE 2.11

(A) Structure of dual-electric 2 traction drive, (B) reliability block diagram of dual-electric 2 traction drive.

5.2 Availability of comparative analyses

Let us use the same strategy for evaluation the availability of full-electric traction drive topologies. Note, due to more free space to install an additional amount of electric battery cells in full-electric embodiments, the battery storage of full-electric traction drive topologies designated as LBS in full-electric topologies. The corresponding Lz-transforms for the LBS is expressed by:

$$Lz\{g^{LBS}(t)\} = p_1^{LBS}(t)z^1 + p_2^{LBS}(t)z^0$$

Then, taking the dual-electric 1 (DE1) traction drive as example to show the availability modeling process by Lz-transform for full-electric traction drives. There are two subsystems working in parallel, each consisting of an electric energy converter (EEC) and an electric motor (EM) in this propulsion system. Using the composition operator $\Omega_{f_{ser}}$, we obtain the Lz-transform $Lz\{g^{EE}(t)\}$ for each subsystem, where the powers of z are calculated as the minimum of the powers of the corresponding terms:

$$\begin{aligned} Lz\{g^{EE}(t)\} = \Omega_{fser}(Lz\{g^{EEC}(t)\}, Lz\{g^{EM}(t)\}) &= \Omega_{fser}(p_1^{EEC}(t)z^1 + p_2^{EEC}(t)z^0, p_1^{EM}(t)z^1 + p_2^{EM}(t)z^0) \\ &= p_1^{EEC}(t)p_1^{EM}(t)z^1 + (p_1^{EEC}(t)p_2^{EM}(t) + p_2^{EEC}(t))z^0 \end{aligned}$$

Using the notations:

$$\begin{cases} P_1^{EE}(t) = p_1^{EEC}(t)p_1^{EM}(t) \\ P_2^{EE}(t) = p_1^{EEC}(t)p_2^{EM}(t) + p_2^{EEC}(t) \end{cases}$$

the resulting Lz-transform for the electric energy converter and electric motor subsystem is as follows:

$$Lz\{g^{EE}(t)\} = P_1^{EE}(t)z^1 + P_2^{EE}(t)z^0$$

Using the composition operator $\Omega_{f_{par}}$ for the two subsystems, connected in parallel, the Lz-transform $Lz\{g^{EES}(t)\}$ for the whole electric energy converter and electric motor subsystems results in:

$$Lz\{g^{EES}(t)\} = \Omega_{fpar}(Lz\{g^{EE}(t)\}, Lz\{g^{EE}(t)\})$$

Using the notations:

$$\begin{cases} P_1^{EES}(t) = \{P_1^{EE}(t)\}^2 \\ P_2^{EES}(t) = 2 \cdot P_1^{EE}(t)P_2^{EE}(t) \\ P_3^{EES}(t) = \{P_2^{EE}(t)\}^2 \end{cases}$$

the resulting Lz-transform for the completely electric energy converter and electric motor subsystems is as follows:

$$Lz\{g^{EES}(t)\} = P_1^{EES}(t)z^2 + P_2^{EES}(t)z^1 + P_3^{EES}(t)z^0$$

Using the composition operator $\Omega_{f_{ser}}$ for subsystems and other elements connected in series, where the powers of z are calculated as the minimum of the powers of the corresponding terms, so the Lz-transform $Lz\{G^{DE1}(t)\}$ for the whole dual-electric 1 propulsion system is defined as follows:

$$\begin{aligned} Lz\{G^{DE1}(t)\} &= \Omega_{f_{ser}}\left(Lz\{g^{LBS}(t)\}, Lz\{g^{EES}(t)\}, Lz\{g^{GB}(t)\}, Lz\{g^{P}(t)\}\right) \\ &= \Omega_{f_{ser}}\begin{pmatrix} p_1^{LBS}(t)z^1 + p_2^{LBS}(t)z^0, P_1^{EES}(t)z^2 + P_2^{EES}(t)z^1 + P_3^{EES}(t)z^0, \\ p_1^{GB}(t)z^1 + p_2^{GB}(t)z^0, p_1^{P}(t)z^1 + p_2^{P}(t)z^0 \end{pmatrix} \\ &= P_1^{DE1}(t)z^1 + P_2^{DE1}(t)z^0 \end{aligned}$$

where

$$\begin{cases} P_1^{DE1} = p_1^{LBS}(t)\left(P_1^{EES}(t) + P_2^{EES}(t)\right)p_1^{GB}(t)p_1^{P}(t) \\ P_2^{DE1} = p_1^{LBS}(t)\left(P_1^{EES}(t) + P_2^{EES}(t)\right)\left(p_1^{GB}(t)p_2^{P}(t) + p_2^{GB}(t)\right) + \left(p_1^{BSb}(t)P_3^{EES}(t) + p_2^{BSb}(t)\right) \end{cases}$$

According to the whole system's Lz-transform, it is possible to obtain the availability function for the constant demand level. For the dual-electric 1 propulsion system, the availability model has the following form:

$$\begin{aligned} A^{DE1}(t) = P_1^{DE1} &= p_1^{LBS}(t)\left(P_1^{EES}(t) + P_2^{EES}(t)\right)p_1^{GB}(t)p_1^{P}(t) \\ &= p_1^{LBS}(t)p_1^{EEC}(t)p_1^{EM}(t)p_1^{GB}(t)p_1^{P}(t)\{p_1^{EEC}(t)p_2^{EM}(t) + p_2^{EEC}(t) + 1\} \end{aligned} \quad (2.16)$$

Similarly, the availability models of single-line electric (SLE) and dual-electric 2 (DE2) propulsion systems can be easily obtained through Markov models and Lz-transform approach, which are shown as follows:

$$A^{SLE}(t) = p_1^{LBS}(t)p_1^{EEC}(t)p_1^{EM}(t)p_1^{P}(t) \quad (2.17)$$

$$A^{DE2}(t) = p_1^{LBS}(t)p_1^{EEC}(t)p_1^{EM}(t)p_1^{P}(t)\{p_1^{EEC}(t)p_1^{EM}(t)p_2^{P}(t) + p_1^{EEC}(t)p_2^{EM}(t) + p_2^{EEC}(t) + 1\} \quad (2.18)$$

Therefore, the comparative results of availability between full-electric traction drive topologies are shown in Fig. 2.12.

Comparing Figs. 2.6, 2.7, and 2.12, although the availability of hybrid-electric traction drives increases with the DoH, the dual-electric 2 traction drive owns the optimal performance, which certifies the advantage and significance of the electrification of vehicles from the perspective of reliability and system safety.

6. Sensitivity analysis of system elements

The sensitivity analysis is used to verify the impact of input parameters changes on a given model output. The sensitivity analysis of reliability measures are used to study the effects of different parameters on the system reliability and mean time to failure. The investigation of the impact of changing different failure rates—for different components in multistate system (MSS)—is often important for practical engineering. Based on this reliability, a researcher or an engineer can make appropriate decisions for MSS reliability improvement.

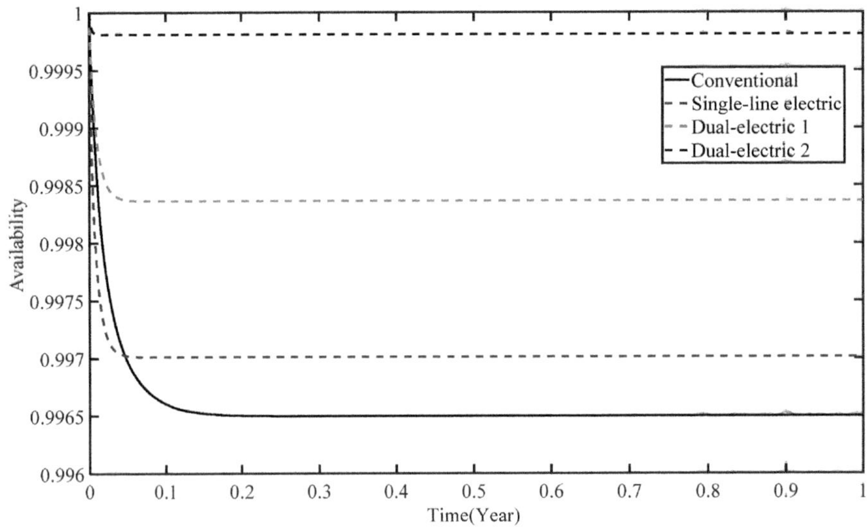

FIGURE 2.12

Availability comparison between full-electric and conventional traction drive topologies.

A most common approach for a sensitivity analysis is changing one factor at a time (OAT) [17], to ascertain the effect occurring on the system output. OAT includes the following steps: (a) assuming one input variable, keeping the rest unchanged; (b) returning the variable to its normal value, and repeating the same process for each of the other parameters. Sensitivity then obtained by monitoring changes in the output by using partial derivatives.

The detailed and theoretical description of the element of sensitivity assessment method is shown in Ref. [18]. The sensitivity measure of availability corresponding to failure rate changes can be calculated as follows:

$$S^{A^m}_{a^{(j)}_{lk}} = \frac{\partial A^m}{\partial a^{(j)}_{lk}} \tag{2.19}$$

where A^m is the availability of each propulsion system, $m = C, SH, PH, CH, SLE, DE1, DE2$. $a^{(j)}_{lk}$ is transition intensity for the transition from state l to state k in the multistate element$j$, $j = FT, GTE, EG, S, GB, P, EEC, EM, SBS, LBS$. The transition intensity matrix a^i $(i = FT, GTE, EG, S, GB, P, EEC, EM, SBS, LBS)$ for each element is shown as follows:

$$a^i = \begin{bmatrix} -\lambda_i \lambda_i \mu_i \\ -\mu_i \end{bmatrix}$$

In Fig. 2.13, the graph of the conventional topology's steady-state availability is presented as functions of the fuel tank failure rate λ_{FT}, the gas turbine engine failure rate λ_{GTE}, the gearbox failure

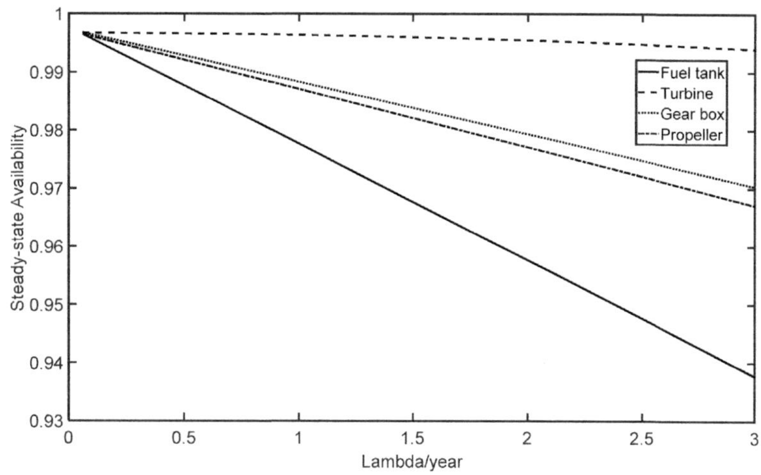

FIGURE 2.13

Graphs of convergent availability as functions of failure rates λ_{FT}, λ_{GTE}, λ_G, and λ_P for conventional helicopter propulsion system.

rate λ_G, and the propeller failure rate λ_P, since the availability of any traction drive topologies converges quickly. Note, that when one failure rate is changing, the rest failure rates are keeping its nominal value, that is, the OAT principle.

The sensitivity measures are calculated by using expression (Eq. 2.19). Values for all HEH propulsion systems are shown as follows (four decimal places) (Table 2.2).

The sensitivity analysis results show the effect of failure rate changes of various components on the availability of the helicopter traction drive. Moreover, the results show that with the current state of the art and repair capabilities, structural and functional redundancy can effectively improve the fault tolerance of the propulsion system of a hybrid-electric helicopter. It will be very useful for determining the safety-critical component and designing the optimal traction drive topology of a helicopter. In conclusion, the dual-electric 2 topology is the optimal structure for the helicopter's traction drive, which performs best in the comparisons of availability between different structures. Moreover, the component sensitivity analysis shows that this traction drive topology provides a very strong fault tolerance.

7. Conclusion

In the current chapter, the results of analysis show that the traction drive of the hybrid-electric and full-electric helicopter have a higher availability than the conventional alternative. The methods presented in this chapter are based on using *Lz*-transform, which is well formalized and which is suitable for practical application in reliability engineering for a real-world MSS's reliability measure analysis.

Based on the availability comparisons between the conventional and the hybrid-electric traction drive topologies of the helicopter, it was concluded that the advancements in multipower source technologies have sufficient potential to provide significant improvements considering the helicopter's

Table 2.2 Sensitivity measures of each propulsion system's element (year^{-2}).

	Conventional	Serial hybrid	Parallel hybrid	Combined hybrid	Single-line electric	Dual-electric 1	Dual-electric 2
Fuel tank	−0.02	0	0	0			
Gas turbine engine	−0.001	0	0	0			
Gearbox	−0.009	0	−0.009	0		−0.009	
Propeller	−0.01	−0.01	−0.01	−0.01	−0.01	−0.01	−0.0002
Shaft				0			
Electric generator		0		0			
Electric energy converter		−0.017	0	0	−0.0017	0	0
Electric motor		−0.008	0	0	−0.008	0	0
Small battery storage		0	0	0			
Large battery storage					−0.0023	−0.0023	−0.0023

reliability and system safety. In addition, the comparative results between full-electric and conventional propulsion system show that using redundancy technology is useful to improve the system availability. The sensitivity analysis shows the fault tolerance of different topologies of traction drive. The comprehensive consideration of all analyses indicates that the dual-electric 2 topology is the optimal structure for a helicopter's traction drive from the point of view of reliability.

However, the fact is that at the current level of technological limitations of battery cells, fuel cell, and other energy storages in weight, size, and energy density characteristics, it is not quite easy to practically implement a full-electric version of the helicopter. Then the dependency of availability of the hybrid-electric propulsion system from the DoH shows that with the increase of *DoH*, the availability can improve, since the electrical part of the propulsion system is much more reliable and easier to repair than the mechanical part. Therefore, in order to improve the availability of the entire traction drive, it is advisable to increase the ratio of electric flight time or to use more reliable gas turbine engines.

References

[1] I. Bolvashenkov, J. Kammermann, H.-G. Herzog, Electrification of helicopter: actual feasibility and prospects, in: Proc. of 13th IEEE Vehicle Power and Propulsion Conference (VPPC'17), 11th–14th December 2017, Belfort, France, 2017, pp. 1–6.

[2] J.J. Kammermann, Potential Analysis of Electrical Drive Trains According to Application Requirements, Dissertation, Department of Electrical and Computer Engineering, Technical University of Munich (TUM), Munich, Germany, 2019, https://doi.org/10.14459/2019md1451565.
[3] M. Cameretti, A. Del Pizzo, L. Di Noia, M. Ferrara, C. Pascarella, Modeling and investigation of a turboprop hybrid electric propulsion system, Aerospace 5 (2018) 123.
[4] D. Buecherl, I. Bolvashenkov, H.-G. Herzog, Verification of the optimum hybridization factor as design parameter of hybrid electric vehicles, in: Proc. of IEEE Vehicle Power and Propulsion Conference, 7th–10th September 2009, Dearborn, MI, USA, 2009, pp. 847–851.
[5] I. Bolvashenkov, H.-G. Herzog, I. Frenkel, L. Khvatskin, A. Lisnianski, Safety-Critical Electrical Drives: Topologies, Reliability, Performance, Springer, 2018.
[6] I. Bolvashenkov, I. Frenkel, J. Kammermann, H.-G. Herzog, Comparison of the battery energy storage and fuel cell energy source for the safety-critical drives considering reliability and fault tolerance, in: Proc. of IEEE International Conference on Information and Digital Technologies (IDT), 5th–7th July 2017, Žilina, Slovakia, 2017, pp. 63–70.
[7] I. Bolvashenkov, J. Kammermann, H.-G. Herzog, Reliability assessment of a fault tolerant propulsion system for an electrical helicopter, in: Proc. of IEEE 12th International Conference on Ecological Vehicles and Renewable Energies (EVER'17), 11th–13th April 2017, Monaco, 2017, pp. 1–6.
[8] I. Bolvashenkov, J. Kammermann, H.-G. Herzog, W.B. Zeng, Reliability evaluation of non-repairable propulsion systems of hybrid-electric helicopter with different level of hybridization, in: Proc. of IEEE 14th International Conference on Ecological Vehicles and Renewable Energies (EVER'19), 8th–10th May 2019, Monaco, 2019.
[9] I. Frenkel, L. Khvatskin, A. Lisnianski, Availability assessment for aging refrigeration system by using Lz-transform, J. Reliab. Stat. Stud. 5 (2) (2012) 33–43.
[10] I. Frenkel, L. Khvatskin, A. Lisnianski, *Lz*-transform application to availability assessment of the air conditioning system with rental equipment working under seasonal weather conditions, J. Inf. Contr. Manag. Syst. 12 (2) (2014) 133–140.
[11] A. Lisnianski, I. Frenkel, Recent Advances in System Reliability, Springer, 2012.
[12] A. Lisnianski, I. Frenkel, Y. Ding, Multi-state System Reliability Analysis and Optimization for Engineers and Industrial Managers, Springer, London, 2010.
[13] A. Lisnianski, *Lz*-transform for a discrete-state continuous-time Markov process and its applications to multi-state system reliability, in: A. Lisnianski, I. Frenkel (Eds.), Recent Advances in System Reliability. Signatures, Multi-State Systems and Statistical Inference, Springer, London, 2012, pp. 79–95.
[14] I. Ushakov, A universal generating function, Soviet Journal of Computer and System Sciences 24 (1986) 37–49.
[15] R de Vries, M.T. Brown, R. Vos, A preliminary sizing method for hybrid-electric aircraft including aero-propulsive interaction effects, in: 2018 Aviation Technology, Integration, and Operations Conference, AIAA AVIATION Forum, (AIAA 2018-4228), 2018.
[16] J.L. Felder, NASA electric propulsion system studies, in: EnergyTech 2015, Cleveland, OH, USA, November 30–December 2, 2015, 2015.
[17] A. Saltelli, K. Chan, E.M. Scott, Sensitivity Analysis, Wiley, Chichester, 2009.
[18] A. Lisnianski, I. Frenkel, L. Khvatskin, On sensitivity analysis of aging multi-state system by using *Lz*-transform, Reliab. Eng. Syst. Saf. 166 (2017) 99–108.

CHAPTER

3 General forms of Bivariate survival functions with reliability applications

Jerzy K. Filus[1], Lidia Z. Filus[2]

[1]*Department of Mathematics and Computer Science, Oakton Community College, Des Plaines, IL, United States;*
[2]*Department. of Mathematics, Northeastern Illinois University, Chicago, IL, United States*

1. Introduction

In this chapter we present a new method for constructing of bivariate probability distributions given any arbitrary pair of the marginals. Usually, whenever facing this type of problem, one is going to seek a proper **copula** [1] in order to "connect" the marginals into the bivariate cdf. For the same problem of the construction we present an **alternative method** for finding a bivariate survival (reliability) function given two marginal survival functions, by means of the defined joiners (see also Refs. [2,3]).

This chapter presents a general theory and methods of construction of any bivariate probability distributions. As shown in Ref. [3] (see Section 3.2, especially Eq. (3.16)) this method is more general than that based on the parameter dependence [4]. As shown, the latter can be regarded as special case of the "joiner method."

As for the applications of joiner method, we mainly stress **reliability** applications of the theory since bivariate distributions can naturally be addressed as possible stochastic models for system component lifetimes, where two-component system reliability structures may either be series or parallel.

As for the theoretical part of this chapter, new methods of model construction, especially in reliability settings (but not only), in many cases are easier, more natural, and more efficient than those by means of copulas. It also allows construction of a wide spread of new classes of bivariate distributions as well as to reinterpret (in Section 3 of this chapter) some existing in literature classes in the light of newly created theory.

The seven known classes of bivariate models as analyzed in Section 3 of the work have all (possibly with exception of the first Gumbel class) reliability or similar roots.

In Section 2, the key point of constructions and the associated theory is the possibility of presenting any bivariate survival function S(x, y) of a random vector (X, Y) having the marginal survival functions, say $S_1(x)\ S_2(y)$ in a simple form of the product $S_1(x)\ S_2(y)\ J(x, y)$, where the "dependence function" J(x, y), called "joiner" [3], is to be found for the sake of the construction. Possibly, in most of the cases, given two marginal distributions, it is easier to find a joiner to model underlying bivariate data than an equivalent copula to construct the proper bivariate cdf. Realize at this point that most of

The Handbook of Reliability, Maintenance, and System Safety through Mathematical Modeling.
https://doi.org/10.1016/B978-0-12-819582-6.00003-4

the copulas rather reflect a mathematical relation, and to find an association between such a relation and the corresponding physical phenomena, may be difficult. In the case of joiners such an association is more natural since, typically, the joiners, as determining stochastic dependences, relate closely to the dependences' description by the Aalen version [5] of the Cox model [6]. It's to be mentioned here that our "joiner theory" takes its roots in the Aalen approach to stochastic dependences [2].

Needless to say, that the theory, outlined in this work, gained more generality and its present formulation became independent from the Aalen ideas.

A significant part of the theory, as the methods of finding proper joiners, given pairs $[S_1(x), S_2(y)]$ of marginals, is to find general analytical criterions for "candidate functions" J(x, y) to fit to the marginals [7]. That is because not all "joiners" work for given fixed marginals.

A solution to this problem is its general formulation we sketched in Ref. [3]. However, the general criterion obtained in Ref. [3], for the case that the second mixed derivative of the product $S_1(x)\ S_2(y)\ J(x,\ y)$ may not exist, regardless of its importance for the theory, is not always easy to apply in practical cases. Thus, in this work, we limit ourselves to the cases we named "continuous" which are applicable to, possibly, all two-component system reliability bivariate models. In these cases the product $S_1(x)\ S_2(y)\ J(x, y)$ is expressible in exponential form [by means of the (existing) hazard rates], given by Eq. (3.4) in the following section. In turn, the candidate for the joiner J(x, y) is represented by another continuous function Ψ(x, y) (see Eq. (3.3)) and the second mixed derivative of $S_1(x)\ S_2(y)\ J(x, y)$ always exists. The main sufficient condition for the joiner (given the two marginals) is nonnegativity of the, existing in the continuous case, second mixed derivative.

This condition (as the condition for the corresponding continuous function Ψ (x, y)) takes on the form of some nontrivial partial integral inequality (Eq. (3.5)). An important part of the whole theory is the theory of this inequality. As shown, some important solutions of it are found immediately. Many solutions can be found by solving analytically well treatable partial integral equations derived from the basic inequality (Eq. (3.5)). The general problem of finding all the solutions of Eq. (3.5) remains, however, open. Other necessary conditions for the joiners to fit the given marginals, or just to be any legitimate joiners, are formulated too (see Ref. [3]).

Some new bivariate models based on the method of joiners were obtained.

Probably the most important underlying fact of the created theory is the achieved **universality** of the joiner representation (Eq. (3.1)). By this we mean that every bivariate survival function is expressible in this simple product form.

An important fact is that the marginals can belong to any class of univariate distributions (including discrete), and the two underlying classes need not be the same.

As for the **applications** of the obtained or obtainable bivariate probability distributions outside reliability settings, there is a wide range of practical situations in which particular models or general methods can be applied. Often the same models as the reliability models may be applied to biomedical or econometric problems and possibly also to meteorology whenever two dependent random quantities (future temperature and air pressure, for example) are of scientific interest.

In association with the applications, especially with modeling reliability of systems, it is a second kind of representation of the bivariate survival (reliability) functions that involves univariate **baseline** (as in a sense a "company" to marginal) survival functions.

The baseline distributions $B_1(x), B_2(y)$ for the bivariate models S(x, y) usually appear during the modeling procedure in situations where there are no physical interactions between tested units.

For example, such a situation one encounters in reliability, when the system components are tested independently of each other outside of the system.

The baseline representation of the same, here considered, survival function also takes on a product form, namely: $S(x, y) = B_1(x)\,B_2(y)\,K(x, y)$, where $B_1(x)$, $B_2(y)$ are the baseline survival functions and K(x, y) (that we call "system function") corresponds to "*physical interactions*" between the "units" (components, for example) after they are put into the system. In general, but not always, the two representations are different and the baseline distributions are not the same as the marginals.

Usually, the baselines are "simpler" than the marginals. The relation between factors of the two representations is described closely in Proposition 1.

Section 3 of this chapter contains applications of the presented theory to seven well-known classes of bivariate distributions. This kind of analysis is a new approach to bivariate survival function investigations. As for the conclusions of the analysis, realize that for the first considered distribution (first Gumbel bivariate exponential) marginal and baseline representations are identical, so the baselines and the marginals are identical too.

This is not the case for any other among the seven distributions we consider. In one case (see Ref. [8]) determination of laboratory conditions and the corresponding baselines was kind of difficult, even if we found some "formal way out" from the situation.

As for the obtained joiners and system functions in those seven examples, some are nice and simple (specially in cases described in Sections 3.1 and 3.3 and also, to a degree, in Sections 3.6 and 3.7 of Section 3) while others, with exception of Section 3.5, are a little problematic.

Anyway, application of the new tools to existing models (at least to some of them) turns out to be promising.

A **direct application** (to reliability and other areas) may be found for three classes of bivariate models obtained by use of the presented methods. The models given by Eqs. (3.7–3.9) for any marginals, determined by two arbitrary failure (hazard) rates $\lambda_1(x), \lambda_2(y)$, seem to, our best knowledge, be new ones.

2. General approach

1. Let X, Y be any two random variables (in **reliability** settings they will mostly be considered as system component lifetimes or strengths) and let $S_1(x) = P(X \geq x)$, $S_2(y) = P(Y \geq y)$ be their survival (reliability) functions, respectively, also called in the following text (probability) "distributions". Actually, there is no restriction on the classes of probability distributions the functions $S_1(x)$, $S_2(y)$ belong to, in particular the two classes may be distinct. However, it is tacitly assumed that both quantities represented by the random variables X, Y describe some "physical" phenomena [in the text below, by the word "*physical*" we do not necessarily mean reality as a subject of physics (i.e., this part of science) but also any "reality" which "exists" outside of describing it mathematics such as sociometrical, psychometrical, econometrical (financial, for example), or other measurable phenomena that could be met in various theoretical considerations] that take place in some common "environment" (so they can be subjected to a variety of possible causality relations). In reliability settings such "environments" represent systems [technical equipment, for example] composed of components, say c_1, c_2 that are characterized by the quantities X, Y.

In absence of such a common environment (when objects, say c_1, c_2, operate in mutual physical isolation) the considered two random variables are always stochastically independent. Consider an arbitrary joint survival function $S(x, y) = P(X \geq x, Y \geq y)$ of the random variables X, Y.

The main thesis of this work is that every such bivariate distribution can be reduced to the same **universal form**, i.e.,

$$S(x, y) = S_1(x)\, S_2(y)\, J(x, y), \tag{3.1}$$

where the function J(x, y), which determines all the stochastic dependences of X and Y, is called the joiner (see Ref. [3]).

Obviously, $J(x, y) = 1$ for all (x, y) if and only if the random variables X, Y are independent.

The joiner is simply defined by

$$J(x, y) = S(x, y)/S_1(x)\, S_2(y). \tag{3.2}$$

If for some x or y either $S_1(x) = 0$ or $S_2(y) = 0$ then the joiner is undefined but then we set in Eq. (3.1) $S(x, y) = 0$. The form Eq. (3.1) is universal for all bivariate survival functions such that the underlying two random variables X, Y take on finite real values only. For that point it is enough to substitute J(x, y) from Eq. (3.2) into Eq. (3.1). So the universality of Eq. (3.1) is simply equivalent to the common (always true) arithmetic identity.

2. As follows directly from Eq. (3.2) every bivariate distribution S(x, y) uniquely determines the corresponding joiner. On the other hand, a properly chosen joiner J(x, y) determines the corresponding bivariate distribution S(x, y) provided both the marginals $S_1(x)$, $S_2(y)$ are given. Thus, characterization of bivariate survival function by the joiner is similar to the characterization of bivariate distribution function $P(X < x, Y < y)$ by **copula** [1].

However, given a cdf $P(X < x, Y < y)$, finding a corresponding copula is not as straightforward as finding joiner (from Eq. (3.2)) for a given bivariate S(x, y). Anyway, the joiner and the copula characterizations of bivariate distributions are competitive to each other (see Arnold [7]).

The main task, both in theory and applications (such as reliability), is, given (any) fixed two survival functions $S_1(x)$, $S_2(y)$, to find bivariate survival functions S(x, y) such that

$$S_1(x) = S(x, 0),\ S_2(y) = S(0, y)$$

In theoretical considerations one may be interested in the whole class of such bivariate distributions (given the two marginals) and eventually in the relation between the pairs $(S_1(x), S_2(y))$ and the corresponding classes of the bivariate S(x, y), while in applications one mostly seeks for some specific subclasses or single bivariate "models" that are consistent with the given data. In the latter case, statistical methods are indispensable.

Let us, for a while, concentrate on the theoretical part as this is also indispensable for good applications such as reliability problems. The main question [7] to be answered is:

Given any fixed pair of distributions $(S_1(x), S_2(y))$, what necessary and sufficient conditions must be satisfied by functions J(x, y) in order that the products $S_1(x)\, S_2(y)\, J(x, y)$ are legitimate survival functions?

This problem was considered in more detail in [7]. Here, let us shortly present the main results.

Suppose a function $S(x, y) = S_1(x)\, S_2(y)\, J(x, y)$ (to be tested) is a candidate for (any) bivariate survival function. Consider only the "continuous case," i.e., the case when both the continuous hazard rates $\lambda_1(x), \lambda_2(y)$ corresponding to the distributions $S_1(x), S_2(y)$ exist. Moreover, suppose the (tested) function J(x, y) has the following representation:

$$J(x,y) = \exp\left[-\int_0^x \int_0^y \Psi(t,u)dt\, du\right] \tag{3.3}$$

where the function Ψ(t, u) is integrable (so all the integrals exist and are finite) and continuous.

The case above described, that we call the "continuous case," is the most important especially in applications including the reliability applications.

[Other, more general, cases that may, between others, contain discrete type distributions $S_1(x), S_2(y)$ we sketched in Ref. [3] where instead of the second derivatives (applied in the following equation) second "differential quotients" (possibly with no limits) were analyzed.]

Now, we have the following representation of the candidate function (for a survival function):

$$S_1(x)\, J(x,y)\, S_2(y) = \exp\left[-\int_0^x \lambda_1(t)dt - \int_0^x \int_0^y \Psi(t,u)dt\, du - \int_0^y \lambda_2(u)du\right] \tag{3.4}$$

In this "continuous case," as described above, the second mixed derivative $\partial^2/\partial x\, \partial y$ of either side of Eq. (3.4) exists and, by the assumed continuity of the functions $\lambda_1(x), \lambda_2(y), \Psi(x,y)$, equals to the derivative $\partial^2/\partial y \partial x$.

Under all those assumptions that still are general enough for application purposes the necessary and sufficient condition for either side of Eq. (3.4) to be a legitimate bivariate survival function is nonnegativity of the derivative $\partial^2/\partial x\, \partial y\{S_1(x)\, J(x,y)\, S_2(y)\}$.

Calculate that derivative from the right-hand side of Eq. (3.4) and set it to be nonnegative. After simplifications the so obtained inequality takes on the form:

$$\left[\lambda_1(x) + \int_0^y \psi(x,u)\, du\right] \times \left[\lambda_2(y) + \int_0^x \psi(t,y)\, dt\right] \geq \psi(x,y). \tag{3.5}$$

Given (in advance), the two hazard rates $\lambda_1(x), \lambda_2(y)$, representing the marginal distributions $S_1(x), S_2(y)$, the necessary and sufficient condition for J(x, y), yielding the product $S_1(x)\, J(x,y)\, S_2(y)$ a legitimate survival function, is condition Eq. (3.5) as satisfied by the, equivalent to hypothetical joiner, function Ψ(x, y) at all points (x, y).

Here, recall that there is one to one relationship Eq. (3.3) between J(x, y) and Ψ(x, y). In the sense of this relationship we may talk about "satisfying Eq. (3.5) by J(x, y), given $S_1(x), S_2(y)$."

Suppose that for all (x, y) we have $\psi(x, y) \geq 0$, which is the condition equivalent to the negative stochastic dependence of the X and Y.

Then all solutions of the inequality

$$\lambda_1(x)\lambda_2(y) \geq \psi(x,y) \tag{3.6}$$

are also solutions of inequality Eq. (3.5).

The reverse is not true as, even if $\psi(x, y) \geq 0$, not all solutions of Eq. (3.5) are solutions of Eq. (3.6).

Nevertheless, the (sufficient) condition Eq. (3.6), by its simplicity, is very convenient for finding new specific models and classes of models for reliability and other applications.

Thus, by Eq. (3.6), we immediately find the following bivariate survival functions:

$$S(x,y) = \exp\left[-\int_0^x \lambda_1(t)\,dt - a\int_0^x\int_0^y \lambda_1(t)\,\lambda_2(u)\,dt\,du - \int_0^y \lambda_2(u)\,du\right], \tag{3.7}$$

where arbitrary "marginals" $\lambda_1(x)$, $\lambda_2(y)$ are given in advance and the constant coefficient "a" satisfies $0 \le a \le 1$.

The specific class (as indexed only by one additional parameter a) of the models Eq. (3.7) is a simple version of the general model Eq. (3.4) which is "universal" (but only) in the continuous case.

The only task to adopt (whenever that's the case) model Eq. (3.7) to a given set of data is to estimate and then verify parameter a. First, of course, the marginal "distributions" $\lambda_1(x), \lambda_2(y)$ must be estimated.

The simple and, still, general (by the generality of the functions $\lambda_1(x), \lambda_2(y)$) model Eq. (3.7) seems to be potentially very important and possibly even "basic" for reliability and other (such as biomedical or econometrical) applications.

Moreover, class Eq. (3.7) can easily be extended upon recognition that the coefficient "a" can be "made" to be dependent on x and y, so it may "become" any (say, continuous) function a(x, y) such that $0 \le a(x, y) \le 1$.

Two simple examples of such functions can immediately be given.

As the first class of such functions we propose the functions: $a(x, y) = \exp[-cxy]$, where $c \ge 0$ is a real constant. In this case the number of (unknown) parameters is the same as in model Eq. (3.7) since the original parameter $0 \le a \le 1$ in Eq. (3.7) is only replaced by the parameter $c \ge 0$.

For more generality the foregoing class of the functions may further be extended by introducing an additional nonnegative parameter "α" so that one obtains a wider class of the functions:

$a(x,y) = \exp[-cx^\alpha y^\alpha]$. When $\alpha = 1$ we arrive at the previous class of the models.

In the case of an arbitrary α, model Eq. (3.7) can be "rewritten" into the form:

$$S(x,y) = \exp\left\{-\int_0^x \lambda_1(t)\,dt - \exp[-cx^\alpha y^\alpha]\int_0^x\int_0^y \lambda_1(t)\,\lambda_2(u)\,dt\,du - \int_0^y \lambda_2(u)\,du\right\} \tag{3.8}$$

Using the most general notation "a(x, y)" one obtains as the widest (in this case) class of the distributions:

$$S(x,y) = \exp\left[-\int_0^x \lambda_1(t)\,dt - a(x,y)\int_0^x\int_0^y \lambda_1(t)\,\lambda_2(u)\,dt\,du - \int_0^y \lambda_2(u)\,du\right]. \tag{3.9}$$

For all other (continuous) models inequality Eq. (3.6) is not uniformly satisfied and $\psi(x, y)$ must be found from inequality Eq. (3.5). Solving the latter may occur a mathematically interesting task as inequality Eq. (3.5) may in some cases be replaced by analytically well treatable (partial) integral equations (see more on that in Ref. [3]) whose solutions are also solutions of Eq. (3.5). Methods of the equations solving may involve a relatively advanced theory of integral equations. The last subject we omitted as being out of main string of our investigations.

3. As noted above, any survival function S(x, y) and not only that satisfying the continuous case's assumptions can be represented by Eq. (3.1), where $S_1(x), S_2(y)$ are the marginal distributions. As it was described in Ref. [3], besides the marginal factors representation Eq. (3.1) there is also "baseline factors representations" valid for, possibly, all survival functions. These representations, like the previously considered marginal, have the product form as follows:

$$S(x, y) = B_1(x)\, B_2(y)\, K(x, y), \tag{3.10}$$

where $B_1(x), B_2(y)$ are "baseline distributions" of (the same) bivariate S(x, y).

The baseline distributions (in general, different from marginal) of a given S(x, y) typically arise in various modeling procedures met in applications. For example, in reliability applications when one considers lifetime of a two-component series or parallel system the baselines $B_1(x), B_2(y)$ may be considered as reliability functions (distributions) of the component lifetimes when tested in "laboratory conditions" where each component is tested in an absence of the other. This physical isolation implies stochastic independence of the components' lifetimes, say, T_1, T_2 whose distributions are $B_1(x)$, $B_2(y)$. The random variables T_1, T_2 are (in general) different from X, Y. The random variables X, Y describe component (marginal) lifetimes when the components work within the system. In the system physical interactions between the components take place. To describe such mechanisms more specifically let us consider, as an example, the situation associated with the construction of the classical Freund [9] model:

Example 1. Two airplane engines attached on the same wing are tested separately of each other in laboratory conditions under the regular load (stress) that is typically present during the regular "quiet flight" with no other engine failure. In these idealized work conditions the engines' **lifetimes $\mathbf{T_1}$, $\mathbf{T_2}$** are independent and have exponential distributions (as the "baseline distributions") with the constant failure (hazard) rates, say α_1, α_2. The failure rates are statistically estimated from data obtained during the laboratory conditions testing. Once statistically "the same" (although physically different) engines are put to work into real flight conditions (to the system) they "influence" each other in the following way. If during the flight either of the two engines [say, i-th engine, i = 1, 2] fails then the remaining j-th engine [j = 1, 2 and j ≠ i] starts to work under an increased stress (load), and, from that moment on, its failure rate, say, α'_j is assumed to be still constant but significantly higher than the previous α_j. This "second" j-th engine works alone until it, eventually, fails too. The now dependent lifetimes X, Y of the engines when they work during the flight (in the system) are different from the independent T_1, T_2 when in the laboratory conditions. The task is to find the joint reliability (survival) function of the random vector (X, Y). It was, eventually, found in 1961 by Freund [9].

Concluding Example 1, the baseline distributions for the Freund model are the exponential survival functions of the (initial) random variables T_1, T_2, while the distributions of X and Y are the marginal given in Section 3.4. In the case of the Freund model, the baseline and marginal distributions are different although this is not always the case for other models. The differences and, in some cases, identity of representations Eq. (3.1) and Eq. (3.10) will be considered in the following sections.

The relationships between the two representations can be formulated by the following Proposition 1.

Proposition 1. Suppose for a given survival function $S(x, y) = S_1(x)\, S_2(y)\, J(x, y)$, a baseline distribution representation $S(x, y) = B_1(x)\, B_2(y)\, K(x, y)$ for $S(x, y)$ exists too.

Then the following relations hold:

$$S_1(x) = B_1(x)\, K(x, 0),$$

$$S_2(y) = B_2(y)\, K(0, y).$$

Moreover,

$$J(x, y) = K(x, y)/K(x, 0)\, K(0, y).$$

The proof of this Proposition (under the name "Theorem 1") as well as more discussion of the baseline representation case one can find in Ref. [3].

In this work we rather concentrate on illustrations of the two (universal) representations based on well-known, from literature, bivariate models. Before that analysis notice the following results formulated.

Proposition 2.

1. The necessary condition for a function $J^*(x, y)$ to be the joiner (i.e., $J^*(x, y) = J(x, y)$) of some bivariate distribution $S(x, y)$ is that: $J^*(0, y) = 1$, for each y and $J^*(x, 0) = 1$, for every x.
2. Suppose there is given representation Eq. (3.10) for some bivariate distribution $S(x, y)$.

If $K(0, y) = K(x, 0) = 1$ then $K(x, y) = J(x, y)$ and the baseline distributions are the same as the marginals.

If this is not the case then the marginals and the joiner of $S(x, y)$ can be obtained from $K(x, y)$ by Proposition 1.

The full justification for part 1 of Proposition 2 can be found in Ref. [3]. Part 2 is a consequence of Proposition 1.

The ("physical") dependence function **K(x, y)** we propose to call **"system function"** as it analytically *describes* the (physical) situation that takes place after the two initially independent objects (such as the two engines of the plane considered in Example 1) is "installed" within a system. Unlike that, the joiner **J(x, y)** may be considered as **"stochastic dependence function."**

3. Analyses of some common bivariate models

All the models we now consider are in the form of bivariate survival functions:

$$S(x, y) = P(X \geq x, Y \geq y).$$

3.1 Gumbel Case

The first classic bivariate model we consider is the first bivariate exponential **Gumbel**.

Distribution [10] is given by the following formula:

$$S(x, y) = \exp[-\lambda_1 x - axy - \lambda_2 y]. \tag{3.11}$$

It's easy to find out that distribution Eq. (3.11) represents the above considered "continuous case" subject to scheme Eq. (3.4), where the hazard (failure) rates λ_1 , λ_2 are constant and $\psi(x, y) = a$, which is a constant satisfying inequality Eq. (3.6) in specific form: $0 \le a \le \lambda_1$, λ_2. Recall, all constants are always continuous functions. In this "Gumbel Case" the baseline ("initial") distributions

$$S_1(x) = \exp[-\lambda_1 x], S_2(y) = \exp[-\lambda_2 y]$$

overlap with the marginals $S_1(x) = S(x, 0)$ and $S_2(y) = S(0, y)$.

Another reason for that statement is the following.

$K(x, y) = \exp[-axy]$ and $K(0, y) = K(x, 0) = 1$ implies that $K(x, y) = J(x, y)$.

Resuming, in the Gumbel Case Eq. (3.11) both baseline distributions are marginal and the system function $K(x, y)$ is the joiner. However, such properties do not always hold. For that see the following "Freund case."

3.2 Freund Case

Illustration of this case from the viewpoint of a particular application was provided in Example 1. However, in the general case, the "after first failure" hazard rate α'_j is not necessarily higher (but different) from the original α_j. The following formula for the bivariate reliability (survival) function of the Freund model can be found in Ref. [11]. Since we will work with our own reparameterization of the original ([11]) formula we first rewrite it and then give our version, so that the reader will be able to compare the two.

To be consistent with the notation adopted here we use the symbols x, y, X, Y, S(x, y), P( ) instead of x_1, x_2, X_1, X_2, $F^{-}_{X1, X2}(x_1, x_2)$, Pr( ), respectively, as applied in Ref. [11].

Thus, the [11] version (page 356) of the Freund bivariate survival function with our choice of symbols, given above, is:

$$S(x, y) = P(X > x, Y > y)$$

$$= 1/\gamma_2\{\alpha_1 \exp[-\gamma_2 x - \alpha'_2 y](\alpha_2 - \alpha'_2)\exp[-(\alpha_1 + \alpha_2)y]\} \quad \text{for } 0 \le x \le y,$$

and

$$= 1/\gamma_1\{\alpha_2 \exp[-\gamma_1 y - \alpha'_1 x] + (\alpha_1 - \alpha'_1)\exp[-(\alpha_1 + \alpha_2)x]\} \quad \text{for } 0 \le y \le x, \tag{3.12}$$

where $\gamma_i = \alpha_1 + \alpha_2 - \alpha'_i$ $(i = 1, 2)$.

Our version of Eq. (3.12) one obtains upon the substitution $\delta_i = \alpha'_i - \alpha_i$, for $i = 1, 2$.

Now, we have the Freund reliability (survival) function in the form:

$$S(x, y) = P(X > x, Y > y)$$

$$= \exp[-\alpha_1 x]\exp[-\alpha_2 y][1/(\alpha_1 - \delta_2)]\{\alpha_1 \exp[-\delta_2(y - x)] - \delta_2 \exp[-\alpha_1(y - x)]\} \quad \text{for } 0 \le x \le y,$$

and

$$= \exp[-\alpha_1 x]\exp[-\alpha_2 y][1/(\alpha_2 - \delta_1)]\{\alpha_2 \exp[-\delta_1(x - y)] - \delta_1 \exp[-\alpha_2(x - y)]\} \quad \text{for } 0 \le y \le x. \tag{3.13}$$

It is easy to note that formula (3.13) is a specific form of the baseline representation Eq. (3.10) of the bivariate survival function, where $B_1(x) = \exp[-\alpha_1 x]$ and $B_2(y) = \exp[-\alpha_2 y]$ are the survival functions of the components (for example, the plane engines from Example 1) when they are operating in physical separation of each other (laboratory conditions). In this case the system function K(x, y) is given by:

$$K(x, y) = [1/(\alpha_1 - \delta_2)]\ \{\alpha_1 \exp[-\delta_2(y - x)] - \delta_2 \exp[-\alpha_1(y - x)]\} \quad \text{for } 0 \le x \le y,$$

and

$$= [1/(\alpha_2 - \delta_1)]\ \{\alpha_2 \exp[-\delta_1(x - y)] - \delta_1 \exp[-\alpha_2(x - y)]\} \quad \text{for } 0 \le y \le x. \quad (3.14)$$

Note that in this case K(x, x) = K(y, y) = 1.

Immediate application of this fact is in the following.

Suppose that X and Y describe the lifetimes of a two-component system which has series reliability structure. Denote the system's (as the whole) lifetime by Z.

Then, as it is very well known, this system reliability function is given as $P(Z > z) = P(X > z, Y > z)$, where $z = x = y$.

As it follows directly from Eq. (3.14) (and upon substituting $x = y = z$ into Eq. (3.13)) we obtain the known formula:

$$P(Z > z) = P(X > z, Y > z) = \exp[-\alpha_1 z]\ \exp[-\alpha_2 z] = \exp[-(\alpha_1 + \alpha_2)z]. \quad (3.15)$$

This relation agrees with the fact, just mentioned above, that the common factors $\exp[-\alpha_1 x]$ and $\exp[-\alpha_2 y]$ in Eq. (3.13) determine the baseline distributions $B_1(x)$, $B_2(y)$ of the joint survival function S(x, y). These were obtained after the components were independently testified in the laboratory conditions.

In this "Freund case," however, the baselines are different from the marginals in Eq. (3.13).

The marginals are immediately obtained from Eq. (3.13) by substituting $S_1(x) = P(X > x) = S(x, 0)$ and $S_2(y) = P(Y > y) = S(0, y)$.

Thus, we obtain:

$$S_1(x) = \exp[-\alpha_1 x]\ [1/(\alpha_2 - \delta_1)]\ \{\alpha_2 \exp[-\delta_1 - x] - \delta_1 \exp[-\alpha_2 x]\} \quad (3.16)$$

for $0 \le y \le x$, and here, of course $y = 0$.

If $0 \le x \le y$ then $y = 0$ implies $x = 0$ and, therefore, we have:

$$S_1(x) = 1, \quad (3.16^*)$$

but, in this case, x takes on only one value $x = 0$ since $0 \le x \le y$.

Similarly, we obtain:

$$S_2(y) = \exp[-\alpha_2 y]\ [1/(\alpha_1 - \delta_2)]\ \{\alpha_1 \exp[-\delta_2 y] - \delta_2 \exp[-\alpha_1 y]\} \quad (3.17)$$

for $0 \le x \le y$, and $x = 0$.

Since if $0 \le y \le x$ then $x = 0$ implies $y = 0$ and we have:

$$S_2(y) = 1 \quad (3.17^*)$$

for $0 \le y \le x$ and, thus, Eq. (3.17∗) holds for $y = 0$ only.

The foregoing four Eqs. (3.16)–(3.17∗) fully determine both marginals of Freund's bivariate survival function Eq. (3.13). It is easy to see that Eq. (3.14) implies that $S_1(x) = B_1(x)\ K(x,0)$ and $S_2(y) = B_2(y)\ K(0,y)$, which agrees with the first part of Proposition 1.

As one can see, the Freund model's marginals as obtained above are pretty far from being exponential. However, this model may still be considered "exponential" because both baseline distributions are exponential and the so understood idea of "*exponentiality*" lies at the bottom of all applications associated with this model (see Example 1). These facts indicate the role of baseline distributions in bivariate models as being in a sense, "competitive" to marginals.

As for the joiner J(x, y) for the Freund bivariate, given by Eq. (3.13), one can obtain it from the second part of Proposition 1 as J(x, y) = K(x, y)/K(x, 0) K(0, y).

Thus, in case $0 \le x \le y$, from Eq. (3.14) we obtain:

$$J(x,y) = [1/(\alpha_1 - \delta_2)]\ \{\alpha_1 \exp[-\delta_2(y-x)] - \delta_2 \exp[-\alpha_1(y-x)]\}$$
$$/\{[1/(\alpha_1 - \delta_2)]\ \{\alpha_1 \exp[\delta_2 x)] - \delta_2 \exp[\alpha_1 x)]\ \}[1/(\alpha_1 - \delta_2)]\ \{\alpha_1 \exp[-\delta_2 y] - \delta_2 \exp[-\alpha_1 y]\}\}$$

and after some simplification we have:

$$J(x,y) = (\alpha_1 - \delta_2)\{\alpha_1 \exp[-\delta_2(y-x)] - \delta_2 \exp[-\alpha_1(y-x)]\}$$
$$/\{\alpha_1 \exp[\delta_2 x)] - \delta_2 \exp[\alpha_1 x)]\}\{\alpha_1 \exp[-\delta_2 y] - \delta_2 \exp[-\alpha_1 y]\} \quad \text{when } 0 \le x \le y \tag{3.18}$$

Similarly, we obtain:

$$J(x,y) = (\alpha_2 - \delta_1)\{\alpha_2 \exp[-\delta_1(x-y)] - \delta_1 \exp[-\alpha_2(x-y)]\}$$
$$/\{\alpha_2 \exp[-\delta_1 x] - \delta_1 \exp[-\alpha_2 x]\}\{\alpha_2 \exp[\delta_1 y)] - \delta_1 \exp[\alpha_2 y)]\} \text{ when } 0 \le y \le x \tag{3.18*}$$

Notice that the necessary condition for J(x, y) (as given by (3.18) and (3.18∗)), to be (any) joiner (namely: J(x, 0) = J(0, y) = 1), is satisfied.

Returning to Example 1 realize that the parameters $\delta_i = \alpha_i' - \alpha_j\ (j = 1,2)$ [which determine the increment (or decrement) of the hazard rate of any j-th component as a result of the other component's failure] are, in turn, numerical characteristics of the underlying stochastic, as well as physical, dependences. Thus, if $\delta_1 = \delta_2 = 0$ (i.e., if the components are insensitive to the other component's status of operation) then under substituting these zero-values into Eq. (3.14) one obtains that always K(x, y) = 1 and so the joint survival function Eq. (3.10) as well as its particular form Eq. (3.13) becomes:

$$S(x,y) = B_1(x)\ B_2(y) = \exp[-\alpha_1 x]\ \exp[-\alpha_2 y]. \tag{3.19}$$

The same result one obtains by realizing that $\delta_1 = \delta_2 = 0$ also implies that J(x, y) = 1 and, therefore, from Eq. (3.1), one obtains:

$$S(x,y) = S_1(x)\ S_2(y). \tag{3.20}$$

Nevertheless, Eq. (3.19) and Eq. (3.20) are consistent since, when $\delta_1 = \delta_2 = 0$, from Eq. (3.16)–Eq. (3.17∗) we obtain

$$S_1(x) = B_1(x) \text{ and } S_2(y) = B_2(y).$$

Remark. As a conclusion from the above one can say that, in the case of the Freund model, when $\delta_1 = \delta_2 = 0$, not only the lifetimes X, Y are stochastically independent but also the components are physically independent (for example, they may be physically isolated from each other). The latter is not always the case for other models. Stochastic independence does not exclude a physical dependence. However, in general, physical independence (for example, when the components are in laboratory conditions as described in Example 1) implies stochastic independence and, therefore, by the logical transposition rule, stochastic dependence implies (or "indicates at") some "physical" dependence.

Analytically, by "physical independence" of the random quantities X, Y one may agree to understand the fact that K(x, y) = 1 for all x, y, whenever the two baseline distributions of a given bivariate exist. Thus, the physical independence is a stronger property than the stochastic independence.

[Recall at this point again that, in this work, by "*physical dependence*" we mean not only purely "physical actions" that may be considered in a framework of physics (as science) but any dependence that takes place "outside of mathematics" (which, as a tool, describes it) such as biological, psychological, sociological, financial, and other possible.]

3.3 Marshall and Olkin bivariate case

The physical genesis of this classical stochastic model [12] is even simpler than in the Freund case.

Consider two parallel system components, say, c_1, c_2, that are subject to three different kinds of fatal shocks. The shocks form Poisson processes with rates $\lambda_1, \lambda_2, \lambda_{12}$. The first shock of the first process causes failure of component c_1, the second causes failure of c_2, while the third with the rate λ_{12} "destroys" both components. So, the j-th component (j = 1, 2) can either fail alone with the rate λ_j, or together with the other component with the rate λ_{12}. Thus, time to failure, say T_j, of each component c_j (j = 1, 2) alone, is exponential with parameter λ_j, and time to failure of both the components is also exponential with parameter λ_{12}. The stochastic model of the system (with parallel reliability structure) whose components c_1, c_2 are subjected to the above described mechanism of failure is the following bivariate survival function (known as "MOBED", i.e., the Marshall–Olkin bivariate exponential distribution, see Ref. [11]):

$$S(x, y) = P(X \geq x, Y \geq y) = \exp[-\lambda_1 x - \lambda_2 y - \lambda_{12} \max(x, y)]. \tag{3.21}$$

In other words:

$$S(x, y) = \exp[-\lambda_1 x - (\lambda_2 + \lambda_{12})y] \quad \text{if } 0 \leq x \leq y$$

$$= \exp[-(\lambda_1 + \lambda_{12})x - \lambda_2 y]. \quad \text{if } 0 \leq y \leq x. \tag{3.21*}$$

From the foregoing formulas we immediately obtain both marginal distributions:

$$S_1(x) = \exp[-(\lambda_1 + \lambda_{12})x] \text{ and } S_2(y) = \exp[-(\lambda_2 + \lambda_{12})y]. \tag{3.22}$$

These are the distributions of the random lifetimes X. Now, we **complete the analysis** of the Marshall and Olkin model applying the theory built in this work and in Refs. [2,3]. First, we obtain the joiner for Eq. (3.21) and Eq. (3.21∗).

According to definition Eq. (3.2) we have for $0 \leq x \leq y$:

$$J(x, y) = S(x, y)/S_1(x)\, S_2(y)$$

$$= \exp[-\lambda_1 x - \lambda_2 y - \lambda_{12} y]/\exp[-(\lambda_1 + \lambda_{12})x]\, \exp[-(\lambda_2 + \lambda_{12})y]$$

$$= \exp[-\lambda_{12}\, x] \text{ and for } 0 \leq y \leq x:$$

$$= \exp[-(\lambda_1 + \lambda_{12})x - \lambda_2 y]/\exp[-(\lambda_1 + \lambda_{12})x]\, \exp[-(\lambda_2 + \lambda_{12})y]$$

$$= \exp[\lambda_{12} y]. \text{ In other words we have obtained: } \quad J(x, y) = \exp[\lambda_{12} \min(x, y)]. \tag{3.23}$$

Notice the remarkable simplicity of the joiner in the MOBED case. Obviously, in this case, the property $J(x, 0) = J(0, y) = 1$ is preserved. The important part of the analysis in the MOBED case is the discussion of the nonnegative parameter λ_{12} which, given the parameters λ_1, λ_2, is a measure of both stochastic dependence of the lifetimes X, Y and physical dependence of the corresponding components c_1, c_2.

It follows directly from Eq. (3.23) that the condition $\lambda_{12} = 0$ makes the joiner Eq. (3.23) equal to 1 which means stochastic independence.

Consider now the underlying physical dependence and the associated baseline representation of the considered model. Thus, in laboratory conditions the components are separated from each other by either being in two places remote each from other or operate in disjoint periods of time or are separated in any other similar sense. If this happens the probability of failures at the same time instant is evidently zero, and the part λ_{12} of each component's hazard rate reduces to zero.

On the other hand, from the condition $\lambda_{12} = 0$ follows physical independence since, as it is assumed, no other system dependences or structures are present. Thus, if $\lambda_{12} = 0$, we obtain from Eq. (3.22) the baseline distributions of the components' (in isolation) lifetimes T_1, T_2:

$B_1(x) = \exp[-\lambda_1 x]$ and $B_2(y) = \exp[-\lambda_2 y]$. This implies that, from Eq. (3.10) and Eq. (3.21), for the system function one obtains:

$$K(x, y) = \exp[-\lambda_{12} \max(x, y)] \tag{3.24}$$

that is,

$$K(x, y) = \exp[-\lambda_{12} y] \quad \text{for } 0 \leq x \leq y \text{ and}$$

$$K(x, y) = \exp[-\lambda_{12}\, x] \quad \text{for } 0 \leq y \leq x$$

Since, when $\lambda_{12} \neq 0$ we have $K(x, y) \neq 1$ and thus the components are physically dependent. This means that the parameter λ_{12} is a measure of both physical and stochastic dependences. Thus, when $\lambda_{12} = 0$, both stochastic and physical independences occur.

Finally realize that in MOBED case, unlike in the Gumbel case, $B_1(x) \neq S_1(x)$ and $B_2(y) \neq S_2(y)$ and therefore $K(x, y) \neq J(x, y)$.

Moreover, the (independent) lifetimes T_1, T_2 in laboratory conditions that have $B_1(x)$, $B_2(y)$ distributions are different from the (dependent) lifetimes X, Y [having the marginal distributions $S_1(x), S_2(y)$ given by Eq. (3.22)] of the components when they work "in the system," i.e., are subject to common failures.

3.4 Singpurwalla and Youngren bivariate exponential model

This bivariate distribution [8] serves as a stochastic model for reliability of a two-component parallel system working in a random environment subjected to (one) shot-noise stochastic process. The situation is somewhat related to the shock model of Marshall and Olkin described above. The difference is that the shocks are, in general, not fatal but may damage any of the two components with a positive probability.

Besides, Marshall and Olkin (M&O) considered three distinct homogeneous Poisson processes of the fatal shocks with constant rates $\lambda_1, \lambda_2, \lambda_{12}$ such that first affects component c_1, second c_2, and the third affects both the components.

In the Singpurwalla and Youngren model [8] they consider one homogeneous Poisson process (with a constant positive rate m) of nonfatal shocks, where each shock affects both components, each with the same positive probability. The authors first derive the general model of the situation and then, under pretty strong additional assumptions, they obtain, as a special case of a more general model, the bivariate exponential distribution:

$$P(X > x, Y > y)$$

$$= [(1 - mx + my)/(1 + mx + my)]^{1/2} \exp[-my] \quad \text{for } x \le y \quad \text{and}$$

$$= [(1 - my + mx)/(1 + m(x + y))]^{1/2} \exp[-mx] \quad \text{for } y \le x. \tag{3.25}$$

As it is assumed, both component lifetimes, when in laboratory conditions, have the same exponential distribution with parameter λ. This parameter when the components are in the system (i.e., under regular stress) turns into the parameter $m = \lambda/b$, where a positive number b characterizes the system conditions. These conditions relay on the fact that the components are subjected to the shocks ("fatal", under the adopted by authors' specific analytic assumptions) that form a Poisson process with constant positive parameter m.

The marginals as obtained from Eq. (3.25) by setting $y = 0$ and then $x = 0$ are:

$$P(X > x) = 1 \quad \text{for } x \le y$$
$$\text{and} \tag{3.26}$$
$$P(X > x) = \exp[-mx] \quad \text{for } y \le x.$$

$$P(Y > y) = \exp[-my] \quad \text{for } x \le y$$
$$\text{and} \tag{3.26*}$$
$$P(Y > y) = 1 \quad \text{for } y \le x.$$

Having the marginals one can write the survival function Eq. (3.25) in form Eq. (3.1) as follows:

$$S(x,y) = S_1(x)S_2(y)J(x,y)$$

$$= 1 \cdot \exp[-my] \cdot [(1 - mx + my)/(1 + mx + my)]^{1/2} \quad \text{for } x \le y \tag{3.27}$$

and

$$= \exp[-mx] \cdot 1 \cdot [(1 - my + mx)/(1 + m(x + y))]^{1/2} \quad \text{for } y \le x. \tag{3.27*}$$

From (3.27) and (3.27∗) one immediately obtains the joiner of Eq. (3.25) in the form:

$$J(x,y)$$

$$= [(1 - mx + my)/(1 + mx + my)]^{1/2} \text{ for } x \le y \text{ and } [(1 - my + mx)/(1 + mx + my)]^{1/2} \text{ for } y \le x. \tag{3.28}$$

At this point recall the authors' remark [8] in the last two lines of their paper:

"*We have not been successful in being able to analytically compute the joint moments of* (3.6) [corresponding to Eq. (3.25) in our paper] *and are thus unable to comment on the nature of the dependence between L_1 and L_2 [here, X and Y].*"

As one can see such a calculation of second mixed moments for Eq. (3.25) is really very hard if at all possible. So, finding a numerical characterization (correlation coefficient, for example) of an underlying stochastic dependence may not be available analytically. However, the functional characterization, as given by the joiner Eq. (3.28), is directly at hand.

As for the stochastic dependence between X and Y first realize that for the joiner Eq. (3.28) we always have

$$0 < J(x,y) < 1 \quad \text{for all nonzero } x, y. \tag{3.29}$$

Recall that the latter condition is not always satisfied by joiners and thus Eq. (3.29) brings some specific information about the character of the dependence we are interested in. As second information that one can derive from the joiner Eq. (3.28) on the stochastic dependence is the minimal value, say $\gamma = \gamma(m)$, of our joiner in the closed quadrant (of the plane): $x \ge 0$, $y \ge 0$.

It's easy to see that this minimum γ is a function of m and always takes on a positive value. This, similarly as some correlation coefficients, satisfy: $0 < \gamma < 1$ and may be considered as numerical measure of the dependence. Even more interesting is the value 1 - γ which measures "how far from 1 can J(x, y) be, as a function of m." Recall, condition $J(x, y) = 1$ for any x, y characterizes independence.

Another, fairly good "indicator of dependence" could be determined by the average value of the integral over the plane's quadrant. Namely if

$$L_1 = \text{Lim}_{x\to y, y\to\infty}(2/xy)\int_0^x dt \int_x^\infty [(1 - mt + mu)/(1 + mt + mu)]^{1/2} du \quad \text{for } x \le y$$

and (3.30)

$$L_2 = \text{Lim}_{x\to y, y\to\infty}(2/xy)\int_0^y du \int_x^\infty [(1 - mu + mt)/(1 + mt + mu)]^{1/2} dt \quad \text{for } y \le x$$

then as indicator we may adopt the value $L = L_1 + L_2$.

One can easily see that $0 < L < 1$.

Again, since condition $J(x, y) = 1$ (for all x, y) is equivalent to independence (and in this case the value of L would be equal to 1), we propose the "magnitude of dependence" to be measured as the distance D of the indicator L from 1, as $D = 1 - L$.

The so defined quantity D is a function of m so that one can analyze the sensitivity of D with respect to changes in m. The integrals in Eq. (3.30) may eventually be calculated numerically for a set of values of m to tabulate this function.

As for the baseline representation of Eq. (3.25) the case of Singpurwalla and Youngren distribution is kind of special. Namely, the laboratory conditions for the considered system's components only rely on the requirement that $b = 1$ for the dependence coefficient $m = \lambda/b$. Thus, the baseline distributions can be obtained simply by substituting in Eq. (3.26) and Eq. (3.26∗) λ everywhere in place of m.

Unfortunately, the so obtained random variables, say, T_1, T_2 (having the distributions Eq. (3.26), Eq. (3.26∗) upon the specific value of $m = \lambda$) are still stochastically dependent having the joint distribution Eq. (3.25) for $m = \lambda$.

This analytical property has its roots in the physical fact that the Poisson process of shocks with the "regular" rate $m = \lambda$ still affects both the components, and therefore induces the stochastic dependence. It cannot be said that in the so created "laboratory conditions" the components physically operate in separation of each other. So it is not in agreement with our former definition of "laboratory conditions" as mutual isolation of the components.

A possible way out from the difficulty in determining baseline distributions and the corresponding system function K(,) is the observation that the considered shot-noise stochastic process of the shocks is the only reason for the component failures, and, on the other hand, it is the only source of stochastic dependence of the lifetimes X, Y.

At this point, we may extend the values of the parameter m from positive to nonnegative ($m \geq 0$ and $m \neq \lambda/b$) and redefine "laboratory conditions" as corresponding to the case $m = 0$ (no shocks).

Therefore, in total absence of the shot-noise process the components (still "operating") are stochastically independent, but then they are "absolutely reliable."

This means that their failure (hazard) rates are constantly equal to zero (as zero is the rate of the corresponding fictious "zero Poisson stochastic process" of the "shocks"), i.e., $m = \lambda = 0$.

In the so defined "zero-laboratory conditions" the baseline survival functions satisfy:

$$B_1(x) = 1, \quad \text{for all } x \text{ and } B_2(y) = 1, \quad \text{for all } y \tag{3.31}$$

which means "both components never fail."

In terms of the random variables (that now we extend) this means that

$$P(X = \infty) = 1 \text{ and } P(Y = \infty) = 1 \tag{3.32}$$

which, formally, is a well-defined mathematical fact.

One would obtain statistical confirmation of this fact if from a sufficiently large sample of either component after a sufficiently long time none of the tested components failed.

Obviously, Eq. (3.31) and Eq. (3.32) only express the mathematical idealization of physical facts reflected by possible statistical observations of "no failures."

Now, the baseline (universal) representation Eq. (3.10) for the Singpurwalla and Youngren bivariate distribution Eq. (3.25) takes the form:

$$S(x, y) = P(X > x, Y > y)$$

$$= 1 \cdot 1 \cdot [(1 - mx + my)/(1 + mx + my)]^{1/2} \exp[-my] \quad \text{for } x \leq y < \infty \tag{3.33}$$

and

$$= 1 \cdot 1 \cdot [(1 - my + mx)/(1 + m(x + y))]^{1/2} \exp[-mx] \quad \text{for } y \leq x < \infty$$

As one sees, in this case, the system function K(x, y) is identical with the whole survival (reliability) function given by Eq. (3.25) for some m, while the baseline distributions (different from the marginal) are "trivial".

The total system conditions are then determined only by one positive number m which here may be considered as both the physical and stochastic dependences characterization after eventual installing the components into the system.

If still $m = 0$ then Eq. (3.33) reduces to

$$S(x, y) = 1 \cdot 1$$

and the "system conditions" reduce to the laboratory conditions according to their definition given above.

Remark. As mentioned above, for the Singpurwalla and Youngren's exponential model the marginal distributions given by Eq. (3.26) and Eq. (3.26∗) are not exponential. However, one can say they are "generated" by exponentials.

3.5 Arnold and Strauss bivariate exponential

This bivariate model's [13] survival (reliability) function has the following form:

$$S(x, y) = \theta(\delta) \exp[-\beta x - \gamma y - \beta\gamma\delta xy]/\{(1 + \delta\beta x)\,(1 + \delta\gamma y)\,\theta(\delta/(1 + \delta\beta x)(1 + \delta\gamma y))\} \tag{3.34}$$

where

$$\theta(\delta) = \delta \exp[-1/\delta]/[-\mathrm{Ei}(1/\delta)], \tag{3.35}$$

and $-\mathrm{Ei}(u) = \int_u^\infty [e^{-w}/w]dw$.

Setting in Eq. (3.34) $y = 0$ and then also $x = 0$ one obtains the marginals:

$$S_1(x) = \theta(\delta) \exp[-\beta x]/(1 + \delta\beta x)\,\theta(\delta/(1 + \delta\beta x)\} \tag{3.36}$$

$$S_2(y) = \theta(\delta) \exp[-\gamma y]/\{(1 + \delta\gamma y)\,\theta(\delta/(1 + \delta\gamma y))\} \tag{3.37}$$

respectively.

Taking the quotient $S(x, y)/S_1(x)\,S_2(y)$ by applying Eqs. (3.34), (3.36), and (3.37) one obtains the joiner of the considered model in the form:

$$J(x, y) = [1/\theta(\delta)] \exp[-\beta\gamma\delta xy]\{\theta(\delta/(1 + \delta\beta x))\,\theta(\delta/(1 + \delta\gamma y))/\theta(\delta/(1 + \delta\beta x)\,(1 + \delta\gamma y))\}. \tag{3.38}$$

At first, joiner Eq. (3.38) may seem to be complicated and "unreadable". However, after a closer look, it reveals itself as a pretty interesting measure of stochastic dependence. Realize, it is the product of two factors. The first factor $[1/\theta(\delta)]\exp[-\beta\gamma\delta xy]$ contains the variables x, y only in the form of (cross) arithmetic product "constant times **xy**" which strongly suggests its application as a measure of some "correlation".

On the other hand the remaining factor:

$$\{\theta(\delta/(1+\delta\beta x))\,\theta(\delta/(1+\delta\gamma y))/\theta(\delta/(1+\delta\beta x)\,(1+\delta\gamma y))\}$$

resembles a function in the form, say, $\beta(t, u) = \alpha(t, u)/\alpha(t)\,\alpha(u)$ which may turn out to be useful in calculations. Notice, the function $\beta(t, u)$ resembles definition Eq. (3.2) of any joiner.

Anyway, the joiner Eq. (3.38) seems to be interesting and the two underlying factors may indicate two distinct sources of dependence.

Remark. At this point recall that the joint distribution Eq. (3.34) (see [13]) was conditionally determined, i.e., the original "data" at the beginning of the construction were two (exponential) conditional densities: $f_{X|Y}(x|y)$ and $f_{Y|X}(y|x)$ [13]. Since they were introduced independently of each other they may be the sources of two kinds of dependences as suggested by the existence of the two factors in the joiner.

Of course, the above comments are suggestions only aimed to encourage readers to investigate dependence by means of the joiner Eq. (3.38). This requires more investigations which unfortunately are out of scope of this work.

Finally, realize too that this "complicated" joiner Eq. (3.38) satisfies both the necessary conditions for joiners:

$$J(x,0) = J(0,y) = 1.$$

As for the **baseline representation** of Eq. (3.34) realize that in this model the parameter which determines stochastic dependence is δ. The case $\delta = 0$ is equivalent to independence. Even as Eq. (3.35) and therefore also Eq. (3.34) is undefined for $\delta = 0$ it is not difficult to find out that the limit of $\theta(\delta)$ as $\delta \to 0$ exists and equals 1. We may then slightly redefine the function $\theta(\delta)$ by setting

$$\theta(0) = 1.$$

Now upon setting $\delta = 0$ in Eq. (3.34) one simply obtains:

$$S(x,y) = \exp[-\beta x]\,\exp[-\gamma y], \tag{3.39}$$

which means that stochastic independence of the baseline random variables, say, T_1, T_2, may be interpreted as the underlying system's component lifetimes in laboratory conditions. [Realize, however, that at the beginning of the modeling process, as performed in Ref. [13], one only had at one's disposal the conditional densities $f_{X|Y}(x|y)$ and $f_{Y|X}(y|x)$ and not the baselines.]

The baseline distributions (of T_1, T_2) are then the exponentials:

$$B_1(x) = \exp[-\beta x] \ \text{and} \ B_2(y) = \exp[-\gamma y].$$

They, evidently, differ from the marginal distributions (of X and of Y, respectively), which are not exponential. However, the underlying initial conditional distributions are both exponential.

As for numerical "measure" of the dependence between X and Y [in Eq. (3.34)] one has at one's disposal the parameter δ.

A more precise description of "physical (and not only stochastic) interactions" between the components when they are put into the system provides the "system function":

$$K(x, y) = S(x, y)/B_1(x)\,B_2(y) \tag{3.40}$$

which also contains the parameter δ.

Eq. (3.40) is other version of the baseline representation Eq. (3.10) of bivariate survival functions. One can easily obtain it from Eq. (3.34).

Recall, in the considered case, K(x, y) is not the joiner.

Remark. In above we have found the two baseline distributions simply by setting the parameter δ to zero. However, in this and other similar cases (not considered in this chapter), finding the baselines may, possibly, be based on the observation that simple exponential survival functions $\exp[-\beta x]$, $\exp[-\gamma y]$ are factors in Eq. (3.34). Such a method, however, not necessarily determines the baselines uniquely.

It is possible that, in some cases, there might be more than one pair of baseline distributions. In particular, this may happen when a considered survival function is the arithmetic product of several survival functions with respect to the same arguments. Then there may be more than one scalar parameter (like the δ in the example above) a set of which governs the dependence.

If such a situation takes place then "physically" more than one "two-component system" would be present (all such systems physically are composed of the same two units), and more than one mathematical description would be included as factors in the same formula for S(x, y).

Resuming this idea, baseline representation Eq. (3.10) may not be unique but, nevertheless, may still be considered as "universal". In contrast, the marginal representation Eq. (3.1) can be viewed as both universal and unique.

3.6 Oakes bivariate frailty model

This model [14] is a bivariate extension of the univariate frailty model [11].

Its survival function has the following form:

$$S(x, y) = P(X \geq x, Y \geq y) = \int \{B_1(x)\,B_2(y)\}^w\, dG(w), \tag{3.41}$$

where $B_1(x)$, $B_2(y)$ are some baseline survival functions, w is any realization of a positive random variable W, and G(w) is a distribution function (often the gamma) of W. This bivariate distribution easily finds applications to two-component system reliability, but mostly is used in modeling biomedical phenomena.

The marginals of Eq. (3.41) takes the form:

$$S_1(x) = \int \{B_1(x)\}^w\, dG(w), \tag{3.42}$$

$$S_2(y) = \int \{B_2(y)\}^w\, dG(w). \tag{3.43}$$

Thus, in this case, the joiner has a bit complicated general form as containing integrals:

$$J(x, y) = \int \{B_1(x)\,B_2(y)\}^w\, dG(w) \Big/ \left\{ \int \{B_1(x)\}^w\, dG(w) \right\} \left\{ \int \{B_2(y)\}^w\, dG(w) \right\} \tag{3.44}$$

However, the baseline representation of Eq. (3.41) looks much better.

First, realize that Eq. (3.41) can be factored as:

$$\int \{B_1(x)\,B_2(y)\}^{w}\,dG(w) = B_1(x)^{r}\,B_2(y)^{r}\int \{B_1(x)\,B_2(y)\}^{w-r}\,dG(w), \tag{3.45}$$

where $0 \le r \le w$.

The factored form Eq. (3.45) may suggest that, for any fixed r ($0 < r \le w$), any pair of survival functions $B_1(x)^r$ and $B_2(x)^r$ may, possibly, be considered as the baselines. If so, then the "baseline representation" Eq. (3.45) would not be unique. This would mean that, unlike the marginal factors representation, the baseline factors representation universality (in general) is "in danger." This problem we leave open. However, in the considered case of Eq. (3.41), and according to the common meaning of the notion "frailty" (here, as a change [governed by w] in reliability of an object affecting the residual lifetime distribution), the "baseline situation" (when no change is present) is described by the conditional distribution given $W = 1$.

One can express such "baseline situation" by:

$$P(X \ge x, Y \ge y | W = 1) = B_1(x)\,B_2(y),$$

where for the "system function" we have: $K(x, y) = 1$, for all x, y (see Eq. (3.10)).

The latter indicates that for all associated practical problems, the "true" baseline distributions (i.e., when the two objects are tested in isolation from each other, which stands for the unique situation) are only $B_1(x)$, $B_2(y)$. Therefore, in Eq. (3.45) the only value of the variable r that corresponds to baseline factors representation is $r = 1$.

This representation is then unique and takes on the form:

$$P(X \ge x, Y \ge y) = B_1(x)\,B_2(y)\int \{B_1(x)\,B_2(y)\}^{w-1}\,dG(w). \tag{3.46}$$

Of course, in this case the system function satisfies: $K(x,y) = \int \{B_1(x)\,B_2(y)\}^{w-1}\,dG(w)$ and thus, for the Oakes bivariate frailty model we have: $K(x,y) \ne J(x,y)$.

In all the, above considered, examples either the marginal or the baseline distributions were exponential. In the next case, the marginal and baseline distributions are different, but all four distributions are Weibullian.

3.7 A bivariate Weibull case

Kotz et al. (in Ref. [10], page 408, B) present an example of a bivariate Weibull distribution, that we write in the following form of a survival function:

$$S(x,y) = P(X \ge x, Y \ge y) = \exp\left\{-\left(\lambda_1 x^{\beta} + \lambda_2 y^{\beta}\right)^{\gamma}\right\}, \tag{3.47}$$

where all the four parameters are positive reals.

A slightly different version of Eq. (3.47) in the application to reliability of two-component system is considered in Hougaard [15].

For the marginals of Eq. (3.47) we obtain:

$$S_1(x) = \exp\left[-\left(\lambda_1 x^{\beta\gamma}\right)\right], \quad S_2(y) = \exp\left[-\left(\lambda_2 y^{\beta\gamma}\right)\right]. \tag{3.48}$$

Evidently, both distributions Eq. (3.48) are Weibullian with common shape parameter $\beta\gamma$.

From Eq. (3.47), Eq. (3.48) we obtain the joiner J(x, y) in the form:

$$J(x,y) = \exp\left[-\left(\lambda_1 x^\beta + \lambda_2 y^\beta\right)^\gamma + \lambda_1 x^{\beta\gamma} + \lambda_2 y^{\beta\gamma}\right]. \tag{3.49}$$

This joiner looks proper and its meaning as dependence measure is clear. Especially, this meaning is clear when $\gamma = 2$ and also when γ is any other positive integer. Thus, for $\gamma = 2$, we obtain:

$$J(x,y) = \exp\{-2\lambda_1\lambda_2 x^\beta y^\beta\}. \tag{3.50}$$

Look at the simplicity of the above.

Suppose X and Y in Eq. (3.47) describe the lifetimes of the two system components.

Notice, it is natural to consider parameter γ in Eq. (3.47) as the numerical measure of the strength of physical interactions among the components when they operate within system. The only value of γ that makes the random variables X, Y independent is $\gamma = 1$. But, in that case we may conclude that the only reason they are stochastically independent (there is no other such situation) is physical separation of the components. Therefore, the (baseline) distributions of the components when in laboratory conditions are:

$$B_1(x) = \exp\left[-\lambda_1 x^\beta\right] \text{ and } B_2(y) = \exp\left[-\lambda_2 y^\beta\right]. \tag{3.51}$$

The baselines Eq. (3.51) are then different from the marginals Eq. (3.48), but they are nicely related by the parameter γ and the baseline and the marginal distributions are identical when $\gamma = 1$. On the other hand, baselines can (as they actually should) be factored out from S(x, y), given by Eq. (3.47).

Namely, based on Eq. (3.47), we have:

$$S(x,y) = \exp\left[-\lambda_1 x^\beta\right]\exp\left[-\lambda_2 y^\beta\right]\exp\left[\left(\lambda_1 x^\beta + \lambda_2 y^\beta\right)\left(1 - \left(\lambda_1 x^\beta + \lambda_2 y^\beta\right)^{\gamma-1}\right)\right] \tag{3.52}$$

and, therefore, for the system function K(,) we obtain:

$$K(x,y;\gamma) = \exp\left[\left(\lambda_1 x^\beta + \lambda_2 y^\beta\right)\left(1 - \left(\lambda_1 x^\beta + \lambda_2 y^\beta\right)^{\gamma-1}\right)\right] \tag{3.53}$$

Notice, that for $\gamma = 1$ we have K(x, y; 1) = 1, so the system function K(,) as a physical dependence measure, and the numerical measure of dependence γ agree in a good sense. Nevertheless, the function K(;) seems to carry more information on the dependence than the single number γ.

An additional information as brought by K(x, y; γ) is the physical dependence magnitude when the events X = x and Y = y happen. Perhaps, an interesting characterization of this type of systems (i.e., with a given type of *physical* mechanisms) would be the rate of change in K(x, y; γ) with respect to γ, possibly expressed in terms of derivative $\partial/\partial\gamma$ K(x, y; γ) or other derivatives with respect to parameters. This subject is, however, out of scope of this work.

References

[1] A. Sklar, Fonctions de repartition a n dimensions et leurs marges, vol. 8, Publications de l'Institut de Statistique de l'Universite de Paris, 1959, pp. 229–231.

[2] J.K. Filus, L.Z. Filus, The Cox-Aalen models as framework for construction of bivariate probability distributions, universal representation, J. Stat. Sci. Appl. 5 (April 2017) 56–63.

[3] J.K. Filus, L.Z. Filus, A general (universal) form of multivariate survival functions in theoretical and modeling aspect of multicomponent system reliability analysis, in: M. Ram, H. Pham (Eds.), Advances in Reliability Analysis and its Applications, Springer Series in Reliability Engineering. Springer, Cham, 2020, pp. 319–342.

[4] J.K. Filus, L.Z. Filus, A method for multivariate probability distributions construction via parameter dependence, Commun. Stat. Theor. Methods 42 (Issue 4) (2013) 716–721.

[5] O.O. Aalen, A linear regression model for the analysis of the life times", Stat. Med. 8 (1989) 907–925.

[6] D.R. Cox, Regression models and life tables (with discussion), J. Roy. Stat. Soc. B 74 (1972) 187–220.

[7] B.C. Arnold, Private Communication, December 2017.

[8] N.D. Singpurwalla, M.A. Youngren, Multivariate distributions induced by dynamic environments, Scand. J. Stat. 20 (1993) 251–261.

[9] J.E. Freund, A bivariate extension of the exponential distribution, J. Amer. Statist. Assoc. 56 (1961) 971–977.

[10] E.J. Gumbel, Bivariate exponential distributions, J. Am. Stat. Assoc. 55 (1960) 698–707.

[11] S. Kotz, N. Balakrishnan, N.L. Johnson, Continuous Multivariate Distributions, second ed., vol. 1, J. Wiley & Sons, Inc, New York, 2000.

[12] A.W. Marshall, I. Olkin, A generalized bivariate exponential distribution, J. Appl. Probab. 4 (1967) 291–303.

[13] B.C. Arnold, D. Strauss, Bivariate distributions with exponential conditionals, J. Am. Stat. Assoc. 83 (1988) 522–527.

[14] D. Oakes, Bivariate survival models induced by frailties, J. Am. Stat. Assoc. 84 (1989) 487–493.

[15] P. Hougaard, A class of multivariate failure time distribution, Biometrika 73 (1986) 671–678.

CHAPTER

4

Reliability analysis of cutting system of sugar industry using intuitionistic fuzzy Lambda–Tau approach

Dinesh Kumar Kushwaha, Dilbagh Panchal, Anish Sachdeva

Department of Industrial and Production Engineering, Dr. B. R. Ambedkar National Institute of Technology, Jalandhar, Punjab, India

1. Introduction

Cutting system, which is an important functionary subsystem of a sugar mill industry, not only chops and cuts the sugarcane but also feeds the tandem mill with small pieces of cane. Due to its importance for cutting the sugarcane in small pieces, its availability plays a key role for achieving the objective of maximum production. Even a small failure of a component/subsystem stops the supply of juice to the juice heater and cuts the supply of bagasse (fuel) to the furnace of boiler [1]. The considered process industry must ensure long-term availability of its system/subsystem/components to fulfill the functional requirements and for uninterrupted supply of the products (sugar) to the market. In order to achieve the aforesaid goal, it is necessary that all the components of system/ subsystem/component must have an optimum maintenance schedule for its long-run availability. In the current scenario, the concept of reliability-based maintenance (RBM) is gaining strength for achieving high availability and maintainability aspects of the complex industrial systems. For developing the RBM, it is essential to analyze the failure behavior of the system using the collected operational failure and repair time data of the considered system. The collection of failure and repair time–based data is a difficult task for the system analyst. Also, dealing with vague data is a great problem for studying the correct failure behavior of the system [2]. Considering this limitation of handling the vagueness in the collected data Intuitionistic fuzzy set (IFS) concept based reliability tool plays an important role in analyzing and studying the correct failure dynamics of the complex repairable Industrial system. So, considering the merits of IFS theory in analyzing the failure behavior of the considered system the current research work presents the application of IFS theory–based Lambda–Tau approach for analyzing the reliability parameters of the sugar mill industry.

The Handbook of Reliability, Maintenance, and System Safety through Mathematical Modeling.
https://doi.org/10.1016/B978-0-12-819582-6.00004-6

2. Literature background

In the past many researchers have studied the failure dynamics of the various industrial systems with the implementation of different mathematical concept–based reliability approaches. The application of Chapman–Kolmogorov birth–death process–based mathematical modeling has been widely used by many researchers for studying the failure behavior of the real industrial systems and was helpful in developing the planned maintenance schedule. Kumar et al. [3] studied the failure dynamics of crystallization system in a sugar mill industry using Chapman–Kolmogorov birth–death equations–based availability and reliability parameters. Again Kumar et al. [4] incorporated Chapman–Kolmogorov birth–death based differential equations and applied its solution on a desulphurization system of a urea fertilizer plant for studying its failure behavior. Arora and Kumar [5] expounded the application of this modeling for analyzing the reliability parameters of ash handling system of a coal-fired thermal power plant. The limitation of Markovian approach lied in the fact that it considers only crisp failure/repair time data and vagueness of data were not considered at all. The literature reported that this shortcoming has been overcome by fuzzy set theory and considered the vagueness and uncertainties of collected data and hence prevents the biasness of the analysis results. Fuzzy methodology (FM) concept has been incorporated within the reliability-based tools and techniques and was implemented by various authors in different areas. Knezevic and Odoom [6] developed fuzzy Lambda–Tau approach with Petri Net modeling for tabulating the various reliability parameters of the industrial system. Based on the tabulated reliability parameters the system failure behavior has been studied. Sharma et al. [7] presented a framework using FM approach to compute various reliability parameters of a paper mill industry to study the behavior analysis of the considered system. Komal et al. [8] used RAM estimation of press and washing units of a paper mill by using Genetic Algorithms based Lambda–Tau (GABLT) technique. In this work, RAM parameters were calculated by Lambda–Tau approach but fuzzy membership function of theses parameters was computed by Genetic Algorithm. Verma and Kumar [9] proposed a vague Lambda–Tau approach for reliability analysis of gas turbine system. Trapezoidal vague set has been incorporated in the work to calculate the failure rate and repair time at different spreads.

Panchal and Kumar [10,11] applied FM for carrying risk analysis of the power-generating unit of a medium size coal-fired thermal power plant using fuzzy along with PN modeling. Panchal and Kumar [10,11] proposed an integrated framework for behavior analysis of a water treatment plant of a coal-fired thermal power plant. The behavior analysis of compressor house unit of a medium size coal thermal power plant and various reliability parameters have been reported [17]. Panchal et al. [2] expounded the application of FM-based integrated framework for studied RAM and risk issues of a chlorine gas plant. In these proposed model FM has been used which considers the highest level of confidence of domain experts as one which entails that there is zero degree of hesitation between membership and nonmembership function values which has direct effect on the correctness of the analysis result. To overcome this drawback for higher degree of correctness of analysis result IFS theory–based concept is of supreme importance and considers level of confidence of domain experts which lies between [0,1]. In the literature IFS theory concept–based reliability analysis has been reported for studying the failure behavior of pulping unit of a paper mill [12].

From the reviewed literature it has been found that the application of IFS theory–based Lambda–Tau approach has not yet been reported for studying the failure behavior cutting system of a sugar mill industry located in northern part of India.

3. Notions of intuitionistic fuzzy set theory and Lambda–Tau approach

3.1 Notions of Intuitionistic fuzzy theory

The basic notions of IFS theory used in the current work are discussed as follows.

3.1.1 Intuitionistic fuzzy set

An IFS in a universal set S is a set of triplets and is given by

$$\breve{A} = \{\langle x, \mu_{\bar{A}}(x), \vartheta_{\bar{x}}(x)\vartheta x \in S\} \tag{4.1}$$

where $\mu_{\bar{A}}(x)$ is a membership function and $\vartheta_{\bar{A}}(x)$ is a nonmembership function which is given by

$$\mu_{\bar{A}} : S \rightarrow [0\ 1],\ x \in S \rightarrow \mu_{\bar{A}}(x) \rightarrow [0\ 1] \tag{4.2}$$

$$\vartheta_{\bar{A}} : S \rightarrow [0\ 1],\ x \in S \rightarrow \vartheta_{\bar{A}}(x) \rightarrow [0\ 1] \tag{4.3}$$

The above two equations must satisfy the following condition to qualify for an IFS

$$\mu_{\bar{A}}(x) + \vartheta_{\bar{A}}(x) = 1 \text{ for all } x \in S.$$

3.1.2 α Cut of IFS

α Cut of IFS has two abbreviated terms and is symbolically represented by a $P^{(\alpha)}$ and $P^{(1-\alpha)}$, and is given for membership function as

$$P^{(\alpha)} = \{x \in S:\ \mu_{\bar{A}}(x) \geq \alpha\} \tag{4.4}$$

For nonmembership function it is defined as

$$P^{(1-\alpha)} = \{x \in S: 1 - \vartheta_{\bar{A}}(x) \geq \alpha\} = \{x \in S:\ \vartheta_{\bar{A}}(x) \leq 1 - \alpha\} \tag{4.5}$$

where α is in the range of $0 \leq \alpha \leq 1$. The α cut is shown graphically in Fig. 4.1 for both membership and nonmembership functions.

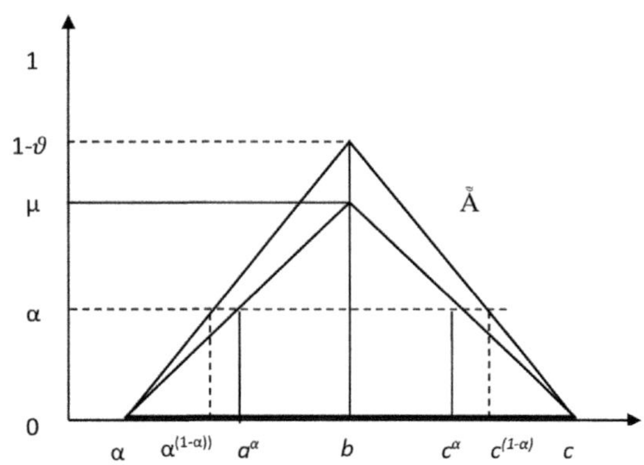

FIGURE 4.1

α cut of IFS $\tilde{\breve{A}}$.

3.1.3 Triangular intuitionistic fuzzy number

The triangular intuitionistic fuzzy number is defined as:

$$\mu_{\bar{A}}(x) = \begin{cases} \mu\left(\dfrac{x-a}{b-a}\right); a \le x \le b \\ \mu; \\ \mu\left(\dfrac{c-x}{c-b}\right); b \le x \le c \end{cases}$$

$$0 \quad \text{otherwise} \tag{4.6}$$

$$1-\vartheta_{\bar{A}}(x) = \begin{cases} 1-\vartheta\left(\dfrac{x-a}{b-a}\right); a \le x \le b \\ 1-\vartheta; \\ 1-\vartheta\left(\dfrac{c-x}{c-b}\right); b \le x \le c \end{cases}$$

$$0 \quad \text{otherwise} \tag{4.7}$$

3.2 Intuitionistic fuzzy set Lambda–Tau approach

Lambda–Tau approach is an effective tool developed by Knezevic and Odoom in 2001. It is used to compute the various reliability indices of real industrial systems under uncertainty, thereby helpful in analyzing their failure behavior. Furthermore, failure behavior is studied by the reliability parameters calculated by reliability formulae. Beside these formulae, AND and OR expressions are used to find the value of the top place. AND expression is used for series combination whereas OR is used for parallel combinations. In the past few years this approach has been incorporated by various researchers for studying the failure behavior of their systems [2,8,10,11,13–16].

The various steps of IFS Lambda–Tau approach are discussed as follows:

Step-1. Develop PN model for representing series–parallel combination of various equipment of the considered system.

Step-2. Failure rate and repair time data have been collected for various equipment/subsystem from different source such as maintenance logbook, computer database, and reliability engineer.

Step-3. Fuzzify failure and repair time data for considering the vagueness of the collected data.

Step-4. Model the series -parallel arrangement of the considered system using AND/OR gate transition expressions Eqs. (4.8–4.15) for both membership and non-membership functions.

For membership function

AND transition expression

$$\lambda^{\alpha\mu} = \left[\prod_{i=1}^{n}\left\{(\lambda_{i2}-\lambda_{i1})\frac{\alpha_\mu}{\mu_i}+\lambda_{i1}\right\}\cdot\sum_{j=1}^{n}\left[\prod_{\substack{i=j\\ \overline{i\neq 1}}}^{n}\left\{(\tau_{i2}-\tau_{i1})\frac{\alpha_\mu}{\mu_i}+\tau_{i1}\right\}\right]\alpha,\right.$$
$$\left.\prod_{i=1}^{n}\left\{-(\lambda_{i3}-\lambda_{i2})\frac{\alpha_\mu}{\mu_i}+\tau_{i3}\right\}\cdot\sum_{j=1}^{n}\left[\left[\prod_{\substack{i=1\\ \overline{i\neq j}}}^{n}-(\tau_{i3}-\tau_{i2})\frac{\alpha_\mu}{\mu_i}+\tau_{i3}\right]\right]\right] \tag{4.8}$$

$$\tau^{\alpha\mu} = \frac{\prod_{i=1}^{n}\left\{(\tau_{i2}-\tau_{i1})\frac{\alpha_\mu}{\mu_i}+\tau_{i1}\right\}}{\sum_{i=j}^{n}\left[\prod_{\substack{i=1\\ i\neq j}}^{n}\left\{-(\tau_{i3}-\tau_{i2})\frac{\alpha_\mu}{\mu_i}+\tau_{i3}\right\}\right]},\frac{\prod_{i=1}^{n}\left\{-(\tau_{i3}-\tau_{i2})\frac{\alpha_\mu}{\mu_i}+\tau_{i3}\right\}}{\sum_{j=1}^{n}\prod_{\substack{i=1\\ i\neq j}}^{n}\left\{(\tau_{i2}-\tau_{i1})\frac{\alpha_\mu}{\mu_i}+\tau_{i1}\right\}} \tag{4.9}$$

OR transition expression

$$\lambda^{\alpha\mu} = \sum_{i=1}^{n}\left\{(\lambda_{i2}-\lambda_{i1})\frac{\alpha_\mu}{\mu_i}+\lambda_{i1}\right\},\sum_{i=1}^{n}\left\{-(\lambda_{i3}-\lambda_{i2})\frac{\alpha_\mu}{\mu_i}+\lambda_{i3}\right\} \tag{4.10}$$

$$\tau^{\alpha\mu} = \frac{\left[\sum_{i=1}^{n}\left[\left\{\lambda_{i2}-\lambda_{i1}\right)\frac{\alpha_\mu}{\mu_i}+\lambda_{i1}\right\}\cdot\left\{(\tau_{i2}-\tau_{i1})\frac{\alpha_\mu}{\mu_i}+\tau_{i1}\right\}\right]}{\sum_{i=1}^{n}\left\{-(\lambda_{i3}-\lambda_{i2})\frac{\alpha_\mu}{\mu_i}+\lambda_{i3}\right.},$$
$$\frac{\sum_{i=1}^{n}\left[\left\{-(\lambda_{i3}-\lambda_{i2})\frac{\alpha_\mu}{\mu_i}+\lambda_{i3}\right\}\cdot\left\{(\tau_{i3}-\tau_{i2})\frac{\alpha_\mu}{\mu_i}+\tau_{i3}\right\}\right]}{\sum_{i=1}^{n}\left\{(\lambda_{i2}-\lambda_{i1})\frac{\alpha_\mu}{\mu_i}+\lambda_{i1}\right\}} \tag{4.11}$$

For nonmembership function

AND transition expression

$$\lambda^{(\alpha_{1-\vartheta})} = \left[\prod_{\substack{i=1\\ \overline{i\neq j}}}^{n}\left\{(\lambda_{i2}-\lambda_{i1})\frac{\alpha_\vartheta}{1-\vartheta_i}+\lambda_{i1}\right\}\cdot\sum_{j=1}^{n}\left[\prod_{\substack{i=1\\ \overline{i\neq j}}}^{n}\left\{(\tau_{i2}-\tau_{i1})\frac{\alpha_\vartheta}{1-\vartheta_i}+\tau_{i1}\right]\alpha\right. \tag{4.12}$$
$$\left.\prod_{i=1}^{n}\left\{-(\lambda_{i3}-\lambda_{i2})\frac{\alpha_\vartheta}{1-\vartheta_i}+\tau_{i3}\right\}\cdot\sum_{j=1}^{n}\left[\prod_{i=1}^{n}-(\tau_{i3}-\tau_{i2})\frac{\alpha_\vartheta}{1-\vartheta_i}+\tau_{i3}\right]\right]$$

$$\tau^{(\alpha_{1-\vartheta})} = \frac{\prod_{i=1}^{n}\left\{(\tau_{i2}-\tau_{i1})\frac{\alpha_{\vartheta}}{1-\vartheta_i}+\tau_{i1}\right\}}{\sum_{i=j}^{n}\left[\prod_{\substack{\neq i=1\\ i\neq j}}^{n}\left\{-(\tau_{i3}-\tau_{i2})\frac{\alpha_{\vartheta}}{1-\vartheta_i}+\tau_{i3}\right]}, \frac{\prod_{i=1}^{n}\left\{-(\tau_{i3}-\tau_{i2})\frac{\alpha_{\vartheta}}{1-\vartheta_i}+\tau_{i3}\right\}}{\sum_{j=1}^{n}\prod_{\substack{i=1\\ i\neq j}}^{n}\left\{(\tau_{i2}-\tau_{i1})\frac{\alpha_{\vartheta}}{1-\vartheta_i}+\tau_{i1}\right\}} \tag{4.13}$$

OR transition expression

$$\lambda^{(\alpha_{1-\vartheta})} = \sum_{i=1}^{n}\left\{(\lambda_{i2}-\lambda_{i1})\frac{\alpha_{\vartheta}}{1-\vartheta_i}+\lambda_{i1}\right\}, \quad \sum_{i=1}^{n}\left\{-(\lambda_{i3}-\lambda_{i2})\frac{\alpha_{\vartheta}}{1-\vartheta_i})+\lambda_{i3}\right\} \tag{4.14}$$

$$\tau^{(\alpha_{1-\vartheta})} = \frac{\left[\sum_{i=1}^{n}\left[\left\{\lambda_{i2}-\lambda_{i1})\frac{\alpha_{\vartheta}}{1-\vartheta_i}+\lambda_{i1}\right\}\cdot\left\{(\tau_{i2}-\tau_{i1})\frac{\alpha_{\vartheta}}{1-\vartheta_i}+\tau_{i1}\right\}\right]\right]}{\sum_{i=1}^{n}\left\{-(\lambda_{i3}-\lambda_{i2})\frac{\alpha_{\vartheta}}{1-\vartheta_i}+\lambda_{i3}\right.}, \frac{\sum_{i=1}^{n}\left[\left\{-(\lambda_{i3}-\lambda_{i2})\frac{\alpha_{\mu}}{\mu_i}+\lambda_{i3}\right\}\cdot\left\{(\tau_{i3}-\tau_{i2})\frac{\alpha_{\mu}}{\mu_i}+\tau_{i3}\right\}\right]}{\sum_{i=1}^{n}\left\{(\lambda_{i2}-\lambda_{i1})\frac{\alpha_{\vartheta}}{1-\vartheta_i}+\lambda_{i1}\right\}} \tag{4.15}$$

Step-5. Tabulate reliability parameters (for different degree of membership in an interval of 0–1) for the system using expression as shown in Table 4.1.

Step-6. Using center of area method as represented by Eq. (4.16) defuzzify the fuzzified reliability parameter values.

$$u* = \frac{\int_{u_1}^{u_2}\mu_{out}\cdot(u)\cdot du}{\int_{u_1}^{u_2}\mu_{out}\cdot(u)\cdot du} \tag{4.16}$$

Table 4.1 Various Reliability parameters.

Reliability indices	Expressions
Mean time to failure	$\mathrm{MTTFs} = \frac{1}{\lambda_s}$
Mean time to repair	$\mathrm{MTTRs} = \frac{1}{\mu_s}$
Mean time between failure	$\mathrm{MTTFs} + \mathrm{MTTR_s}$
Reliability	$\mathrm{Rs} = e^{-\lambda s\cdot t}$
Availability	$\mathrm{As} = \frac{\mu_s}{\mu_s+\lambda_s} + \frac{\lambda_s}{\mu_s+\lambda_s}\cdot e - (\mu_s + \lambda_s)t$
Expected number of failures	$(\mathrm{ENOF}) = \frac{\lambda_S\cdot\mu_S\cdot t}{\mu_S+\mu_S} + \frac{\lambda_S^2}{(\mu_s+\lambda_s)^2}\left[1 - e^{-(\mu_S\cdot\lambda_S\cdot t}\right]$

4. Case study

A cutting system of a sugar mill industry has been considered as case study in this work, which is located in northern part of India. The sugar mill has a capacity of crushing sugarcane 2700 tonne/day. Cutting system, one of the important functionary subsystems of feeding system, has been considered in the present work and the schematic diagram of the considered system is represented in Fig. 4.2.

4.1 System description

Unloader ($i = 1$)**:** The function of unloader is to unload the sugarcane from truck and bullock cart and load the sugarcane supply system. There is only one unloader connected in series configuration with the cutting system.

Cutters ($i = 2, 3$): It consists two cutters which are arranged in series configuration used to cut the supplied sugarcane in desired pieces.

Crusher ($i = 4$): Crusher comprises of tandem mill, used to crush the small chopped cane and extract the juice from it. It is arranged in the series configuration with the considered system.

4.2 Application of intuitionistic fuzzy set Lambda–Tau approach

The complex series–parallel arrangement with AND/OR symbol of the considered system was represented with PN model as shown in Fig. 4.3.

Failure and repair time values of each subsystem/component have been collected from expert feedback, maintenance logbook, computer database, etc., as shown in Table 4.2.

Using Eqs. (4.6)–(4.7) the collected values have been converted into triangular fuzzy number (TFN) at various spreads ($\pm$ 15% , $\pm$25% , $\pm$50%) for considering the vagueness of the raw data. Furthermore, using Eqs. (4.8)–(4.15) for AND/OR gate transitions (for both membership and nonmembership function) modeling as per the considered system's series–parallel arrangement (Fig. 4.3) for the top event has been developed and the value of degree of acceptance ($\mu = 0.6$) and degree of rejection($\vartheta = 0.2$) has been considered on the basis of expert feedback. Here, considering the system for heavy process industry the mission time was considered as $t = 168$ h. Using the TFN in the modeling, various reliability parameters were tabulated as per the expression shown in Table 4.1. Fuzzified values with left and right spreads for $\pm$ 15%, $\pm$ 25%, and $\pm$ 50% were tabulated in the interval of 0–1 for the various reliability parameters for membership

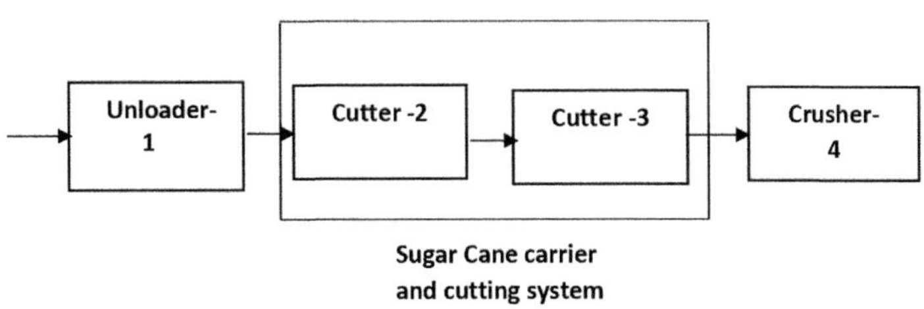

FIGURE 4.2

Schematic diagram of cutting system.

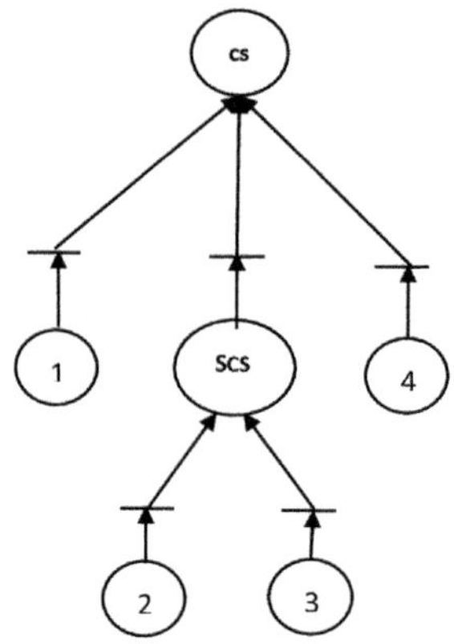

FIGURE 4.3

PN model of cutting system.

Table 4.2 Failure and repair time data.

Components	Unloader	Cutter	Crusher
Failure rate (h)	0.0042	0.0056	0.0056
Repair time (h)	3	4	3

function, nonmembership function, and Lambda–Tau approach as shown in Tables 4.3–4.5, respectively. Similarly, fuzzified values for ± 25% and ± 50% were also to be tabulated. Due to space limitation these values were not shown here.

The defuzzified values of various parameters are shown in Table 4.6.

5. Result discussion

It has been observed from Table 4.6 that failure rate of the current system shows increasing trend with increase in spread values. Failure rate is increased by 0.4% with increase in spread from ± 15% to ± 25% and 0.5% with the increase in spread from ±25% to 50% for value of membership function. Same trend is followed for nonmembership function value too. Furthermore, failure rate is increased by 0.2386% with increase in spread from ± 15% to ± 25% and 0.6061% with increase in spread from ±25% to 50%. Likewise, other reliability parameters such as repair time, mean time between failure (MTBF), availability, and reliability have increasing trend with increase in spread values. Since repair time shows increase in its trend and therefore the availability of the system in the analysis results shows decreasing trend. Availability is decreased by 0.00703% with increase in spread from ± 15% to ± 25% and by 0.0521% with the increase in spread from ±25% to 50% for

Table 4.3 Reliability parameters at ±15% spread values for membership function.

	Left spread					Right spread				
DOF	Failure rate	Repair time	MTBF	Reliability	Availability	Failure rat rate	Repair time	MTBF	Reliability	Availability
1.0	0.023100	4.750370	53.120563	0.041786	0.953130	0.018900	2.601818	43.674390	0.020634	0.901117
0.9	0.022575	4.414279	51.706589	0.038258	0.948201	0.019425	2.812287	44.652371	0.022537	0.909378
0.8	0.022050	4.100526	50.369184	0.035028	0.942873	0.019950	3.036984	45.680748	0.024615	0.917081
0.7	0.021525	3.807393	49.102696	0.032071	0.937123	0.020475	3.276951	46.762769	0.026885	0.924254
0.6	0.021000	3.533333	47.902066	0.029364	0.930925	0.021000	3.533333	47.902066	0.029364	0.930925
0.5	0.020475	3.276951	46.762769	0.026885	0.924254	0.021525	3.807393	49.102696	0.032071	0.937123
0.4	0.019950	3.036984	45.680748	0.024615	0.917081	0.022050	4.100526	50.369184	0.035028	0.942873
0.3	0.019425	2.812287	44.652371	0.022537	0.909378	0.022575	4.414279	51.706589	0.038258	0.948201
0.2	0.018900	2.601818	43.674390	0.020634	0.901117	0.023100	4.750370	53.120563	0.041786	0.95313
0.1	0.018375	2.404630	42.743907	0.018892	0.892267	0.023625	5.110714	54.617436	0.045639	0.957685

Table 4.4 Reliability parameters at ±15% spread values for nonmembership function.

	Left spread					Right spread				
DOF	Failure rate	Repair time	MTBF	Reliability	Availability	Failure rat rate	Repair time	MTBF	Reliability	Availability
1.0	0.021788	3.951483	49.72740	0.033517	0.940053	0.020213	3.154990	46.21484	0.025725	0.920732
0.9	0.021394	3.737147	48.79655	0.031372	0.935617	0.020606	3.339461	47.04207	0.027484	0.925967
0.8	0.021000	3.533333	47.90207	0.029364	0.930925	0.021000	3.533333	47.90207	0.029364	0.930925
0.7	0.020606	3.339461	47.04207	0.027484	0.925967	0.021394	3.737147	48.79655	0.031372	0.935617
0.6	0.020213	3.154990	46.21484	0.025725	0.920732	0.021788	3.951483	49.72740	0.033517	0.940053
0.5	0.019819	2.979421	45.41876	0.024078	0.915206	0.022181	4.176967	50.69668	0.035809	0.944244
0.4	0.019425	2.812287	44.65237	0.022537	0.909378	0.022575	4.414279	51.70659	0.038258	0.948201
0.3	0.019031	2.653155	43.91432	0.021094	0.903237	0.022969	4.664152	52.75956	0.040874	0.951934
0.2	0.018638	2.501620	43.20338	0.019744	0.896768	0.023363	4.927383	53.85821	0.043670	0.955453
0.1	0.018244	2.357304	42.51840	0.018480	0.889959	0.023756	5.204835	55.00542	0.046656	0.958767

Table 4.5 Reliability parameters at ±15% spread values for Lambda–Tau approach.

	Left spread					Right spread				
DOF	Failure rate	Repair time	MTBF	Reliability	Availability	Failure rat rate	Repair time	MTBF	Reliability	Availability
1.0	0.021000	3.533333	47.90207	0.029364	0.930925	0.021000	3.533333	47.90207	0.029364	0.930925
0.9	0.020685	3.377466	47.21140	0.027850	0.926981	0.021315	3.695562	48.61481	0.030959	0.934699
0.8	0.020370	3.227683	46.54190	0.026415	0.922860	0.021630	3.864447	49.35057	0.032642	0.938308
0.7	0.020055	3.083721	45.89275	0.025053	0.918557	0.021945	4.040302	50.11038	0.034416	0.941758
0.6	0.019740	2.945333	45.26315	0.023762	0.914065	0.022260	4.223461	50.89533	0.036286	0.945054
0.5	0.019425	2.812287	44.65237	0.022537	0.909378	0.022575	4.414279	51.70659	0.038258	0.948201
0.4	0.019110	2.684361	44.05973	0.021375	0.904491	0.022890	4.613136	52.54540	0.040337	0.951205
0.3	0.018795	2.561347	43.48458	0.020274	0.899395	0.023205	4.820434	53.41309	0.042529	0.954071
0.2	0.018480	2.443048	42.92633	0.019229	0.894086	0.023520	5.036606	54.31110	0.044841	0.956803
0.1	0.018165	2.329276	42.38443	0.018237	0.888555	0.023835	5.262114	55.24096	0.047278	0.959406

Table 4.6 Defuzzified values of various reliability parameters.

Reliability parameters		Defuzzified value (15% spread)	Defuzzified value (25% spread)	Defuzzified value (50% spread)
Failure rate	1	0.000382	0.000637	0.001274
	2	0.000970	0.001274	0.003234
	3	0.001323	0.002205	0.004410
Repair time	1	35.582620	69.559560	284.300907
	2	87.964550	164.605838	569.346026
	3	119.34570	221.471894	742.751258
MTBF	1	50.394407	53.222738	66.759946
	2	49.263355	51.126922	61.120856
	3	48.654908	50.084716	58.882400
Reliability	1	0.036310	0.045318	0.090523
	2	0.033554	0.039899	0.074009
	3	0.032169	0.037484	0.068584
Availability	1	0.932604	0.926044	0.881307
	2	0.928245	0.919739	0.872609
	3	0.925202	0.914852	0.863689

1→ Membership function values; 2→Nonmembership function values; 3→ Lambda–Tau approach values.

value of membership function. It is declined by 0.00916% with increase in spread from ±15% to ±25% and 0.0512% with increase in spread from ±25% to 50% for nonmembership values. Increase of failure rate and repair time shows increased unavailability of system which means increase in sudden failure of components and thereby loss and interruption in production.

6. Conclusion

The considered has been modeled by intuitionistic fuzzy Lambda–Tau approach for enhancing the reliability. IFS theory has been used to handle the vagueness of the collected data. Various reliability parameters were computed to study the failure dynamics of the system. The major merits of the technique lie in the fact that it has considered the degree of hesitation as 0.2 between degree of acceptance and rejection. Eventually, theses parameters facilitate the reliability engineer to plan a proper maintenance schedule for the cutting system of a sugar industry also the behavior analysis of the system could be done.

References

[1] D. Kumar, J. Singh, I.P. Singh, Availability of the feeding system in the sugar industry, Microelectron. Reliab. 28 (6) (1988) 867–871.

[2] D. Panchal, A.K. Singh, P. Chatterjee, E.K. Zavadskas, M. Keshavarz-Ghorabaee, A new fuzzy methodology-based structured framework for RAM and risk analysis, Appl. Soft Comput. 74 (2019) 242–254.

[3] D. Kumar, P.C. Pandey, J. Singh, Process design for a crystallization system in the urea fertilizer industry, Microelectron. Reliab. 31 (5) (1991) 855–859.

[4] S. Kumar, N.P. Mehta, D. Kumar, Steady state behaviour and maintenance planning of a desulphurization system in a urea fertilizer plant, Microelectron. Reliab. 37 (6) (1997) 949–953.

[5] N. Arora, D. Kumar, Stochastic analysis and maintenance planning of the ash handling system in the thermal power plant, Microelectron. Reliab. 37 (5) (1997) 819–824.

[6] J. Knezevic, E.R. Odoom, Reliability modelling of repairable systems using Petri nets and fuzzy Lambda–Tau methodology, Reliab. Eng. Syst. Saf. 73 (1) (2001) 1–17.

[7] R.K. Sharma, D. Kumar, P. Kumar, Predicting uncertain behavior of industrial system using FM—a practical case, Appl. Soft Comput. 8 (1) (2008) 96–109.

[8] Komal, S.P. Sharma, D. Kumar, RAM analysis of repairable industrial systems utilizing uncertain data, Appl. Soft Comput. 10 (2010) 1208–1221, 2010.

[9] M. Verma, A. Kumar, Y. Singh, Vague reliability assessment of combustion system using Petri nets and vague lambda-tau methodology, Eng. Comput. 30 (5) (2013) 665–681.

[10] D. Panchal, D. Kumar, Stochastic behaviour analysis of power generating unit in thermal power plant using fuzzy methodology, Opsearch 53 (1) (2016) 16–40.

[11] D. Panchal, D. Kumar, Integrated framework for behavior analysis in a process plant, J. Loss Prev. Process. Ind. 40 (2016) 147–161.

[12] H. Garg, Performance and behavior analysis of repairable industrial systems using Vague Lambda–Tau methodology, Appl. Soft Comput. 22 (2014) 323–338.

[13] S.P. Sharma, D. Kumar, Stochastic behavior analysis of the press unit in a paper mill using GABLT technique, Int. J. Int. Comput. Cybern. 2 (3) (2009) 574–593.

[14] S.P. Sharma, N. Sukavanam, N. Kumar, A. Kumar, Reliability analysis of complex robotic system using Petri nets and fuzzy lambda-tau methodology, Eng. Comput. 27 (3) (2010) 354–364.

[15] S.P. Sharma, D. Kumar, A. Kumar, Behavior prediction of washing system in a paper industry using GA and fuzzy lambda–tau technique, Appl. Math. Model. 36 (6) (2012) 2614–2626.

[16] M. Verma, A. Kumar, A novel general approach to evaluating the reliability of gas turbine system, Eng. Appl. Artif. Intell. 28 (2014) 13–21.

[17] D. Panchal, D. Kumar, Reliability analysis of CHU system of coal fired thermal power plant using fuzzy λ-τ approach, Procedia Eng. 97 (2017) 2323–2332.

CHAPTER

5 Game theoretic modeling and dependability analysis of small cell relays under bandwidth spoofing attack in 5G wireless communication network

K.C. Lalropuia, Vandana Gupta

Department of Operational Research, University of Delhi, Delhi, India

1. Introduction

Recent years have seen a dramatic increase in the demands of users in the sector of wireless communication networks (WCNs). In fact, mobile data traffic are expected to increase 100 times in the near future. This motivated the development of ultra-high data rate such as fifth-generation (5G) network that is designed to be fully interconnected with the possibilities of limitless and quicker access to the information [1–4]. Several technologies, such as massive MIMO (multiple-input and multiple-output), device-to-device communication, small cells (femtocells, picocells, and microcells), millimeter wave communication, etc., are incorporated in the existing ones to achieve 5G service requirements [5–8]. This in turn poses a number of challenges in coverage region, energy efficiency, spectrum utilization, cost, latency, data rate, and security [5,9,10]. Out of all the technologies mentioned above, small cells have become the integral part of 5G networks, which help in the improvement of network coverage, spectral and energy efficiency, and reduction of transmission power [1–3,5]. Small cells are low-powered cellular radio access nodes operating in licensed and unlicensed spectrum covering 10 meters to a few kilometers. A number of small cells can be deployed around a macrocell base station (MBS) in which small cells act as relay nodes and these are called small cell relays (SCRs) [11,12]. Generally, a relay network is employed when the transmission range of the communicating devices is greater than the distance between them. End users (EUs) that are on the edge of a macrocell coverage or those EUs that are not able to communicate properly with MBS due to weak signals employ small cells as relays for effective communications [5,9,11]. Although small cells have become promising technologies toward the goal of 5G networks, their deployment renders the network highly susceptible to various kinds of attacks [13–16]. For example, eavesdropping is a serious and common threat in small cell networks [3,12,13,17], and bandwidth spoofing attack is possible during communication between SCRs and an MBS [5]. In a heterogeneous network with dense small cells, one of the attackers' targets could be a privacy issue such as location

The Handbook of Reliability, Maintenance, and System Safety through Mathematical Modeling.
https://doi.org/10.1016/B978-0-12-819582-6.00005-8

information [14]. Authentication in small cell–based smart grids is also a current issue in 5G network [15]. Further, denial-of-service (DoS) attacks can be launched in small cell networks [18,19].

Relay can be of half-duplex (HD) or full-duplex (FD) [2]. The secrecy performance of FD relay is better than that of HD relay if it is possible to supress the self-interference [20]. Transmission in FD relay is exposed to the risk of eavesdropping both in the uplink and downlink [21]. For a decode and forward relay, Lin et al. [22] evaluated maximum achievable secrecy rate with power constraints on the source and energy harvesting relay by using some optimization techniques. The energy harvesting can help in achieving energy efficiency goal. However, such approach makes the network susceptible to security attacks [2]. Based on graph theoretical techniques, Zhang et al. [23] proposed cooperative anti-eavesdropping strategies for physical layer security enhancement in 5G large-scale relay networks. Fang et al. [24,25] focused on eavesdropping attacks and proposed a Stackelberg game model to capture the interactions between the source and the multiple relays in order to defend signals forwarded by the relays against multiple eavesdroppers.

Devi et al. [9] presented security attacks on SCRs in 5G network in which an intrusion detection system (IDS) is implemented using Adaptive neuro-fuzzy inference system. The authors [9] focused on the IDS for detecting intrusion in a security attack on the SCRs. However, the methods for responding to the malicious activities and the attacker's probable strategies are not considered. Gandotra et al. [2] proposed a 5G network architecture in which SCRs are employed for attaining required data rate as well as saving the battery lifetime. However, the proposed architecture is susceptible to spoofing attacks [2]. Gupta et al. [5] discussed bandwidth spoofing attack in a multistage 5G WCN based on a prisoner dilemma game.

In a bandwidth spoofing attack, the attacker intrudes into the network masquerading as a legitimate user in order to procure the bandwidth allocated to SCRs. In case of bandwidth distributed DoS attack, attackers disrupt the network infrastructure operation by causing congestion that is carried out by increasing the total amount of traffic or the total amount of packets [33]. However, a bandwidth spoofing attack is different from a bandwidth distributed DoS attack in that the attacker tries to acquire the bandwidth by sending bandwidth requests to the MBS other than causing traffic congestion to disrupt the communication [5,34]. In case of successful attack, the bandwidth is obtained by the attacker resulting in the unavailability of the network. When the attacker carries out bandwidth spoofing attack during communication between SCRs and an MBS, there is a very high chance that the game being played at one stage transforms into a different game at the next stage if the IDS fails to detect the intrusion. Further, the game that is to be played next could depend on the previous game played as well as the actions taken by both the attacker and the defender. Hence, we incorporate these aspects in SCRs attack and develop a novel game theoretic model based on stochastic game. In a stochastic game, players play from a set of game elements and the game elements are played in a sequence of stages. The transition from one game element to another is probabilistic and it depends on the current game element and the actions taken by the players.

From the above discussion, it is clear that if the bandwidth that is allocated to the SCR is acquired by the attacker, the SCR will not be available to the EUs for communication. Therefore, it is important to consider the dependability issues due to the bandwidth spoofing attack such as availability and reliability of the SCR network. Dependability analysis evaluates system capability of avoiding service failures that can result in significant damages or losses. Dependability includes various measures to assess system performances such as reliability, availability, integrity, maintainability, security, etc. [26]. A survey on concepts and methodologies for evaluating system dependability is performed in

Ref. [27]. This paper [27] presents a summary of how these techniques can be employed to assess system security. Dependability analysis has widely been applied in critical infrastructure systems, communication systems, and aircraft and defense systems [28]. In this chapter, we also perform dependability analysis of SCRs under bandwidth spoofing attack by formulating a state-based stochastic model with continuous-time Markov chain (CTMC) as the underlying stochastic process. We then obtain the dependability attributes such as mean time to failure (*MTTF*, a measure of reliability) and steady state availability (*SSA*) of the SCR. Reliability is used to measure the continuous delivery of correct service (or the average time to failure), and availability is concerned with readiness for correct service [29]. The main contributions of this chapter are outlined as below:

- A novel game theoretic model is proposed based on a two-person zero-sum stochastic game to capture the interactions between an attacker and the network defender when a bandwidth spoofing attack is carried out during the communication between an MBS and SCRs in a 5G WCN.
- The attacker's optimal strategies are determined, thereby predicting the attacker's behavior in the process of the bandwidth attack. Further, the optimal defense strategies are computed for responding to the attacker's malicious activities. The feasibility of the proposed model is demonstrated with the help of numerical illustrations.
- Dependability measures such as *MTTF* and *SSA* of an SCR network in the presence of a bandwidth spoofing attack are obtained based on a CTMC. In addition, a numerical illustration on how the reliability and availability of the network are affected by the bandwidth spoofing attack is presented. It is shown that both the reliability and the availability of the network decrease as the rate of attack increases.

The rest of the chapter is organized as follows: In Section 2, an overview of a stochastic game is given. Section 3 covers the proposed SCR attack model, determination of optimal strategies for both the attacker and the defender, and numerical illustrations of how the attacker and the defender are motivated by the game parameters. Section 4 discusses various dependability measures of the SCR under attack. Model validation via simulation results is also presented in Section 4.2 and discussion of the results is given in Section 5 and the conclusion follows in Section 6.

2. An overview of stochastic game

Game theory has become a very useful tool in the area of network security [30]. In this section, we first describe a stochastic game and then present the proposed game model in the next section. A two-person zero-sum stochastic game consists of "q" game elements, Γ_r, $r = 1,2, \ldots,q$, and the two players—I and II play the games in a sequence of stages [31]. The transition from the game being played to the next game element is probabilistic and it depends on both the current game element and the actions taken by the players. Each game element Γ_r, $r = 1,2, \ldots,q$ can be represented by an $m_r \times n_r$ matrix (γ_{ij}^r) with entries given by

$$\gamma_{ij}^r = a_{ij}^r + \sum_{s=1}^{q} \phi_{ij}^{rs} \Gamma_s \tag{5.1}$$

where $\phi_{ij}^{rs} \geq 0$, and

$$\sum_{s=1}^{q} \phi_{ij}^{rs} < 1. \tag{5.2}$$

Eq. (5.1) states that if ith pure strategy is chosen by player I and jth pure strategy is chosen by player II in the game element Γ_r, then there is a payoff a_{ij}^r from player I and II. In addition, the players will play the sth game element next with a transition probability φ_{ij}^{rs} and the game will terminate with probability $\phi_{ij}^{r0} = 1 - \sum_{s=1}^{q} \phi_{ij}^{rs}$. Condition (2) states that, in each stage, there is always a positive probability that the game will terminate. This implies that the probability that the game will continue indefinitely is 0 and therefore all expected payoffs will be finite.

In rth game element, the player I's stationary mixed strategy is an m_r vector, $\mathbf{x^r}$ given by $\mathbf{x^r} = \left(x_1^r, x_2^r, \ldots, x_{m_r}^r\right), r = 1, 2, 3, \ldots, q$, such that

$$\sum_{k=1}^{m_r} x_k^r = 1, \quad \text{and } x_k^r \geq 0 \tag{5.3}$$

In other words, x_k^r is the probability that the player I chooses his/her kth pure strategy, assuming that he/she is playing the rth game element Γ_r.

Similarly, the player II's stationary mixed strategy is an n_r vector, $\mathbf{y^r}$ given by $\mathbf{y^r} = \left(y_1^r, y_2^r, \ldots y_{n_r}^r\right), r = 1, 2, 3 \ldots, q$, such that

$$\sum_{l=1}^{n_r} y_l^r = 1, \quad \text{and } y_l^r \geq 0 \tag{5.4}$$

For a given pair of strategies, an expected payoff ω_r can be computed for any $r = 1, 2, \ldots, q$, on the assumption that the game element at the first stage is Γ_r. Therefore, a value vector $\omega = (\omega_1, \omega_2, \ldots, \omega_q)$ can be determined for a pair of strategies. If the value vector is to exist, the value component ω_r must be able to replace the game element Γ_r and thus we must have

$$\omega_r = val(B_r) \tag{5.5}$$

where $val(B_r)$ represents the value of the matrix game B_r, and B_r denotes an $m_r \times n_r$ matrix (b_{ij}^r) having elements of the form

$$b_{ij}^r = a_{ij}^r + \sum_{s=1}^{q} \phi_{ij}^{rs} \omega_s \tag{5.6}$$

Therefore, a set of stationary optimal strategies can be obtained for each player at the end from the optimal mixed strategies for the matrix game B_r, $r = 1, 2, \ldots, q$.

3. Proposed game model

In this chapter, we focus on a "bandwidth spoofing attack" that is carried out by an attacker during the process of communication between an MBS and an SCR. The n SCRs are stationed around the MBS as shown in Fig. 5.1. A bandwidth spoofing attack takes place in medium access control layer, and it can be described as follows [5,32]: In the attack scenario given in Fig. 5.1, the attacker knows the uplink/downlink traffic between an SCR and the MBS. The communication between the SCR and the MBS

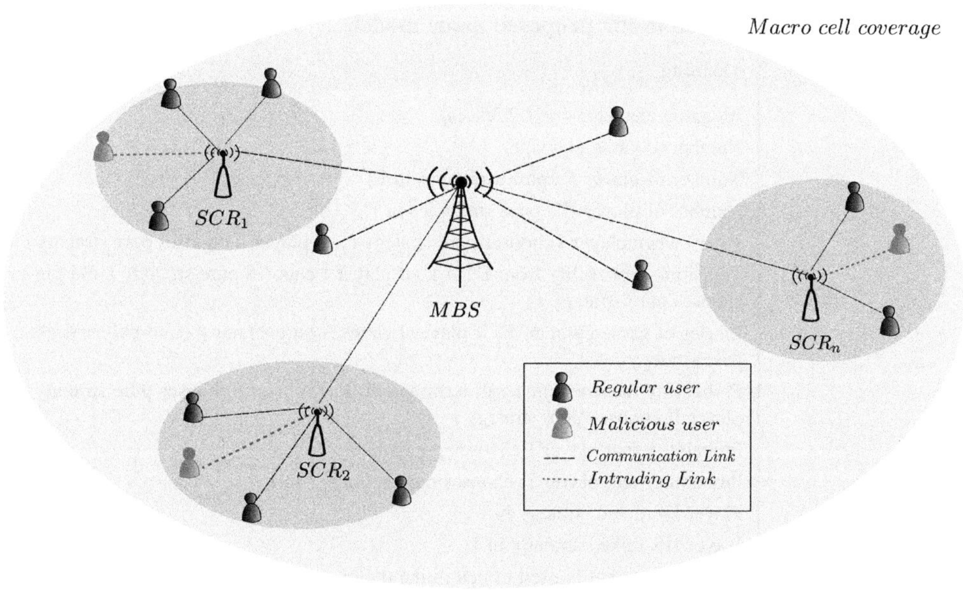

FIGURE 5.1

A scenario of small cell relay (SCR) attack in a fifth-generation (5G) wireless communication network (WCN).

occurs in three phases. In phase 1, the operation of ranging is performed by the MBS. Once it has been done, the SCR can send a request to the server, and this request is sent with the help of the MBS (uplink) in phase 2. In phase 3, the server responds to the particular application from the MBS (downlink) to the SCR. For this process, a bandwidth is required, which is to be allocated by the MBS to the SCR. During this communication process, an attacker masquerading as a regular mobile user can penetrate the system to carry out a bandwidth spoofing attack. Thus, the attacker tries to procure the bandwidth allocated to the SCR by sending unnecessary bandwidth requests to the MBS. If the bandwidth is obtained by the attacker, the system will be unavailable, thereby causing a DoS attack. In a bandwidth distributed DoS attack, attackers disrupt the network infrastructure operation by causing congestion that is carried out by increasing the total amount of traffic or the total amount of packets [33]. However, in this chapter, we consider a bandwidth spoofing attack in which an attacker (a malicious EU), instead of intending to cause traffic congestion, tries to acquire the bandwidth by sending bandwidth requests to the MBS [5,34]. In this section, a stochastic game model is formulated to capture the interactions between the attacker and the system defender. In this bandwidth spoofing attack, a "mobile user" who pretends to be a regular user is the "attacker" and an "intrusion detection system (IDS)" is the "network defender." The unavailability issue due to the DoS attack is addressed in the next section. Some of the notations used in the proposed game model are given in Table 5.1.

Now, we assume that the SCR under the bandwidth spoofing attack undergoes different states and these are explained as follows:

Table 5.1 Some notations used in the proposed game model.

Notation	Meaning
Γ_r	rth game element, $r = 1,2,3, \ldots,q$
q	Number of game elements
m_r	Number of player I's pure strategies in Γ_r
n_r	Number of player II's pure strategies in Γ_r
a^r_{ij}	Payoff when player I chooses pure strategy i and player II chooses pure strategy j in Γ_r
ϕ^{rs}_{ij}	Transition probability from Γ_r to Γ_s if player I chooses pure strategy i and player II chooses pure strategy j
γ^r_{ij}	Entries of game element Γ_r if player I chooses pure strategy i and player II chooses pure strategy j
ϕ^{r0}_{ij}	Probability that the game will terminate in Γ_r if player I chooses pure strategy i and player II chooses pure strategy j
x^r_k	Probability that player I chooses pure strategy k in Γ_r
y^r_l	Probability that player II chooses pure strategy l in Γ_r
$\mathbf{x}^\mathbf{r}$	Player I's mixed strategy in Γ_r
$\mathbf{y}^\mathbf{r}$	Player II's mixed strategy in Γ_r
B_r	Matrix game that is used to determine the expected payoff
ω_r	Expected payoff if the game begins with Γ_r
ω	Expected payoff vector
b^r_{ij}	Value of γ^r_{ij} obtained by replacing Γ_s with its value
a	Attacker's action "attack"
τ	Attacker's action "not-attack" or defender's action "not-monitor"
m	Defender's action "monitor"
α	Hardware failure rate of SCR
β_{rs}	Attacker's accumulated attack intensity from state r to state s
ψ_{rs}	System administrator's effort intensity from state r to state s
$v(r, s_1, s_2)$	Payoff corresponding to a strategy profile (s_1, s_2) in Γ_r
λ_{ij}	Transition rate from state i to state j in CTMC

CTMC, *continuous-time Markov chain;* SCR, *small cell relay.*

- *Free state* (N): Initially, the SCR is assumed to be available when requested, and this state is denoted by N as shown in Fig. 5.2. In this state N, a certain level of data volume is attained in the uplink.
- *Intermediate* (I): After a period of time, the SCR enters intermediate state I from the state N. The SCR is said to be in the intermediate state when data volume in the uplink is below a threshold value which is fixed at some value (or average value) [35].
- *Busy state* (B): The SCR makes a transition from the intermediate state to the busy state (B) when data volume in the uplink is above the threshold value. The peak value of the data volume is attained in this state [35].
- *Penetrated state* (P):The attacker who is pretending to be a regular user intrudes into the network when the SCR is either in state I or B, and this intrusion renders the network penetrated state (P).

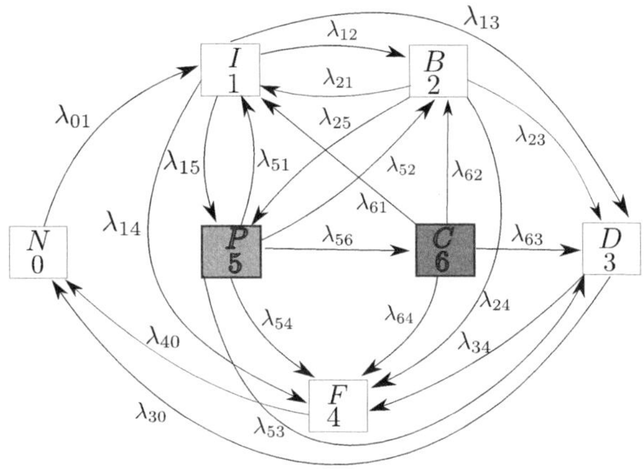

FIGURE 5.2

State transition diagram of the small cell relay (SCR) network with identified game elements.

The attacker in this state can perform either "attack" or "not-attack" action. If the attacker performs "attack" and the IDS fails to detect the attack (i.e., the IDS monitors the system but does not detect the intrusion), then he/she will capture the bandwidth allocated to the SCR, and hence the SCR will make a transition to the compromised state (C) from the state P which is indicated by the directed arc from P to C in Fig. 5.2. However, if the IDS discovers the malicious activity, then by taking appropriate actions, the network is restored to the previous state either at state I or state B. Further, the SCR being in the penetrated state will move back from the state P to its former state either at state I or state B if the attacker withdraws (i.e., "not-attack") from his/her attack intention.

Here the action "attack" refers to the malicious action taken by the attacker when he/she tries to procure the bandwidth allocated to the SCR by the MBS, whereas "not-attack" represents the action as a result of which the attacker intruding into the system as a regular user withdraws from the attack intention. Note that the attacker is said to be successful either when he/she chooses the action "attack" and the system IDS takes the action "not-monitor" or when the IDS monitors the network but does not detect any adversarial activity. In this case, the attacker procures the bandwidth allocated to the SCR and hence gets a positive reward or payoff. However, if the IDS performs "monitor" and detects the malicious actions of the attacker, then a necessary action is taken to remove the threat or to make sure that the system is secure from the attack. Thus, in this case, the attacker loses the bandwidth (or gets a negative payoff or cost). Moreover, the "monitor" refers to an action when the IDS monitors if there is any malicious activity in the system and generates an alarm to the system administrator in case of intrusion detection [36]. As a result of successful defense, positive payoff will be assigned to the defender. On the other hand, the action "not-monitor" will result in a negative payoff or cost if the attacker performs harmful activities.

- *Compromised state* (C):In the compromised state, the attacker tries to maximize his/her payoff by continuing the "attack" process in order to capture the whole SCR network. In that case, the SCR will go from the compromised state (C) to the failed state (F) and the EUs will not be able to access the SCR for communications. On the other hand, if the IDS detects the malicious activities, then by performing necessary actions, the SCR is brought back from the state C to the previous state—I or B as indicated by the directed arcs from C to I and C to B as shown in Fig. 5.2.

- *Down state* (D): As the SCR is battery powered, its battery will be depleted at a certain rate. Moreover, in the presence of the attacker, a part of its energy (battery) will be consumed while resisting the intrusion at the SCR [2]. Further, relaying is not possible when its energy level is below a threshold level [2]. If the SCR is in any one of the states—I, B, P, and C—then it will move to the down state (D) due to the energy depletion, which is specified by the directed arcs from the states I, B, P, and C to the state D, respectively. The SCR will move from the down state (D) to the initial state (N) as soon as the battery is revived, and this is represented by the directed arc from D to N.
- *Failed state* (F): In addition to the transition from the state C to the state F described above, the SCR can go to the failed state at any time due to the hardware failure when the SCR is in any one of the states such as I, B, P, C, and D. This condition is indicated by the directed arcs from the states I, B, P, C, and D to the state F, respectively.

In this model, hardware failure of the SCR is defined to be the failure other than the one caused by battery shortage. The SCR will make a transition to the free state (N) from the failed state F if recovery action is performed on the failed SCR. This transition is indicated by the directed arc from F to N.

Note that in the proposed model, the down state (D) is different from the failed state (F). The SCR will be in the down state (D) when its battery level is below a threshold value in which relaying is not possible, whereas the SCR can move to the failed state (F) at any time due to the hardware failures of the SCR.

To simplify the analysis, we label each of the states defined above (free, intermediate, busy, penetrated, compromised, down, and failed) states numerically using 0, 1, 2, 3, 4, 5, and 6, respectively as shown in Fig. 5.2. The regular users within the network coverage of the SCR under attack communicate with the SCR at random point of times and the attacker is assumed to intrude into the system according to a Poisson process. Further, the time taken by the attacker to carry out the bandwidth spoofing attack, the occurrence of hardware failure, and the repair completion time can be characterized by random variables, and we assume that these times follows exponential distributions. Thus, we model the network lifetime based on a CTMC and λ_{ij} denotes the transition rate from state i to state j in the context of CTMC as shown in Fig. 5.2. The transition rates from the state N to both the state F and the state D are relatively so small that they are negligible [35], and therefore no arc is drawn from N to F and D. Further, the SCR undergoes different states and due to the actions taken by the attacker and the IDS, the two states such as the penetrated state (P) and the compromised state (C) are transitioned into different states with certain probabilities.

To capture this system dynamics, stochastic game is the appropriate technique in which the attacker and network IDS are the two players. Based on the above discussions, there are various states undergone by the SCR in the presence of the attacker; however, the intruder is interested only in the penetrated state (P) and the compromised state (C). Thus, the stochastic game consists of the two game elements and is given by $\Gamma = \{\Gamma_P, \Gamma_C\}$. Further, the game is assumed to begin from the game element Γ_P.

To predict the attacker behavior, the game is analyzed mainly from the attacker's perspective. For this, a two-person zero-sum stochastic game model is formulated as given below:

The proposed game model can be characterized by the five tuple:

$$\langle Z, \Gamma, S, \phi, \nu \rangle$$

where

- The set of players: $Z = \{$attacker, defender (network IDS)$\}$.
- The set of game elements: $\Gamma = \{\Gamma_B\ \Gamma_C\}$, where Γ_P and Γ_C correspond to the penetrated and the compromised state, respectively.
- $S = S_1 \times S_2$, where $S_1 = \{$attack (a), not-attack (τ)$\}$ and $S_2 = \{$not-monitor (τ), monitor (m)$\}$ are the action sets of the attacker and the defender, respectively.
- The state transition probability function: $\phi : \Gamma \times S \times \Gamma \mapsto [0,1]$.
- The payoff function: $\nu : \Gamma \times S \mapsto \mathbb{R}$, where $\mathbb{R}$ is the set of real numbers.

In each of the game elements, such as Γ_P and Γ_C, the attacker and the defender are assumed to take actions from S_1 and S_2, respectively. In the rth game element Γ_r, let the attacker's mixed strategy be denoted by $\mathbf{x^r} = (x_a^r, x_\tau^r)$, i.e., $\mathbf{x^r}$is a probability distribution on S_1 such that $x_a^r + x_\tau^r = 1$. Also let the defender's mixed strategy be represented by $\mathbf{y^r} = (y_\tau^r, y_m^r)$, i.e., $\mathbf{y^r}$ is a probability distribution on S_2 such that $1y_\tau^r, y_m^r = 1$. Further, let ϕ_{ij}^{rs} denote the transition probability that the game element Γ_r moves to the game element Γ_s when the ith pure strategy is chosen by the attacker and the jth pure strategy is chosen by the defender in the rth game element Γ_r. If the IDS fails to detect the attacker activities, the game moves from Γ_P to Γ_C. In moving from Γ_r to Γ_s, there exists an accumulated attack intensity β_{rs} if the action (a) is always chosen by the attacker. One way of determining this accumulated attack intensity is to depend on a subjective assesment of security experts and/or empirical data obtained from intrusion experiments [37], and estimating these rates is beyond the scope of this chapter. Let ψ_{rs} be the effort intensity of the system administrator from a system state r to a state s, and α denotes the hardware failure rate of the SCR. Therefore, the probability with which the game element Γ_P makes transition to the game element Γ_C can be computed based on the system's Markov properties [37]:

$$\phi_{ij}^{PC} = \begin{cases} \dfrac{\beta_{PC}}{\beta_{PC} + \psi + \alpha + \mu}, & \text{if } i = a \text{ and } j = \tau \\ 0, & \text{otherwise} \end{cases} \tag{5.7}$$

where β_{PC} is the accumulated attack intensity from the system state P to the state C, $\psi = \psi_{PI} + \psi_{PB}$ is the system administrator's effort intensity from the system state P to the state I or from the state P to the state B, α is the hardware failure rate of the SCR, and μ is the energy (battery) depletion rate of the SCR. Here ψ, α, and μ are the rates of the other events that may interfere the attack.

Note that the transition probability from a game element Γ_r to a game element Γ_s is denoted byϕ_{ij}^{rs}, and we assume that $\phi_{ij}^{PP} = 0, \phi_{ij}^{CP} = 0, \phi_{ij}^{CC} = 0$. Thus, the entries of each of the game elements Γ_P and Γ_C can be expressed as below:

$$\gamma_{ij}^{P} = \begin{cases} a_{ij}^{P} + \phi_{ij}^{PC}\Gamma_C, & \textit{if } i = a \textit{ and } j = \tau \\ a_{ij}^{P}, & \textit{otherwise} \end{cases} \tag{5.8}$$

$$\gamma_{ij}^{C} = a_{ij}^{C} \textit{ for all } i \textit{ and } j \tag{5.9}$$

3.1 Predicting attacker behavior

In this section, we predict the attacker's behavior in terms of the optimal strategies (optimal attack probabilities) in each of the penetrated and the compromised state, respectively. In addition, we obtain the best defense strategies for responding to the attack strategies. Note that predicting the attacker behavior is to determine the optimal strategies of the attacker both in the game elements Γ_P and Γ_C. The optimal mixed strategies of the attacker and the defender in the game element Γ_r are denoted by $\mathbf{x}^{\mathbf{r}^*}$ and $\mathbf{y}^{\mathbf{r}^*}$, respectively. These optimal mixed strategies can be obtained by determining

$$\max_{\mathbf{x}^{\mathbf{r}}} \min_{\mathbf{y}^{\mathbf{r}}} E(\mathbf{x}^{\mathbf{r}}, \mathbf{y}^{\mathbf{r}}),\ for\ all\ \Gamma_r \in \Gamma \tag{5.10}$$

where

$$E(\mathbf{x}^{\mathbf{r}}, \mathbf{y}^{\mathbf{r}}) = \sum_{i \in S_1} \sum_{i \in S_2} x_i^r \gamma_{ij}^r y_j^r,\ for\ all\ \Gamma_r \in \Gamma. \tag{5.11}$$

and the probability distributions $\mathbf{x}^{\mathbf{r}} = (x_a^r, x_\tau^r)$ and $\mathbf{y}^{\mathbf{r}} = (y_\tau^r, y_m^r)$ are the mixed strategies of the attacker and the defender, respectively, in the game Γ_r.

The value of the matrix game B_r denoted by val(B_r) can be obtained by

$$val(B_r) = E(\mathbf{x}^{\mathbf{r}^*}, \mathbf{y}^{\mathbf{r}^*})\ for\ all\ \Gamma_r \in \Gamma \tag{5.12}$$

Thus, based on the following algorithm [31,38], we determine the sets of the optimal strategies of the attacker and the defender such as $\mathrm{x}^* = \{x^{P^*}, x^{C^*}\}$ and $\mathrm{y}^* = \{y^{P^*}, y^{C^*}\}$, where $x^{P^*} = (x_a^{P^*}, x_\tau^{P^*}), x^{C^*} = (x_a^{C^*}, x_\tau^{C^*})$, $y^{P^*} = (y_\tau^{P^*}, x_m^{P^*})$, and $y^{C^*} = (y_\tau^{C^*}, y_m^{C^*})$.

3.1.1 Algorithm 1

Input: $\langle Z, \Gamma, S, \phi, u \rangle$

Output: $\mathbf{x}^*$ and $\mathbf{y}^*$

Initialize the payoff value vector $\omega = (\omega_B\ \omega_C) = (0{,}0)$;

1. Form the matrix $B_C = B_C = \left(b_{ij}^C\right) = \left(\gamma_{ij}^C\right) = \left(a_{ij}^C\right)$;
2. Determine x^{C^*} and B_C using Eq. (5.10);
3. Put $\omega_C = val(B_C) = E(x^{C^*}, y^{C^*})$;
4. Determine b_{ij}^P by replacing Γ_C in Eq. (5.8) with ω_C and then form B_P;
5. Determine x^{P^*} and y^{P^*} in B_P using Eq. (5.10);

Return $\mathrm{x}^* = \{x^{P^*}, x^{C^*}\}$& $\mathrm{y}^*\{y^{P^*}, y^{C^*}\}$.

3.2 Illustration of the attacker behavior

Note that the proposed model is a two-person zero-sum game model, and in each of the game elements, such as Γ_P and Γ_C, we assume that the attacker takes actions from the action set $S_1 = \{$attack (a), not-attack (τ)$\}$ and the defender takes actions from the action set $S_2 = \{$monitor (m), not-monitor (τ)$\}$. In the game element Γ_P, in order to illustrate how the attacker would carry out the attack and correspondingly how the defender would respond to the malicious activities, we consider the attacker's payoffs corresponding to the four strategy profiles such as (a, τ), (a, m), (τ, τ), and (τ, m), and these are given in the payoff matrix V_P:

$$V_P = \begin{array}{c} \\ \text{attack}(a) \\ \text{not}-\text{attack}(\tau) \end{array} \begin{array}{c} \text{not-monitor}(\tau) \quad \text{monitor}(m) \\ \left[\begin{array}{cc} 1 & \nu(P,a,m) \\ \nu(P,\tau,\tau) & 0 \end{array} \right] \end{array} \tag{5.13}$$

Further, in the game element Γ_C, the payoffs of the attacker corresponding to the different strategy profiles are represented by the entries of the payoff matrix V_C:

$$V_C = \begin{array}{c} \\ \text{attack}(a) \\ \text{not}-\text{attack}(\tau) \end{array} \begin{array}{c} \text{not-monitor}(\tau) \quad \text{monitor}(m) \\ \left[\begin{array}{cc} 2 & \nu(C,a,m) \\ \nu(C,\tau,\tau) & 0 \end{array} \right] \end{array} \tag{5.14}$$

Here $\nu(r,s_1,s_2)$denotes the payoff corresponding to the strategy profile $(s_1, s_2) \in S$ in the game element Γ_r, $r \in \{P, C\}$, i.e., $\nu(r,s_1,s_2)$is the payoff obtained by the attacker when the attacker chooses the action "$s_1 \in S_1$" and the defender chooses the action "$s_2 \in S_2$" in the game element Γ_r. The other payoffs corresponding to the different strategy profiles are also represented by the corresponding entries of the matrices V_P and V_C.

When the attacker intrudes into the SCR as a regular user, the two players, the attacker, and the network defender interact in the penetrated state (P), a positive payoff (reward) 1 is awarded to the attacker if he/she chooses "*attack* (a)" and the defender chooses "*not-monitor* (τ)." Moreover, a payoff value 0 is assigned to the attacker if he/she chooses "*not-attack* (τ)" and the defender chooses "*monitor* (m)." Thus, corresponding to the profiles of strategies (a, τ) and (τ, m), the attacker's payoffs are 1 and 0, respectively, as shown in the payoff matrix V_P. Note that the payoff values themselves are not important, but it is their relative values that will reflect how the attacker behaves in the system.

Further, we have the payoff matrix V_C as given above. When the SCR is in the compromised state (C), the attacker persists in attacking the system trying to maximize his/her payoffs. Therefore, in the compromised state, the malicious activity is assumed to have more severe impact on the SCR than it does in the penetrated state (P) if the attacker chooses the "*attack* (a)" and the defender chooses "*not-monitor* (τ)." Therefore, a payoff value 2 is assigned to the attacker corresponding to the action profile (a,τ) in V_C.

Based on the two parameters such as $\nu(r,\tau,\tau)$ and $\nu(r,a,m)$in the payoff matrix V_r where $r \in \{P, C\}$, the attacker will determine how to carry out his/her attack in the game element Γ_r and note that a negative reward denotes the cost he/she has to pay against the system defender. In fact, the cost incurred when the malicious action is detected will be a demotivating factor for the attacker. Hence, to characterize the attacker behavior in the SCR attack, the proposed model is solved using Algorithm 1 by setting the game parameters $\nu(r,\tau,\tau)$ and $\nu(r,a,m)$in the range of negative payoffs (costs) such as between -9 and 0. Thus, we obtain the optimal mixed strategies (probabilities) $x_a^{P^*}$ and $x_a^{C^*}$ of the attacker in the game element Γ_P and Γ_C, respectively, for different values of the game parameters. Fig. 5.3 depicts the graph of optimal attack probabilities $x_a^{P^*}$ versus $\nu(P,a,m)$and $\nu(P,\tau,\tau)$ in the penetrated state (P). Fig. 5.4 also shows the attacker's optimal mixed strategies in the compromised state (C). Further, solving Algorithm 1, we also obtain the optimal defense strategies (optimal probabilities) of the defender in the penetrated (P) and compromised state (C), and these are plotted as shown in Figs. 5.5 and 5.6, respectively. It can be observed from the graphs that actions are carried out by the attacker and the defender on the basis of the costs $\nu(r,\tau,\tau)$and $\nu(r,a,m)$ where $r \in \{P, C\}$.

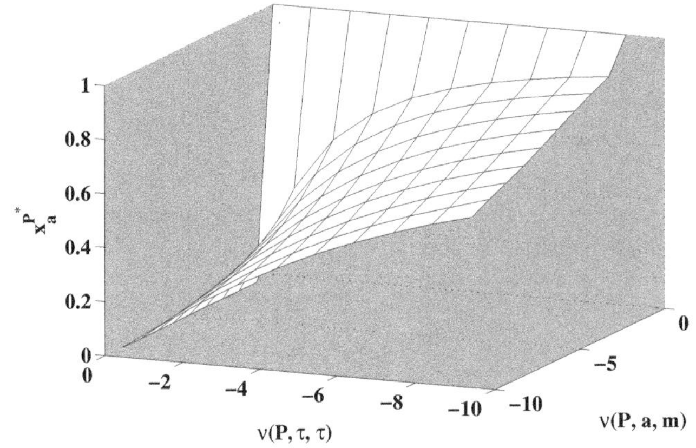

FIGURE 5.3

Attacker's optimal probability in the penetrated state $x_a^{P^*}$ versus withdrawal cost $\nu(P, \tau, \tau)$ and detection cost $\nu(P, a, m)$.

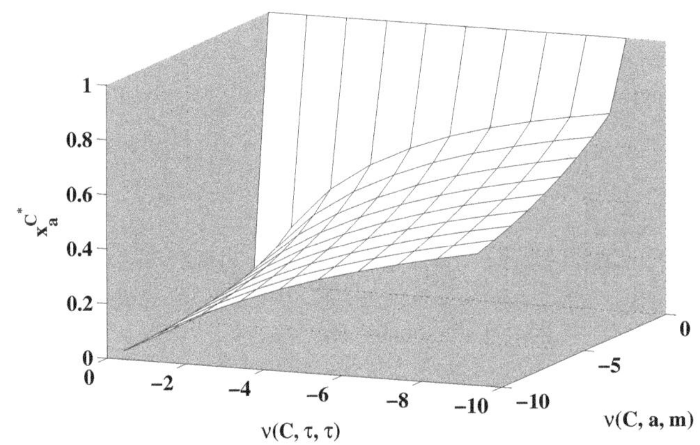

FIGURE 5.4

Attacker's optimal probability in the compromised state $x_a^{C^*}$ versus withdrawal cost $\nu(C, \tau, \tau)$ and detection cost $\nu(C, a, m)$.

It can be observed from the results in Table 5.2 that the motivation of the attacker (i.e., the attack probabilities $x_a^{P^*}$ and $x_a^{C^*}$) increases as the cost of detection $\nu(r,\ a,\ m)$ decreases.

Thus, we observe from the results in Table 5.3 that the attack probabilities $x_a^{P^*}$ and $x_a^{C^*}$ increase as the cost of withdrawal from the attack intention $\nu(r,\ \tau,\ \tau)$ increases.

It is clear from the outcomes in Table 5.4 that the defender's optimal defense probabilities (i.e., $y_m^{P^*}$ and $y_m^{C^*}$) increase as the cost of detection $\nu(r, a, m)$ decreases. This is a reasonable result because the optimal attack probability increases as the cost of detection decreases as shown in the previous results (*see* Table 5.2), and hence the defense probability is also expected to increase in order to respond to the malicious activities.

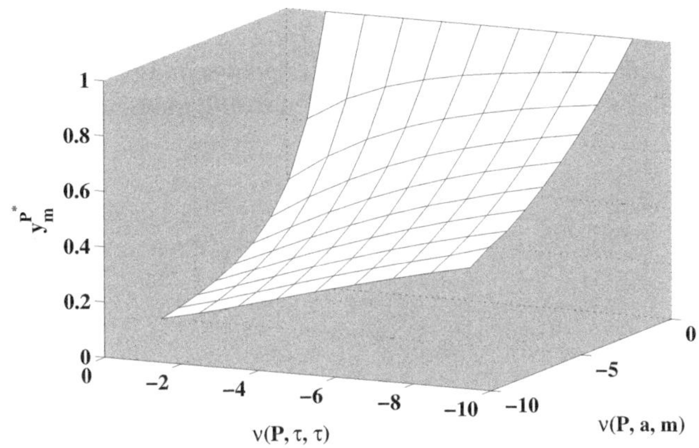

FIGURE 5.5

Defender's optimal probability in the penetrated state y_m^{P*} versus attacker's withdrawal cost $\nu(P, \tau, \tau)$ and detection cost $\nu(P, a, m)$.

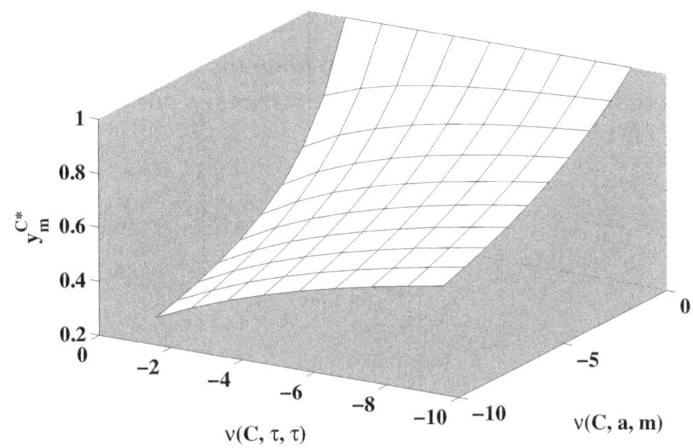

FIGURE 5.6

Defender's optimal probability in the compromised state y_m^{C*} versus attacker's withdrawal cost $\nu(C, \tau, \tau)$ and detection cost $\nu(C, a, m)$.

We can observe from the results in Table 5.5 that the defenders' defense probability increases as the cost of withdrawal increases. Again, this is the expected result because the optimal attack probability increases as the withdrawal cost increases as shown in the previous results (*see* Table 5.3), and therefore the defense probability is also expected to increase to respond to the attacker's malicious actions.

Table 5.2 Attacker behavior when $\nu(r, \tau, \tau) = -4$. Here $\nu(r, \tau, \tau)$ denotes the payoff corresponding to the profile of strategy (not-attack [τ], not-monitor [τ]). Also $\nu(r, a, m)$ denotes the payoff corresponding to the profile of strategy (attack [a], monitor [m]) in the game element Γ_r where $r \in \{P, C\}$.

$\nu(r, a, m)$ (cost of detection)	$x_a^{P^*}$	$x_a^{C^*}$
−9	0.342	0.267
−8	0.368	0.286
−7	0.402	0.308
−6	0.440	0.333
−5	0.484	0.364
−4	0.536	0.400
−3	0.595	0.444
−2	0.662	0.500
−1	0.733	0.571

Note that $x_a^{P^*}$ and $x_a^{C^*}$ represent the optimal attack probabilities in the penetrated (P) and the compromised (C) state, respectively.

Table 5.3 Attacker behavior when $\nu(r, a, m) = -4$. Here $\nu(r, a, m)$ denotes the payoff corresponding to the profile of strategy (attack (a), monitor (m)). Also $\nu(r, \tau, \tau)$ denotes the payoff corresponding to the profile of strategy (not-attack (τ), not-monitor (τ)) in the game element Γ_r where $r \in \{P, C\}$.

$\nu(r, \tau, \tau)$ (Cost of withdrawal)	$x_a^{P^*}$	$x_a^{C^*}$
−1	0.183	0.143
−2	0.331	0.250
−3	0.446	0.333
−4	0.536	0.400
−5	0.605	0.455
−6	0.660	0.500
−7	0.704	0.538
−8	0.740	0.571
−9	0.769	0.600

4. Dependability analysis of small cell relay under DoS attack

In the previous section, a stochastic game model is proposed to capture the interactions between the system defender and the attacker who performs bandwidth spoofing attack during the communication between an SCR and an MBS. If the attacker successfully exploits the bandwidth that is allocated to the SCR, then the SCR will not be available for the EUs which are under the network coverage of the

Table 5.4 Defense probabilities when $\nu(r, \tau, \tau) = -4$. Here $\nu(r, \tau, \tau)$ denotes the payoff corresponding to the profile of strategy (not-attack (τ), not-monitor (τ)). Also $\nu(r, a, m)$ denotes the payoff corresponding to the profile of strategy (attack (a), monitor (m)) in the game element Γ_r where $r \in \{P, C\}$.

$\nu(r, a, m)$ (Cost of detection)	$y_m^{P^*}$	$y_m^{C^*}$
−9	0.231	0.400
−8	0.260	0.429
−7	0.296	0.462
−6	0.340	0.500
−5	0.395	0.545
−4	0.464	0.600
−3	0.554	0.667
−2	0.669	0.750
−1	0.817	0.857

Note that $y_m^{P^*}$ and $y_m^{C^*}$ represent the optimal defense probabilities in the penetrated (P) and the compromised (C) state, respectively.

Table 5.5 Defense probabilities when $\nu(r, a, m) = -4$. Here $\nu(r, a, m)$ denotes the payoff corresponding to the profile of strategy (attack (a), monitor (m)). Also $\nu(r, \tau, \tau)$ denotes the payoff corresponding to the profile of strategy (not-attack (τ), not-monitor (τ)) in the game element Γ_r where $r \in \{P, C\}$.

$\nu(r, \tau, \tau)$ (Cost of withdrawal)	$y_m^{P^*}$	$y_m^{C^*}$
−1	0.267	0.429
−2	0.338	0.500
−3	0.405	0.556
−4	0.464	0.600
−5	0.445	0.636
−6	0.560	0.667
−7	0.598	0.692
−8	0.630	0.714
−9	0.658	0.733

SCR. This leads to a DoS attack. Therefore, it is important to study the dependability of the network under the DoS attack, and hence in this section, we determine the dependability measures of the network such as *availability* and *reliability* of the SCR in the presence of a DoS attack. When the attacker intrudes into the SCR to carry out a bandwidth spoofing attack (a DoS attack), the system undergoes various states such as penetrated state (P), compromised state (C), down state (D), and failed state (F) depending on the actions taken by both the attacker and the network defender. Further, the system can end up in the down state (D) or the failed state (F) due to hardware failure and/or the energy depletion of the SCR. Thus, the system makes transitions from one state to another as shown in Fig. 5.2. and the system stays at any one of these states and then moves to another state after a certain period of time (called sojourn time of the process). As described in the previous section, the sojourn times of the process are assumed to be exponentially distributed. Therefore, the SCR attack model can be characterized based on a CTMC having the state space $\Omega = \{0, 1, 2, 3, 4, 5, 6\}$ and the transition rates λ_{kl}, where $k, l \in S$. We can now obtain the dependability measures of the SCR under the bandwidth spoofing attack such as *MTTF* (or reliability) and *SSA*. For this, we consider a time homogeneous CTMC represented by $\{X(t), t \geq 0\}$ with state space Ω, where $X(t)$ denotes the state of the system at time t and $X(t) \in \Omega$.

The steady state probabilities π_k, $k \in \Omega$, can be computed by solving the following system of equations [39]:

$$\pi \mathbf{Q} = 0 \tag{5.15}$$

and

$$\sum_{k \in \Omega} \pi_k = 1 \tag{5.16}$$

where $\mathbf{Q}$ is the transition rate matrix.

Recall that β_{kl} denotes the accumulated attack intensity, and ψ_{kl} is the system administrator's effort intensity from the system state k to the state l. Also, λ_{kl} denotes the transition rate from the system state k to the state l in the context of CTMC, and α and μ denote the hardware failure rate and the SCR energy (battery) depletion rate, respectively. Further, let θ_k be the IDS's probability of detecting the intrusion in the system state k, where $k \in \{P, C\}$. In the previous section, we know that the optimal mixed strategies $\mathbf{x}^* = \{x^{P^*}, x^{C^*}\}$ and $\mathbf{y}^* = \{y^{P^*}, y^{C^*}\}$ of the attacker and the defender can be computed from Algorithm 1. Thus, we compute these optimal strategies to determine the transition rate matrix $\mathbf{Q}$ of the underlying CTMC of the proposed SCR attack model given in Fig. 5.2. For example, in the penetrated state (P), the IDS monitors the system with the optimal defense probability $y_m^{P^*}$ with probability of successful detection θ_5. Further, the system administrator effort from the state 5 to the state 1 is ψ_{51} and hence the rate at which the system makes the transition from the state 5 (penetrated state [P]) to the state 1 (intermediate state [I]) is given by $\lambda_{51} = \psi_{51}\theta_5 y_m^{P^*}$. On the other hand, in the state P, the IDS fails to detect the intrusion with probability $(1 - \theta_5)$ and the attacker performs attack with the optimal attack probability $x_a^{P^*}$ along with the accumulated attack intensity β_{56}. Therefore, the system makes transition from the state 5 (penetrated state [P]) to the state 6 (compromised state [C]) with rate $\lambda_{56} = \beta_{56}(1 - \theta_5)x_a^{P^*}$. Similarly, we can derive the other transition rates of the transition rate matrix. Note also that $\lambda_{13} = \mu$ because the energy (battery) of the SCR is depleted at the rate μ from the state 1 (intermediate state [I]) to the state 3 (down state [D]) and $\lambda_{14} = \alpha$ since the hardware failure of the SCR occurs at the rate α from the state 1 (intermediate state [I]) to the state 4 (failed state [F]). In a similar way, we can derive the other transition rates involving hardware failures and/or energy (battery) depletion. Thus, the transition rate matrix $\mathbf{Q}$ is given by

$$\begin{pmatrix} \lambda_{00} & \lambda_{01} & 0 & 0 & 0 & 0 & 0 \\ 0 & \lambda_{01} & \lambda_{12} & \mu & \alpha & \lambda_{15} & 0 \\ 0 & \lambda_{21} & \lambda_{22} & \mu & \alpha & \lambda_{25} & 0 \\ \lambda_{30} & 0 & 0 & \lambda_{33} & \alpha & 0 & 0 \\ \lambda_{40} & 0 & 0 & 0 & \lambda_{44} & 0 & 0 \\ 0 & \psi_{51}\theta_5 y_m^{P^*} & \psi_{52}\theta_5 y_m^{P^*} & \mu & \alpha & \lambda_{55} & \beta_{56}(1-\theta_5)x_a^{P^*} \\ 0 & \psi_{61}\theta_6 y_m^{C^*} & \psi_{62}\theta_6 y_m^{C^*} & \mu & \beta_{64}(1-\theta_6)x_a^{C^*}+\alpha & 0 & \lambda_{66} \end{pmatrix} \tag{5.17}$$

4.1 Numerical illustration

Availability: Availability of a system is characterized by the probability that the system is working properly when needed for use. In the proposed SCR attack model, a CTMC is used as the underlying stochastic process to characterize the lifetime of the SCR. Therefore, we obtain the steady state probabilities of the proposed model using SHARPE software [40]. Note that in the proposed model, the SCR network is available in the states 0, 1, 2, and 5. Thus, we determine *SSA* of the SCR under attack as

$$SSA = \pi_0 + \pi_1 + \pi_2 + \pi_5 \tag{5.18}$$

where π_k is the steady state probability that the network under attack is in state k. Based on the parameter values in Table 5.6 and setting $\lambda_{40} = 0.001$ we get the value of *SSA* as $SSA = 0.697$. Note that the parameter values are chosen for the purpose of numerical illustration only.

MTTF: This is a common measure for the reliability of a system and is the average length of time that the system is expected to be in operation. To obtain the *MTTF* of the SCR, we remove the arc from the failed state (*F*) to the normal state (*N*) in Fig. 5.2, resulting in an absorbing Markov chain in which the state *F* is the absorbing state. We then compute the *MTTF* of the SCR under bandwidth spoofing

Table 5.6 Parameters and their values (per hour).

Parameter	Value	Parameter	Value
λ_{01}	0.50000	λ_{25}	0.00040
λ_{12}	0.00060	λ_{51}	0.00094
λ_{21}	0.00060	λ_{52}	0.00094
λ_{13}	0.00020	λ_{54}	0.00003
λ_{34}	0.00003	λ_{64}	0.00300
λ_{23}	0.00020	λ_{53}	0.00020
λ_{14}	0.00003	λ_{63}	0.00040
λ_{24}	0.00003	λ_{62}	0.00220
λ_{30}	0.00500	λ_{61}	0.00220
λ_{56}	0.05950	λ_{15}	0.00040

attack by solving the absorbing CTMC. For example, on implementing the Algorithm 1 with $\nu(r,a,m) = -9$ and $\nu(r,\tau,\tau) = -9$, $r \in \{P,C\}$, we get $x_a^{P^*} = 0.595$, $x_a^{C^*} = 0.450$, $y_m^{P^*} = 0.405$, and $y_m^{C^*} = 0.550$. Substituting, these parameter values in the sixth and seventh row of the rate matrix Q with $\theta_5 = 0.7$ and $\theta_6 = 0.8$, $\beta_{51} = 0.0033$, $\psi_{52} = 0.0033$, $\psi_{61} = 0.005$, $\psi_{62} = 0.005$, $\beta_{56} = 0.33$, and $\beta_{64} = 0.03$, and using the parameter values in Table 5.6, we obtain the *MTTF* as *MTTF* = 6112.339 h.

4.2 Model validation

In this section, the proposed SCR attack model is validated via simulation results. The performance measures of the network under attack such as *MTTF* (or reliability) and *SSA* are simulated using MATLAB. The simulation is performed based on the parameter values presented in Table 5.6. In order to study how the network reliability and availability are affected by the DoS attack, we compute these measures when the rate of successful attack λ_{56} varies from 0.25 to 3.0. As shown in Figs. 5.7 and 5.8, the simulation results corresponding to the *MTTF* and the *SSA* are plotted in dashed lines. The analytical results are plotted in thick lines. As expected, it is found from the graphs that both the reliability (*MTTF*) and the availability (*SSA*) decrease as the rate of attack λ_{56} increases. Further, it is observed from these figures that the simulation results are very close to the analytical results, thereby validating the proposed model.

5. Result discussion

From the proposed game model, we can obtain the optimal strategies for both the attacker and the defender in the event of bandwidth spoofing attack. From Table 5.2, it can be seen that the attack probabilities ($x_a^{P^*}$ and $x_a^{C^*}$) decreases as the cost of detection $\nu(r,a,m)$ increases. This implies that

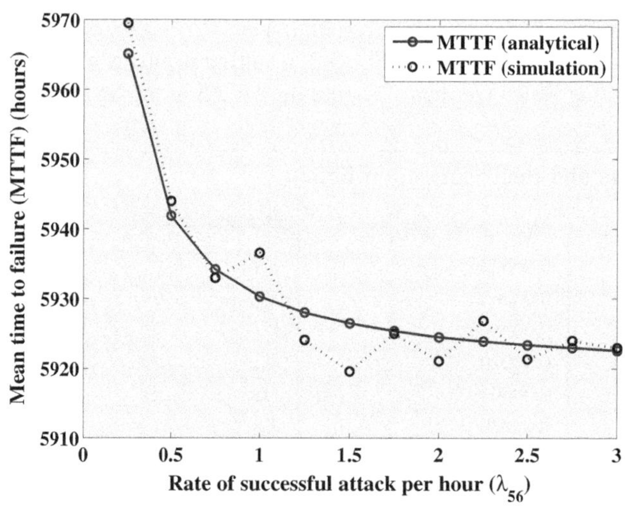

FIGURE 5.7

Mean time to failure (MTTF) versus rate of successful attack (λ_{56}).

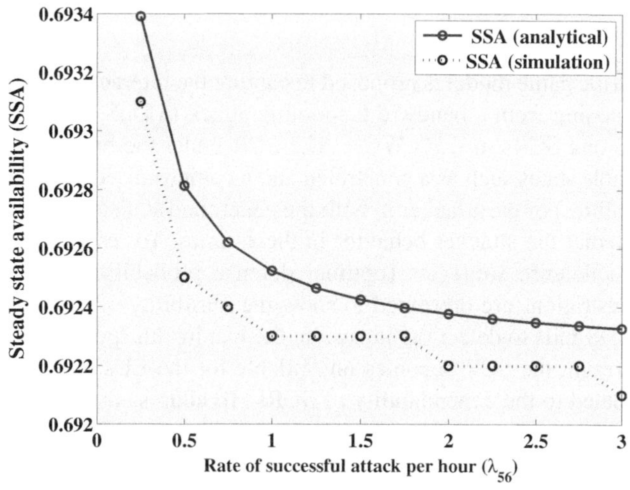

FIGURE 5.8

Steady state availability (SSA) versus rate of successful attack (λ_{56}).

attacking cost is a demotivating factor for the attacker to carry out the attack. Further, the attack probabilities in the penetrated state are observed to be greater than that of the compromised states (i.e., $x_a^{P^*} > x_a^{C^*}$). This shows that it is very important to defend the SCR under attack in the penetrated state to prevent the system from transitioning into the compromised state.

It can also be observed from Table 5.3 that the attack probabilities $x_a^{P^*}$ and $x_a^{C^*}$ increase as the cost of withdrawal from the attack intention $\nu(r, \tau, \tau)$ increases. In fact, the attack probabilities are higher in the penetrated state than that of the compromised state (i.e., $x_a^{P^*} > x_a^{C^*}$). This signifies that the attacker is motivated to persist in performing the attacks both in the penetrated and the compromised state in order to maximize the attacker's payoffs.

On the other hand, Table 5.4 shows that the defender's optimal defense probabilities (i.e.,$y_m^{P^*}$ and $y_m^{C^*}$) decrease as the cost of detection $\nu(r, a, m)$ increases. Further, it can be observed from Table 5.5 that the defenders' defense probability increases as the cost of withdrawal increases. In addition, the optimal defense probabilities are larger in the penetrated state than that of the compromised state (i.e., $y_m^{P^*} < y_m^{C^*}$ and $y_m^{P^*} < y_m^{C^*}$). These observations indicate that the cost of performing monitoring activities is a demotivating factor for the defender to defend the system and the defense action against the adversarial attacks should be taken effectively in the compromised state as compared to the penetrated state.

Further, the bandwidth spoofing attack makes the network unavailable if the attack is not successfully defended. As we can see from Section 4, the SSA and the MTTF of the system under attack can be derived. It is found from the graphs in Figs. 5.7 and 5.8 that both the reliability (*MTTF*) and the availability (*SSA*) decrease as the rate of attack λ_{56} increases. This implies that higher the number of successful penetration, greater the chance of disrupting the system.

6. Conclusion

In this chapter, a stochastic game model is proposed to capture the interactions between an attacker and the network defender arising from a bandwidth spoofing attack (a DoS attack) during the communication between a MBS and SCRs in a 5G WCN. The SCR under the bandwidth spoofing attack undergoes certain vulnerable states such as a penetrated and a compromised state. The optimal strategies (optimal attack probabilities) of the attacker in both the penetrated state and the compromised state are obtained in order to predict the attacker behavior in these states. To respond to the attacker's adversarial attacks, the best defense strategies (optimal defense probabilities) are also determined. In addition, numerical illustrations are presented to show the feasibility of the proposed model.

Further, when the IDS fails to detect the intrusion, the bandwidth spoofing attack is carried out by the attacker, and as a result, the SCR becomes unavailable for the EUs. Therefore, it is important to deal with the issues related to the dependability of SCRs. To address these issues, the lifetime of the SCR under the bandwidth spoofing attack is characterized using a CTMC. The dependability measures such as *MTTF* and *SSA* of the SCR under the attack are obtained based on the CTMC. Numerical results are also discussed to show how the reliability and the availability of the SCR are affected by the bandwidth attack. The numerical results demonstrate that these dependability measures decrease as the rate of attack increases. Moreover, the viability of the proposed model is verified using simulation results. The results and the analysis of the proposed models are expected to be helpful not only for designing and developing a robust defense mechanism for SCR security but also for paving the way in the modeling aspects of network security in 5G WCN.

Acknowledgment

One of the authors (KC Lalropuia) is grateful to the University Grant Commission (UGC), India, for granting him financial support through the CSIR UGC Junior Research Fellowship (UGC Ref. No.: 1031).

References

[1] Y. Zou, Intelligent interference exploitation for heterogeneous cellular networks against eavesdropping, IEEE J. Sel. Area. Commun. 36 (7) (2018) 1453–1464.

[2] P. Gandotra, R.K. Jha, A survey on green communication and security challenges in 5G wireless communication networks, J. Netw. Comput. Appl. 96 (2017) 39–61.

[3] Y. Zou, M. Sun, J. Zhu, H. Guo, Security-reliability trade-off for distributed antennas systems in heterogeneous cellular networks, IEEE Trans. Wireless Commun. 17 (12) (2018) 8444–8456.

[4] N. Panwar, S. Sharma, A.K. Singh, A survey on 5G: the next generation of mobile communication, Phys. Commun. 18 (2) (2016) 64–84.

[5] A. Gupta, R.K. Jha, P. Gandotra, Bandwidth spoofing and intrusion detection system for multistage 5G wireless communication network, IEEE Trans. Veh. Technol. 67 (1) (2018) 628–632.

[6] T.S. Rappaport, Y. Xing, G.R. MacCartney, A.F. Molisch, E. Mellios, J. Zhang, Overview of millimeter wave communications for fifthgeneration (5G) wireless networks - with a focus on propagation models, IEEE Trans. Antenn. Propag. 65 (12) (2017) 6213–6230.

[7] K. Xiao, W. Li, M. Kadoch, C. Li, On the secrecy capacity of 5G MmWave small cell networks, IEEE Wireless Communications 25 (4) (2018) 47–51.

[8] N. Al-Falahy, O.K. Alani, Milimetre wave frequency band as a candidate sprectrum for 5G architecture: a survey, Phys. Commun. 32 (2019) 120–144.

[9] R. Devi, R.K. Jha, A. Gupta, S. Jain, P. Kumar, Implementation of intrusion detection system using adaptive neuro fuzzy inference system for 5G wireless communication network, Int. J. Electron. Commun. 74 (2017) 94–106.

[10] A. Roy, S. Midya, K. Majumder, S. Phadikar, Enhancing QoS in 5th generation Het-Net via synergistic TVWS spectrum sharing for distributive adaptive small cells, Phys. Commun. 36 (2019).

[11] C. Song, Massive-MIMO enabled FDD wireless backhaul small-cell relay networks: AF protocol based designs with low channel estimation and feedback complexity, IEEE Access 6 (2018) 31050–31064.

[12] N. Nguyen, C. Kundu, Hien, Q. Ngo, T.Q. Duong, B. Canberk, Secure-full duplex small cell networks in a spectrum sharing environment, IEEE Access 4 (2016) 3087–3099.

[13] Y. Wu, A. Khisti, C. Xiao, K. Wong, A survey of physical layer security techniques for 5G wireless networks and challenges ahead, IEEE J. Sel. Area. Commun. 36 (4) (2018) 679–695.

[14] D. Fang, Y. Qian, R.Q. Hu, Security for 5G mobile wireless networks, IEEE Access 6 (2017) 4850–4874.

[15] M.A. Ferrag, L. Maglaras, A. Argyriou, D. Kosmanos, H. Janicke, Security for 4G and 5G cellular networks: a survey of existing authentication and privacy-preserving schemes, J. Netw. Comput. Appl. 101 (2018) 55–82.

[16] D. Fang, Y. Qian, R.Q. Hu, Security analysis for interference management in heterogeneous networks, Ad Hoc Netw. 84 (2019) 1–8.

[17] A. Babaei, A.H. Aghvami, A. Shojaeifard, K. Wong, Full-duplex small-cell networks: a physical-layer security perspective, IEEE Trans. Commun. 66 (2018) 3006–3021.

[18] V.G. Vassilakis, H. Mouratidis, E. Panaousis, I.D. Moscholios, M.D. Logothetis, Security Requirements Modelling for Virtualized 5G Small Cell Networks, 24th International Conference on Telecommunications (ICT), 2017.

[19] R. Roman, J. Lopez, M. Mambo, Mobile edge computing Fog et al.: a survey and analysis of security threats and challenges, Future Generat. Comput. Syst. 78 (2) (2018) 680–698.

[20] G. Chen, Y. Gong, P. Xiao, J.A. Chambers, Physical layer network security in the full-Duplex relay system, IEEE Trans. Inf. Forensics Secur. 10 (3) (2015) 574–583.

[21] D. Wang, B. Bai, W. Chen, Z. Han, Energy efficient secure communication over decode-and-forward relay channels, IEEE Trans. Commun. 63 (3) (2015) 892–905.

[22] Y. Lin, J. Huang, H. Chen, L. Chen, F. Zhang, Y. Xiang, A. Xu, X. Guo, B. Chen, Resource allocation for multicarrier secure communications in energy harvesting decode-and-forward relay network, in: Proceeding of the 11th Conference on Industrial Electronics and Applications, ICIEA), 2016.

[23] C. Zhang, J. Ge, Z. Xia, H. Du, Graph theory based cooperative transmission for physical-layer security in 5G large - scale wireless relay networks, IEEE Access 5 (2017) 21640–21649.

[24] H. Fang, L. Xu, X. Wang, Coordinated multiple-relay based physicallayer security improvement a single-leader multiple follower Stackelberg game scheme, IEEE Trans. Inf. Forensics Secur. 13 (1) (2018) 197–209.

[25] H. Fang, L. Xu, K.R. Choo, Stackelberg game-based relay selection for physical layer security and energy efficiency enhancement in cognitive radio networks, Appl. Math. Comput. 296 (2017) 153–167.

[26] A. Avizienis, J.C. Laprie, B. Randell, C. Landwehr, Basic concepts and taxonomy of dependable and secure computing, IEEE Trans. Dependable Secure Comput. 1 (2014) 11–33.

[27] D.M. Nicol, W.H. Sanders, K.S. Trivedi, Model-based evaluation: from dependability security, IEEE Trans. Dependable Secure Comput. 1 (1) (2004) 48–65.

[28] R. Zeng, Y. Jiang, C. Lin, X. Shen, Dependability analysis of control centre networks in smart grid using stochastic petri nets, IEEE Trans. Parallel Distr. Syst. 23 (9) (2012) 1721–1730.

[29] A. Avizienis, J.C. Laprie, B. Randell, Fundamental concepts of dependability, in: Proceedings of the 3rd IEEE Information Survivability Workshop, 2000, pp. 7–12.
[30] M.H. Manshaei, Q. Zhu, T. Alpcan, T. Bascar, J.P. Hubaux, Game theory meets network security and privacy, ACM Comput. Surv. 45 (3) (2013).
[31] G. Owen, Game Theory, third ed., Academic Press, 2001.
[32] K.C. Lalropuia, V. Gupta, A Bayesian game model and network availability model for small cells under denial of service (DoS) attack in 5G wireless communication network, Wireless Network 26 (2020) 557–572.
[33] M. Geva, A. Herberg, Y. Gev, Bandwidth distributed denial of service, IEEE Security & Privacy 12 (1) (2014) 54–61.
[34] F. Siddiqui, S. Zeadally, S. Fowler, Broadband wireless technologies, in: N. Chilamkurti, et al. (Eds.), Next Generation Wireless Technology: 4G and beyond, 2013, pp. 71–103.
[35] Z. Liao, J. Liang, C. Feng, Mobile relay deployment in multihop relay networks, Comput. Commun. 112 (2017) 14–21.
[36] A.A. Ramki, A. Rasoolzadegan, A.J. Jafari, A systematic view on intrusion detection based on the Hidden Markov model, Stat. Anal. Data Min. 11 (2018) 111–134.
[37] K. Sallhammar, B.E. Helvik, S.J. Knapskog, On stochastic modeling for integrated security and dependability evaluation, J. Network. 1 (2006) 31–42.
[38] S. Shen, R. Han, L. Guo, W. Li, Q. Cao, Survivability evaluation towards attacked WSNs based on stochastic game and continuous-time Markov chain, Appl. Soft Comput. 12 (2012) 1467–1476.
[39] K.S. Trivedi, Probability and Statistics with Reliability, Queuing, and Computer Applications, second ed., John Wiley & Sons, 2001.
[40] R.A. Sahner, K.S. Trivedi, A. Puliafito, Performance and Reliability Analysis of Computer Systems: An Example Based Approach Using Sharpe Software Package, Kluwer Academic Publishers, Massachusetts, USA, 1996.

CHAPTER

Standbys provisioning in machine repair problem with unreliable service and vacation interruption

6

Chandra Shekhar[1], Shreekant Varshney[2], Amit Kumar[1]

[1]*Department of Mathematics, Birla Institute of Technology and Science Pilani, Pilani Campus, Pilani, Rajasthan, India;*
[2]*Department of Mathematics, ICFAITech, Faculty of Science and Technology, ICFAI Foundation for Higher Education (IFHE), Hyderabad, India*

1. Introduction

In a real-time scenario, due to the rapid advancement in science and engineering technologies, the machining systems have now become an essential requirement of daily life. There are many examples of machinery-based service systems, to be specific, production systems, computer and communication systems, transportation and supply chain management, flexible manufacturing systems, etc., which are cited as real-time industrial systems. The industrial systems update with the advancement of digitization, technology, and become sophisticated and complex. The service quality and performance of any machining system is highly influenced by the variability of processing times, randomness of repair times, and amount of random failures, which are some of the common critical factors among all manufacturing and commercial industries. Though, machines are unreliable as their units and capability of working decrease significantly with the passes of time. It leads to unexpected failure of the machining system, costly repair of units, and expensive replacement of units that cause the loss of production in any industrial management system. Therefore, the complexity of machining systems, as well as the costs caused by the unpredictable failure of their units, attracts the attention of researchers and system analysts to maintain the market or acceptable value of any business industry. Industrial systems necessitate the high reliability and availability of the machining system. For the mathematical modeling of reliability-based machine repair problems through a queueing-theoretical approach, several research articles and books (cf. [1–6], and [7]) have been presented by many of the researchers.

In the mid-20th century, the queueing-based telephonic communication problems, developed by the Danish mathematician A. K. Erlang, laid the foundation for the development and implementation of queueing problems in real-time scenarios. Henceforth, during the last century, queueing problems remained more popular and emerged as the most prominent and active research area among scholars and researchers. However, because of the increasing complexity of many stochastic machining and service systems, classical queueing theory, which was once quite successful in modeling of telephone systems, appears to be insufficient today. To overcome the constraints of classical queueing theory, the vacation queueing models were introduced and developed as an extension of the previous theory in the

The Handbook of Reliability, Maintenance, and System Safety through Mathematical Modeling.
https://doi.org/10.1016/B978-0-12-819582-6.00006-X

1970s. In a vacation queueing model, the server is allowed to take a vacation after the service completion instant rather than continuously serving the newly arriving customers after spending some idle time. The provision of opting the vacation by the server makes the queueing system more flexible in terms of finding the optimal operating conditions of the system with a minimum associated expected cost. Henceforth, the waiting line problems or merely the vacation queueing models fascinate considerable attention of system analysts, decision-makers, and researchers and become an active and popular area in the research and development (R&D) field. To analyze and study the vacation queueing models, the optimal operating policies in Markovian and non-Markovian environment has been used by many researchers and scientists (cf. [8–18] and [19]).

In the past, the concept of queueing theory has been applied to many industrial problems with several basic queueing terminologies such as impatient, feedback, breakdown, batch arrival/service, the retrial of customers, and many more. In these models, generally, it is assumed that the service provided by the service provider is successful and satisfactory. Conversely, this proposition may not always be correct in many real-time customer-service-based management problems. Therefore, a new terminology, unreliable service, has been introduced in the queueing literature (cf. [20]) to ascertain whether the ongoing service of the customer has been completed or not satisfactorily. It can generally be happened because of some external interference, i.e., the service is interrupted neither by the server's fault nor by the customer arrival.

In the forthcoming sections, we delineate the effect of working vacation (WV) and vacation interruption (VI) policy on some reliability characteristics of the machining system with unreliable service. We classify our study in three parts: (i) machine repair problem (MRP), (ii) MRP with WV and VI, and (iii) MRP with VI with unreliable service. For the analysis purpose, we formulate the Chapman-Kolmogorov differential-difference equations. Next, to calculate the transient state probability distribution, we employ the Runge-Kutta method of fourth-order since it is an arduous task to derive the closed-form expressions of the mean-time-to-failure *MTTF* and reliability of the machining system analytically. For a better understanding from the mathematical point of view, the matrix structures using different levels and phases of the quasi-birth and death (QBD) process for all three developed models are also provided.

2. Machine repair problem

2.1 Notations

M: Number of operating units in the system
S: Number of standby units in the system
m: Minimum number of units required in short mode
λ: Mean arrival rate
μ_b: Mean repair rate during busy period of the repairman

In queueing literature, there are many basic models related to real-time Markov modeling. To the specific, the machine repair model is a typical example of a finite population queueing model. In the machine repair model, units represent the population of prospective customers, a failure of the unit corresponds to an arriving customer, and the repairman who provides the repair to the failed units is known as server. Subsequently, for the mathematical modeling, we deal with a finite population MRP consisting of M identical operating units and S warm standby units under the care of a single repairman. When an active operating unit fails, it is immediately replaced by an available standby unit with the negligible switchover time. At the moment when state of standby unit changes to operating state, the failure characteristics of the standby unit become as same as an operating unit. In the normal working of machining system, there is a requirement of M operating units in the system. But the machining system can also continue functioning in degraded manner even when the number of operating units in the system is at least m. The (M,m) machining system operates in normal-short mode depending on the number of operating units in the system; therefore, maximum $K = M + S - m + 1$ units are allowed to fail. Moreover, we assume that the time-to-failure of an operating unit as well as a standby unit are independently and exponentially distributed random variates with parameters λ and ν $(0 < \nu < \lambda)$, respectively. Similarly, the time-to-repair a failed unit is exponentially distributed with parameter μ_b. Henceforth, the state-dependent failure rate of units is represented as

$$\lambda_n = \begin{cases} M\lambda + (S-n)\nu; & 0 \le n < S \\ (M+S-n)\lambda; & S \le n < K \\ 0; & \text{otherwise} \end{cases}$$

In the last few decades, countless research papers on performance characteristics of the machining system have been reviewed using a queuing-theoretic approach to deal MRP with essential and/or optional terminologies (cf. [21–40] and [41]). More recently, a fault-tolerant machining system with the random failure events, common cause failure, and imperfection is investigated (cf. [42,43]). For the comparative and optimal analysis, they provided a numerical simulation of several test experiments in different graphs and tables.

In this chapter, we present reliability-based analysis of MRP with different machining variants. For that purpose, we define the states of the machining system at time instant t using the fundamental law of Markov chain as:

$J(t) \equiv$ State of the repairman/server.

$N(t) \equiv$ Number of failed units in the machining system.

Therefore, $X(t) = \{(J(t), N(t)); t >= 0\}$ represents the continuous-time Markov chain (CTMC) on state space

$$\Theta \equiv \{(0, \text{n});\ n = 0, 1, \ldots, K-1\} \cup \{K\}$$

where, K is the failure state of the machining system.

Hence, the Markov chain $\{X(t); t >= 0\}$ is irreducible. Also, since the state space Θ is finite, the Markov chain is positive recurrent. To get view of transitions between the precedence states of the machining system, the state transition diagram for a basic machine repair model is provided in Fig. 6.1.

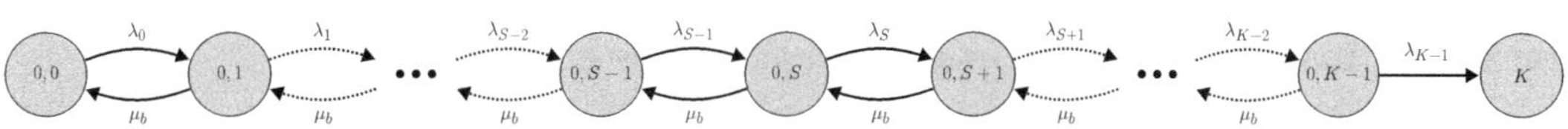

FIGURE 6.1

State transition diagram of a basic machine repair model.

Now, by balancing the input and output flows in Fig. 6.1 and using the fundamental law of QBD process, the governing Chapman-Kolmogorov differential-difference equations of the studied model are derived as follows:

(i) When there is no failed unit in the machining system and the repairman is idle

$$\frac{dP_{0,0}(t)}{dt} = -\lambda_0 P_{0,0}(t) + \mu_b P_{0,1}(t) \tag{6.1}$$

(ii) When there are n failed units in the machining system and the repairman is busy in working state

$$\frac{dP_{0,n}(t)}{dt} = -(\lambda_n + \mu_b)P_{0,n}(t) + \lambda_{n-1}P_{0,n-1}(t) + \mu_b P_{0,n+1}(t); \quad 1 \le n \le K-2 \tag{6.2}$$

(iii) When out of K, maximum allowed failed units, total $(K-1)$ units are failed and the repairman is busy in working state

$$\frac{dP_{0,K-1}(t)}{dt} = -(\lambda_{K-1} + \mu_b)P_{0,K-1}(t) + \lambda_{K-2}P_{0,K-2}(t) \tag{6.3}$$

(iv) The state where the machining system fails completely

$$\frac{dP_K(t)}{dt} = \lambda P_{0,K-1}(t) \tag{6.4}$$

Further, for employing the matrix method to determine the transient state probabilities, the system of differential-difference equations Eqs. (6.1)−(6.4) can be represented into the matrix form $\mathbf{DY} = \mathbf{Q}_1\mathbf{Y}$, wherein, $\mathbf{Y}$ is a column vector of all time-dependent probabilities of dimension $K+1$, $\mathbf{DY}$ is the derivative of the column vector $\mathbf{Y}$, and $\mathbf{Q}_1$ is the block-square transition matrix of order $K+1$, which is generated by using the tridiagonal characteristics of matrix algebra. The block structure of the transition rate matrix $\mathbf{Q}_1$ is partitioned as

$$\mathbf{Q}_1 = \begin{bmatrix} \mathbf{B}_{00}^1 & \mathbf{B}_0^1 & \mathbf{0} \\ \mathbf{C}_1^1 & \mathbf{A}_1^1 & \mathbf{B}_1^1 \\ \mathbf{0} & \mathbf{C}_2^1 & \mathbf{A}_2^1 \end{bmatrix}$$

where $\mathbf{B}_{00}^1$ is the scalar matrix and $\mathbf{B}_0^1$, $\mathbf{C}_1^1$ & $\mathbf{B}_1^1$ are the row and column vectors of order $K-1$, respectively. Similarly, $\mathbf{A}_1^1$ is the tridiagonal square matrix of order $K-1$ and $\mathbf{C}_2^1$ & $\mathbf{A}_2^1$ are the zero vectors. The structures of these block submatrices are given as

$$\mathbf{B}_{00}^1 = [-\lambda_0]$$

$$\mathbf{B}_0^1 = [\lambda_0, 0, 0, \cdots, 0]$$

$$\mathbf{B}_1^1 = [0, 0, \ldots, 0, \lambda_{K-1}]^T$$

$$\mathbf{C}_1^1 = [\mu_b, 0, 0, \ldots, 0]^T$$

and

$$\mathbf{A}_1^1 = \begin{bmatrix} u_1^1 & v_1^1 & 0 & \dots & 0 & 0 \\ w_2^1 & u_2^1 & v_2^1 & \dots & 0 & 0 \\ 0 & w_3^1 & u_3^1 & \dots & 0 & 0 \\ \vdots & \vdots & \vdots & \vdots & \vdots & \vdots \\ 0 & 0 & 0 & \dots & u_{K-1}^1 & v_{K-1}^1 \\ 0 & 0 & 0 & \dots & w_K^1 & u_K^1 \end{bmatrix}$$

where

$$u_n^1 = \begin{cases} -(\lambda_n + \mu_b); & 1 \le n \le K-1 \\ 0; & \text{otherwise} \end{cases}$$

$$v_n^1 = \begin{cases} \lambda_n; & 1 \le n \le K-2 \\ 0; & \text{otherwise} \end{cases}$$

$$w_n^1 = \begin{cases} \mu_b; & 2 \le n \le K-1 \\ 0; & \text{otherwise} \end{cases}$$

In the process of analyzing the efficiency and working quality of any service/machining system, performance measures play a vital role. These measures may either be qualitative or quantitative and helps the system engineers and decision-makers to rank the complex machining/service systems. The following are some essential queueing performance measures that necessarily be required to investigate the machine repair model.

- Expected number of failed units in the machining system at time t

$$E_N(t) = \sum_{n=0}^{K-1} nP_{0,n}(t) + KP_K(t) \tag{6.5}$$

- Probability that the repairman is idle at time t

$$P_I(t) = P_{0,0}(t) \tag{6.6}$$

- Failure frequency of the machining system at time t

$$FF(t) = \lambda_{K-1}P_{0,K-1}(t) \tag{6.7}$$

- Throughput of the machining system at time t

$$\tau_P(t) = \sum_{n=1}^{K-1} \mu_b P_{0,n}(t) \tag{6.8}$$

- Reliability of the machining system at time t

$$R_Y(t) = 1 - P_K(t) \tag{6.9}$$

- *MTTF* of the machining system

$$MTTF = \int_0^{\infty} R_Y(t)dt \tag{6.10}$$

The numerical simulation has been done in the following section for the sensitivity analysis of abovementioned performance characteristics of MRP.

3. Working vacation and vacation interruption

3.1 Notations

μ_b: Repair rate of the repairman during busy mode
μ_v: Repair rate of the repairman during WV period
$1/\theta$: Mean vacation time of the repairman
T: Threshold value for VI

In the queueing literature, the vacation queueing models have emerged as intensive research topics in recent years. Though from the literature, it is observed that the existing queueing models mainly focused on the maintenance optimization, but the reliability modeling where the repairman take a sequence of vacations was less studied. The concept of working vacation (WV) policy was first conceptualized in 2002 by Servi and Finn [8], inspired by the WDM optical access network using multiple wavelengths. In the machining systems, during the WV period, the repairman continues rendering the intended repair to the failed units rather than terminating the repair as in vacation in general. However, in real-time congestion problems, in spite better than complete vacation, the assumption of WV still seems more restrictive. Therefore, to overcome this limitation, in 2007 Li and Tian [44] proposed a vacation interruption (VI) policy for a single service provider in Markovian environment. Under VI policy, during the WV period if the repairman finds more failed units in the system waiting for repair than prespecified threshold after the service completion instant, the repairman immediately terminates his vacation and resumes the regular working attribute. Several studies have been done in the context of parametric and optimal analysis of a machining system with WV and VI by many researchers (cf. [45–51] and [52]).

The basic assumptions considered previously for MRP queueing model are also assumed for the study of queueing model involved in MRP with WV and VI. Let the repair time of the failed units in WV follow an exponential distribution with parameter μ_v and the WV time of the repairman also follow the exponential distribution with mean time $1/\theta$. We assume that the interarrival times, repair times in both busy and WV states, and vacation times are mutually independent. Let, $J(t)$ denotes the state of the repairman at time t, and $N(t)$ represents the total number of failed units in the machining system at time t. Therefore, the possible states of the repairman are characterized as follows:

$$J(t) = \begin{cases} 0; \text{ The repairman is in a working vacation period at time instant } t \\ 1; \text{ The repairman is in a normal busy period at time instant } t \end{cases}$$

Clearly, $\{J(t), N(t)\}$for $t \geq 0$ is a CTMC with the state space

$$\Theta \equiv \{(0,0)\} \cup \{(j,n); j=0,1 \ \& \ n=1,2,\dots,K-1\} \cup \{K\}$$

where, K is the state representing that the machining system fails completely.

Using the fundamental law of probability and balancing the transitions between adjacent states in Fig. 6.2 of an MRP with WV and VI policies, the governing differential-difference equations are developed as follows:

(i) When there is no failed unit in the machining system and the repairman is on WV

$$\frac{dP_{0,0}(t)}{dt} = -\lambda_0 P_{0,0}(t) + \mu_0 P_{0,1}(t) + \mu_b P_{1,1}(t) \tag{6.11}$$

(ii) When there are n failed units in the machining system and the repairman is on WV

$$\frac{dP_{0,n}(t)}{dt} = -(\lambda_n + \mu_v + \theta)P_{0,n}(t) + \lambda_{n-1}P_{0,n-1}(t) + \mu_v P_{0,n+1}(t); \quad 1 \leq n \leq T-1 \tag{6.12}$$

(iii) States of the system at which the WV of the repairman is interrupted

$$\frac{dP_{0,n}(t)}{dt} = -(\lambda_n + \mu_v + \theta)P_{0,n}(t) + \lambda_{n-1}P_{0,n-1}(t); \quad T \leq n \leq K-1 \tag{6.13}$$

(iv) The state of the system after which the busy repairman takes a vacation

$$\frac{dP_{1,1}(t)}{dt} = -(\lambda_1 + \mu_b)P_{1,1}(t) + \mu_b P_{1,2}(t) + \theta P_{0,1}(t) \tag{6.14}$$

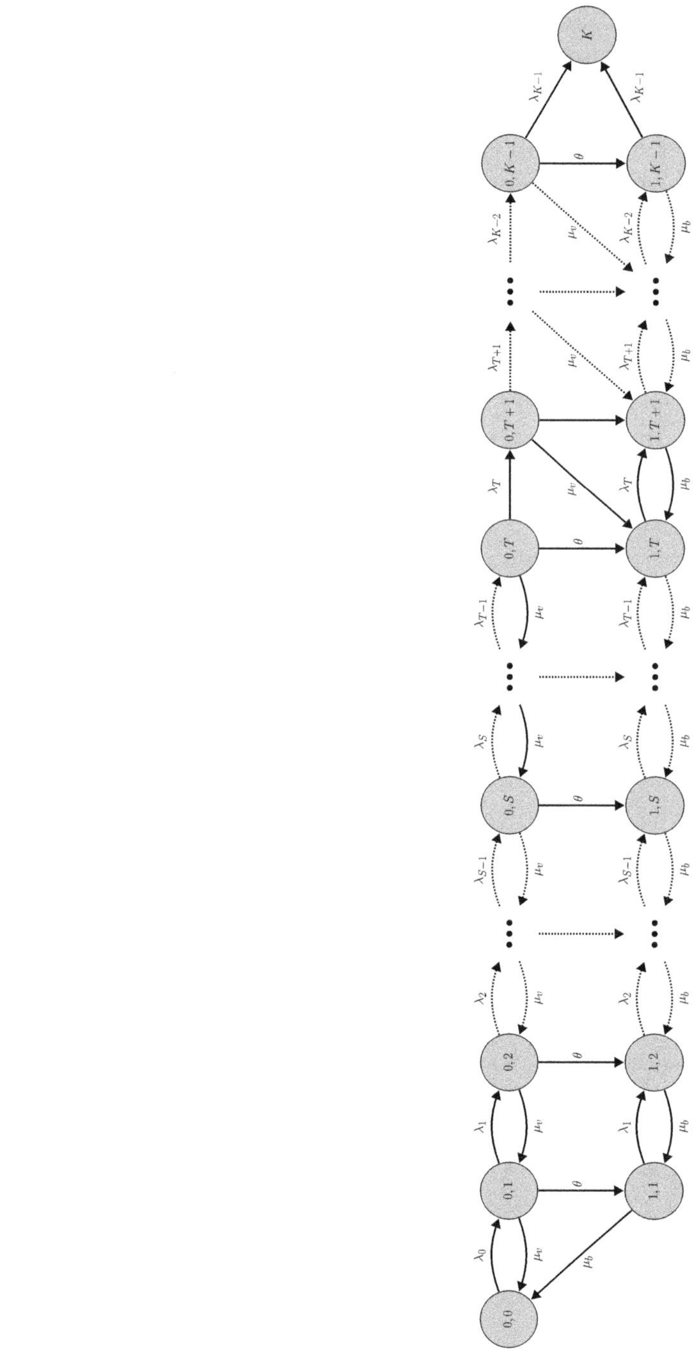

FIGURE 6.2

State transition diagram of a machine repair model with working vacation and vacation interruption.

(v) When there are n failed units in the machining system and the repairman is on regular busy mode

$$\frac{dP_{1,n}(t)}{dt} = -(\lambda_n + \mu_b)P_{1,n}(t) + \lambda_{n-1}P_{1,n-1}(t) + \mu_b P_{1,n+1}(t) + \theta P_{0,n}(t); \quad 2 \le n \le T-1 \tag{6.15}$$

$$\frac{dP_{1,n}(t)}{dt} = -(\lambda_n + \mu_b)P_{1,n}(t) + \lambda_{n-1}P_{1,n-1}(t) + \mu_b P_{1,n+1}(t) + \theta P_{0,n}(t) + \mu_v P_{0,n+1}(t); \quad T \le n \le K-2 \tag{6.16}$$

(vi) When $(K-1)$: out-of-K, maximum allowed, units are failed in the machining system during the busy period of the repairman

$$\frac{dP_{1,K-1}(t)}{dt} = -(\lambda_{K-1} + \mu_b)P_{1,K-1}(t) + \lambda_{K-2}P_{1,K-2}(t) + \theta P_{0,K-1}(t) \tag{6.17}$$

(vii) The state of the machining system at which the system fails completely

$$\frac{dP_K(t)}{dt} = \lambda_{K-1}P_{0,K-1}(t) + \lambda_{K-1}P_{1,K-1}(t) \tag{6.18}$$

The generator matrix, denoted by $\mathbf{Q}_2$, is the composition of block matrices obtained by the corresponding transitions between adjacent states of the machining system. The structure of the generator matrix is expressed as follows:

$$\mathbf{Q}_2 = \begin{bmatrix} \mathbf{B}_{00}^2 & \mathbf{B}_0^2 & \mathbf{0} & \cdots & \mathbf{0} & \mathbf{0} & \mathbf{0} & \cdots & \mathbf{0} & \mathbf{0} & \mathbf{0} \\ \mathbf{C}_1^2 & \mathbf{A}_1^2 & \mathbf{B}_1^2 & \cdots & \mathbf{0} & \mathbf{0} & \mathbf{0} & \cdots & \mathbf{0} & \mathbf{0} & \mathbf{0} \\ \mathbf{0} & \mathbf{C}_2^2 & \mathbf{A}_2^2 & \cdots & \mathbf{0} & \mathbf{0} & \mathbf{0} & \cdots & \mathbf{0} & \mathbf{0} & \mathbf{0} \\ \vdots & \vdots & \vdots & \ddots & \vdots & \vdots & \vdots & \ddots & \vdots & \vdots & \vdots \\ \mathbf{0} & \mathbf{0} & \mathbf{0} & \cdots & \mathbf{A}_{T-1}^2 & \mathbf{B}_{T-1}^2 & \mathbf{0} & \cdots & \mathbf{0} & \mathbf{0} & \mathbf{0} \\ \mathbf{0} & \mathbf{0} & \mathbf{0} & \cdots & \mathbf{C}_2^2 & \mathbf{A}_T^2 & \mathbf{B}_T^2 & \cdots & \mathbf{0} & \mathbf{0} & \mathbf{0} \\ \mathbf{0} & \mathbf{0} & \mathbf{0} & \cdots & \mathbf{0} & \mathbf{C}_3^2 & \mathbf{A}_{T+1}^2 & \cdots & \mathbf{0} & \mathbf{0} & \mathbf{0} \\ \vdots & \vdots & \vdots & \ddots & \vdots & \vdots & \vdots & \ddots & \vdots & \vdots & \vdots \\ \mathbf{0} & \mathbf{0} & \mathbf{0} & \cdots & \mathbf{0} & \mathbf{0} & \mathbf{0} & \cdots & \mathbf{A}_{K-2}^2 & \mathbf{B}_{K-2}^2 & \mathbf{0} \\ \mathbf{0} & \mathbf{0} & \mathbf{0} & \cdots & \mathbf{0} & \mathbf{0} & \mathbf{0} & \cdots & \mathbf{C}_3^2 & \mathbf{A}_{K-1}^2 & \mathbf{B}_{K-1}^2 \\ \mathbf{0} & \mathbf{0} & \mathbf{0} & \cdots & \mathbf{0} & \mathbf{0} & \mathbf{0} & \cdots & \mathbf{0} & \mathbf{0} & \mathbf{0} \end{bmatrix}$$

where, the block matrix $\mathbf{B}_{00}^2$ is a scalar matrix and $\mathbf{B}_0^2, \mathbf{C}_1^2$ &$\mathbf{B}_{K-1}^2$ are the row and column vectors of order 2, respectively. The rest of all subblock matrices are the square matrices of order 2. The vector and matrix representations of these matrices are viewed as

$$\mathbf{B}_{00}^2 = [-\lambda_0], \quad \mathbf{B}_0^2 = [\lambda_{0,}0], \quad \mathbf{C}_1^2 = [\mu_v, \mu_b]^T, \quad \mathbf{B}_{K-1}^2 = [\lambda_{K-1}, \lambda_{K-1}]^T$$

$$\mathbf{A}_n^2 = \begin{bmatrix} -(\lambda_n + \mu_v + \theta) & \theta \\ 0 & -(\lambda_n + \mu_b) \end{bmatrix}; \quad 1 \le n \le K-1$$

$$\mathbf{B}_n^2 = \begin{bmatrix} \lambda_n & 0 \\ 0 & \lambda_n \end{bmatrix}; \quad 1 \le n \le K-2$$

$$\mathbf{C}_2^2 = \begin{bmatrix} \mu_v & 0 \\ 0 & \mu_b \end{bmatrix} \quad \& \quad \mathbf{C}_3^2 = \begin{bmatrix} 0 & \mu_v \\ 0 & \mu_b \end{bmatrix}$$

The closed-form expressions for the expected number of failed units in the machining system $E_N(t)$, probability that the repairman is on WV P_{WV} (t), probability that the vacation of the repairman is interrupted $P_{VI}(t)$, reliability of the machining system $R_Y(t)$, etc., are executed in the following manner:

- Expected number of failed units in the machining system at time t

$$E_N(t) = \sum_{j=0}^{1} \sum_{n=1}^{K-1} nP_{j,n}(t) + KP_K(t) \tag{6.19}$$

- Probability that the repairman is idle at time t

$$P_I(t) = P_{0,0}(t) \tag{6.20}$$

- Probability that the repairman is in normal working mode at time t

$$P_B(t) = \sum_{1}^{K-1} P_{1,n}(t) \tag{6.21}$$

- Probability that the repairman is on WV at time t

$$P_{WV}(t) = \sum_{n=0}^{T} P_{0,n}(t) \tag{6.22}$$

- Probability that the vacation of the repairman is interrupted at time t

$$P_{VI}(t) = \sum_{n=T+1}^{K-1} P_{0,n}(t) \tag{6.23}$$

- Failure frequency of the machining system at time t

$$FF(t) = \lambda_{K-1} P_{0,K-1}(t) + \lambda_{K-1} P_{1,K-1}(t) \tag{6.24}$$

- Throughput of the machining system at time t

$$\tau_p(t) = \sum_{n=1}^{K-1} \mu_v P_{0,n}(t) + \sum_{n=1}^{K-1} \mu_b P_{1,n}(t) \tag{6.25}$$

- Reliability of the machining system at time t

$$R_Y(t) = 1 - P_K(t) \tag{6.26}$$

- *MTTF* of the machining system

$$MTTF = \int_0^{\infty} R_Y(t)dt \tag{6.27}$$

4. MRP with WV, VI, and unreliable service

4.1 Notations

μ_b: Repair rate of the repairman during busy mode, successful or not
μ_v: Repair rate of the repairman during WV period, successful or not
β_1: The rate of a successful repair
β_2: The rate of an unsuccessful repair

In this section, we choose the queueing terminology and assumptions as same as in previous sections along with the service failure. The random occurrence of the service failure is neither because of the server as it would appear in unreliable server queueing models (cf. [53–60] and [61]), nor by the arrivals' fault as it would be in several interruption models (cf. [56,62–69] and [70]). The waiting and/or in-service arrival do not abandon (due to a high amount of impatient) from the system, and we preserve the First Come First Serve (FCFS) protocol. We assume that the random service failures occur due to external shocks, environmental forces. The failed units, for which primary repair remains incomplete, continue to strive for successful repair until it is entirely successful.

Additionally, neither the server nor the arrival knows whether the repair was successful or not until the service time is completed, at which instant we hypothesize a quality check to take place, which determines that whether the service is complete and successful or not.

Let, $N(t)$ be the number of failed units in the machining system at time instant t, and $J(t)$ be the state of the repairman at time instant t. Then, there exist a total four possible states of the repairman, which are characterized as follows:

$$J(t) = \begin{cases} 0; \text{ The repairman is in a WV period at time instant } t \\ 1; \text{ States which represents the check points during WV period} \\ 2; \text{ The repairman is in a busy period at time instant } t \\ 3; \text{ States which represents the check points during the busy period} \end{cases}$$

Then, $\{(J(t), N(t)); t \geq 0\}$ becomes a CTMC with state space

$$\Theta \equiv \{(0,0)\} \cup \{(j,n); j=0,1,2,3 \text{ and } n=1,2,\cdots,K-1\} \cup \{K\}$$

Fig. 6.3 represents the state transition diagram of an MRP with multiple WV, VI policy, and unreliable service of the repairman. For mathematical modeling, repair times of the failed unit follow exponential distribution with parameters μ_v and μ_b when repairman is on vacation or busy, respectively. After repair, the service is checked where checking time also follows exponential distribution with parameters β_1 and β_2 when repair is successful and not successful, respectively.

The governing set of differential-difference equations is given as follows:

(i) State on which the repairman is idle

$$\frac{dP_{0,0}(t)}{dt} = -\lambda_0 P_{0,0}(t) + \beta_1 P_{1,1}(t) + \beta_1 P_{3,1}(t) \tag{6.28}$$

(ii) When the repairman is in a WV state

$$\frac{dP_{0,n}(t)}{dt} = -(\lambda_n + \mu_v + \theta)P_{0,n}(t) + \lambda_{n-1}P_{0,n-1}(t) + \beta_1 P_{1,n+1}(t) + \beta_2 P_{1,n}(t); \quad 1 \leq n \leq T-1$$

$$\frac{dP_{0,T}(t)}{dt} = -(\lambda_T + \mu_v + \theta)P_{0,T}(t) + \lambda_{T-1}P_{0,T-1}(t) + \beta_2 P_{1,T}(t) \tag{6.30}$$

(iii) When the WV of the repairman is interrupted

$$\frac{dP_{0,n}(t)}{dt} = -(\lambda_n + \mu_v + \theta)P_{0,n}(t) + \lambda_{n-1}P_{0,n-1}(t) + \beta_2 P_{1,n}(t); \quad T+1 \leq n \leq K-1 \tag{6.31}$$

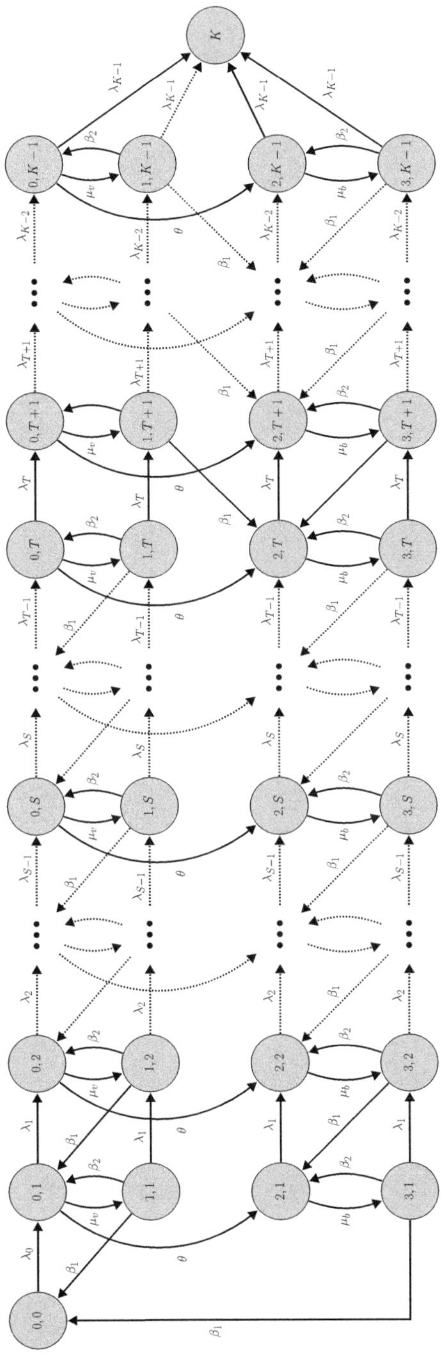

FIGURE 6.3

State transition diagram of a machine repair model with working vacation interruption and unreliable service.

(iv) Check points immediately after the repair is rendered during WV mode

$$\frac{dP_{1,1}(t)}{dt} = -(\lambda_1 + \beta_1 + \beta_2)P_{1,1}(t) + \mu_v P_{0,1}(t) \tag{6.32}$$

$$\frac{dP_{0,n}(t)}{dt} = -(\lambda_n + \beta_1 + \beta_2)P_{0,n}(t) + \lambda_{n-1}P_{1,n-1}(t) + \mu_v P_{0,n}(t); \quad 2 \le n \le T \tag{6.33}$$

$$\frac{dP_{0,n}(t)}{dt} = -(\lambda_n + \beta_1 + \beta_2)P_{0,n}(t) + \lambda_{n-1}P_{1,n-1}(t) + \mu_v P_{0,n}(t); \quad T+1 \le n \le K-1 \tag{6.34}$$

(v) States representing that the repairman is on regular busy mode of his service

$$\frac{dP_{2,1}(t)}{dt} = -(\lambda_1 + \mu_b)P_{2,1}(t) + \theta P_{0,1}(t) + \beta_2 P_{3,1}(t) + \beta_1 P_{3,2}(t) \tag{6.35}$$

$$\frac{dP_{2,n}(t)}{dt} = -(\lambda_n + \mu_b)P_{2,n}(t) + \lambda_{n-1}P_{2,n-1}(t) + \theta P_{0,n}(t) + \beta_2 P_{3,n}(t) + \beta_1 P_{3,n+1}(t); \quad 2 \le n \le T-1 \tag{6.36}$$

$$\frac{dP_{2,n}(t)}{dt} = -(\lambda_n + \mu_b)P_{2,n}(t) + \lambda_{n-1}P_{2,n-1}(t) + \theta P_{0,n}(t) + \beta_1 P_{1,n+1}(t) + \beta_2 P_{3,n}(t) + \beta_1 P_{3,n+1}(t); \quad T \le n \le K-2 \tag{6.37}$$

$$\frac{dP_{2,K-1}(t)}{dt} = -(\lambda_{K-1} + \mu_b)P_{2,K-1}(t) + \lambda_{K-2}P_{2,K-2}(t) + \theta P_{0,K-1}(t) + \beta_2 P_{3,K-1}(t) \tag{6.38}$$

(vi) Check points immediately after the repair is rendered during regular busy period of the repairman

$$\frac{dP_{3,1}(t)}{dt} = -(\lambda_1 + \beta_1 + \beta_2)P_{3,n}(t) + \mu_b P_{2,1}(t) \tag{6.39}$$

$$\frac{dP_{3,n}(t)}{dt} = -(\lambda_n + \beta_1 + \beta_2)P_{3,n}(t) + \lambda_{n-1}P_{3,2}(t) + \mu_b P_{2,n}(t); \quad 2 \le n \le K-1 \tag{6.40}$$

(vii) The state when the system fails completely

$$\frac{dP_K(t)}{dt} = \lambda_{K-1}P_{0,K-1}(t) + \lambda_{K-1}P_{1,K-1}(t) \tag{6.41}$$

Now, using the lexicographic sequence of the states of the machining system, the structure of the generator matrix is represented as

$$
\mathbf{Q}_3 = \begin{bmatrix}
\mathbf{B}_{00}^3 & \mathbf{B}_0^3 & \mathbf{0} & \cdots & \mathbf{0} & \mathbf{0} & \mathbf{0} & \cdots & \mathbf{0} & \mathbf{0} & \mathbf{0} \\
\mathbf{C}_1^3 & \mathbf{A}_1^3 & \mathbf{B}_1^3 & \cdots & \mathbf{0} & \mathbf{0} & \mathbf{0} & \cdots & \mathbf{0} & \mathbf{0} & \mathbf{0} \\
\mathbf{0} & \mathbf{C}_2^3 & \mathbf{A}_2^3 & \cdots & \mathbf{0} & \mathbf{0} & \mathbf{0} & \cdots & \mathbf{0} & \mathbf{0} & \mathbf{0} \\
\vdots & \vdots & \vdots & \ddots & \vdots & \vdots & \vdots & \ddots & \vdots & \vdots & \vdots \\
\mathbf{0} & \mathbf{0} & \mathbf{0} & \cdots & \mathbf{A}_{T-1}^3 & \mathbf{B}_{T-1}^3 & \mathbf{0} & \cdots & \mathbf{0} & \mathbf{0} & \mathbf{0} \\
\mathbf{0} & \mathbf{0} & \mathbf{0} & \cdots & \mathbf{C}_2^3 & \mathbf{A}_T^3 & \mathbf{B}_T^3 & \cdots & \mathbf{0} & \mathbf{0} & \mathbf{0} \\
\mathbf{0} & \mathbf{0} & \mathbf{0} & \cdots & \mathbf{0} & \mathbf{C}_3^3 & \mathbf{A}_{T+1}^3 & \cdots & \mathbf{0} & \mathbf{0} & \mathbf{0} \\
\vdots & \vdots & \vdots & \ddots & \vdots & \vdots & \vdots & \ddots & \vdots & \vdots & \vdots \\
\mathbf{0} & \mathbf{0} & \mathbf{0} & \cdots & \mathbf{0} & \mathbf{0} & \mathbf{0} & \cdots & \mathbf{A}_{K-2}^3 & \mathbf{B}_{K-2}^3 & \mathbf{0} \\
\mathbf{0} & \mathbf{0} & \mathbf{0} & \cdots & \mathbf{0} & \mathbf{0} & \mathbf{0} & \cdots & \mathbf{C}_3^3 & \mathbf{A}_{K-1}^3 & \mathbf{B}_{K-1}^3 \\
\mathbf{0} & \mathbf{0} & \mathbf{0} & \cdots & \mathbf{0} & \mathbf{0} & \mathbf{0} & \cdots & \mathbf{0} & \mathbf{0} & \mathbf{0}
\end{bmatrix}
$$

where, the block subvectors and matrices have following representation.

$$\mathbf{B}_{00}^3 = [-\lambda_0], \quad \mathbf{B}_0^3 = [\lambda_0, 0, 0, 0], \quad \mathbf{C}_1^3 = [0, \beta_1, 0, \beta_1]^T$$

$$\mathbf{B}_{K-1}^3 = [\lambda_{K-1}, \lambda_{K-1}, \lambda_{K-1}, \lambda_{K-1}]^T$$

$$
\mathbf{A}_n^3 = \begin{bmatrix}
-(\lambda_n + \mu_v + \theta) & \mu_v & \theta & 0 \\
\beta_2 & -(\lambda_n + \beta_1 + \beta_2) & 0 & 0 \\
0 & 0 & -(\lambda_n + \mu_b) & \mu_b \\
0 & 0 & \beta_2 & -(\lambda_n + \beta_1 + \beta_2)
\end{bmatrix}; \quad 1 \le n \le K-1
$$

$$
\mathbf{B}_n^3 = \begin{bmatrix}
\lambda_n & 0 & 0 & 0 \\
0 & \lambda_n & 0 & 0 \\
0 & 0 & \lambda_n & 0 \\
0 & 0 & 0 & \lambda_n
\end{bmatrix}; \quad 1 \le n \le K-2
$$

$$
\mathbf{C}_2^3 = \begin{bmatrix}
0 & 0 & 0 & 0 \\
\beta_1 & 0 & 0 & 0 \\
0 & 0 & 0 & 0 \\
0 & 0 & \beta_1 & 0
\end{bmatrix}
$$

and

$$\mathbf{C}_3^3 = \begin{bmatrix} 0 & 0 & 0 & 0 \\ 0 & 0 & \beta_1 & 0 \\ 0 & 0 & 0 & 0 \\ 0 & 0 & \beta_1 & 0 \end{bmatrix}$$

To examine the performance of reliability characteristics, namely, reliability of the machining system, *MTTF*, of the governing mathematical model, we delineate the closed-form expressions in terms of the transient state probabilities of various states. Following are some critical performance measures of the machining system which are depicted as:

- Expected number of failed units in the machining system at time t

$$E_N(t) = \sum_{j=0}^{3} \sum_{n=1}^{K-1} nP_{j,n}(t) + KP_K(t) \tag{6.42}$$

- Probability that the repairman is idle at time t

$$P_I(t) = P_{0,0}(t) \tag{6.43}$$

- Probability that the repairman is in normal working mode at time t

$$P_B(t) = \sum_{n=1}^{K-1} P_{2,n}(t) \tag{6.44}$$

- Probability that the repairman is on WV at time t

$$P_{WV}(t) = \sum_{n=0}^{T} P_{0,n}(t) \tag{6.45}$$

- Probability that the vacation of the repairman is interrupted at time t

$$P_{VI}(t) = \sum_{n=T+1}^{K-1} P_{0,n}(t) \tag{6.46}$$

- Failure frequency of the machining system at time t

$$FF(t) = \sum_{j=0}^{3} \lambda_{K-1} P_{j,K-1}(t) \tag{6.47}$$

- Throughput of the machining system at time t

$$\tau_p(t) = \sum_{n=1}^{K-1} \mu_v P_{0,n}(t) + \sum_{n=1}^{K-1} \mu_b P_{1,n}(t) \tag{6.48}$$

- Reliability of the machining system at time t

$$R_Y(t) = 1 - P_K(t) \tag{6.49}$$

- *MTTF* of the machining system

$$MTTF = \int_0^{\infty} R_Y(t)dt \tag{6.50}$$

5. Special cases

The studied models are advanced MRPs with many unique queueing terminologies. Relaxing one or more assumptions, our studied models match the specific cases of the published models available in the existing literature. The special cases validate our modeling, methodology, and results. Some of them are as follows:

Case 1: On considering the facility of cold standby units in place of warm standby units $\nu = 0$, and reliable service $\beta_1 = \mu_b$, $\beta_2 = 0$, our machine repair model results match with special case of MRP in article [71].
Case 2: If $K \to \infty$, $\beta_1 = \mu_b$, $\beta_2 = 0$, the studied queueing model gives similar results as in Ref. [13] for *M*/*M*/1 queue with WV and VI.
Case 3: If $T = K \to \infty$, our model reduces to single-server queueing system with WV and unreliable service (cf. [72]).
Case 4: If $T = K \to \infty$, $\theta \to \infty$, and $\mu_v = 0$, the model reduces to *M*/*M*/1 queueing system with unreliable service (cf. [20]).
Case 5: In the case of infinite capacity queueing system $K \to \infty$, on removing the assumptions of VI, i.e., $T = K$ and unreliable service $\beta_1 \to \infty$, our model is equivalent to the model proposed by Ref. [8] in which he introduced the notion of WV in single-server queueing system.

6. Cost analysis

In this section, we formulate an expected total cost function of the machining system to develop the cost optimization problem and calculate optimal system design parameters, which helps system analysts and engineers in decision-making.

6.1 Steady-state analysis

In this section, the steady-state analysis at equilibrium is performed to examine the optimal operating policy of the developed machine repair model with unreliable service and VI. In steady-state, i.e., in equilibrium ($t \rightarrow \infty$), the state probability distribution of the machining system is defined as follows:

$$P_{0,0} = \lim_{t \rightarrow \infty} Pr[J(t) = 0, N(t) = 0]$$
$$P_{j,n} = \lim_{t \rightarrow \infty} Pr[J(t) = j, N(t) = n];\ j = 0, 1, 2, 3\ \&\ n = 1, 2, \cdots, K-1$$

and

$$P_K = \lim_{t \rightarrow \infty} P_K(t)$$

Now, using the state transition diagram in Fig. 6.3, governing differential-difference equations, or the matrix formulation $\mathbf{Q}_3\mathbf{Y} = \mathbf{0}$ for the Markov machine repair model with standbys provisioning, unreliable service, and VI, the steady-state probability distribution can be easily demonstrated by employing the matrix method. Further, for the optimal analysis, the expected total cost function is also formulated in the following section using intrinsic performance measures, which incur some cost.

6.2 Cost function

For the optimal analysis, the system design parameters, namely, μ_v (repair rate during WV) and μ_b (repair rate during the busy period), are taken into consideration. The main objective of our intuition is to exhibit the optimal repair rates, say, μ_v^* and μ_b^* respectively, for minimizing the incurred expected total cost in operating the machining system. The system engineers and decision-makers have to identify the states of the machining system, which incur some costs. Following are some cost elements associated with different performance measures and states of the system that are considered and defined as follows:

$C_h \equiv$ Holding cost for each failed unit present in the machining system
$C_b \equiv$ Cost associated with the regular busy state of the repairman
$C_{wv} \equiv$ Cost associated with the WV state of the repairman
$C_{vi} \equiv$ Cost incurred with the VI of the repairman
$C_i \equiv$ Fixed cost for the idle state of the repairman
$C_1 \equiv$ Associated cost for providing the repair with rate μ_b
$C_2 \equiv$ Associated cost for providing the repair with rate μ_v

Using the concept of queueing-theoretic approach and the abovementioned cost elements, we formulate the cost function as follows:

$$TC(\mu_v, \mu_b) = C_h E(N) + C_b P_B + C_{wv} P_{WV} + C_{vi} P_{VI} + C_i P_I + C_1 \mu_b + C_2 \mu_v \tag{6.51}$$

The cost optimization (minimization) problem of the described model involved in MRP with WV, VI, and unreliable service can be mathematically represented as an unconstrained problem as follows:

$$TC\left(\mu_v^*, \mu_b^*\right) = \min_{(\mu_v, \mu_b)} TC(\mu_v, \mu_b) \tag{6.52}$$

The expected total cost function is an implicit function of cost elements and performance measures, which depend on state probabilities derived from governing system of equations that are delineated in terms of rates. The expected total cost function is too complex to get optimal value via theory of calculus since first derivative is not evaluative directly, gradient method, and any other well-known optimization techniques. The direct search method is too time-taken computational technique to get any useful results. In the next section, we employ the metaheuristic technique, particle swarm optimization (PSO), which depends on the theory of survival of the fittest or nature-inspired behavior in swarm for existence.

6.3 Particle swarm optimization

In this section, the PSO algorithm is employed to deal with the established cost optimization problem. The concept of PSO algorithm is first proposed by American social psychologist James Kennedy and engineer Russell C. Eberhart [73], having inspired by the social behavior of birds flocking/fish schooling. The PSO algorithm works utilizing the phenomena of exploration and exploitation with a population of particles in the feasible search space. In PSO, each particle has its inherent velocity and position by which it moves randomly within the search space. Also, the movement of every particle is influenced by its local/personal best (p-best) and global/common best (g-best) positions in the solution space. Suppose that $\mathbf{V}_i$ and $\mathbf{S}_i$ are the velocity and the position vector of the ith particle, respectively. The velocity component is updated in iteration $t+1$ from the information at iteration t using the following formula:

$$\mathbf{V}_i^{t+1} = \mathbf{V}_i^t + \kappa_1 \varphi_1 \left(\mathbf{G}^* - \mathbf{S}_i^t\right) + \kappa_2 \varphi_2 \left(\mathbf{S}_i^{*(t)} - \mathbf{S}_i^t\right) \tag{6.53}$$

wherein, κ_1 and κ_2 are the learning coefficients having standard value 2 for each, φ_1 and φ_2 are two random vectors having the range (0,1). $\mathbf{G}^*$ and $\mathbf{S}_i^{*(t)}$ are the global and local best of the ith particle. The position updating formula for the ith particle is characterized as

$$\mathbf{S}_i^{t+1} = \mathbf{S}_i^t + \mathbf{V}_i^{t+1} \tag{6.54}$$

To control the overshooting and undershooting in exploration and exploitation among particles, there is a requirement of an inertia function $\omega_2(t)$, which is introduced by Ref. [74] in the PSO algorithm. So, the improved velocity updating formula is given by

$$\mathbf{V}_i^{t+1} = \omega_2 \mathbf{V}_i^t + \kappa_1 \varphi_1 \left(\mathbf{G}^* - \mathbf{S}_i^t\right) + \kappa_2 \varphi_2 \left(\mathbf{S}_i^{*(t)} - \mathbf{S}_i^t\right) \tag{6.55}$$

The standard value of the inertia function $\omega_2(t)$ has been found in the literature to be between 0.5 and 0.9. The pseudocode of the PSO algorithm is given below.

Particle Swarm Optimization: Pseudocode

Input: Input parameters, population size, learning parameter.

Output: Approximate the solution (μ_1^*, μ_2^*) and compute the value of cost function

$$TC\left(\mu_1^*, \mu_2^*\right)$$

Step 1: Population Initialization: find the positions $\mathbf{S}_n$ of n particles.
Step 2: Find $\mathbf{G}^*$ (common best) from $\{TC(\mathbf{S}_1), TC(\mathbf{S}_2), \ldots, TC(\mathbf{S}_n)\}$.
while ($t <$ MaxGeneration) or (stop criterion)
for loop over all the n particles and all d dimensions
Step 3: Find new velocity vector for the ith particle $\mathbf{V}_i^{t+1}$.
Step 4: Find new positions for the ith particle $\mathbf{S}_i^{t+1} = \mathbf{S}_i^t + \mathbf{V}_i^{t+1}$.
Step 5: Evaluate cost function at new positions $\mathbf{S}_i^{t+1}$.
Step 6: Find the current best for each particle $\mathbf{S}^*{}_i$.
end for
Step 7: Update global best $\mathbf{G}^*$.
$t \rightarrow t + 1$
end while
Step 8: Output final results $\mathbf{S}^*{}_i$ and $\mathbf{G}^*$.
Step 9: Output optimal value of the cost function: TC^* at $\mathbf{G}^*$.

7. Numerical results

The prime goal of the present chapter is to get and to understand the qualitative and perceptible performance of the developed machining system using several reliability-based performance measures. The expressions of some other queue-based performance measures are also provided for the straightforward comparative analysis. For validation of formulation and methodology, we establish numerical simulations through various numerical experiments with three studied models.

Model 1: Machine repair problem (MRP)
Model 2: MRP with WV and VI
Model 3: MRP with WV, VI, and unreliable service

For that purpose, we fix the default values of the system parameters as $M = 10$, $S = 5$, $m = 2$, $T = 8$, $\lambda = 0.3$, $\nu = 0.1$, $\mu_b = 3.0$, $\mu_v = 1.0$, $\theta = 6.0$, $\beta_1 = 5$, and $\beta_2 = 1$. To determine the state probability distribution numerically, we employ the Runge-Kutta method of fourth-order and develop the code in MATLAB (2018b) since it is not possible to derive the analytical expressions of governing state probabilities.

In Figs. 6.4–6.9, the variation of reliability of the machining system $R_Y(t)$ is explored wrt the increasing values of different system parameters for all the developed models. From each figure, we observe that initially, the reliability of the machining system is constant, but after some time, it continuously decreases, which is the intuitively apparent result. We plot three different figures simultaneously to compare the findings of the studied models and to show the decrements in the reliability function with increasing values of time t.

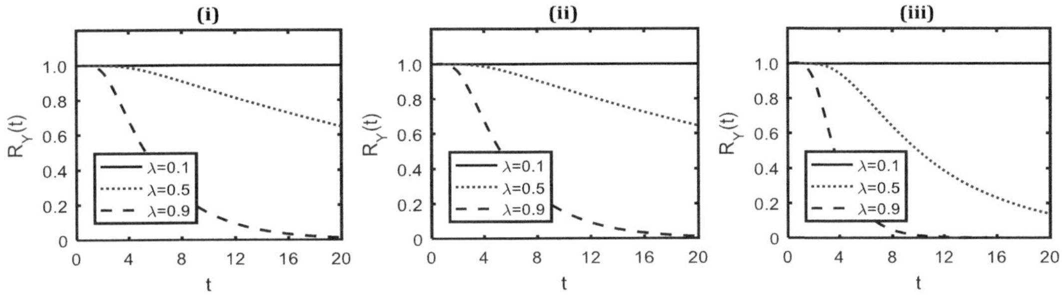

FIGURE 6.4

Variation of reliability of the machining system wrt failure rate of operating units λ for (i) Model 1, (ii) Model 2, and (iii) Model 3.

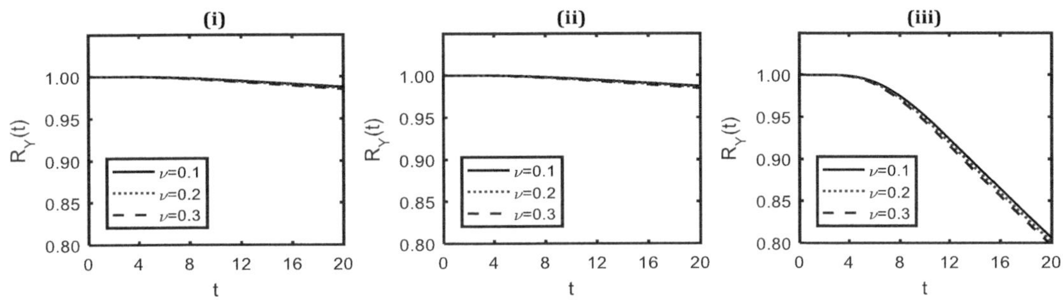

FIGURE 6.5

Variation of reliability of the machining system wrt failure rate of standby units ν for (i) Model 1, (ii) Model 2, and (iii) Model 3.

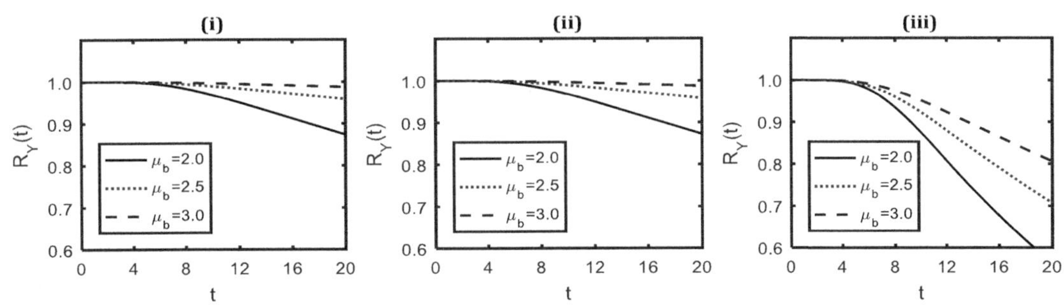

FIGURE 6.6

Variation of reliability of the machining system wrt repair rate μ_b for (i) Model 1, (ii) Model 2, and (iii) Model 3.

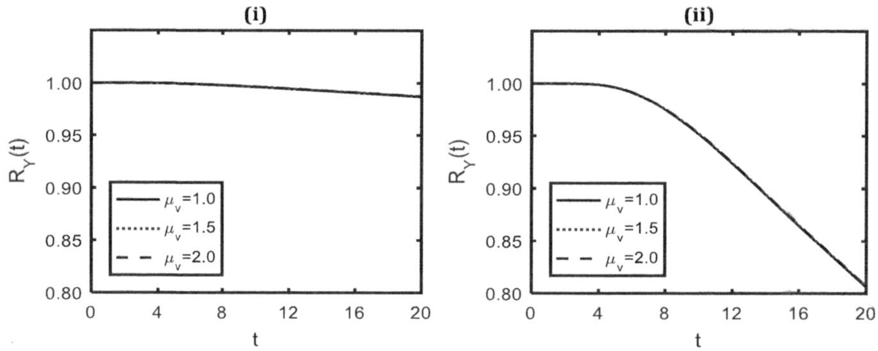

FIGURE 6.7

Variation of reliability of the machining system wrt repair rate μ_v for (i) Model 2 and (ii) Model 3.

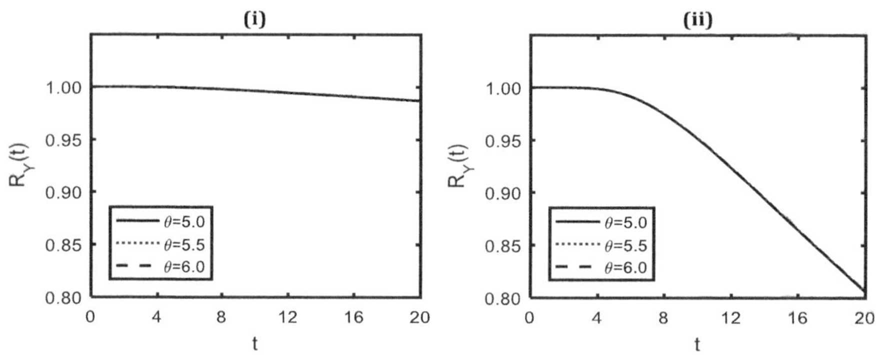

FIGURE 6.8

Variation of reliability of the machining system wrt vacation rate θ for (i) Model 2 and (ii) Model 3.

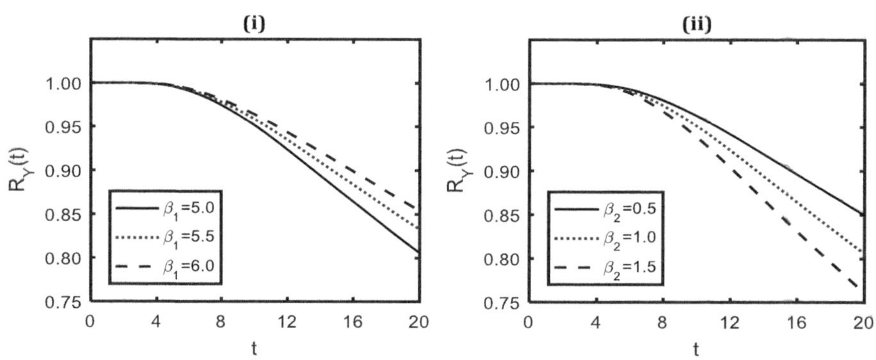

FIGURE 6.9

Variation of reliability of the machining system wrt (i) rate of successful service β_1 and (ii) rate of unsuccessful service β_2 for Model 3.

In Figs. 6.4 and 6.5, it is noted that the reliability of the machining system decreases with an increase in the failure rate of the operating units (λ) and standby unit (ν), respectively, which follow obvious trend. This also prompts that for higher values of λ, the reliability of the machining system decreases more rapidly. It is intuitively anticipated that the reliability of the machining system can be increased by maintaining an appropriate level of repair rate (μ_b). The results in Fig. 6.6 validate the hypothesis of the repair rate and demonstrate that the reliability of the machining system increases with the increasing value of μ_b. The effects of WV on the reliability of the machining system are depicted in Figs. 6.7 and 6.8.

Fig. 6.9i shows that the high rate of successful service attempt increases the $R_Y(t)$, which is expected for any machining or service system. But the reverse trend is depicted in Fig. 6.9ii, which indicates that the cumulative value of unsuccessful attempts reduces the reliability of the machining system. It is recommended that proper preventive measures should be opt to avoid the failure of unit and optimal corrective maintenance strategies should be established to maintain the desired level of reliability at minimum expected cost.

The variation of the *MTTF* of the machining system for different system parameter values is shown in Figs. 6.10–6.12. In each figure, we portray comparative bar graph for the studied model(s). The *MTTF* decreases extensively for the higher values of failure rate (λ). It also decreases gradually wrt ν and β_2, and remains more or less constant wrt μ_v, θ. It also prompts that better corrective measures are always necessitate since the *MTTF* is an increasing function for higher values of repair rate μ_b. Also, it appears that the value of *MTTF* is very less for the third model, which includes the concept of unreliable service of the repairman. It prompts that perfect corrective measure is always important.

For the optimal analysis in a steady-state condition, we delineate a surface plot and contour plot of the expected total cost function (Eq. 6.51) in Fig. 6.13 for combined values of machining system design parameters μ_b and μ_v besides the default system parameters value in Figs. 6.4–6.12 and unit cost elements as $C_h = 80$, $C_b = 30$, $C_{wv} = 20$, $C_{vi} = 20$, $C_i = 10$, $C_1 = 60$, and $C_2 = 7$ to validate its convex nature. The shape of these plots confirms that the developed expected total cost function is very much convex in nature.

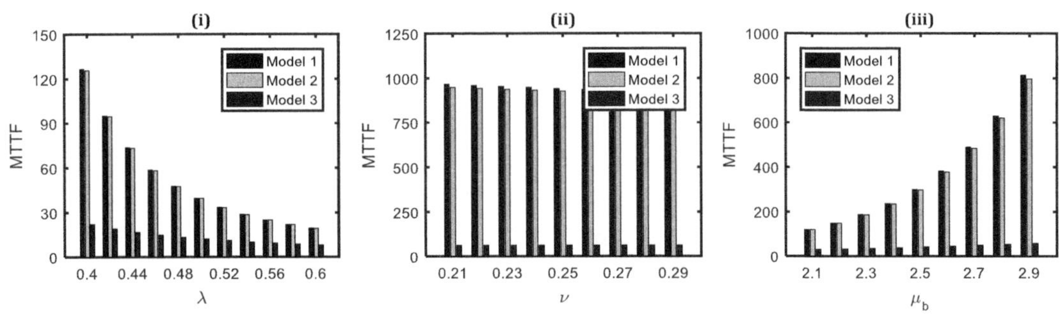

FIGURE 6.10

Effect of different system parameters on mean time to failure of the machining system in Models 1, 2, and 3.

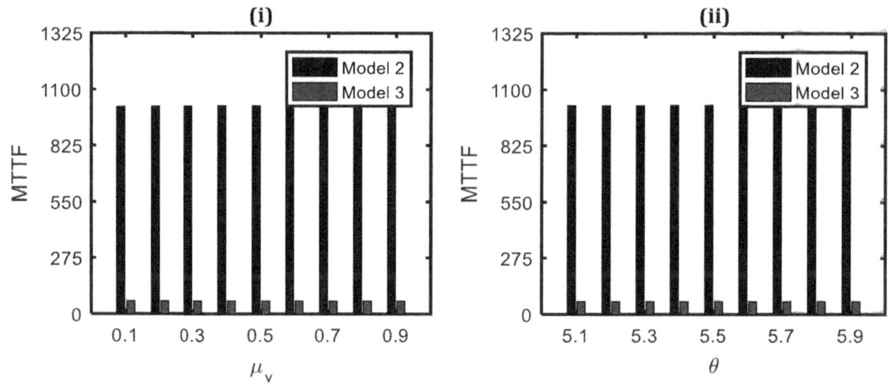

FIGURE 6.11

Effect of different system parameters on mean time to failure of the machining system in Models 2 and 3.

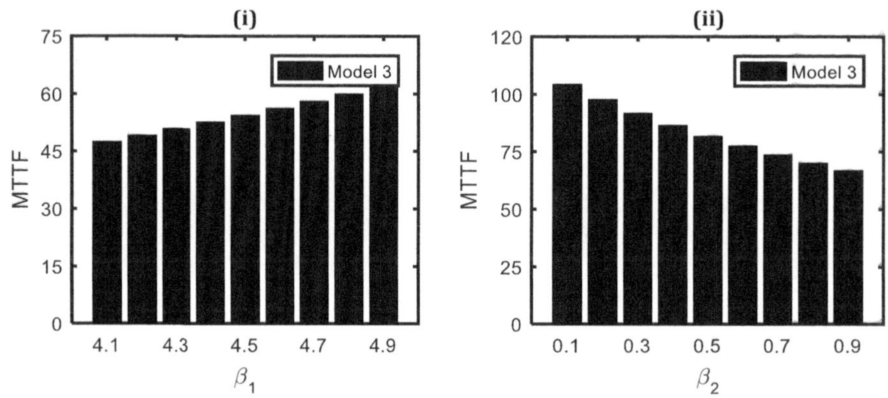

FIGURE 6.12

Effect of different system parameters on mean time to failure of the machining system in Model 3.

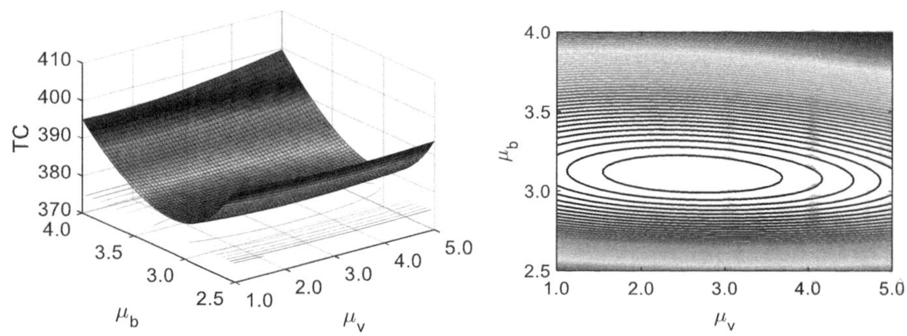

FIGURE 6.13

Surface plot of the expected total cost of the machining system wrt pair of system design parameters.

To achieve the numerical solution of the governing cost optimization problem (Eq. 6.52), we implement the swarm intelligence-based global optimization technique, PSO. For that purpose, we fix the default values of system parameters as $M = 10$, $S = 5$, $m = 2$, $T = 8$, $\lambda = 0.1$, $\nu = 0.01$, $\theta = 3.0$, $\beta_1 = 3.0$, and $\beta_2 = 0.5$ along with the unit cost elements associated with performance measures and states of the machining system as $C_h = 80$, $C_b = 30$, $C_{wv} = 20$, $C_{vi} = 20$, $C_i = 10$, $C_1 = 60$, and $C_2 = 7$. We range the lower and upper bounds of both the decision variables μ_v and μ_b as (0, 8). The default values for parameters of PSO algorithm are set as $\kappa_1 = 2$, $\kappa_2 = 2$, and $\omega_2 = 0.5$. The random vectors $\boldsymbol{\varphi}_1$ and $\boldsymbol{\varphi}_2$ take the values of their elements between 0 and 1.

For the abovementioned default values of the system parameters, some selected generations of the PSO algorithm are provided in the feasible domain for the illustrative purpose in Fig. 6.14. With the help of these generations, we depict the optimal combination of decision parameters μ_v and μ_b along with the optimal expected cost of the machining system.

Because the PSO algorithm is a generation and agent-based stochastic optimization technique, we easily examine that in the first generation, all the search particles (solution points) are randomly distributed in the whole feasible domain. After that, as the generation passes, i.e., in generations 25, 50, and 100, they approach closer and closer to the converging results in a significant manner by exploring and exploiting the whole feasible region and show its capability to converge to the optimal results

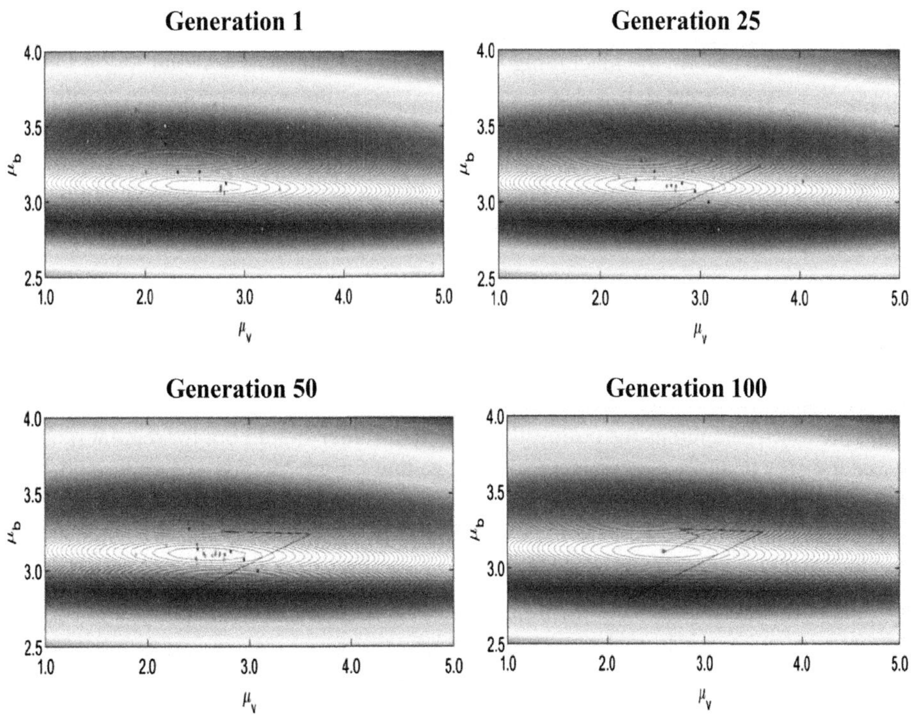

FIGURE 6.14

Several generations of the PSO algorithm in order to find the optimal pair (μ_v,μ_b).

within a reasonable time interval. It implies the robustness of the PSO algorithm and confirms its applicability for all such numerical experiments. Using the PSO algorithm, we achieve the coordinates of the best particle as $[\mu_v^*, \mu_b^*] = [2.585448, 3.105217]$ along with the minimal expected cost of the machining system $TC^* = 378.073039$.

The numerical simulation for different combinations of system parameters and cost elements is performed by developing several test instances, and results are tabulated in Tables 6.1–6.3. For each test instance, we execute numerical experiments by employing the PSO algorithm for 50 search particles, 100 generations, and 20 runs. All the results obtained in each run of the PSO algorithm are mutually independent of each other. Besides, for the validation of the research findings and to show the robust nature of PSO algorithm, we utilize the concept of statistical parameters, namely, mean ratio and the maximum ratio of the optimal cost (*TC*) among all runs of the PSO algorithm. From Tables 6.1–6.3, for all test instances, we examine that the mean ratio $\left[\frac{TC}{TC^*}\right]$ and max ratio $\left[\frac{TC}{TC^*}\right]$ ranges [1.000000000000, 1.000012673892] and [1.000000000000, 1.000038018489], respectively. It signifies the searching characteristic of the PSO algorithm to move toward the best position.

Table 6.1 prompts clearly that for more operating units, better repair rates require. It allows the WV repair rate little less, which signifies the nonworking attitude of the server in vacation mode. The apparent result of the VI threshold is also revealed in Table 6.1, i.e., for high threshold value, a high WV repair rate requires. Table 6.2 summarizes the optimal design specifications for different rates involved in the studied model. For a higher failure rate of units, λ, ν, higher working repair rate is required. In the optimal design, the machining system does not prefer WV repair. For a long vacation time, a lower WV repair rate is necessary since the machining system stabilizes with time. The substantial effect of the unreliable service on the optimal design parameters is also clearly depicted. Table 6.3 tabulates the results of optimal design parameters with variation in the incurred cost for different states of the machining system. The results give a glance for the design of the machining system under the constraints of resource or budget.

8. Discussion

In a nutshell, we recommend following notable points from the studied model:

- For the predictive maintenance, proper modeling, methodology, and analysis are required.
- For better preventive maintenance policy, the system designer should opt for regular maintenance check, redundancy under budget constraints, etc., so that failure of units or machining system can be delayed.
- For just-in-time corrective maintenance policy, a prompt repair facility should be made available with some budgetary constraints.

In short, the optimal design of the machining system is required from installation to operation, the operation to repair, and repair to replacement.

Table 6.1 Optimal values of (μ_v^*, μ_b^*) along with minimum cost of the machining system TC^* using PSO algorithm for system thresholds.

(M, S, m, T)	μ_v^*	μ_b^*	TC^*	Mean$\left\{\frac{TC}{TC^*}\right\}$	Max$\left\{\frac{TC}{TC^*}\right\}$	CPU time
(10, 5, 2, 8)	2.585448	3.105217	378.073039	1.000000000281	1.000000000804	438.712
(12, 5, 2, 8)	2.013138	3.757219	457.574207	1.000000000034	1.000000000097	530.994
(14, 5, 2, 8)	0.435713	4.432288	545.939766	1.000000000015	1.000000000068	578.153
(10, 4, 2, 8)	2.387868	2.962138	367.620676	1.000000000457	1.000000001329	396.839
(10, 6, 2, 8)	2.706459	3.222075	386.475177	1.000000001501	1.000000004322	459.041
(10, 5, 3, 8)	2.585425	3.105148	378.071183	1.000000003709	1.000000011101	275.330
(10, 5, 4, 8)	2.585130	3.104612	378.055332	1.000000000686	1.000000002005	246.331
(10, 5, 2, 5)	2.539595	3.106588	378.053896	1.000000001082	1.000000001913	297.634
(10, 5, 2, 6)	2.572589	3.105605	378.067681	1.000000000182	1.000000000333	341.115

Table 6.2 **Optimal values of (μ_v^*, μ_b^*) along with minimum cost of the machining system TC^* using PSO algorithm for system rates.**

$(\lambda, \nu, \theta, \beta_1, \beta_2)$	μ_v^*	μ_b^*	TC^*	Mean$\left\{\frac{TC}{TC^*}\right\}$	Max$\left\{\frac{TC}{TC^*}\right\}$	CPU time
(0.10, 0.1, 0.3, 0. 3.0, 0.5)	2.585448	3.105217	378.073039	1.000000000281	1.000000000804	438.712
(0.11, 0.01, 3.0, 3.0, 0.5)	2.350304	3.402750	415.536376	1.000000000574	1.000000001090	438.655
(0.12, 0.01,3.0, 3.0, 0.5)	1.889981	3.696572	454.370538	1.000000000392	1.000000000948	311.395
(0.10, 0.03, 3.0, 3.0, 0.5)	2.071489	3.183738	394.941291	1.000000001246	1.000000002151	425.438
(0.10, 0.05, 3.0, 3.0, 0.5)	1.349724	3.245301	410.590482	1.000000000487	1.000000001337	527.494
(0.10, 0.01, 3.5, 3.0, 0.5)	1.160355	3.137190	377.035219	1.000000006007	1.000000010261	514.017
(0.10, 0.01, 3.25, 3.0, 0.5)	1.874644	3.122627	377.698759	1.000000000445	1.000000001233	317.015
(0.10, 0.01, 3.0, 2.5, 0.5)	1.219253	3.363463	420.133113	1.000012673892	1.000038018489	324.824
(0.10, 0.01, 3.0, 2.75, 0.5)	2.037311	3.224146	396.721953	1.000000000146	1.000000000425	320.377
(0.10, 0.01, 3.0, 3.0, 0.75)	2.125596	3.268074	393.522662	1.000000000013	1.000000000115	320.601
(0.10, 0.01, 3.0, 3.0, 1.0)	1.612174	3.424405	408.300535	1.000000000289	1.000000000172	353.761

Table 6.3 Optimal values of (μ_v^*, μ_b^*) along with minimum cost of the machining system TC^* using PSO algorithm for system incurred costs.

$(C_h, C_b, C_{wv}, C_{vi}, C_i, C_1, C_2)$	μ_v^*	μ_b^*	TC^*	Mean$\left\{\frac{TC}{TC^*}\right\}$	Max$\left\{\frac{TC}{TC^*}\right\}$	CPU time
(80, 30, 20, 20, 10, 60, 7)	2.585448	3.105217	378.073039	1.000000000281	1.000000000804	438.713
(85, 30, 20, 20, 10, 60, 7)	3.049352	3.154306	387.879518	1.000000000683	1.000000001908	385.228
(75, 30, 20, 20, 10, 60, 7)	2.082261	3.052302	367.882500	1.000000004205	1.000000012601	432.220
(80, 35, 20, 20, 10, 60, 7)	2.693672	3.112864	379.709848	1.000000001744	1.000000004899	352.091
(80, 40, 20, 20, 10, 60, 7)	2.800584	3.120475	381.332747	1.000000004705	1.000000013472	382.103
(80, 30, 25, 20, 10, 60, 7)	2.608424	3.103066	378.396733	1.000000000130	1.000000000197	345.326
(80, 30, 30, 20, 10, 60, 7)	2.631280	3.100929	378.719856	1.000000003277	1.000000012483	336.438
(80, 30, 20, 15, 10, 60, 7)	2.585449	3.105216	378.073036	1.000000000223	1.000000000517	354.679
(80, 30, 20, 25, 10, 60, 7)	2.585460	3.105217	378.07304	1.000000000005	1.000000000010	512.927
(80, 30, 20, 20, 5, 60, 7)	2.714117	3.110244	376.730988	1.000000000037	1.000000000107	289.431
(80, 30, 20, 25, 15, 60, 7)	2.455805	3.100195	379.399740	1.000000000450	1.000000001342	290.850
(80, 30, 20, 25, 10, 65, 7)	2.698389	3.025877	393.397104	1.000000000686	1.000000001058	290.158
(80, 30, 20, 25, 10, 70, 7)	2.801896	3.954549	408.345142	1.000000000524	1.000000001381	291.181
(80, 30, 20, 25, 10, 60, 8)	1.333921	3.125021	380.013859	1.000000000125	1.000000000365	289.993
(80, 30, 20, 25, 10, 60, 9)	0.263702	3.137975	380.801008	1.000000000000	1.000000000000	291.469

9. Conclusion

In general, the permanent repair facility deteriorates the performance and service quality of any machining/service system due to exhaust work, wear or tear, more idleness, etc. To reduce the wastage of valuable resources, we use some critical queueing terminologies like WV, VI, and unreliable service of the repairman in our modeling and develop different models. To show the dynamical behavior of developed models and the comparative analysis among them, we use several concepts of reliability theory and queueing-theoretic approach. For that purpose, using the fundamental law of transition between adjacent states, the Chapman-Kolmogorov differential-difference equations are developed for each model, and corresponding matrix structures in terms of block matrices are also provided. Moreover, to show the variability of reliability of the machining system and *MTTF*, several plots are provided, and numerical simulation has been performed for the illustrative purpose.

With the observations of transient analysis, the research scientists, decision-makers, and engineers can conclude that the reliability of the machining system, *MTTF*, can be significantly improved by increasing the standby components and the repair rates of the repairman. As a conclusive remark, the findings of the reliability measures of the machining system reveal that the utilization of WV policy is more beneficial for the system analysts and engineers instead of employing unreliable service altogether. Numerical simulations and optimal analysis for multiple combinations of default data sets of system parameters and cost elements affirm that such queueing methodologies may be appropriate for many commercial and manufacturing industries. From a future perspective, one can extend this work to general and hyperexponential service times rather than exponential repair times of the failed units.

Acknowledgment

The second author (SV) extends his sincere thanks to funding agency CSIR, New Delhi, India, for the financial grant SRF/NET (09/719(0068)/2015-EMR-I). Also, all authors are supported by DST FIST (India) grant SR/FST/MSI-090/2013(C).

References

[1] D.R. Cox, H.D. Miller, The Theory of Stochastic Processes, Methuen & Co Ltd, London, UK, 1965.
[2] U.N. Bhat, Elements of Applied Stochastic Processes, John Wiley & Sons, New York, 1972.
[3] R.B. Cooper, Introduction to Queueing Theory, MacMillan, New York, 1972.
[4] J. Medhi, Stochastic Processes, John Wiley & Sons, 1982.
[5] D. Gross, C.M. Harris, Fundamentals of Queueing Theory, Wiley Series in Probability and Statistics, 1998.
[6] J. Medhi, Stochastic Models in Queueing Theory, Elsevier, 2002.
[7] D. Gross, J.F. Shortle, J.M. Thompson, C.M. Harris, Fundamentals of Queueing Theory, John Wiley & Sons, 2008.
[8] L.D. Servi, S.G. Finn, *M/M/1* queues with working vacations (*M/M/1/WV*), Perform. Eval. 50 (1) (2002) 41–52.
[9] Y. Baba, Analysis of a *GI/M/1* queue with multiple working vacations, Oper. Res. Lett. 33 (2) (2005) 201–209.

[10] N. Tian, Z. Ma, M. Liu, The discrete time *Geom/Geom/*1 queue with multiple working vacations, Appl. Math. Model. 32 (12) (2008) 2941–2953.

[11] C.H. Lin, J.C. Ke, Multi-server system with single working vacation, Appl. Math. Model. 33 (7) (2009) 2967–2977.

[12] D.Y. Yang, K.H. Wang, C.H. Wu, Optimization and sensitivity analysis of controlling arrivals in the queueing system with single working vacation, J. Comput. Appl. Math. 234 (2) (2010) 545–556.

[13] J. Li, N. Tian, Performance analysis of a *GI/M/*1 queue with single working vacation, Appl. Math. Comput. 217 (10) (2011) 4960–4971.

[14] N. Selvaraju, C. Goswami, Impatient customers in an *M/M/*1 queue with single and multiple working vacations, Comput. Ind. Eng. 65 (2) (2013) 207–215.

[15] D. Guha, A.D. Banik, On the renewal input batch-arrival queue under single and multiple working vacation policy with application to epon, INFOR Inf. Syst. Oper. Res. 51 (4) (2013) 175–191.

[16] D. Guha, V. Goswami, A.D. Banik, Equilibrium balking strategies in renewal input batch arrival queues with multiple and single working vacation, Perform. Eval. 94 (2015) 1–24.

[17] P. Rajadurai, M.C. Saravanarajan, V.M. Chandrasekaran, A study on *M/G/*1 feedback retrial queue with subject to server breakdown and repair under multiple working vacation policy, Alex. Eng. J. 57 (2) (2018) 947–962.

[18] W.M. Kempa, M. Kobielnik, Transient solution for the queue-size distribution in a finite-buffer model with general independent input stream and single working vacation policy, Appl. Math. Model. 59 (2018) 614–628.

[19] C. Shekhar, S. Varshney, A. Kumar, Optimal control of a service system with emergency vacation using bat algorithm, J. Comput. Appl. Math. 364 (2020) 112332.

[20] J. Patterson, A. Korzeniowski, *M/M/*1 model with unreliable service, Int. J. Stat. Probab. 7 (1) (2018) 125–136.

[21] E. Balagurusamy, K.B. Misra, Reliability calculation of redundant systems with nonidentical units, Microelectron. Reliab. 15 (2) (1976) 135–138.

[22] K. Trivedi, J.B. Dugan, R. Geist, M. Smotherman, Hybrid reliability modeling of faulttolerant computer systems, Comput. Electr. Eng. 11 (2–3) (1984) 87–108.

[23] J. Sztrik, B.D. Bunday, Machine interference problem with a random environment, Eur. J. Oper. Res. 65 (2) (1993) 259–269.

[24] R. Levantesi, A. Matta, T. Tolio, Performance evaluation of continuous production lines with machines having different processing times and multiple failure modes, Perform. Eval. 51 (2–4) (2003) 247–268.

[25] L. Haque, M.J. Armstrong, A survey of the machine interference problem, Eur. J. Oper. Res. 179 (2) (2007) 469–482.

[26] I. Dimitriou, C. Langaris, A repairable queueing model with two-phase service, start-up times and retrial customers, Comput. Oper. Res. 37 (7) (2010) 1181–1190.

[27] J.C. Ke, C.H. Lin, Maximum entropy approach to machine repair problem, Int. J. Serv. Oper. Inf. 5 (3) (2010) 197–208.

[28] S. Lv, D. Yue, J. Li, Transient reliability of machine repairable system, J. Inf. Comput. Sci. 7 (13) (2010) 2879–2885.

[29] M. Jain, G.C. Sharma, R.S. Pundhir, Some perspectives of machine repair problems, Int. J. Eng. 23 (3) (2010) 253–268.

[30] Q. Wu, S. Wu, Reliability analysis of two-unit cold standby repairable systems under Poisson shocks, Appl. Math. Comput. 218 (1) (2011) 171–182.

[31] J.E. Ruiz Castro, Q.L. Li, Algorithm for a general discrete k-out-of-$n : G$ system subject to several types of failure with an indefinite number of repairpersons, Eur. J. Oper. Res. 211 (1) (2011) 97–111.

[32] M. Nourelfath, E. Châtelet, N. Nahas, Joint redundancy and imperfect preventive maintenance optimization for series-parallel multi-state degraded systems, Reliab. Eng. Syst. Saf. 103 (2012) 51–60.

[33] M.A. El Damcese, M.S. Shama, Reliability and availability analysis of a standby repairable system with degradation facility, Int. J. Res. Rev. Appl. Sci. 16 (3) (2013) 501–507.

[34] C.E. Wells, Reliability analysis of a single warm-standby system subject to repairable and nonrepairable failures, Eur. J. Oper. Res. 235 (1) (2014) 180–186.

[35] C.C. Kuo, S.H. Sheu, J.C. Ke, Z.G. Zhang, Reliability-based measures for a retrial system with mixed standby components, Appl. Math. Model. 38 (19–20) (2014) 4640–4651.

[36] J.C. Ke, T.H. Liu, C.H. Wu, An optimum approach of profit analysis on the machine repair system with heterogeneous repairmen, Appl. Math. Comput. 253 (2015) 40–51.

[37] T.C. Yen, W.L. Chen, J.Y. Chen, Reliability and sensitivity analysis of the controllable repair system with warm standbys and working breakdown, Comput. Ind. Eng. 97 (2016) 84–92.

[38] J. Hu, Z. Jiang, H. Wang, Preventive maintenance for a single-machine system under variable operational conditions, Proc. Inst. Mech. Eng. Part O J. Risk Reliab. 230 (4) (2016) 391–404.

[39] D.Y. Yang, Y.D. Chang, Sensitivity analysis of the machine repair problem with general repeated attempts, Int. J. Comput. Math. 95 (9) (2018) 1761–1774.

[40] P. Perez Gonzalez, V. Fernandez Viagas, M.Z. García, J.M. Framinan, Constructive heuristics for the unrelated parallel machines scheduling problem with machine eligibility and setup times, Comput. Ind. Eng. 131 (2019) 131–145.

[41] C. Shekhar, N. Kumar, M. Jain, A. Gupta, Reliability prediction of computing network with software and hardware failures, Int. J. Reliab. Qual. Saf. Eng. (2019) 2040006.

[42] C. Shekhar, A. Kumar, S. Varshney, Load sharing redundant repairable systems with switching and reboot delay, Reliab. Eng. Syst. Saf. (2019) 106656.

[43] C. Shekhar, A. Kumar, S. Varshney, S.I. Ammar, *M/G/*1 fault-tolerant machining system with imperfection, J. Ind. Manag. Optim. (n.d), pp. 433–441.

[44] J. Li, N. Tian, The *M/M/*1 queue with working vacations and vacation interruptions, J. Syst. Sci. Syst. Eng. 16 (1) (2007) 121–127.

[45] L. Tao, Z. Liu, Z. Wang, The *GI/M/*1 queue with start-up period and single working vacation and Bernoulli vacation interruption, Appl. Math. Comput. 218 (8) (2011) 4401–4413.

[46] S. Gao, Z. Liu, An *M/G/*1 queue with single working vacation and vacation interruption under Bernoulli schedule, Appl. Math. Model. 37 (3) (2013) 1564–1579.

[47] L. Tao, L. Zhang, X. Xu, S. Gao, The *GI/Geo/*1 queue with Bernoulli-schedulecontrolled vacation and vacation interruption, Comput. Oper. Res. 40 (7) (2013) 1680–1692.

[48] P. Vijayalaxmi, J. Kanithi, Impatient customer queue with Bernoulli schedule vacation interruption, Comput. Oper. Res. 56 (2015) 1–7.

[49] T. Li, L. Zhang, S. Gao, Performance of an *M/M/*1 retrial queue with working vacation interruption and classical retrial policy, Adv. Oper. Res. 2016 (2016).

[50] K. Li, J. Wang, Y. Ren, J. Chang, Equilibrium joining strategies in *M/M/1* queues with working vacation and vacation interruptions, Oper. Res. 50 (3) (2016) 451–471.

[51] P. Vijayalaxmi, S. Indira, Ant colony optimisation for impatient customer queue under *N*-policy and Bernoulli schedule vacation interruption, Int. J. Math. Oper. Res. 10 (2) (2017) 167–189.

[52] P. Rajadurai, V.M. Chandrasekaran, M.C. Saravanarajan, Analysis of an unreliable retrial *G*-queue with working vacations and vacation interruption under Bernoulli schedule, Ain Shams Eng. J. 9 (4) (2018) 567–580.

[53] M. Jain, C. Shekhar, V. Rani, *N*-policy for a multi-component machining system with imperfect coverage, reboot and unreliable server, Prod. & Manuf. Res. 2 (1) (2014) 457–476.

[54] M. Jain, C. Shekhar, S. Shukla, Machine repair problem with an unreliable server and controlled arrival of failed machines, Opsearch 51 (3) (2014) 416–433.
[55] M. Zhang, Q. Liu, An *M/G/1 G*-queue with server breakdown, working vacations and vacation interruption, Opsearch 52 (2) (2015) 256–270.
[56] C.D. Liou, Markovian queue optimisation analysis with an unreliable server subject to working breakdowns and impatient customers, International J. Syst. Sci. 46 (12) (2015) 2165–2182.
[57] C.C. Kuo, J.C. Ke, Comparative analysis of standby systems with unreliable server and switching failure, Reliab. Eng. & Syst. Saf. 145 (2016) 74–82.
[58] G. Choudhury, M. Deka, A batch arrival unreliable server delaying repair queue with two phases of service and Bernoulli vacation under multiple vacation policy, Qual. Technol. & Quantit. Manag. 15 (2) (2018) 157–186.
[59] F.M. Chang, T.H. Liu, J.C. Ke, On an unreliable-server retrial queue with customer feedback and impatience, Appl. Math. Model. 55 (2018) 171–182.
[60] T. Jiang, B. Xin, Computational analysis of the queue with working breakdowns and delaying repair under a Bernoulli-schedule-controlled policy, Commun. Stat. Theory & Methods 48 (4) (2019) 926–941.
[61] A. Nazarov, J. Sztrik, A. Kvach, T. Bérczes, Asymptotic analysis of finite-source *M/M/*1 retrial queueing system with collisions and server subject to breakdowns and repairs, Annals Oper. Res. 277 (2) (2019) 213–229.
[62] Y. Wang, J. Guo, A.A. Ceder, G. Currie, W. Dong, H. Yuan, Waiting for public transport services: queueing analysis with balking and reneging behaviors of impatient passengers, Transp. Res. Part B: Methodol. 63 (2014) 53–76.
[63] R. Kumar, A single-server markovian queuing system with discouraged arrivals and retention of reneged customers, Yugosl. J. Oper. Res. 24 (1) (2016) 1–9.
[64] A.A. Bouchentouf, L. Yahiaoui, On feedback queueing system with reneging and retention of reneged customers, multiple working vacations and Bernoulli schedule vacation interruption, Arab. J. Math. 6 (1) (2017) 1–11.
[65] D.Y. Yang, Y.Y. Wu, Analysis of a finite-capacity system with working breakdowns and retention of impatient customers, J. Manuf. Syst. 44 (2017) 207–216.
[66] A.A. Bouchentouf, A. Messabihi, Heterogeneous two-server queueing system with reverse balking and reneging, Opsearch 55 (2) (2018) 251–267.
[67] Q. Wang, B. Zhang, Analysis of a busy period queuing system with balking, reneging and motivating, Appl. Math. Model. 64 (2018) 480–488.
[68] R. Kumar, S. Sharma, Transient analysis of an *M/M/c* queuing system with balking and retention of reneging customers, Commun. Stat. Theory & Methods 47 (6) (2018) 1318–1327.
[69] B.K. Som, A stochastic feedback queuing model with encouraged arrivals and retention of impatient customers, in: Advances in Analytics and Applications, Springer, 2019, pp. 261–272.
[70] A.A. Bouchentouf, M. Cherfaoui, M. Boualem, Performance and economic analysis of a single server feedback queueing model with vacation and impatient customers, Opsearch 56 (1) (2019) 300–323.
[71] B. Liu, L. Cui, Y. Wen, J. Shen, A cold standby repairable system with working vacations and vacation interruption following markovian arrival process, Reliab. Eng. Syst. Saf. 142 (2015) 1–8.
[72] J. Patterson, A. Korzeniowski, *M/M/*1 model with unreliable service and a working vacation, Int. J. Stat. Probab. 8 (2) (2019) 1–10.
[73] J. Kennedy, R. Eberhart, Particle swarm optimization (PSO), in: Proc. IEEE International Conference on Neural Networks, Perth, Australia, 1995, pp. 1942–1948.
[74] Y. Shi, R. Eberhart, A modified particle swarm optimizer, in: 1998 IEEE International Conference on Evolutionary Computation Proceedings. IEEE World Congress on Computational Intelligence (Cat. No. 98TH8360), IEEE, 1998, pp. 69–73.

CHAPTER

7 Methods of modeling the maintenance of a steam turbine based on condition assessment

Zdravko N. Milovanović[1], Ljubiša R. Papić[2], Valentina Z. Janičić Milovanović[3], Snježana Z. Milovanović[4], Svetlana R. Dumonjić-Milovanović[5], Dejan Lj. Branković[1]

[1]*Department of Hydro and Thermal Engineering, University of Banja Luka, Faculty of Mechanical Engineering, Banja Luka, Republic of Srpska, Bosnia and Herzegovina;* [2]*DQM Research Center, Čačak, Serbia;* [3]*Routing Ltd., Banja Luka, Republic of Srpska, Bosnia and Herzegovina;* [4]*Department of Materials and Structures, University of Banja Luka, Faculty of Architecture, Civil Engineering and Geodesy, Banja Luka, Republic of Srpska, Bosnia and Herzegovina;* [5]*Partner Engineering Ltd., Banja Luka, Republic of Srpska, Bosnia and Herzegovina*

1. Introduction

Most authors who deal with maintenance and reliability of technical systems consider maintenance in such a way that after corrective maintenance (CM) or preventive maintenance (PM), the system returns to a state defined: "as good as new," which means that maintaining is perfect, or in the state "as bad as old" which means that maintenance is minimal [1]. In practice, perfect maintenance is indeed possible in some systems where only one structural component is concerned, while minimal repairs are possible where the system is made up of many nondominant components interchangeable with new components of the same type. Results of practice to date will show that the maintenance results are not in these extremes, but are often hidden somewhere in the middle. Therefore, perfect and minimal maintenance, as such, is rarely realized in practice [2].

There are several methods that try to provide a solution to the problem of maintaining technical systems and each has its own specific goals and objectives. We can distinguish several characteristic methods: condition-based maintenance (CBM), reliability-centered maintenance (RCM), lean maintenance (LM), total costs life cycle strategy (TCLCS), and total productive maintenance (TPM) [3–6]. New opportunity for modeling maintenance in a more realistic way is provided by imperfect maintenance, which is a step further in the theory of reliability and maintenance of technical systems. The imperfect maintenance method has multiple models that are classified by the author (Wang and Pham, 2006). Some of the authors of these methods are Nakagawa and Yasui [7]; Wang and Pham [8,9]; Brown and Proschan [10]; Block, Borges, and Savits [11]; and a number of other authors. To allow these models of imperfect maintenance to give the results for a particular

The Handbook of Reliability, Maintenance, and System Safety through Mathematical Modeling.
https://doi.org/10.1016/B978-0-12-819582-6.00007-1

technical system it is necessary to know: the parameters of a given system (connection of elements within the system, failure rates, etc.), the characteristics of the maintenance system, and deciding which maintenance tasks to optimize, corrective or preventive, or both.

Maintenance as such can be divided into one of three principles into three main categories: corrective maintenance (CM), preventive maintenance (PM), and combined maintenance (COM). Corrective maintenance is the maintenance that takes place when the system fails. According to some authors, CM is defined as repair. According to MIL-STD-721B [2], CM represents all activities that take place, as a result of a failure, to restore the system to a predefined state. It is obvious that CM is performed at unpredictable time intervals since the failure time of each component is unknown. CM typically takes three steps: diagnosing a problem, repairing and/or replacing a broken component, and verifying a repair. Preventive maintenance is a maintenance that takes place when the system is in operation, i.e., when it is operational. According to MIL-STD-721B, PM designates all activities performed as an attempt to restore the system by conducting systematic inspections, and detecting and preventing failures that are still under development. Combined maintenance is a model of a combination of the previously presented two maintenance models (preventive and corrective).

Preventive maintenance models are most often based on the cost criterion and less on the certainty criterion. The optimization criterion against which to choose the best variant of a maintenance strategy is expressed as: the minimum cost (which is obtained by comparing direct and indirect costs generated by different maintenance concepts) and the maximum profit or other business indicators (obtained by comparing economic effects), differences obtained, and subtracted for different maintenance concepts) [12].

According to the second principle, maintenance can also be classified according to the degree to which the operating state of the system returns to the previous ones by applying maintenance actions, as follows [13]:

- Perfect repair or perfect maintenance (Perfect repair or maintenance, where the system run like as a new and has a law of distribution of time between failures and intensity cancellation of such a new system. The general repair of an engine is an example of a perfect maintenance. The replacement of cancelled components is considered as a perfect repair);
- Minimal repair or minimal maintenance (System is put into the right situation, but the intensity of failures is like before the failure that was removed. The system is in state of "as bad as old." The change of punctured tires on the car is an example of minimal repair. The tire was replaced but the car as system has the same distribution of time between failures as before);
- Imperfect maintenance or imperfect repair (System repair by this principle puts it in the state of a not very new system (as good as new) but still a little newer. It is usually considered to be a nonperfect maintenance that puts the system in an operational state somewhere between "as good as new" and "as bad as old." Obviously, imperfect maintenance is essentially a general repair that can involve two very extreme cases: minimal and perfect maintenance);
- Worse repair or maintenance worse (Represents non intentional worsening failure intensity of the system. The system is leading into a state more time age than that which is current but still the system does not lead immediately to state failure. It means that the poor maintenance puts the system in a worse state than it was before of this maintenance actions);
- Worst repair or worst maintenance (System is not intentionally lead to the failure by using of some of the maintenance actions [1,2]).

In this case the corrective maintenance and preventive maintenance may be in all four levels previously shown (perfect, minimal, worse, and the worst).

The desire for a profitable business, most often in an open market environment, forces a rethinking of existing ones and the search for new methods of maintaining technical facilities. In choosing the appropriate maintenance method, documentation on the previous behavior of the existing plant (or its equivalent in the case of designing a new one), i.e., a database or database of types of failures, their frequency, and the life of individual parts or entire installations, proved particularly useful. With the reduction of production costs on the one hand, and the preservation of existing resources on the other, it also monitors the efficiency of maintenance within the overall business of the system as a whole. According to rough estimates, the funds allocated annually in order to maintain the existing facilities for companies within the electric power sector is from 5 up to 20 % of total costs, depending on the age of the plant and the percentage of implemented reconstruction, rehabilitation and modernization of the same. Continuous demand for reducing production costs requires the introduction of certain measures to increase the efficiency of plant management and maintenance, which also requires the realization of certain preconditions, such as: continuous collection and sorting of data on the plant and plant components in order to increase reliability and reduce mandatory maintenance measures, procurement of appropriate tools (applications) for better organization of management and maintenance, transparency of operations, presentation of all costs, and assessment of the justification of individual investments (benefit—cost analysis) [14].

Since failures are an integral part of the normal operation of the turbine plant and its auxiliary equipment, further analysis is necessary. Most failures on equipment are developed gradually, and the time for the development of the cause of failure depends on a number of causes (turbines work parameters, the environment exploitation conditions, lubrication, material, type of load, stress, etc.). By the analysis of all the data the aim is to bring decision on the process of implementation of the certain activities and actions for maintenance improvement. The final goal is: how to impact on reducing the number and duration of failures in the equipment, and plant stop which is directly connected to the safety of the plant, maintenance costs and the overall business. Attention should also be paid to the importance of staff training for monitoring and diagnosing turbine plants and associated equipment. It is necessary for the management of the plants to understand the importance and the need to train operational personnel for performing professional activities in operation, primarily repair, maintenance, and servicing of the related equipment, in accordance with the applicable legal regulations and instructions given by the equipment manufacturer [15,16].

Modern concepts of steam turbine maintenance, implemented with the use of modern equipment for technical diagnostics and monitoring, in accordance with the instructions from the normative-technical documentation (NTD) and instructions prescribed by the manufacturer, as well as the adopted regulations by the users of the equipment, ensure the safe and reliable operation of the turbine plant and its accessories. The number, then the period of realization, as well as the scope of activities that must be performed during the steam turbine overhaul are closely related to the location and role of the associated power plant within the higher hierarchical power system [17].

2. Terms and definitions

The process of exploitation of a steam turbine system within a condensing thermal power plant is impaired by the degradation of its performance. In order to maintain the level of production of condensation power plants within the electric power system, any negative changes to the system condensation thermal power plant (hence its turbine system) must be prevented by certain countermeasures in order to restore the original operability of a system as a whole. One of the measures is its maintenance [18].

The definition of maintenance as a technological process can be defined as the measures needed to maintain and restore the nominal state of the facility or the procedures related to determining and assessing the current state of technical facilities and systems.

Maintenance technology (MT) is a very broad term, which defines the content and timing of maintenance operations. MT determines the manner of performing a particular maintenance procedure, the order of the activities themselves, the required equipment and tools, as well as a timetable for the implementation of maintenance procedures.

From a technological point of view, maintenance jobs, activities and operations include, in the most common terms, the following works: inspection (status overview, control review, inspection), testing and measurement, auditing and inspection, cleaning, washing and lubrication, corrosion protection (DE conservation, conservation and re-eventual conservation), external monitoring, servicing, technical diagnostics, safety measurements, occupational safety, servicing and finishing, assembly and disassembly, overhaul (small, medium and large, general audit), reconstruction, revitalization and modernization (restoration, corrective and preventive programs, etc.) [12].

Preventive work differs according to time related to the malfunction (operations: condition control and monitoring of degradation of control parameters, finding and elimination of weaknesses, preventive replacement of parts due to aging and wear, preventive adjustments, "care" of the system in the form of washing, cleaning, lubrication, and corrosion protection) as well as remedial work (works performed to restore the malfunctioning function of the technical system to eliminate malfunctions: adjustment; small, medium, and large repairs; parts replacement; revitalization; dismantling and installation, with or without associated adjustments; etc.). In the general case, all activities within the preventive works are performed in cases where an element has passed its useful life (according to constant durability), i.e., when the term plan has been determined in advance, regardless of the wear of the parts (according to a constant date) or on the basis of parameters for maintenance according to the condition (according to the established condition, using the methods of technical diagnostics) [19].

From the aspect of complexity, maintenance works can be classified into two groups: basic or elemental work (cleaning, lubrication, protection, inspection, calibration, adjustment, replacement of parts, and works related to the control and condition of technical diagnostics) and complex works or works of more categories (all other works, which may include basic works). Below are definitions of individual activities related to the maintenance work of a complex steam turbine technical system that provide long-term and reliable operation of turbine equipment, as well as an example of the practical organization of the distribution of 50 MW turbine components.

Maintenance is a complex of operations performed to reflect the operational readiness of steam turbines. They include monitoring of equipment, systematic monitoring of its correctness, control of operating mode, then continuous implementation of the prescribed rules for safe operation during exploitation, as well as elimination of small defects that do not require equipment downtime, as well as reducing the scope of the ongoing overhaul.

Overhaul is a complex of activities performed to establish the correctness or working ability of the equipment. Performing an ongoing overhaul (usually once a year) reduces the need for frequent capital overhauls (usually once every 4 years). On the turbine equipment, during the ongoing repairs, the technical and economic condition of the flow section (sediment deposits, increase in the flow gap, air suction test, etc.) is controlled. The efficiency of the performed works is usually determined by means of certain methods for rapid testing, which not only determine the quality of the performed overhaul, but also carry out regular control of the equipment during the intermittent period.

The overhaul cycle is the smallest interval of operation time of a block unit during which, in accordance with the NTD, the intended forms of repair are performed. The structure of the overhaul cycle determines the order of the different types of overhauls and technical maintenance work of the equipment within a single overhaul cycle.

An unplanned overhaul is an overhaul whose implementation is carried out without a prior plan and is carried out in the event of a failure, which can lead to a complete shutdown of operation at the power plant.

Scheduled overhauls are performed in accordance with the NTD and may be capital, medium, or current in scope. A major overhaul is performed to establish the full equipment resource by replacing or repairing (restoring) all parts in accordance with the NTD. It represents the most extensive form of overhaul and can be type or specialized. Medium overhaul serves to restore the integrity and partial restoration of equipment resources by replacing or repairing and checking technical condition. Medium overhaul in its scope includes partly capital and partly ongoing overhaul of equipment. Ongoing overhaul is used to train the equipment through the replacement of individual parts. It is the least extensive overhaul and can be of the first or second category according to the amount of funds invested.

Starting from the data on embedded materials, their structural dimensions and design and working parameters of work surfaces and failure analyses are performed, with a detailed review of all previous tests performed. On the other hand, knowing the mechanisms of influence on the material under consideration for which there are certain possibilities (identification of pressure and associated temperature, possibility of testing the structure on-site by nondestructive methods, possibility of measuring dimensions, wall thickness or oval shape, computational possibilities of checking the saturation of the material, etc.), it is possible to effectively define the scope and dynamics of testing embedded materials. The extent of testing depends primarily on the knowledge of the main mechanisms of damage (failure records, experientially or on the basis of analogs), such as erosion, corrosion, fatigue, damage, combined mechanisms, etc.

The safety of the energy system and its associated energy equipment is determined by a number of different (by its nature) factors, such as construction, quality of materials used, manufacturing technology, quality of installation, service and exploitation conditions, steam quality, etc. In the process of exploitation, there are cases where complete or partial loss of functional properties occurs. An event that results in a power system shutdown is called a cancellation. The cancellation may be complete (breakdown or shutdown) or partial (downgrade). In doing so, the resulting cancellations can be immediate or gradual. Instantaneous failure is usually characterized by the failure and failure of individual elements or parts of the energy system, which by their function automatically mean complete shutdown of the system, while gradual failure has a temporary change in the state of one or more elements of the plant. Most often these failures are due to the weakening of the material due to working in thermally unfavorable conditions or caused by the removal of the material and the reduction of the walls due to corrosion, erosion, and abrasion. The availability time of a condensing thermal power plant within a hierarchically higher power system is of great importance during the process of its exploitation. Time during which the system is unavailable or downtime is precisely the time during which maintenance is performed and is the total time during which the system is not in an acceptable operational state. In contrast, operating time is the time during which the system operates satisfactorily. System free time is the time when the system is not used. It may or may not be a downtime, and it depends on what condition the system was threatening. The same condensation thermal power plant within the power system can be placed in a state of cold or hot (operating) reserve (Fig. 7.1).

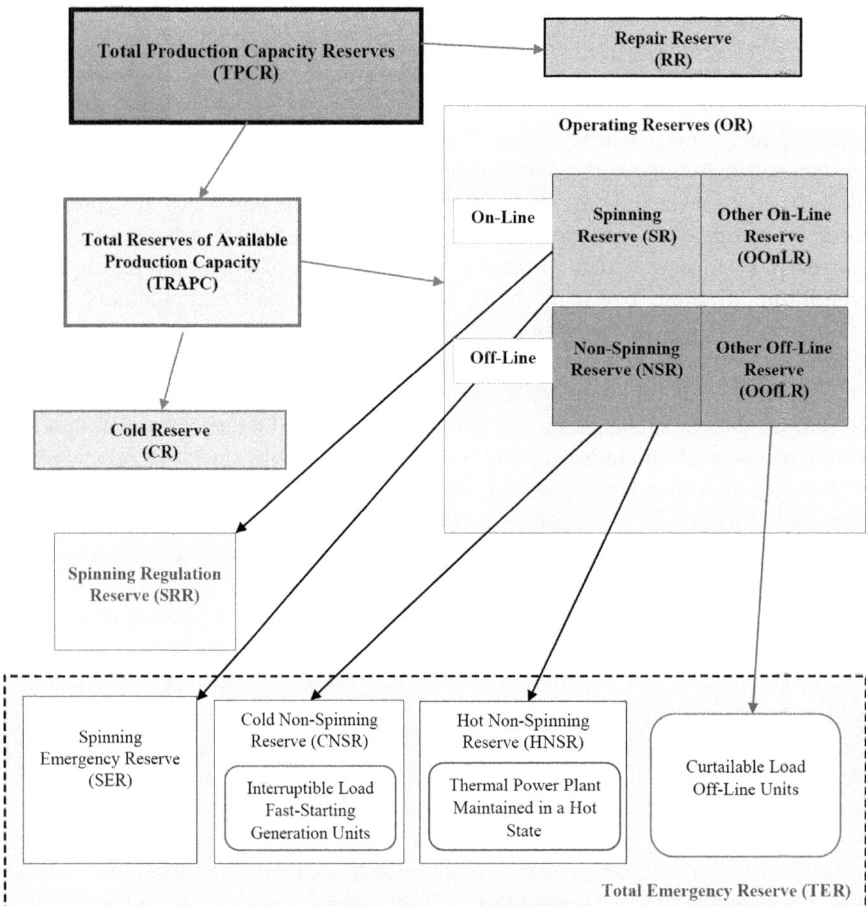

FIGURE 7.1

Schematic representation of production capacity reserves in the power system.

The notion of reserve in production capacity refers to the difference between the installed capacity of the generator and the actual load on the power system. By their nature and purpose of use, the various types of reserves in an electricity system can vary greatly. Existence of production capacity reserves in exploitation is a basic factor that ensures the reliability and safety of system operation. The allocation of production capacity reserves can be carried out according to different principles. For exploitation, the most important division is by the role of certain types of reserves. This division is not sharp because some types of reserves can be used for multipurpose. The total reserve of installed production capacity is the difference between the installed (nominal) power in the system generators and the current load. The total reserve is divided into the overhaul reserve and the total reserve of available production capacity. The overhaul reserve represents that part of the production capacity, which is used to cover the power of the production units that are in the state of planned repairs, repairs, and care. The total reserve of available generation capacity is that portion of the installed system generator power that is used to cover any unforeseen outages due to failures at the generators in operation. It consists of two parts: cold reserves and operational reserves. Cold reserves are made up of generators in older and more expensive thermal power plants, which are capable of being started and used relatively quickly (12 days) to meet the needs of consumers. The operating reserve is part of the production capacity reserve, which is directly used in operation. It consists of a rotating and a nonrotating operating reserve. Rotating reserve is a part of unused production capacity, which is synchronized to the network, which can be used in a very short time (510 min). The two basic components of a rotary reserve are a control reserve (used to compensate for errors between projected and realized system consumption) and a rotating emergency reserve (to cover sudden outages of production capacity from the plant). The nonrotating operating reserve is a backup capacity that is not synchronized to the network but can be quickly put into operation, synchronized, and loaded. The components of this reserve are cold nonrotating operating reserve (made up of fast-start units in hydropower and gas turbine power plants) and warm nonrotating operating reserve (consisting of units in steam power plants that are out of operation but are kept in a warm state, ready for rapid start-up and sync). The total emergency operating reserve is made up of all spare capacity in the operating reserve, except for the regulatory reserve. Some authors classify the operational reserve of production capacities on the basis of the time required for their engagement to the primary reserve (activated for a maximum of 10 minutes) and secondary reserve (activated for a maximum of 20-30 minutes). The first group obviously includes the regulating and emergency rotary reserve, as well as the fast-starting units from the cold reserve, which can be put into operation and loaded for a period of 10 min, and the second units from the cold nonrotating reserve, which require start-up and loading time longer than 10 min. The division of US companies (with even finer diversification) includes: regulating reserves (with the current activation), emergency rotating reserves (activated for a period of 10 min), fast in nonrotating reserves (fast-extracting executable units, which are launched and loaded for 10–30 min), and slow rotating reserves (used for permanent replacement protruding capacity: it can be done in the available time of about 60 min).

By systematic procedures for determining the causes, types, and consequences of failures that may occur, it is necessary to define and specify activities to minimize the catastrophic consequences of failures, especially those related to the means and the environment (preventive engineering). Managing the remaining life of complex technical systems (such as a steam turbine), with the inevitable analysis and specification of its "weakness," today is a multidisciplinary task for a team of experts, whose implementation requires new methods and concepts, as well as appropriate algorithms for methods work. The main tendency in the development of these methods is efficiency, speed, and cost, i.e., obtaining certain numerical values on the basis of which an appropriate and timely decision can be made in the maintenance process (decision optimization).

In addition to estimation, data for determining of reliability can be obtained through calculation and verification or naturally (unforced), through customer experiences, production and other experiences, and through data from relevant service organizations engaged in steam turbine maintenance. If the object under consideration is complex (e.g., a steam turbine system), then the problem of determining reliability is solved if one knows the reliability of the components, or at least their "most critical" parts, their interconnection (structure) and operating conditions (constraints and environmental conditions).

It should be emphasized that the verification of reliability, that is, testing the hypothesis in practice, is carried out at all stages of the development, design, construction, and operation of the facility, and is mainly related to several basic limiting factors—money and time, environmental conditions, and other technical constraints. The reliability verification itself is accompanied by the corresponding mathematical apparatus, with a certain level of confidence in the parameters tested. The inadequate level of reliability during the exploitation of the complex technical system itself and the existence of irrational labor-based investments by eliminating consequences rather than causes clearly indicate the need to harmonize existing methods to achieve optimal reliability and adapt them to the system, with the prior definition and elaboration of the appropriate algorithm.

3. Maintenance conceptions for steam turbine system

The exploitation of a steam turbine as a complex technical system primarily involves the realization of the projected objective function (production of electricity, heat, and process steam), with as little downtime as possible. Raising the level of utilization of the turbine and its auxiliary equipment to a higher level is achieved by quality and by the manufacturer of the equipment prescribed by the operation and maintenance. The basic parameter that characterizes the methods for maintaining them is information about their current technical condition and reliability. The efficiency of the steam turbine system through maintenance is also influenced by the logistics system and administration, which through personal, legal, and fiscal regulation contributes to decision-making. Different models of maintenance of complex technical systems, such as the steam turbine system, are the result of the need for improvement and more realistic monitoring of their safety and reliability. As for the model itself, a decision needs to be made: what maintenance concept will be applied, and what is the optimization criterion (cost, availability, etc.). It can be seen from the above that the starting point for the application of the nonperfect maintenance method is the failure history of a system, which is the result of innate reliability and organization of the maintenance process. Upgrades that make changes possible are provided by the nonperfect maintenance models themselves, with their variations in the form of applied maintenance concepts and the selection of optimization criteria. Further elaboration of the results obtained in terms of making predictive reliability calculations for the next period of use can be performed using computer simulations. The practical use of a reliability model, that individual authors insist on [20,21] or greater use of availability as a criterion instead of cost or at least in combination [1,2], in the literature examples has not been demonstrated in the aviation domain. With their approach to reliability-based maintenance, advocated by both airlines and aircraft manufacturers and military aviation since the early 1960s, problems in the aviation domain make it possible to apply some models of poor maintenance.

3.1 Life cycle of a steam turbine system

Maintenance during the life cycle of a steam turbine system combines a number of supporting activities, ranging from the idea and definition of the concept, the evaluation of their cost-effectiveness, realization, exploitation, and until the system is decommissioned. The training of the maintenance system, through designing on the basis of maintenance, is conditioned by the development of the production forces of the company and aims at extending the working life, with the achievement of more optimal connections of technical, technological, and economic characteristics.

The process of maintaining a steam turbine within a complex technical system of a thermal power plant, as one of the most important parts of the overall production process in the production of electricity, has the task of preventing and eliminating system failures, first of all by rationalizing and optimizing their use and increasing the productivity and economy of consumption in the process of production itself or exploitation.

The life cycle begins when the idea of a new steam turbine is born, and ends when it is withdrawn from use. The main processes that help the system through the life cycle stages are: marketing (specifications), design, production, use, and finally, withdrawal from use. Life cycle analysis is a systematic and analytical approach to identify the resources needed to support the design, production, use, and decommissioning processes. Therefore, useful life analysis is a tool for useful life engineering, whose main goals are to influence useful life design, to identify and quantify total resources related to useful life processes, and to analytically manage useful life process activities.

In other words, engineering life cycle should enable decision-making process, in order to reach the best compromise between investment and providing the necessary resources for the design, production, and use and withdrawal system of the steam turbine from service. In doing so, this approach enables: early and continuous impact on system design from the aspect of life cycle, reduction of life cycle costs of steam turbine systems by limiting the main cost generators over the life cycle, and identification of resources that follow all processes (stages) of steam turbine system service life.

Generally speaking, traditional (sequential) engineering is mainly focused on the performance of the steam turbine system as the main goal, rather than on the development of a general integral approach. Recent knowledge and experience gained in recent decades indicate that the proper exercise of the objective function, i.e., the required degree of competitiveness of the steam turbine, cannot be ensured by investing mainly after their production and reaching the utilization phase, which is most often done. It is much more important for engineers to be sensitive to looking at the consequences of potential errors that can occur during the early stages of steam turbine design and development. This means that engineers should be able to take responsibility for life cycle engineering (competitive, simultaneous engineering), which has previously been largely neglected.

An indispensable part of the useful life is the extension of the working life (revitalization), reconstruction, and modernization of the steam turbine, i.e., the process of extending the working life of these facilities with modernization and reconstruction, with the additional improvement of technical, technological, economic, and environmental acceptability. This procedure, by its structure, is extremely complex and is often compared to the rank of a new steam turbine. Such a systematic and comprehensive process on a part or plant of a steam turbine as a whole is an unavoidable and logical process in the working life of the facility.

The relation between the process of reengineering on the maintenance of a steam turbine with the basic and associated auxiliary equipment, with the aim of realizing the corresponding advantages and improving the reliability of the system, is given through the following characteristic elements: cost analysis related to the maintenance and availability and availability of the steam turbine system (as one of the most important characteristics of efficiency), determining the general aspects related to the motives and justification of revitalization, as well as the scope and defining the most optimal term for realization of this process. In particular, the influence of the reliability and availability characteristics of this system facility on the application of the principle of reengineering through the process of maintaining the steam turbine as a whole, i.e., on the systematic approach to the revitalization of its individual capacities, should be singled out.

The design, development, construction, and operation, while maintaining the equipment and systems of the steam turbine, carries with it a large number of phenomena that can cause damage and endanger the health and life of people directly involved in the main power plant (MPP) of the steam power plant where the steam turbine is installed as well as the wider environment. In short, there is a high risk of adverse events and their consequences. For more complex technical systems, such as thermal power plants, which have a high interdependence of their subsystems and elements, failure of any of them can mean automatic shutdown of the whole system, or reduced power operation (or more often, operation at a technical minimum), which can result in increased operating costs of the system itself, thermal and other overloads, as well as greater damage during system outages. For these reasons, such a complex thermal power system needs to be reliable in the operation. The safety of steam turbines as part of this complex thermal power system can be considered from two aspects. The first and most important aspect is the protection of the operator (human) from injuries during system operation. Another aspect is the protection of the steam turbine itself from damage caused by external causes. Preference is given to operator safety. In doing so, the two aspects are not unconditionally complementary, and an increase in operator safety can be achieved at the expense of the safety of the steam turbine system as a whole.

Any technical system, including steam turbines, even if it performs the function of the target within the tolerances, can be damaged if it is incorrectly handled. The main causes of operator risk include: engaging body parts such as hands in the process of system operation, inattention to the working parts of the system (especially poorly attached units), contact with sharp and abrasive surfaces, the impact of operator static on moving objects, or vice versa.

The sources of risk for steam turbines are diverse and numerous, and in the design phase, the consequences of critical types of failures must be minimized through the prediction of protective devices during the operation of the steam turbine itself. Risks in basic and auxiliary equipment of steam turbines include: shocks, vibrations, corrosion, environment, fire, and contact with high voltage, high-pressure and high-temperature elements, as well as mismanagement (overload or operation below the technical minimum).

Ability to operate basic and auxiliary equipment of the steam turbine without failure in stationary and nonstationary modes of operation, economic and technical convenience for repair of both the elements and the turbine as a whole, restrictions accompanying its exploitation (environment or parent system, environmental protection, financial resources, and other), the possibility of using appropriate type solutions on the basis of analogy with similar facilities, standards for control, and diagnostics—all these are characteristics that do not have a detailed budget and experimentally substantiated base, related to availability and reliability.

On the other hand, with the increasing complexity of technical systems, such as turbine washes, the problem of their optimal functionality arises as an accompanying problem, especially if it is known that such systems can often cause major economic losses or endanger the security of the wider macro-region and the people who serve them.

3.2 Criteria for determining the strategy for maintaining steam turbines

When deciding on the type of maintenance strategy for a particular steam turbine, it is necessary to determine the primary goal (usually minimizing downtime) first. One of the important segments that need to be analyzed when making a decision is the downtime data for each segment (element) of the steam turbine in the previous period and their usability (objectivity). When selecting a maintenance strategy, other parameters should be considered, among which are: the results of the analysis of the importance of equipment, requirements for reliability and availability in the operation of equipment, structure of causes of damage and failures, consequences of occurrence of damage and failures, availability of personnel, and request for providing minimum costs.

On the other hand, the contemporary approach to work organization is based on the experience and knowledge of work organization and on the development of modern scientific disciplines of systems, information, and feedback. The maintenance organization, when not entrusted to specialized companies, as a function of the business technical system in practice and as a process, is implemented as a centralized or decentralized maintenance system. It is often the case that the elements of one of them are also applied to another concept of a maintenance organization (mixed or combined concept).

Market competitions, as well as new legal legislation within EU countries, also pose new challenges to companies within the electricity sector, in order to choose the right procedures for determining the maintenance periods of equipment. The impact of an open (liberalized) market has the effect of rigorously reducing maintenance costs, with priority being given to reliability over availability.

The overall maintenance strategy is based on the "knowledge of the state" of the steam turbine system. In doing so, coordination between short- and long-term maintenance emerges as one of the criteria for selecting a strategy, with the emergence of complex financial transactions that occur when maintaining certain technological entities. It should be noted that in these systems, an optimal maintenance schedule increases their reliability, reduces operating costs, and saves on capital investment in new plants. The right choice of maintenance strategy should ensure the achievement of long-term maintenance goals: high reliability of the plant and the plant as a whole, greater environmental friendliness, improved utility, while reducing maintenance costs, then shortening the duration of the overhaul (annual or capital), as well as extending the period between repairs, and as much maintenance as possible by equipment condition.

After defining the maintenance strategy (maintenance approach), it is also necessary to define the MT, in order to achieve the set goals with an appropriate maintenance strategy. This includes the elaboration of the MT itself, the known principles of failure recording and the repair itself, the identification and diagnosis of various parameters defining the state of the steam turbine system, as well as the definition of certain repair technologies for repairing damaged parts, lubrication and corrosion protection, etc.

The very development of access to appropriate technological processes for the maintenance of the steam turbine system is realized in several stages [16,19]:

- processing of substrates for the design of maintenance technologies (documentation of equipment manufacturers; information on the functioning of equipment; operating conditions—operating mode, environment, personnel, etc.; production capabilities—capacity and quality of work, planned useful life, etc.),
- development of general principles of technology for maintenance of individual parts of technical system of steam turbine (system for detecting failures or weaknesses; methods of condition control and diagnostics of failures—visual, acoustic, ultrasound, etc.; existing methods for elimination of failures and weaknesses in the system; methods for simplest system dismantling and assembly; defined levels of required dismantling (disassembly) to eliminate common turbine failures or planned cleaning; lubrication; modification of individual variable positions or assemblies; ways of testing the functionality of installed systems—test operation and/or operation of the turbine),
- approaches and methods of MT (access "to the equipment, to the facility—steam turbine as a complex technical system," where maintenance comes on the spot and access "from the technical system," where the steam turbine is dismantled and transported to a special location, where the maintainers have all the conditions for the implementation of the required maintenance activities: an individual maintenance method used for major repairs or planned operations after shutdown and preparation of the plant (general overhaul) and an aggregate or group maintenance method, with timing of individual activities),
- elaboration of specifics in MT (different substrates, multiple methods of execution, small number of the same or unified parts of the steam turbine, different manufacturers of similar equipment, years of installation, etc.).

The decision on the type of maintenance is made on the basis of the cost criteria of the company related to maintenance and operation. In this way, it is possible to determine all the most economically acceptable maintenance activities. These activities must be realized by applying the appropriate type of maintenance. The activities of MT related to control and diagnostics in maintenance, the repair technology itself, as well as the activities related to lubrication and corrosion protection of technical systems must not be neglected. The most commonly used methods for repairing broken or worn parts are: welding on, welding, metallization, electrolytic application, electromechanical treatment, metal lock bonding process for fractured parts, bonding techniques, patented repair welding technologies, application of materials to surfaces by plasma spraying and flame spraying, etc.

Each selected type of maintenance solution for the steam turbine as a whole provides the required level of output sizes (efficiency, effectiveness, effect, etc.), characterized by certain costs of the maintenance process, or a share in the total operating costs of the steam turbine. Searching for the best solutions, according to certain criteria, is an optimization of the maintenance system and aims to provide the required level of reliability at minimum maintenance costs. The set of rules that dictate maintenance activities, presented in the form of a maintenance strategy (MS), is defined as:

$$MS = \{\tau_1, MP_{prev.1}(t), MP_{cor.1}(t)\}, \tag{7.1}$$

where the concept of preventive replacements of a simple system is defined (answer to the question: what?), at fixed time intervals τ_1, by the technology and organization that provide the maintainability $MP_{prev.1}$ and $MP_{cor.1}$ from the standpoint of preventive and corrective maintenance procedures (answer to the question: how?).

As stated earlier, maintaining a steam turbine plant can be:

- corrective (applicable only when failure has occurred or an error has occurred during the interservice period),
- preventive (carried out in certain and predetermined periods of time, which characterizes the operation of thermal power plants within the electricity system, as a hierarchically superior system) or
- by condition (for more modern turbine plants, which have the necessary diagnostic equipment for determining the condition of individual installed components of the turbine plant, or the state of the materials from which those components are made).

Certainly, the most convenient and reliable approach is maintenance according to the condition of the plant. Only the maintenance of the condition is based on the control of the technical condition at certain time intervals, regardless of the degree of damage to the component and/or the system as a whole, and after that, depending on the condition, its replacement or repair is performed or the element remains and further in production.

The characteristic size of the technical condition, which characterizes the change of state, is taken to be the most influential size in the process of exploitation, which most often coincides with one of the selected characteristic sizes of the state of the technological process and serves the tasks of automatic control. Monitoring (control) of the technical condition of the system for the purpose of performing maintenance activities can be identified with monitoring the state of the system in order to guide the optimal process. The result of any technical diagnostics must be a decision on the continued usability of the component or system as a whole (part for reinstallation, part for repair and reinstallation, part must be taken out of further use).

In order for any decision to be made at all, it is necessary first of all to know the allowable wear limits and then the other conditions given in Fig. 7.2. The limit of wear and tear is actually the boundary between the operational usability of the element or system and its damage. As mentioned earlier, any change in the state of the system during exploitation is a process of random character, with the moment of transition of the correct to the fault condition occurs as a conditional failure, that is, characterizes the beginning of the fault occurrence (Fig. 7.2).

3.3 Cost–maintenance ratio

Recovery of individual elements of a steam turbine after certain random and deterministic time intervals is a common feature of most maintenance strategies. In doing so, the process of their functioning is broken down into specific cycles, i.e., stochastically equivalent intervals with respect to the duration and associated direct costs of maintenance and indirect costs (damages and losses caused by the occurrence and duration of downtime due to recorded failure) [19]. In this case, the definition of the mathematical expectation (mean value) of maintenance costs (exploitation costs) in a unit of time, with an indefinite time for the element (steam turbine system) to function, takes the form:

$$K = \lim_{t \to \infty} \frac{E(C(t))}{t}. \tag{7.2}$$

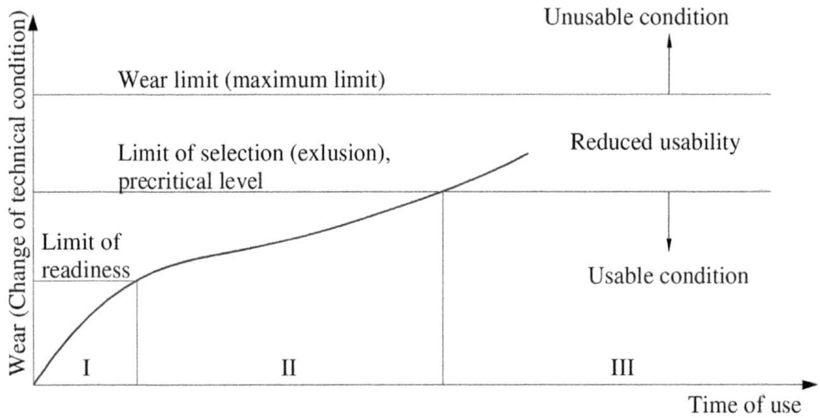

FIGURE 7.2

Boundaries of the technical system for forecasting the behavior of the technical system of the steam turbine.

Costs within the life cycle of a steam turbine are represented by the total costs incurred during its lifetime (Fig. 7.3). Their occurrence is still in the design and development phase of the project documentation, to grow more intensively during the execution itself.

During use and associated maintenance, they have more moderate growth, and they end with the exclusion of the steam turbine from the process of operation and its removal from the microlocation, with the accompanying environmental remediation. Costs also depend on the maintenance interval, that is, on the reliability level of the steam turbine as a whole, and are closely related to the possible risks (Fig. 7.4).

If the intervals between individual maintenance interventions on the steam turbine increase, the direct maintenance costs decrease, with the risk of sudden failures increasing (indirect maintenance

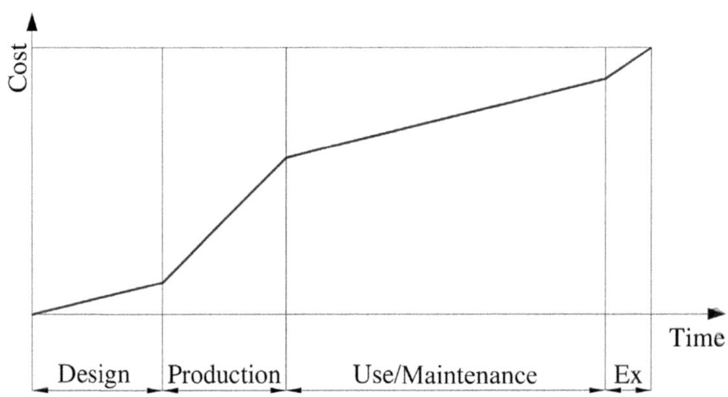

FIGURE 7.3

Impact of parts of life cycle of steam turbine system on costs [12].

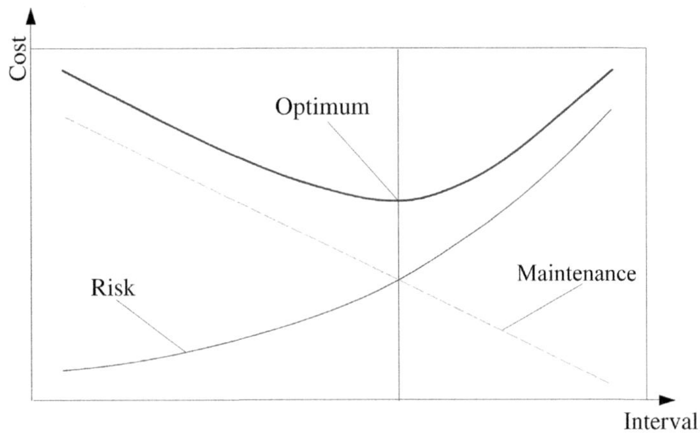

FIGURE 7.4

Outline of the cost–maintenance ratio of the steam turbine system [16].

costs). It should be noted that in energy-process engineering, the intervals depend on some additional requirements that are not directly related to the maintenance of the steam turbine (operation of the hierarchically superior system, ensuring production in the desired period, avoiding overhaul during the winter, tourist season, etc.).

The quality of the work process and the effect of maintenance are directly related to the investment in the maintenance of the steam turbine as a whole. Low investments lead to major negative consequences for complex technical systems, such as a steam turbine, as well as their owner, resulting in corresponding negative financial effects. On the contrary, a high maintenance investment results in a small negative effect on the steam turbine itself.

3.4 Maintenance methods for technical systems

In terms of philosophy or methodological approach to maintenance, there are two schools (lines) that have been attracting the most attention lately: RCM and TPM. The revitalization of steam turbines is the process of rebuilding their parts, bringing the steam turbine with its basic and auxiliary equipment back into a state of further reliable exploitation (at least 15 years). Extension of the service life or life span of a steam turbine plant is usually realized in parallel with the reconstruction and modernization of other plants at the power plant, with an increase in the degree of utility and the level of reliability, as well as an increase in the power of thermal power plants. In this case, in most cases, the revitalization, reconstruction, and modernization of thermal power plants are closely interconnected and most often performed simultaneously.

The primary goal of the system user is to keep the system in a working state for as long as possible. To achieve this, it is necessary for the system to "assist" by performing certain maintenance tasks. Important decisions about the responsibilities, duties, content, and timing of individual maintenance tasks define the maintenance methodology or philosophy.

The very process of planning and implementing the process of revitalization and exploitation of plants within the higher hierarchical system (electrical energy system (EES)) is realized with the aim of achieving a high level of operational safety, which involves defining and detecting possible sources of unreliability, defining measures to eliminate and mitigate their effects, and the most commonly used economic criterion is used as a criterion.

The reconstruction of the thermal power plant changes significantly. The heat scheme is changed; changes are made to changes in the function of operation (e.g., condensing thermal power plant is changed to thermal power plant, technologically regulated subtractions introduced, etc.), as well as changes resulting from the inevitable modernization of the thermal power plant (increase in technical and economic efficiency of production, improvement of exploitation characteristics, increase of degree of beneficial effect (DBE) of individual plants and systems as a whole, increase of reliability and safety in exploitation, etc.).

Good quality maintenance of certain system is chosen according to specifics of the system itself and its position within the hierarchical superior system and its surroundings. During choosing maintenance strategy, methods given in Fig. 7.5 are usually combined.

To get the answer for the question which maintenance strategy to choose in order to make the system economically more effective and the market more competitive, we must take into account the loss in production caused as a result of a malfunction, reducing the cost of maintenance and organizational costs. Increasing efficiency of maintenance can be performed with the help of insights into the state of the system, the information support in decision-making, and optimization of maintenance procedures. After collecting information on the degree and function of fatigue of individual system components and statistics of stochastic failures, we are able to choose a particular maintenance method. The chosen maintenance strategy is selected on the basis of the information obtained from the system empties and defines the type of maintenance, the time required to perform the maintenance, and the maintenance goals themselves.

The purpose of the chosen maintenance strategy, depending on the observed system undergoing maintenance, the overall economic situation of the company, and the selection of maintenance strategies, is to fulfill the target maintenance functions as much as possible (increasing usability of the system, ensuring a certain degree of reliability, optimizing the number of employees, reducing the total cost, etc.). Advantages of this approach are high reliability and maximum utilization of reserve parts of the system, and drawback is dependence of the system on possibilities of conducting adequate measurements and reliability of the diagnosis of the system. Also, as the time to perform the overhaul is not planned anymore but it mostly depends on the state of the system, it is possible that the execution of major operations on the system falls at an inappropriate time, that is, when the market (or signed contracts for the delivery of the final product) requires high availability of the system as a whole.

4. Steam turbine maintenance method according to state

The optimal management of a complex steam turbine system must be based on the evaluation and complex optimization of the reliability indicators, depending on the means of providing them and the hierarchical level of detail as well as the current life cycle phase. For this reason, the optimization process includes basic structural, parametric, and structural solutions related to the steam turbine

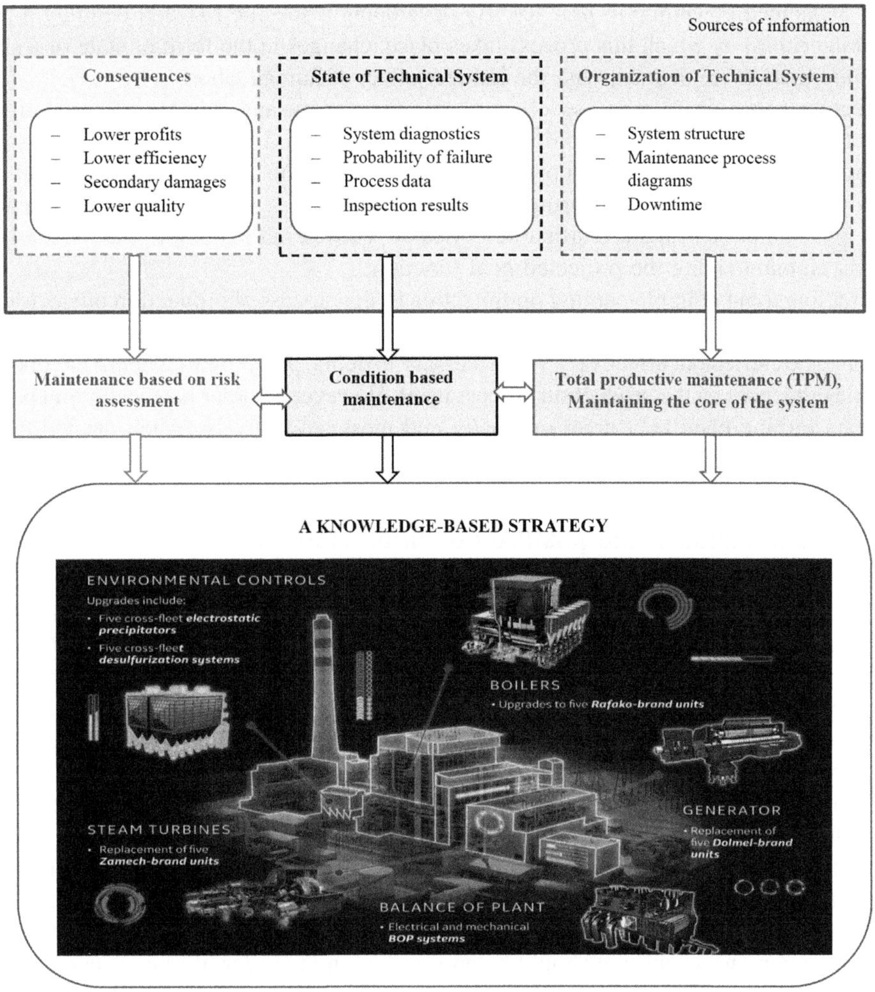

FIGURE 7.5

Information sources for each type of maintenance.

technical system itself and its associated equipment by changing its most important characteristics: efficiency (most often energy), maneuverability, reliability, and cost-effectiveness as a whole. The set of optimization goals is concluded in the overall choice of reliability indicators and possible ways to achieve them, given the already established rules related to the higher hierarchical level of the system (thermal power plant, power system).

After selecting the right strategy for maintaining steam turbines, based on the possible approaches and concepts of maintenance, it is necessary to define the appropriate MT, in order to eliminate the planned or unplanned downtime in practice. By developing a specific MT and defining a particular procedure (algorithm) by which this process takes place, changes in the form or state of a substance, energy, or information are defined, with the indirect effect of human labor.

An effective approach to increase efficiency is the timely and reliable identification of the condition of the equipment using technical diagnostic measures. Using adequate methods of technical diagnostics makes it easier and more reliable to diagnose the condition of individual elements of a steam turbine (rotor, blades, bearings, housing, regulating system, protection systems, foundation, etc.), on the basis of which further activities are carried out leading to the realization of the set goal, that is, maintaining the projected goal function.

The increasing trend of implementing optimization of the process of production of electricity, heat, and process steam by economic criterion aims to reduce costs and increase productivity. In order for optimization to be carried out effectively, it is necessary to define process-influencing factors as well as their possible influences on the production process itself. However, in order for the thermal power plant and the steam turbine plant to operate optimally and most productively (when needed most), it is necessary to have reliable equipment necessary to carry out the process.

4.1 Theoretical basics of the posture condition method

CBM represents the combination of the two: periodical maintenance and maintenance before the failure [12]. The idea is to take advantage of postbreakdown maintenance and maximize the use of spare parts and components while avoiding the disadvantages of periodic maintenance. CBM uses all available methods to determine technical level state of the system and equipment in order to maintain access only when the condition of components of the system falls below the specific critical level. The system condition is determined by certain tests, inspections, diagnosis, measurement, and analysis of measured data [14]. Today, when modern and profitable production rests on market demands, optimal technology, and optimal technological process, from the aspect of techno-economic life of the steam turbine system, the need for intensive revitalization (extension of basic working life), reconstruction, and modernization of existing technological processes and related equipment is of particular importance. This requires further expansion and implementation of modeling, simulation, and optimization methods based on the use of information technologies. Changes in the maintenance system should be approached in the order of importance of the factors that influence the change in the performance indicators of the steam turbine system, as well as the possibilities of application from the aspect of the required investments. The aim is to achieve such a sequence of development steps that provide the greatest effects with the least investment (economic criterion).

The specific tasks of MT are to provide the process of maintenance optimization and to refine the directions for achieving higher quality, reliability, and economy of the steam turbine system and its production itself. The decision on the type and activities of maintenance can also be made based on the costs of the company related to maintenance and operation and the methods of maintenance chosen. The order of development steps to be implemented affects the efficiency and effectiveness of the maintenance system.

Commitment to a maintenance strategy or technology will certainly influence the character, scope, and frequency of maintenance work to be performed in a specific technical system (application of the integrated logistics support (ILS) concept). In doing so, the greatest attention is paid to reducing administrative time through the implementation of necessary changes in the type and form of management and organizational structure and the application of the concept of computerized maintenance management system (CMMS). Another factor in importance is the shortening of logistic times by determining the most optimal levels, methods of managing and allocating spare parts by levels, as well as accelerating material and information flows related to them. Merely improving the quality of work execution requires more changes in the behavior of the equipment repairer and maintainer than the material investment itself. In addition to the introduction of modern overhaul equipment, the use and application of diagnostic analysis methods can further improve the performance indicators of the MT applied.

Condition assessment of the object under consideration is performed on-site by an employee or remotely with the help of software for the remote control and surveillance (monitoring) system. With advances in communications and information technology, equipment for remote surveillance of technical systems became accessible and available to business subjects. Invested means analysis for the equipment for remote monitoring and analysis, thus acquired capabilities and services confirms the cost-effectiveness of investing in such systems. Information and sizes essential for the operation and maintenance of the technical system are necessary to collect continuously during the time of its working life. Such information is collected, locally processed and prepared, and then transferred to the central part with the same for monitoring where further processing of the collected data is performed, with complex methods and algorithms for analyzing, processing, and storing the collected data.

Choosing the right maintenance organization defines the basic elements of a work-sharing structure that will allow all required maintenance tasks to be performed from the aspect of the selected MT. On the basis of all the defined characteristics of the maintenance process of a particular technical system, it is not difficult to conclude that the characteristics of the selected MT depend on the characteristics of the technical system itself, that is, they must also be considered within the characteristics and contents of the "technical factor."

On the other hand, the operation of equipment over a period of time under conditions dictated by process optimization by economic criteria may adversely affect the length of the basic life expectancy for exploitation, primarily due to operation under different mechanical conditions than prescribed (overload, shortening of maintenance period and delay of certain repair planned activities, recommended by the equipment manufacturer, forced system operation, etc.). Such work very often results in increased stresses in the elements of the equipment, especially rotary machines, and leads to a faster degradation of the mechanical condition. This fact is often neglected, which in a long period of time can completely undo the positive results previously achieved by process optimization (increased maintenance and repair costs, as well as losses due to longer delays in the exploitation process). The operation of the turbine under mechanical and process unfavorable conditions generates variable stresses of the material, which leads to damage to some of its assemblies, cracks, and fractures with often catastrophic consequences.

The development of microprocessor devices for complete monitoring and analysis of operation, with the ability to determine the current mechanical state of the steam turbine, have provided a completely different approach to the maintenance of the steam turbine plant, which increasingly applies the maintenance method according to the state (compared to the previously planned maintenance according to a constant date). In doing so, all maintenance activities are carried out only when necessary or when the condition of the turbine requires it.

4.2 Maintenance and decomposition activities of the steam turbine system

Steam turbine plant is a complex system under condensation thermal power plant. It consists of several subsystems (housing with jet blades, a rotor with working blades, foundation, rotor bedding system, sealing system, etc.) or individual parts (components), wherein the condition of each part directly impacts on the complete system condition. In addition, this complex energy object for its function requires the use of auxiliary systems on the thermal power plant (system of regenerative heaters of low and high pressure, a boiler with superheater of fresh steam, the condensation plant, steam lines and other pipelines with associated fittings, control and protection system, lubrication system, etc.). Each of these shared facilities affects with certain behavior weight on the whole system. To select the most suitable approach to maintenance it is necessary to make the proper decomposition of the observed plant and its disassembly into simpler parts from the point of maintenance. Various approaches to decomposition are possible, which is usually performed by expert with a high degree of knowledge about the production process, operation, and maintenance of the observed system.

Basic division on the basis of empirical data for the observed system which can be done is a division of the system according to the priorities (a division of equipment in classes according to the level of necessity for system functioning). The simplest division of the observed system is: first class (important parts for the system functioning) and second class (irrelevant parts or equipment - the system can work for some time without these). A more detailed subdivision of the system into the corresponding classes can be made according to the consequences of a possible failure on the observed system, its parts, or individual equipment. In this case, the parts of the system can be divided into five classes as follows [13]:

- system or parts of the system of which a failure has impact on reduction in security of the plant or it may cause malfunction of other parts of the system (vacuum distortion at the junction of the low pressure turbine plant-condensation);
- system or parts of the system of which failure could lead to a complete standstill of a plant (system control and protection system of turbine installations);
- system or parts of the system of which a failure leads to the reduction of production (high-pressure heater malfunction);
- system or parts of the system of which failure leads to reduction of the efficiency of the main production process (sealing system); and
- system or parts of the system of which failure does not have a direct effect on the production process (damage to the pump at the pumping station with working and reserve pumps).

Such a division of a system has proven to be as a useful approach for plants related to the production, transmission, and distribution of electricity [18].

Results of many researches and analyses of damage to turbines, which are in service, showed susceptibility to damage and more frequent cancellation (a higher probability) of some parts of turbine, while for the other elements this probability is very small. Therefore, the classification of damage to the turbine parts is often made according to the categories of probability of their occurrence: damage with a high probability of occurrence, then damage with a medium probability of occurrence, and damage with a low probability of occurrence. The following are likely to be damaged by exploitation: labyrinth seals, bearings, rotor blades of the first and last stage, nozzles of the first stage, spindles and trays of control valves, condenser tubes, etc. The average probability of damage to the parts usually includes: rotor and other stages blades, turbine shafts, stator blades, valve housings, etc. Steam turbines with a low probability of failure include all remaining turbine components that are not exposed to high temperatures and operate at approximately constant pressures (rotor discs, turbine housing), turbine condenser jacket, ejector, oil cooler, etc.

The previous categorization is of a conditional type and can only serve as a preliminary estimate for the procurement of new spare parts and a prerequisite for defining the positions that need more attention when working. Applying technical diagnostics in the course of performing activities for maintenance of the turbine and introducing systematic monitoring of the condition of parameters and damage for specific conditions of exploitation can determine the categorization of damage of individual parts for a particular turbine. In this way, it is also possible to further reduce the scope of research into the causes of failure of the most critical element (drive) on the turbine, which will result in an adequate reduction of the costs necessary for the research (preliminary or preliminary assessment of the situation). Also, the results obtained in this way can serve for further reconstruction, revitalization, and modernization of individual parts of the structure, that is, the steam turbine system as a whole.

5. Methods of modeling of a steam turbine plant maintenance

There are a number of tasks in energy, for which, apart from the basic indicators, information is needed on the dynamics of their application, as well as on the characteristics of the process and its distribution. The probability functions of the occurrence of certain states of the system can be twofold: *discrete*, that is, intermittent change functions, where variables take a finite number of values and *continuous* or continuous change functions, in which the variables take any value within a given interval. Starting from the fact that the steam turbine, as a unit within the technical system of the condensing power plant, which again forms one of the elements of the power system, operates during the accounting period of time with variable power, and that it is not always available with the highest power, there is a difference between the indicators which is given by the time and energy method.

Indicators determined by certain time method are always more favorable than those of specific energy, so, in order to distinguish them, there should always be determinant *timely* in front of those indicators which are determined by specific time method [18]. The indicators often used in the analysis of the reliability of energy equipment—the coefficient of readiness (operational readiness) and the coefficient of technical efficiency—only reflect complete failures, not partial ones. In the literature, the following supplementary coefficients are often used to reflect these failures, which result in a decrease in electricity production [15]:

- coefficient of relative reduction of the given electricity;
- coefficient of relative deterioration of the economic and technical indicators of the plant;
- coefficient of relative operation of the steam turbine complex within the thermal power plant system;
- coefficient of nominal readiness of the steam turbine, or of the power plant as a whole.

The structure of the technological system, with its internal and external connections, and the reliability characteristics of the individual elements are the basic elements for evaluating the measure of importance of individual elements. When considering issues of importance of elements of a technological system, various questions can usually be asked, the most common are [14]:

- change in the operational readiness of the technological system while changing the operational readiness of certain elements;
- evaluation of the elements most likely to influence the failure of the technological system (steam turbines as a whole);
- assessment of the elements with the highest probability of failure in the failure of the technological system (steam turbines as a whole);
- estimation of the increase in operational readiness of a steam turbine while increasing the operational readiness of its individual components and ancillary equipment;
- assessment and rational allocation of resources when increasing the reliability of individual elements.

Possible objectives for conducting a basic reliability analysis of a steam turbine can be defined as [16,19]:

- assessment and research of the most critical facility, that is, the most critical details of that facility, by setting influential basic and supplementary research criteria and determining the rank of critical plants and summarizing them (determining and comparing them using the ranking method);
- optimizing the means of ensuring reliability, with an analysis of their internal and external links;
- analysis of the interrelation of the requirements for the reliability of the parts, that is, of the system as a whole and the total cost of providing them;
- forecasting the optimal reliability of a steam turbine plant as part of a thermal power plant, that is, a thermal power plant as a component of a power system, and linking it to the tasks of reliability optimization tasks at the power system level, considering the functional relationship between the required reliability curve and the cost curve necessary to realize them.

Starting from the basic technological scheme and the composition of the equipment, as well as the basic flows of the substances participating in the technological process, it is possible to adequately observe the technical maintenance system as a system for managing the reliability and safety of the technical maintenance strategies.

5.1 Data collection

The use of modern and developed equipment for monitoring of the system facilitates the collection of data on the system, where there is certain information about the system that cannot be collected in that way (oil chromatography, general condition of the object, the degree of corrosion, erosion, and abrasion, droplet condensation in the last stages of a low pressure turbine, etc.). The collection of information of such type, due to the complex measurement process or the nature of information, requires the participation of trained and especially skilled staff, capable for measuring and giving an adequate estimation of the object or systems and the estimation of possible impacts on the system as a whole. From the point of method of information gathering there are dynamic data type (dynamic information collected with the help of software for remote monitoring, collection, and transmission of data) and static data type (static information which can be reached on the spot or by measuring or assessment of the expert on the system).

Based on the collected dynamic or static information, decisions are made on maintenance of the system. The information collected is appropriately processed and interpreted; additional analysis of the state of facilities shows that its condition mostly depends on the state of a number of different attributes, which in turn requires decision-making analysis to be used to determine its current state, based on multiple attributes (Multiple Decision Data Analysis (MDDA)), most often combined with the fact (evidential) conclusion (evidential reasoning) [13].

The analysis of data collected by technical diagnostics of the system, the association of appropriate assessments, and decision-making shall be carried out in an objective, reliable, reproducible, and transparent manner. Results of state assessment as one numerical value and with the interval from minimum to maximum values allow easy access to the state of the system and its comparison with other similar or with the state of the same technical system in a different point in time. Critical parts of the system can get high weight coefficients, with the aim of potentiating their impact on the complete state of the steam turbine system, as well as the condensation thermal power plant system and electro energy system as the whole (hierarchically superior systems).

5.2 Testing and determining methodology of the remaining service life of the structural parts of a turbine plant

In the stochastic behavior of complex technical systems with a large number of circuits, subassemblies and their components, such as a steam turbine system, the future state is not determined solely by the initial condition and mode of control. Therefore, methods for estimating optimal reliability on the basis of the economic criterion get their role in the design and planning processes for the design, use, and maintenance of the system, as well as its parts. Also, the application of probability theory and mathematical statistics based on the history of failure data is very important for making lasting decisions in the maintenance system, which enables timely action, with an adequate reduction in maintenance costs. An important step in the safety analysis, and therefore the reliability of the steam turbine system, is the mere standardization of safety, that is, the formulation of security requirements for the turbine system. However, the problem of forming a minimum sufficient set of indicators characterizing the considered property of a particular part of the turbine system has not yet been fully resolved. Depending on the part of the turbine system under consideration, safety or reliability is the result of the superposition of other more "elementary properties," such as: mechanical strength, stability, fire resistance, elasticity, etc.

The existence of potential sources of danger and thus the frequency of hypothetical damage can serve as a universal quantitative characteristic of safety or reliability of all steam turbines [14]. This enables their comparison of different purposes and principles of operation, i.e., "measurement" according to the accident scale of different sources of danger. This poses a risk that characterizes the frequency of occurrence of adverse events per unit of time. In the dictionary of the European Organization for Quality (EOQ), as part of the terms used in the field of integrated quality management (total quality management (TQM)), risk is defined as "a common factor likelihood of the risk and their consequences" [12]. Previously, the basic methods of reliability analysis, as a component of a broader notion of safety, were based on the conservative concept of "absolute safety," which is not adequate to the stochastic nature of the occurrence of failures and disturbances of exploitation, caused most often by changes in exploitation conditions.

On the other hand, in order to avoid the usual differences between the set reliability requirements and their dependence on the fulfillment of operational requirements, special attention should be paid to defining analytical terms and numerical values of the reliability parameters. To accomplish this task, it is necessary to create an appropriate database, linked not only to the system as a whole, but also to the system components, as the basic links in the reliability chain. The failure rate of some of the components of the system depends on many factors (mechanical and thermal overload, environmental impact, operating conditions, method of repair or replacement, human factor influence, etc.). In doing so, the reliability assessment, depending on the purpose and phase of the life cycle of a complex system such as a steam turbine within a thermal power plant or a nuclear power plant, is in principle realized in three basic ways:

- reliability assessment on the principle of similarity of equipment, based on its typing or retrospective analogue information, with correction for new prediction design conditions,
- reliability estimation by the component enumeration method or so-called "rough" calculation of reliability, with the formation of appropriate statistical methods and logical-stochastic models, as well as estimates in incomplete information determination and reliability estimation by the stress analysis method, or so-called "fine" calculation of reliability (characteristics of possible relationships between operating parameters and load),
- estimation of the likelihood of durability parameters and possible deviations of structural elements, expert correction of the characteristics of durability, and resources of details, with the participation of adverse impacts.

The intensive development of stochastic security analysis methods has resulted in the formulation of a set of probabilistic security analysis methods for technical systems. There are different ways of realizing these methods, which, when viewed as a complex whole, can be classified as follows:

- the use and reduction to the model of the probability of failure, frequency, and their duration, which corresponds to the laws of the "switching" Boolean algebra with two basic conditions: full working ability or complete failure,
- methods based on the use of Markov or semi-Markov security models, characterized by multiple states (including reserve states) and the time dependence function of the probability of the condition,
- use of the Weibull distribution, for both the elements and subsystems and the system as a whole and its testing.

Fatigue of the material due to the high temperature and pressure over a long period of time (fatigue over time) is the root cause of the service life of the structural parts of the turbine plant. Certainly, fatigue also occurs as a consequence of their spread over the course of their operation (fatigue in change of spread). Stationary loads from vapor pressure, stationary temperature differences, stationary external loads, existence of centrifugal force, and absolute temperatures during operation—are elements of fatigue due to time (temporal involvement). On the other hand, thermal loads (especially in nonstationary modes of operation: start, stop, change of power), with the participation of the jump load by steam pressure, represent the so-called variable participation. The final degree of fatigue (exhaustion) is obtained according to the hypothesis of linear superposition of damage according to the extended Palmgren–Miner's rule [14]:

$$E_{iscrplj.} = \sum_{ij} \frac{t_{ij}}{t_{Bij}} + \sum_{k} \frac{N_k}{N_{Bk}} = E_{ps} + E_{tl} = E_{ps} + \left(E_{termal}^{start/stop} + E_{jl}^{steam.press} + E_{termal}^{var.power} \right), \quad (7.3)$$

where: t_{ij} is the sum of the operating hours of the part under voltage and temperature; t_{Bij} is adequate time limit of break due to creep; N_k is the number of load cycles with equal stretching ranges; N_{Bk} is the appropriate number of load cycles until cracks occur; E_{tps} is the temporal participation or share; E_{vtl} is the variable participation or share; $E_{termal}^{start/stop}$ is the proportion of thermal load at start or stop; $E_{jl}^{steam.press}$ is the share of jump load by pressure of steam; $E_{termal}^{var.power}$ is share of thermal load due to power change.

Besides, the fatigue limit values E_{lim} are defined as:

- $E_{lim} = 1$—is for the upper surface of the shaft and other structural parts, with uneven stress distribution;
- $E_{lim} = 0.5$—is for the center of the shaft and other structural parts, with a uniform distribution of voltage.

For the operation of steam turbines with the associated equipment to be successful, it is necessary that its vital components are characterized by high reliability in operation. Starting from a common base diagram for all elements of Phase I and III thermal power plants given in Figs. 7.6 and 7.8, a test program for the Phase I turbine subsystem is defined, followed by data processing and analysis, as well as correlation with the phase I data, which determines Phase II for a turbine plant different from the boiler plant within the power unit of the thermal power plant (Fig. 7.7).

The scheme of estimation of the remaining working life according to the recommendations of the manufacturer Siemens AG UB KWU–Mülheim an der Ruhr is shown in Fig. 7.9. Thus, the composition of the drive data must be adapted to the accompanying KWU questionnaire.

When reaching the values E_{lim}, the following control tests are predicted: crack testing (testing of external surfaces visually by endoscope, etc.; depth determination of cracks; and fault localization—Gill method, ultrasound, casing for housing, etc.), measuring of oval shape, and sampling for material testing with methods with destruction.

In order to provide the required reliability in the operation of steam turbines and to avoid possible downtime due to the failure of some of its components, it is necessary to carry out certain maintenance tasks (audits and overhauls) and to monitor the behavior of the material, i.e., monitoring during the operation of the turbine.

The designed service life of the turbines, previously used in the calculations, was 100,000 to 150,000 h, while today it is calculated with a value of 200,000 to 250,000 h (improved materials, more accurate calculation procedures for the structural parts of the turbine, improved and new test methods, as well as the introduction of new criteria and procedures for evaluating embedded material). Extending the service life requires additional assessments due to the need to change the operating mode from basic to peak loads and because of increased requirements regarding the degree of utilization (replacement of worn parts, change of operating mode, setting new intervals in the maintenance

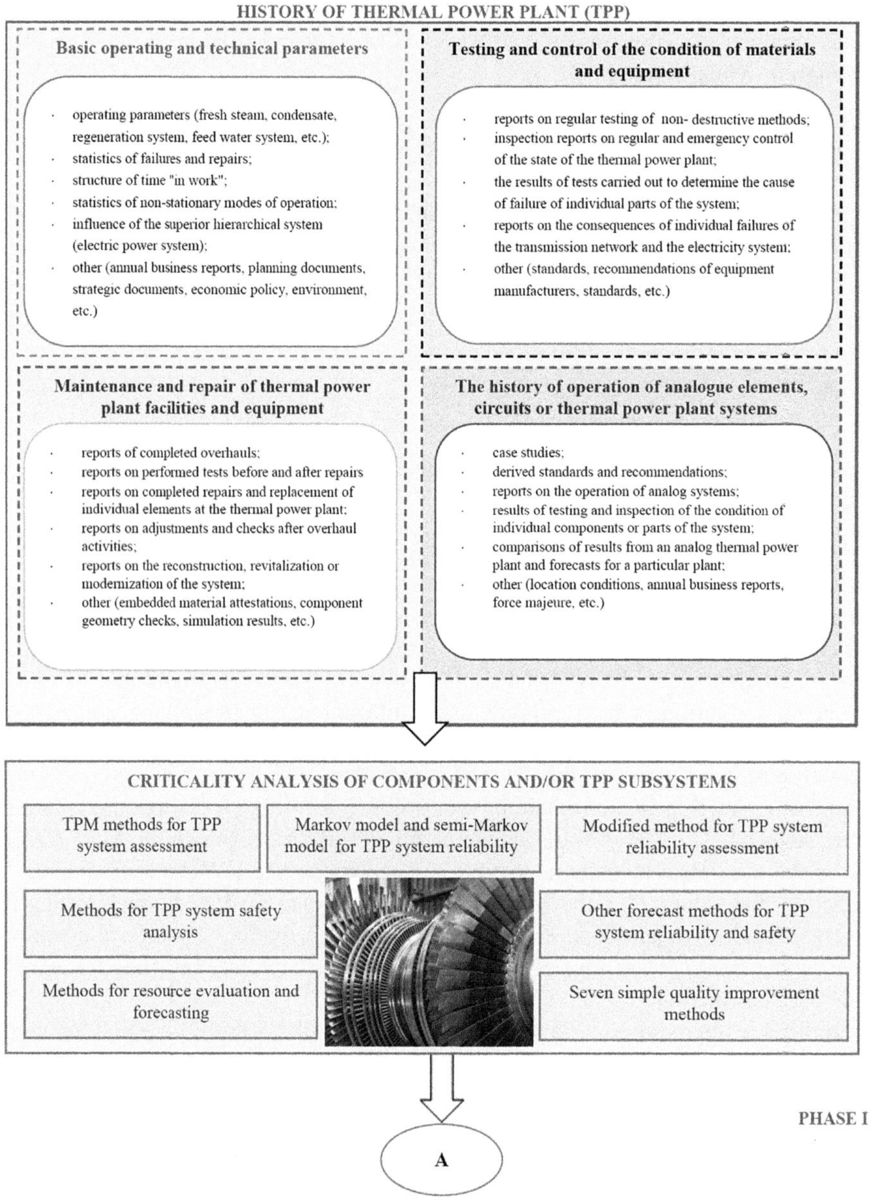

FIGURE 7.6

Display of the level of collection, systematization, and processing of design data and data on the previous history of the thermal power plant, Phase I.

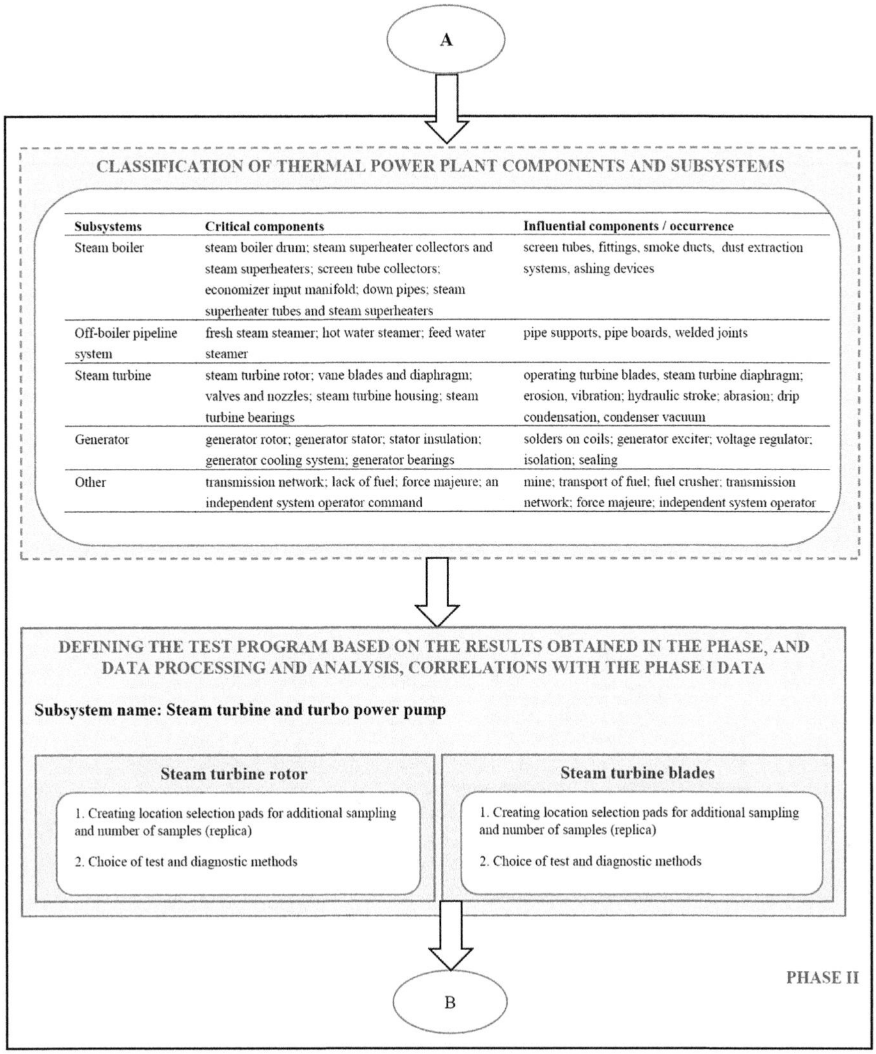

FIGURE 7.7

Definition of test program for Phase I turbine plant subsystem, data processing and analysis, correlation with Phase I and Phase II data.

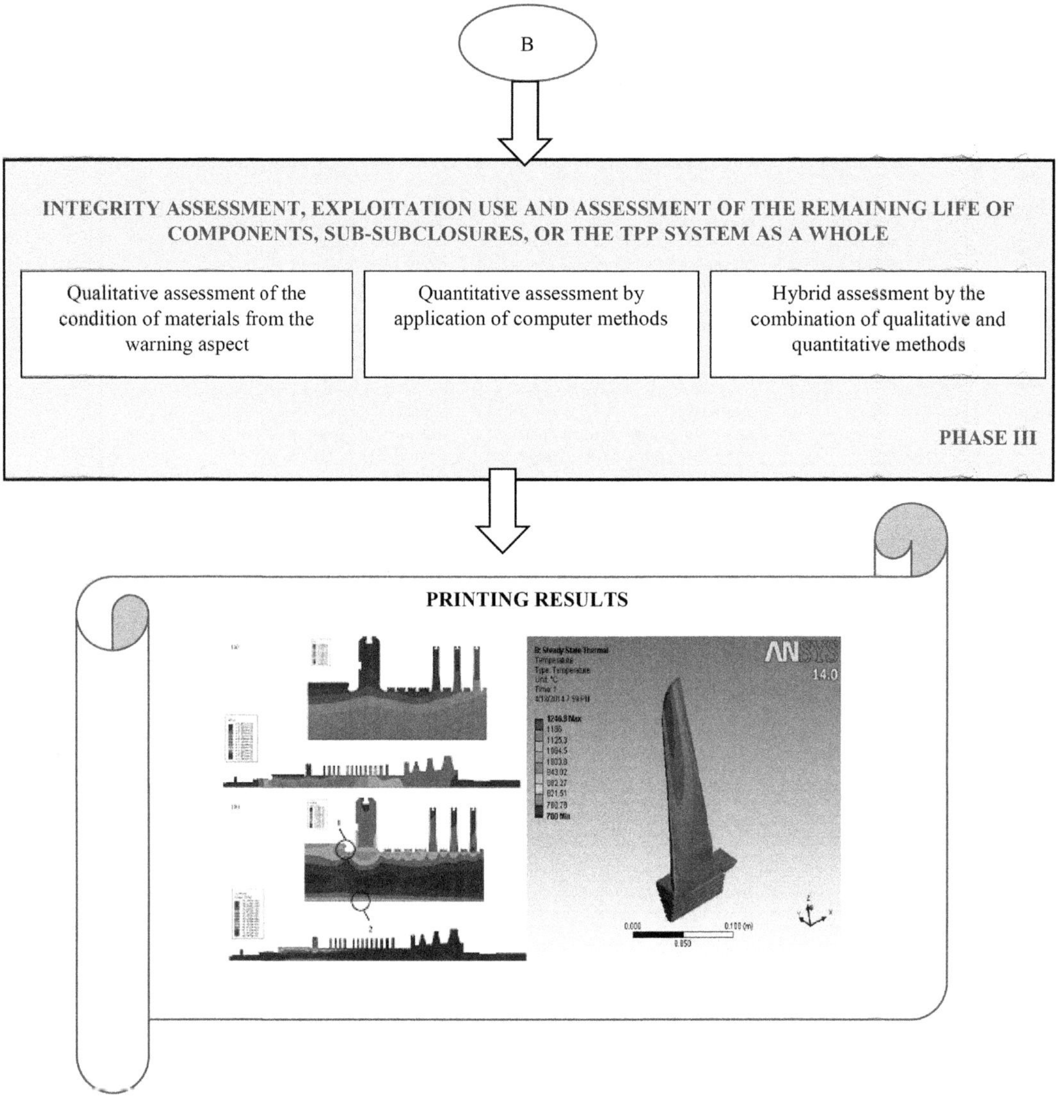

FIGURE 7.8

Assessment of integrity and exploitation usability, phase III.

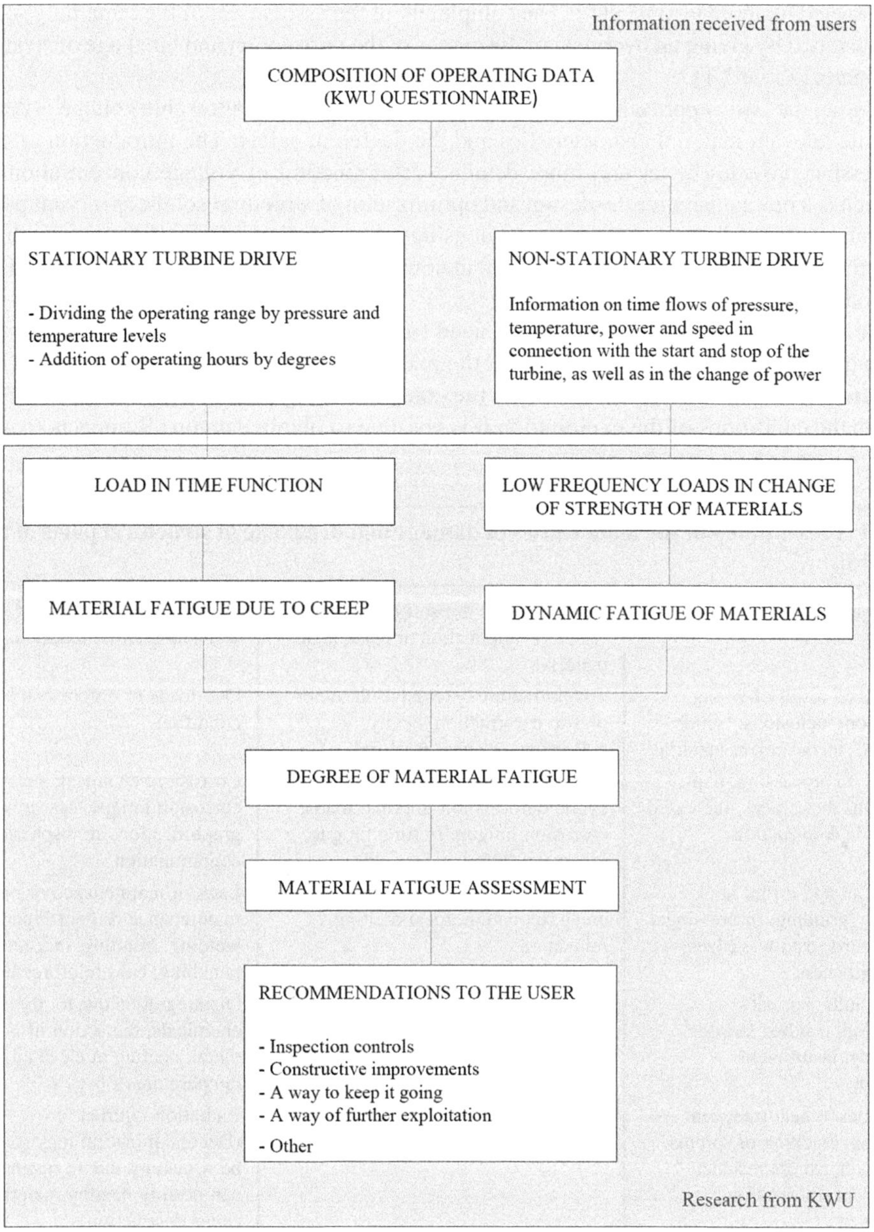

FIGURE 7.9

Scheme of determining residual working life according to Siemens AG UB KWU—Mülheim an der Ruhr.

process, targeted material testing, etc.). The complexity of the assessment of the remaining service life is often illustrated by giving an overview of the causes of the destruction and breakage of steam turbine structural parts (Table 7.1).

In doing so, the basic approach to budgeting does not change. The allowable voltage is determined based on the relevant material characteristics and the degree of safety. The introduction of computer data processing provides better and more detailed determination of voltage concentration (voltage state), which is a prerequisite for the design and optimization of structural solutions of vital parts of the turbine plant (construction of multiarmored housings, construction of welded rotors with reduced voltage at the same diameter, for better behavior at nonstationary thermal loads, determination of blade loads—model of corrosion fatigue life, etc.).

A large number of influential factors that need to be taken into account when assessing the situation relate to the conditions of exploitation, the material, and the calculation procedures selected. It is of particular importance to have data for the commissioning period or a recorded "zero" initial state. From the conditions of the exploitation it is possible to obtain data on all impacts to which the

Table 7.1 Presentation of the main causes of damage and breakage of structural parts of turbine plants [16].

Destruction as a result of processing errors	Destruction as a result of incorrect structural application or inadequate material	Destruction as a result of worsening exploitation conditions
Mistakes as a result of wrong composition: inclusions, "crude impurities," inconvenient material	Tough fracture (excessive elastic or plastic deformation, cavity collection, or shear fracture)	Overloads or unforeseen load conditions
Errors due to processing: flaps, folds, seams, hot cracks, increased local plastic deformation	Fatigue fracture: cyclic loading, cyclic deformation, thermal loads, corrosion fatigue, fretting fatigue, low-cycle fatigue	Corrosion: chemical, stress, corrosion fatigue, cast iron graphitization, atmospheric contamination
Irregularities and errors in machining, grinding, or pressing: grooves, burrs, cavities, edges, cracks, brittleness	High-temperature destruction: creep, oxidation, local melting, relaxation	Lack of maintenance or poor maintenance or poor reparation: welding, grinding, inaccurate hole punching, cold reinforcement
Welding faults: porosity, undercutting, residual stresses, nonwelding, insufficient penetration	Delayed static fractures: hydrogen brittleness, caustic brittleness, slight crack growth stimulated by aggressive environment	Disintegration due to: the effects of chemicals, the action of liquid metal, coating at elevated temperatures
Irregularities in heat treatment: overheating, cracking of sprouts, grain growth, excess residual austenite, decarburization, deposition	The expressed stress concentration inherent in the structure	Radiation injuries (Decontamination may sometimes be necessary due to testing, which can destroy healthy material and cause destruction)
Surface hardening errors: carbide separation along grain boundaries, soft core, wrong thermal cycle	Incorrect voltage analysis or inability to calculate the voltage in the complex part	Accidental (unexpected) circumstances: abnormal operating temperature, strong vibrations, thermal shocks
Errors due to surface treatment: cleaning, coating, chemical diffusion, hydrogen brittleness.		

material was exposed during the exploitation, especially from the aspect of the realized turbine modes, material time characteristics and stress distribution across the component, unexpected material behavior in exploitation, and occurrence of failures (hidden material errors, residual voltages, poor maintenance, etc.).

It should be noted that newer turbines have built-in special systems that prevent uncontrolled start and unwanted power changes. Also, in the design phase, the expected number of starts has already been evaluated, with the projection of the material behavior in the first 150,000 h and 250,000 h, respectively. On the other hand, procedures for testing nondestructive methods have been developed, which very reliably determine the existence of faults even before the turbine plant is commissioned (ultrasound, radiography, magnetic particles, acoustic emission, holography, endoscopy, etc.).

6. Technical diagnostic methods for steam turbine within the complex of thermal power plant technical system

The term technical diagnostics comes from the Greek word diagnosis, which means recognition, conclusion, and evaluation (grading). The first explanations for this term date from the field of medical sciences. Thus, in Webster's vocabulary, diagnostics is defined as the act or process of deciding a problem or disease through symptom examination, careful evaluation, and fact analysis in an attempt to understand or explain something, a decision or opinion formed on the basis of an assessment, or a brief scientific explanation for a taxonomic classification.

Technical diagnostics of steam turbines with overfitting equipment represents all activities performed to evaluate the current state or to provide a forecast of the behavior of a steam turbine system over a period of time. Thus, it uses all available algorithms, rules, and models necessary to determine the state of the system, with a view to predicting a malfunction in a timely manner. This increases the reliability, availability, and efficiency of the steam turbine plant with associated equipment. Since there is still no general concept of forming a diagnostic system at thermal power plants, nor a complex diagnostic supply of thermal power plants in our region, it is necessary to point out the following:

- technical diagnostics is a significant tool for increasing the reliability, economy, and safety in the operation of the steam turbine system with its associated and basic equipment;
- the greatest effect of the application of technical diagnostics is obtained by its harmonization with the methods for the short- and long-term reliability forecasting and its optimization, most commonly by economic criteria;
- basic tasks of technical diagnostics on steam turbine within the system of thermal power plants are usually formulated as:
  - **(a)** predictions and measures to prevent accidents,
  - **(b)** reducing the number and duration of outages through timely forecasting,
  - **(c)** detecting and monitoring the development of the cause of the failure,
  - **(d)** shortening the scope of planned and unplanned overhauls at the expense of improving and applying the methods of technical diagnostics,

(e) prevention or elimination in the process of exploitation of working conditions, which are the generator of damage and occurrence of failures, and
(f) Computer-aided monitoring of the workforce and production efficiency of the steam turbine system within the thermal power plant.

The application of technical diagnostic methods and the suitability for controlling the condition of a steam turbine significantly affect the suitability of maintenance as well as its internal characteristic (its units or elements), or the state of functionality under defined conditions for a specified period of time, assuming that maintenance is performed in accordance with planned and prescribed procedures. This impact is reflected by certain factors, of which the following should be noted: the dynamics and characteristics of failure occurrence, the ability of the components of the steam turbine system, and the thermal power plant system as a whole to maintain working capacity for purpose and safety, with the possibility of assessing the condition of the elements and identifying the causes of failure occurrence.

The possibility of testing and inspection of the system or its integral part (element) in the process of control of the state of the steam turbine system is already planned at the stage of development and design by specifying the elements for performing the main and auxiliary functions at the level of projected sizes (with tolerances), as well as safety elements, occupational safety, fire protection, and environmental protection. Each of the procedures for evaluating the technical system of a steam turbine, comprising the analysis of the test object, the setting of theory and test methods, the development of appropriate algorithms of the test program, and the determination of ways and means for studying the specific properties and characteristics of the technical system as a whole, its specificities affect in various ways the suitability maintenance. Of particular importance is the consideration of the test mode itself within the process of exploitation of this system or at the time of interruption of operation, the way of its realization, and the degree of automation of the database (possibility of application of information technologies).

The development of technical diagnostics on steam turbines went in the direction of realizing the functions that the turbine should provide. Checking the correctness, working ability, and functionality of the turbine plant, along with locating the failure site at the lowest hierarchical level, are the elements on the basis of which the remaining service life or the trend of failure is estimated.

Significant economic effects and reduction of operating costs through timely detection of possible causes of failure of turbine plant components can be achieved through the application of methods and technical diagnostics. In this case, predicting and defining the causes of failure can be achieved during the operation of the turbine plant itself or within the downtime and time for overhaul of the plant and equipment, so that exploitation (operating mode) and overhaul (stationary mode) technical diagnostics are distinguished as integral elements of maintenance according to the state of the turbine plant within the power plant, as a higher hierarchical system. Significant application of technical diagnostics is also in giving the short- and long-term reliability of a steam turbine system with associated equipment and its optimization, most often by economic criteria.

The importance of applying diagnostic methods in the enclosed space, in terms of enhancing the safety of such facilities, is the timely detection of failures on equipment in the area affected by the explosion, with the aim of preventing the occurrence of major accidents, which can further cause the ignition of the explosive atmosphere. In doing so, any overheating of the equipment or part of the equipment is a sign of an error or failure. The most common diagnostic methods that can be used on both electrical and mechanical equipment are certainly vibration and infrared diagnostics. Both of these methods belong to the nondestructive testing (NDT) technique.

6.1 Technical diagnostic methods for the analysis of the condition of steam turbines

Methods for technical diagnostics of steam turbine plants in operating mode include:

(a) Vibrodiagnostic methods, of which the following are distinguished:
- analysis of the total vibration level as a change in the mechanical or process state of the rotary machine,
- spectral Fast Fourier Transform (FFT) and Discrete Fourier Transform (DFT) analysis,
- phase analysis of phase angle of vibration,
- analysis of nonstationary signals or line analysis at variable speed or when analyzing a steam turbine system when starting or stopping,
- vector analysis in nonstationary or stationary modes of operation of the turbine plant,
- analysis in logarithmic scale for detection of defects at the very stage of emergence,
- high-frequency detection (HFD) of total vibrations of high spectrum from 5 to 60 kHz,
- low-frequency detection (LFD) of very low-frequency defects for determining the eccentricity of large turbine rotors,
- orbital or dual analysis for analyzing the operation of sliding turbine bearings,
- SKF's analysis for the detection of failure of bearings and gears in the early stage of formation, as well as for the detection of cavitation, friction, cracks, electrical problems, etc.,
- rolling-element bearing activity monitor (REBAM) analysis for monitoring and evaluation of the condition of rolling bearings on auxiliary turbine equipment,
- operating (for virtual machine movements) and classical (obtaining dynamic characteristics of machine parts) modal analysis,
- trend analysis for online monitoring and diagnostics;

(b) noise monitoring and analysis, by measuring noise levels to determine human and environmental impacts, then analyzing noise emissions from a sound source to determine the state of the machine (total noise level analysis, frequency sound pressure analysis, acoustic emission analysis);

(c) monitoring and analysis of the displacement and propagation parameters of individual elements of the steam turbine, such as the analysis of the position of the rotor sleeve in the bearing, the analysis of relative displacement or axial displacement of the shaft, the analysis of the relative elongation of the shaft, the analysis of the absolute elongation of the turbine housing, the analysis of the eccentricity of the shaft, etc.;

(d) monitoring and analysis of technological (operating) parameters: speed, flow, pressure, temperature, differential pressure on the filter, as well as other parameters, which are the result of technological process of transformation of potential steam energy via kinetic energy into mechanical operation of rotation;

(e) analysis of the temperature field, with the aim of determining the temperature differences in the coupling and connection joints (thermovision), in order to determine the overheating and detection of electrical problems (poor grounding, poor cooling, damage to insulations, phase imbalance of the three-phase system, etc.) and mechanical (malfunction or shaft curvature, valve leakage, etc.) origin;

(f) analysis of the quality of turbine oil and particulate matter through:
- analysis of the physical and chemical properties of the oil (appearance, color and odor, acidity, base number, viscosity, viscosity index, oil water content or percentage of oil drainage, resistance to oxidation of turbine oil),
- analysis of oil contamination through internal contamination (as a result of different types of wear—abrasive, adhesive, cavitation, corrosion, fatigue of the material, etc.) and external contamination (impurities, water, oxidation products, etc.), using:
  - spectrometric analysis (analysis of particles in oil formed by wear),
  - infrared analysis of the physicochemical characteristics of the oils and impurities,
  - particle counting technique, usually in conjunction with ferrography, i.e., analysis of metallic particles in oil for analysis of impurities in oil samples and their ranking by size from 5 to 200;

(g) analysis of the current signal for evaluation and monitoring of the condition of the generator and auxiliary electric motors within the turbine plant and its auxiliary equipment (spectral analysis of the current signal for determining the geometric deviation in the electric motors; detection of damaged rotor rods, damaged stator winding insulation, etc.; analysis of the developed current signals for damaged rotors in electric motors, etc.);

(h) partial discharge analysis or PDA—analysis for monitoring the insulation quality of stator windings on large rotary machines such as steam turbines, generators, electric motors, etc.;

(i) magnetic field analysis for determining the deviation of the rotor axis with respect to the stator and determining the condition of the coils by individual phases on the generators;

(j) analysis of combustion products for internal combustion engines;

(k) analysis of the working medium, through the analysis of water quality at the inlet of the steam boiler and the analysis of condensate at the inlet of the degasser;

(l) corrosion, erosion, and abrasive deposition analysis (Table 7.2).

In stationary diagnostics, the steam turbine must be stopped and dismantled, with each component of the turbine plant and associated equipment separately tested. It is most often performed at major downtime (during the overhaul), with:

- control and analysis of the condition of the materials used for the individual components of the turbine plant and its associated equipment, with sampling of the material in order to determine its structure and mechanical properties;
- optical testing for inaccessible parts (endoscopy, magneto flux, penetrate testing, etc.);
- determination of the voltage state in certain parts of the construction of the steam turbine and its associated equipment (tension metric analysis);
- to determine the condition and characteristics of electrical machines within the turbine plant's auxiliary equipment, special test platforms and transformer shock tests are used.

The major problem with the introduction of technical diagnostics in a complex technical system (such as steam turbine plants), in order to maintain the required level of technical condition and estimate the optimum reliability of work for the future (short- and long-term forecast), is the choice of

Table 7.2 Comparative presentation of materials, type of corrosion, and possible means for monitoring the condition of the steam turbine system.

Construction material	Kind of a corrosive process	Steam turbine monitoring device
Carbon and low alloy steels	Hydrogen brittleness at low and high temperatures	- Hydrogen sensors; - Hydrogen meters
	A significant increase in magnetite at higher temperatures	
	Corrosion due to fatigue of the material	- Analysis using a sound transmitter
Stainless and high alloy steels	Cracking, crack corrosion	- Electrochemical sensors
	Rupture	- Magnetic measurement permeability
Copper and copper alloys	Local corrosion, crack corrosion	- Electrochemical sensors
	Corrosion, erosion	- Measurement of eddy current
Aluminum and aluminum alloys	Local corrosion, hole corrosion, crowning, and crack corrosion	- Electrochemical sensors

methods and instruments for technical diagnostics, given to the aspects of objectivity, uniqueness, repeatability, and specific operating parameters (operation with high values of pressure and temperature close to material fluidity).

Since most technical diagnostic methods are limited to the widespread presence of fault detection methods, the following will be proposed and their classification in terms of their possible applicability to steam turbine components as a power plant subsystem.

Complex technical systems have all the classic methods and diagnostic tools at their disposal, while only a small number of them are specially developed for the diagnostics of equipment at turbine plants within power plants. The formation of a diagnostic system on these systems generally follows their hierarchical level of organization, including the establishment of the lawfulness of changing the system state parameters and the processes that take place in it. The system of technical diagnostics is essentially an accompanying system within these facilities, which includes the selection of diagnostic parameters and establishing their relationships with the system state parameters, characteristics of their change, norms, as well as determining possible state assessment and diagnosis for the system at any given time.

Controlling the current state of the system or its elements (usually the most critical when a failure occurs), defining the lawfulness of a failure in time based on a database, and predicting the behavior of the system in the future are directly related to the development of diagnostic tools and the development of diagnostic devices. Increasing the complexity of technical systems, increasing the dependence of human work on the reliability of these systems, with increasingly stringent requirements for the quality of the implemented processes and safety and environmental protection, implies the application of information theory, the establishment and study of failure methods, and the use of computers in the diagnosis and processing obtained data.

6.2 Maintenance control of current status by condition

The result of any condition control must be a decision on the continued usability of the constituent element or the system as a whole (the element for re-installation, the element for repair and re-installation or the element must be withdrawn from further use). In order for any decision to be made at all, it is necessary first of all to know the permissible limits of wear and then the other conditions necessary for its operation (Fig. 7.2). The limit of wear is, in fact, the boundary between the operational usability of the element or system as a whole and its damage.

Maintenance models by condition are usually classified into two groups: models of current change of technical condition (with application of condition inspection) and models of gradual change of condition (with application of condition control). Inspection control usually only determines the state of the system at a particular moment of exploitation, that is, the definition of the state of the system (one of the states: "in failure" or "in operation"), while the state control is tasked with defining the parameters of the technical system, i.e., determining in what phase the system is in between the previous two extreme states. Thus, maintenance according to state is a very successful way of managing the maintenance processes of technical systems, that is, preventing the occurrence of failures and restoring each system individually (parameter control) or homogeneous group of systems (reliability level control), while providing the maximum system operation time and minimum cost.

From all the above, it is easy to see that technical diagnostics, as an integral part of the maintenance process according to the condition, should determine the technical condition of the system elements within a precisely defined time and with precisely defined limits of accuracy. This can be achieved by using appropriate instrumentation or by sensing the senses of a diagnostic specialist.

Longevity, reliability, and condition control represent three parameters that are related to each other, with whose knowledge we have a fully defined system state. Diagnostics, as a science, deals with the recognition of the technical state of a system, with certain accuracy at a certain point in time. Thus, it uses algorithms, rules, and certain models, which assume the existence of formal descriptions of system elements and their behavior when in operation and in failure (mathematical model of diagnostics). As a result of the technical diagnostics, data are obtained to determine and maintain the operating status, as well as their controls. This implies knowledge of certain criteria of permissible and impermissible conditions.

As mentioned earlier, any change in the state of the system during exploitation is a process of random character, whereby the moment of transition of the correct to the defective state occurs as a conditional failure, that is, characterizes the beginning of the occurrence of a fault. In the general case, when looking for a fault, a test diagnostic (characterized by the possibility of giving special effects) and functional diagnostics (used to check the working ability of system components in the process of their operation) are distinguished.

Measurement methods for determining the technical state of the system, which are a set of special procedures for defining the relationships of some measured quantities, can be absolute (instantaneous reading of the absolute value of a measured size) and realistic (determining the relationship of a measured and predefined other size) direct measurement methods (direct reading of the measured size) and indirect (computational or other determination of the measured size based on the read value of the measurement) measurement methods, contact (measuring instruments in direct contact with the measured medium) and contactless (without physical contact of the measuring instrument with the measured medium), or differential (in which certain sizes are measured separately, and the conclusion about the required measure is made on the basis of a large number of measurement results) and complex (with simultaneous reading of data on multiple parameters) measurement methods.

6.3 Defining the legality of steam turbine failure

The steam turbine failure is defined as the cessation of the ability of any element of the turbine or turbine as a whole to perform the functions for which it was designed. Reduction or loss of working capacity of the technical system during exploitation is due to the effect of various factors (built-in, accidental, or temporal), which change the initial parameters of the system, thus causing different levels of damage. There are several categorizations of cancellations according to the selected criterion. The development of a failure can be viewed through a malfunction (condition change monitoring in time interval) and through a failure (monitoring of the number of those failures over time). In doing so, developed methods are used for monitoring the phenomena with modeling of distribution, all with the aim of obtaining a mathematical model of changing the state of the technical system, thus creating the necessary assumptions for the possibility of making diagnostic decisions. The determination of mathematical models is carried out by a deterministic and stochastic method.

Causes of steam turbine failures and accidents are most commonly divided into three groups: systemic (built-in), random and monotonous or temporal causes of failure. In the initial period of operation of the steam turbine usually occur systemic or built-in system failures. These failures can be: incorrect design (poor construction with defects in construction, errors as a result of incorrect material selection, etc.), design reliability, occurrence of design voltages, manufacturing errors, assembly errors, adjustment errors, heat treatment and residual thermal stresses, failures of technical controls (insufficient control and testing), etc. On the other hand, accidental causes of failure include: unstable environmental conditions, overload or unstable (especially nonstationary) modes of operation (instability of technological parameters), poor handling and maintenance, inadequate control, and instability of structural parameters (load gradient, etc.). Monotonous or weathering causes may include: exploitation mode, maintenance mode, lubrication, fatigue of materials (fatigue processes and changes in material properties), warming, erosion and corrosion of parts, wear of embedded parts, contamination of working medium, regulation and concentration of parts, etc.

6.4 Defining the lawfulness of steam turbine failure

Each handover of both newly constructed and revitalized turbine plants requires certain warranty tests to determine the actual values of the operating parameters and the improvements obtained in the event of the revitalization and modernization of the steam turbine and its associated equipment. This indirectly leads to the conclusion about the quality of the contracted work performed from the aspect of achieving the required operating performance of the steam turbine. Also, standardized normative measurements are occasionally performed, which are an integral part of the monitoring of the steam turbine in operation, in order to determine the value of their degree of usefulness (baseline for the standards in operation). The tests performed can sometimes be a consequence of the orders and instructions received by the thermal inspection, when performing their regular and extraordinary inspections.

Guarantee and normative tests on steam turbine plants are defined by appropriate standards (DIN 1943, GOST 18322-78, and other regulations [21–27]), which define in more detail the operating and other conditions under which the tests are performed, as well as the measures and procedures given for each of the stages of tests:

- preparatory phase for testing (termination of activities, selection of appropriate measuring equipment, and preparation of the place on the steam turbine at which the tests are carried out);
- the test implementation phase, with providing of the standard required operating conditions necessary to start measuring, measuring, and archiving the operating parameters measured values;
- the phase of systematization of the data, their processing, analysis of measurement uncertainty and permissible error thresholds, and analysis of the resulting results, with a conclusion containing their interpretation.

Guarantee tests for new steam turbine plants within a higher hierarchical system (thermal power plant, power plant, boiler plant or nuclear power plant) is performed within six months after the trial operation of a newly built or revitalized steam turbine plant.

In case of delay in the performance of warranty tests, due to the inevitable changes that occurred during the exploitation on the surfaces of the movable and fixed blades, certain corrections of the results obtained for specific steam consumption and specific heat consumption should be made. In most cases, unless otherwise specified in the contracts for the delivery or works on the reconstruction, revitalization, and modernization of the turbine plant, a reduction of 0.1% is made for each month of operation, within the first 8 months after the expiration of the foreseen deadline, or 0.06% for drive each month over the next 12 months, and after expiry of the previous 8 months [28].

The conduct of normative tests has a periodic character. The results obtained from these tests allow the formation of an image of the actual technical condition of the plant before periodic overhauls, thus creating the necessary preconditions for the most reliable planning of the necessary overhauls (preparation of contracts with specialist maintenance companies, timely procurement of spare parts, consumables and necessary specialist equipment, and similar). If these tests are carried out after the overhaul, the results obtained will be the basis for evaluating the quality of the performed overhaul works and for assessing the cost-effectiveness of the plant in the coming period (usually until a new capital overhaul).

7. Measures to reduce damages and increase reliability of steam turbines

Functions to be fulfilled by the system of technical diagnostics are given in the form of certain checks of the technical state of the system, checks of working ability, checks of functionality, location of failure sites at the lowest possible hierarchical level, as well as giving estimates of the remaining time of use or the trend of malfunction occurrence. Undoubtedly, the use of technical diagnostics has opened up new possibilities for managing power plants, thus creating all the preconditions for a significant reduction of corrective and preventive maintenance activities, while maintaining the same or achieving even higher levels of reliability of the plant as a whole. This is of particular importance for energy-processing plants operating within a higher hierarchical system (power system, oil industry, petrochemical plant, etc.) and for which a high level of reliability is required during exploitation and increased protection for both personnel and the environment. To increase the reliability of the turbine, the following measures and recommendations are necessary [29]:

- operation of the turbine with increased fresh steam temperature above the budgeted (nominal) value should be avoided (due to reduced creep speed, several hours of operation during the year with a fresh steam temperature higher than nominal by 10–20°C are allowed);
- due to the possibility of a water impact, the sudden drop in fresh steam temperature below the permissible limit should not be allowed;
- the vapor pressure in the control stages should not exceed 10%–15% of the nominal value for a given amount of steam (a gradual increase in the pressure in the control stages indicates a decrease in the flow cross-section of the stator and rotor blades due to salinity);
- for a given cross-section of the turbine, the permitted upper level of difference of temperature between the lower and upper housings must not exceed 30–50°C (in the case of larger differences in temperature, the case and the rotation of the rotor part in the stator part of the turbine are usually bent);
- the vibration amplitude of the turbine bearings at a speed of 50 s^{-1} should not exceed 15, while the vibration speed should be less than 4.5 mm/s;
- the insertion of sheets of labyrinth seals into the impeller as well as its greater deflection of 0.2 mm (manifested by the occurrence of metallic noise or sparks and resulting in severe damage to the flow part of the turbine) should not be allowed;
- control the intrinsic frequency of the rotor blades in order to prevent its routing by the rows of rotation speed of the turbine (even up to the eighth row, since in the case of long and slender blades with low damping and the eighth rope it can cause cracks and vibration fatigue and fracture of the blades);
- the rotor speed must be controlled (an increase in speed indicates a malfunction of the control system, which may cause the blades to break due to the increase in centrifugal force);
- the axial displacement of the rotor should remain within the permissible range (a larger displacement indicates the wear or melting of the white metal of the axial bearing segments);
- the relative displacement of the rotor relative to the stator must also not exceed the permissible value (the relative position of the rotor and stator changes due to their uneven heating when the turbine is started or when the flow or temperature of the steam is changed);
- a vacuum in the condenser is not recommended less than 0.07 MPa (reducing the vacuum increases the outlet temperature of the turbine housing and changes the relative position of the rotors and the TNP housing, which can cause them to radically interfere);
- the pressure drop in the rear stage of the pressure turbines should remain within the permissible value (rear stage overload may cause diaphragm breakage);
- the oil temperature at the exit of the bearings should not exceed 80°C (at higher temperature turbine oil is more prone to oxidation, with a decrease in viscosity, which again reduces the size of the minimum gap in the bearing);
- for elastic turbine rotors, it is necessary to install the sheets of labyrinth seals in the rotor and to perform ridges on the segments of seals installed in the stator part of the turbine (this design does not lead to permanent deformation of the rotor with increased vibrations and scraping);

- in this way, the introduction of maintenance according to state, with the accompanying application of technical diagnostics and the correct determination of the remaining service life (reliability management), can reduce the number of failures of the steam turbine system. Of course, this should be accompanied by the application of computer technology, as well as databases at both the power plant and the EES level.

8. Result discussion

The assessment result of steam turbine system condition within the condensation power plants is defined and ranked marks with a certain degree of reliability and safety of the correctness of the assigned marks. Based on the condition assessment, a decision on maintenance activities is made in the coming period, so that the main purpose of maintenance is to be implemented: lifting the state of technical steam turbine system at higher level in relation to the level of state defined as critically bad (condition evaluation can be: bad, good enough, average good, very good, and excellent). These ratings correspond to the usual numerical grades from 1 to 5. In order to successfully evaluate the state of a technical system, it is necessary to establish a certain hierarchy among the attributes or parts of the system. If the influence of some attributes cannot be determined we can leave the appropriate degree of uncertainty in the assessment of state attributes, and with that to the whole system (the difference missing in the assessment of the first attribute represents the degree of uncertainty, that is, indeterminacy).

Calculation methods that deviate from the classical orientations in the theory of reliability, where the distribution of failures is not explicitly given, but are based on their dependence on the system of planned overhauls, after which an assessment of the existing situation and the possibility of exploitation for the next period are made. Further progress in improving the reliability assessment, except adjusting the classical methods to the specifics of a complex technical complex, lies in the need to shorten the test time of one or more factors through the selection of an optimal short test plan, the automation ("*online*") of reliability assessment procedures, and its optimization based on selected criteria (most commonly economic criteria). It is also necessary, given the structure of the technical system and the reliability characteristics of the individual elements, to give a measure of importance and ranking of the elements from the aspect of rational allocation of resources while increasing the reliability of each of them. As a result of problem-solving, a list of criticality of the ultimate consequences (effects) of failure is established. Conditions that are necessary to have, to do lists criticality occurred, are knowledge of the working conditions of the system and its structure, and the possession of a data element failure [18]. It should be noted that the methodology of reliability assessment has advanced much in the field of electronics, while in the case of propulsion systems where heterogeneous technologies are represented (mechanical engineering, electronics, energy, etc.), it requires further study in terms of introducing other assumptions (establishing regular maintenance processes) with the introduction of diagnostics, organizing and collecting data on failures, and using existing statistical analyses. The objectives of reliability prediction, that is, the process of determining numerical values for the design ability to meet the set requirements of reliability, are: feasibility assessment, comparison of possible solutions, identification of possible problems, planning of supply and maintenance, identification of data gaps, harmonization in cases of interdependence of parameters, allocation of reliability, and measuring progress in achieving set reliability.

Conducted analyses of the implementation of the maintenance policy by condition versus classic maintenance over a constant time period showed an increase in operational reliability, an extension of service life, and a reduction in the cost of maintaining the turbines. Special contribution from the point of view of the progress made in the quality of maintenance and the availability achieved are the effect and potential reduction of the costs of operation and maintenance of this facility. The first stage of the gradual modernization of the maintenance of the turbine plants is the introduction of maintenance according to the state with parameter control, for which application (for an initial "zero" state) it is necessary to develop a monitoring method and technical diagnostics of the system status, as well as an adequate information system. It should be noted that newer turbines have a high degree of automation, which allows the application and introduction of new diagnostic methods with the existing information system. Further development of preventive maintenance of turbines is maintenance according to the condition with control of the level of reliability, as well as maintenance with the help of expert systems. The determination of the initial projected reliability, then the average level of exploitation reliability, as well as the legality of their change form the basis, on the basis of which further defines the current level of exploitation reliability and makes its prediction (forecast) in the future.

9. Conclusion

To reduce unplanned downtime, prevent damage, and increase reliability in the operation of the steam turbine, strict application of quality assurance regulations during the life of the steam turbine, from the stage of preparation and design to the end of operation and its decommissioning, is necessary. The realization of the projected reliability of the turbine requires strict application of the rules, procedures, and criteria for exploitation, as well as the good training of the operating staff that operates the steam turbine. During the period of exploitation, the condition of both the elements and the turbine as a whole is inevitable. The methodological presentation of the scientific prediction of the state or behavior of the steam turbine system is based on deterministic, stochastic, or most commonly combined deterministic-stochastic prediction methods. The modeling of system behavior is mostly based on specific operational research (for real conditions, with treatment of system functionality only), application of mathematical statistics methods (defining and selecting distributions, estimating observed parameters, test hypotheses, defining ranges, and evaluating characteristics), and application of probability theory methods (different mathematical models). In order to ensure the necessary availability of steam turbines and their reliable long-term operation, it is necessary to carry out adequate maintenance during the exploitation period. Traditional preventative maintenance at a fixed time, since certain elements of the turbine plant as a whole do not cancel in specific time periods, often resulted in the execution of some unnecessary overhaul activities and works, and sometimes delays in certain maintenance operations (occurrence of damage or damage during the interperiod). This has resulted in increased costs and loss of production. A far better approach to preventative maintenance is the introduction of a maintenance policy according to the condition, with the accompanying introduction of technical diagnostic methods and a maintenance information system based on regulated and updated databases.

References

[1] H.Z. Wang, H. Pham, Reliability and Optimal Maintenance, Springer, London, 2006.

[2] H.Z. Wang, H. Pham, Availability and maintenance of series systems subject to imperfect repair and correlated failure and repair, Eur. J. Oper. Res. 174 (3) (2006) 1706−1722.

[3] M. Bengtsson, Standardization issues in condition based maintenance, COMADEM, in: Proceedings of the 16th International Congress, August 27−29, 2003, pp. 651−660.

[4] R.C. Eisenmann, R.C. Eisenmann JR., Machinery Malfunction Diagnosis and Correction, Prentice-Hall, Inc., Saddle River, New Jersey, 1998.

[5] N. Majdandžić, Maintenance Strategies and Maintenance Information Systems, University of Osijek, Faculty of Mechanical Engineering in Slavonski Brod, Slavonski Brod, 1999 (in Croatian).

[6] F.A. Sturm, Efficient Operations, Intelligent Diagnosis and Maintenance, VGB Power Tech Service GmbH, Essen, Germany, 2003.

[7] T. Nakagawa, K. Yasui, Optimum policies for a system with imperfect maintenance, IEEE Trans. Reliab. 36 (5) (1987) 631−633.

[8] Y. Wang, H. Pham, Multi-objective optimization of imperfect preventive maintenance policy for dependent competing risk system with hidden failure, IEEE Trans. Reliab. 60 (4) (2011) 770−781.

[9] Y. Wang, H. Pham, Dependent competing risk model with multiple-degradation and random shock using time-varying copulas, IEEE Trans. Reliab. 61 (1) (2012) 13−22.

[10] M. Brown, F. Proschan, Imperfect maintenance, in: IMS Lecture Notes-Monograph Ser. 2: Survival Analysis, Inst. Math. Statist., Hayward, Calif., 1982, pp. 179−188.

[11] H.W. Block, et al., A general age replacement model with minimal repair, Nav. Res. Logist. 35 (5) (1988) 365−372.

[12] L. Papić, Z. Milovanović, Maintenance and reliability of technical systems, in: DQM Monograph Library Quality and Reliability in Practice, Book 3, Prijevor, 2007 (in Serbian).

[13] D. Blažević, Predicting Maintenance of Technical System by Condition Assessment, PHD thesis, Josip Juraj Strossmazer University of Osijek, Faculty of Electrical Engineering Osijek, Osijek, 2012 (in Croatian).

[14] Z. Milovanović, Optimization of Power Plant Reliability, University of Banja Luka, Faculty of Mechanical Engineering Banja Luka, Banja Luka, 2003 (in Serbian).

[15] Z. Milovanović, Modified Method for Reliability Evaluation of Condensation Thermal Electric Power Plant, Ph.D. thesis, University of Banja Luka, Faculty of Mechanical Engineering Banja Luka, Banja Luka, 2000 (in Serbian).

[16] Z. Milovanović, et al., Sustainable energy planning: technologies and energy efficiency, in: DQM Monograph Library Quality and Reliability in Practice, Book 9, Prijevor, 2017 (in Serbian).

[17] D. Miličić, Z. Milovanović, Monograph of Energy Machines - Steam Turbines, University of Banja Luka, Faculty of Mechanical Engineering Banja Luka, Banja Luka, 2010 (in Serbian).

[18] Z. Milovanović, Monographs: Energy and Process Plants, Tom 1: Thermal Power Plants - Theoretical Foundations, University of Banja Luka, Faculty of Mechanical Engineering Banja Luka, Banja Luka, 2011 (in Serbian).

[19] Z. Milovanović, Monographs: Energy and Process Plants, Tom 2: Thermal Power Plants - Technological Systems, Design and Construction, Exploitation and Maintenance, University of Banja Luka, Faculty of Mechanical Engineering Banja Luka, Banja Luka, 2011 (in Serbian).

[20] Z. Milovanović, D. Miličić, Steam Turbines for Cogeneration Energy Production, Library of Monographs, Energy Machines 3, University of Banja Luka, Faculty of Mechanical Engineering Banja Luka, Banja Luka, 2012 (in Serbian).

[21] P. O'Connor, A. Kleyner, Practical Reliability Engineering, fifth ed., Wiley, 2012.

[22] RD 34.20.581-96 SP ORGRÉS, Methods of Estimation of Technical Condition of Steam Turbine Units before and after Overhaul and during the Period of Overhaul, Moscow, 1998 (in Russian).

[23] D 153.34.1.17.421-98 RD 10-262-98, Typical Instruction for Control and Extension of Service Life of Boiler, Turbine, and Pipelines of Thermal Power Stations, ORGRÉS, Moscow, 1999 (in Russian).

[24] RD 34.17.440-96, Methodical Instructions on the Procedure of Carrying Out Work in the Estimation of the Individual Resource of Steam Turbines and the Extension of the Term of Their Exploitation from above the Park Resource, VTI, Moscow, 1996 (in Russian).

[25] RD 34.17.415-96, Instruction on Ultrasonic Inspection of Fastening of Power-Equipment, SPO ORGRÉS, Moscow, 1995 (in Russian).

[26] Overhaul of Steam Turbines RTM 108.021.55-77, Minerner USSR, Moscow, 1977 (in Russian).

[27] Y.M. Brodov, V.N. Rodin, Overhaul of Steam Turbines, GOU UGTU - UPI, Ekaterinburg, 2002 (in Russian).

[28] I. Smajević, K. Hanjalić, Heat Turbo Machines, Šahinpašić, Sarajevo, 2007 (in Bosnian).

[29] B. Staniša, Revitalization and extension of the service life of the steam turbine 125 MW TPP Plomin I, Opatija, in: Proceedings of the International Congress “Energy and Environment 2002”, 2002, pp. 291–298 (in Croatian).

CHAPTER

Qualitative analysis in the reliability assessment of the steam turbine plant

8

Zdravko N. Milovanović[1], Ljubiša R. Papić[2], Snježana Z. Milovanović[3], Valentina Z. Janičić Milovanović[4], Svetlana R. Dumonjić-Milovanović[5], Dejan Lj. Branković[1]

[1]*Department of Hydro and Thermal Engineering, University of Banja Luka, Faculty of Mechanical Engineering, Banja Luka, Republic of Srpska, Bosnia and Herzegovina;* [2]*DQM Research Center, Čačak, Serbia;* [3]*Department of Materials and Structures, University of Banja Luka, Faculty of Architecture, Civil Engineering and Geodesy, Banja Luka, Republic of Srpska, Bosnia and Herzegovina;* [4]*Routing Ltd., Banja Luka, Republic of Srpska, Bosnia and Herzegovina;* [5]*Partner Engineering Ltd., Banja Luka, Republic of Srpska, Bosnia and Herzegovina*

1. Introduction

The probability that a technical system will successfully enter into operation and perform the required function of the criteria within the tolerance limits for a given period of time and given environmental conditions (operating temperature, pressure, humidity, allowable vibration, noise and shocks, changes in operating mode parameters, etc.) represents the effectiveness of the technical system. The performance indicator is characterized by a single (unit parameter) or several performance characteristics (complex parameter), such as: reliability (the ability of the system to maintain continuous operating capacity within the tolerances during a calendar time period, quantified through indicators: probability of failure-free operation, mean time in operation, failure rate, and failure density), maintenance convenience (ability to prevent and detect failures and damage systems, to restore operating ability and correctness through technical service and technical repairs, quantified through: probability of renewal for a given calendar time period, average renewal time, and the intensity of renewal), durability (the ability of the system to maintain working capacity from the very beginning of its application or exploitation to the transition to limit states where certain delays in the implementation of certain maintenance and maintenance activities are possible and repairs, defined through indicators: medium resource, gamma-percentage resource, mean lifetime, and gamma-percent lifetime), and persistence (the ability of the system to continuously maintain warm reserves, storage, and/or transportation) [1].

The safety of the technical system, especially the thermal power plant (TPP) (pump) and its associated energy equipment, is determined by a number of different factors, such as: construction of the quality of the materials used, manufacturing technology, quality of installation, conditions of service and exploitation, quality of steam, etc. In the process of exploitation, there are cases of complete or partial loss of functional properties, that is, failure of the system, which may be complete (breakdown or shutdown) or partial (decrease in working capacity). In doing so, the resulting

The Handbook of Reliability, Maintenance, and System Safety through Mathematical Modeling.
https://doi.org/10.1016/B978-0-12-819582-6.00008-3

cancellations can be immediate or gradual. In addition to the criteria for assessing the reliability indicators, it is also necessary to define basic and supplementary reliability indicators.

Adequate maintenance of a steam turbine plant within the condensing TPP during the period of exploitation is important for providing the necessary availability and reliability. As certain elements of the turbine plant as a whole do not fail in specific time periods, traditional preventive maintenance at a fixed time often resulted in the execution of certain unnecessary overhaul activities and works, and sometimes delays were observed in certain maintenance operations (occurrence of damage or breakdowns during the overhaul period). This has resulted in increased costs and loss of production [2]. A far better approach to preventative maintenance is the introduction of a maintenance policy according to the condition, with following introduction of technical diagnostic methods and a maintenance information system based on regulated and updated databases. Conducted analyses of the application of the maintenance policy by condition in relation to conventional maintenance over a constant time period showed an increase in operational reliability, an extension of service life, and a reduction in the cost of maintaining the turbines. The initial stage of the gradual modernization of the turbine plants maintenance is the introduction of maintenance according to the state with parameter control, for which application (for an initial "zero" state) it is necessary to develop a monitoring method and technical diagnostics of the system status, as well as an adequate information system. It should be noted that newer turbines have a high degree of automation, which allows the application and introduction of new diagnostic methods with the existing information system. Further development of preventive maintenance of turbines is maintenance according to the condition with control of the level of reliability, as well as maintenance with the help of expert systems. The determination of the initial projected reliability, then the average level of exploitation reliability, as well as the legality of their change represent the basis, on the basis of which the current level of exploitation reliability is further defined and its prediction (forecast) is made in the future.

On the other hand, the first works in the field of reliability date back to 1930, treating the safety and security of civil aviation in England. The United States has been intensively engaged in reliability, especially during the Korean War. It was only after the World War II that reliability was intensified in the fields of military and civilian aviation, armaments, and space exploration. One of the first areas of reliability in which certain mathematical solutions "have been achieved is the area of system service (Hinchin, 1932; Palm, 1943)" [3]. As special mathematical disciplines, the theories of renewal and reliability "were promoted between 1937 and 1952 (Lotka, 1939; Weibul, 1939; Daniels, 1945; Feller, 1947, etc.), and also Gnedenko, Belyaev, and Solovyev, 1965" [1]. Thus, the German mathematician Lusser, while studying German V-1 rocket systems during World War II, laid out the theoretical foundations of serial systems [4]. In 1951, the US Department of Defense formed a group of experts on the reliability of electronic components, which marked the beginning of the development of reliability theory as a separate discipline in the technological sciences. A great contribution to the development of reliability was the adoption of military standards (MIL-STD), which prescribe the essential characteristics of military equipment and weapons. During 1962, National Aeronautics and Space Administration (NASA) and the US Air Force introduced reliability into the development and design procedures of the space program and the military aerospace industry [3]. The acquired knowledge and experience is being transferred to civil aviation. In the early 1970s, the reliability-centered maintenance (RCM) methodology was developed in the aviation industry. The end of the 1970s, after several disasters, was characterized by the use of reliability in the construction of nuclear power plants. The application extends to the chemical and petroleum industries. The study of the possibility of applying the theory of reliability to the ship's energy systems SNAME (Society of Naval Architects and Marine Engineers) came to light in 1971 [5]. A Ship Reliability Investigation Committee (SRIC) was established in Japan in 1981 with the task of investigating the reliability of marine energy system (MES). The result of it is that after 2000, the reliability of MES is handled by VeriSTAR, MOSys, and RAM/SHIPNET [6].

Reliability Centered Maintenance, IEC Draft 56 (Sec.), 317 (1990) defines priority centers of reliability in preventative maintenance technology [7]. They relate to the mathematical modeling of the main reliability functions (failure density, failure rate, and reliability). The centers are ranked according to the priorities of failure occurrence in their work and include recommendations on determining the weak points from the aspect of the reliability theory.

ISO standard 2372 made recommendations for vibration strength values, which are used for a general assessment of machine condition and vibration limit values that are acceptable [8].

Blanchard and Fabrysky give an overview of the analysis of machine systems from the aspect of constructive solutions, plausibility, and convenience of operation [9]. It clearly defines the states of operation of the system, which allow the analysis formation of a failure that is tree for the constituent components of the system. This approach allows the formation of a universal failure tree of any technical system.

Over the last 50 years, many methods of reliability analysis have been developed and put into practice to evaluate the reliability of various technical systems with greater or less complexity. However, their effectiveness, advantages, and disadvantages can vary greatly depending on their application to a particular technical system. The analysis so far has shown that there is no general rule for selecting the best method applicable to assessing the reliability of a particular functional block (Bernard et al., 2007) [10]. The use of these methods for particular applications depends on a variety of factors, including specific requirements and needs, the type of functional block, and the preferences of the persons involved in conducting the specific analysis. In doing so, ease of use and requirements for a particular analyst experience may vary from method to method [11]. For these reasons, international and national standards, as well as the standards of professional associations, provide specific guidelines to support engineers in selecting the appropriate technical method. The basic objective of qualitative reliability analysis is to identify potential functional block failures under consideration, their consequences at the same level, and the cause and effect relationships of the failures, as well as identify possible repair strategies [12]. For qualitative methods, the magnitudes of the likelihood of failure occurrence are estimated by the analyst's subjective judgment or expert opinion. Thus, quantitative methods aim at generating a list of potential failures, with the failure data relying heavily on previous experience or expert opinion. On the other hand, quantitative reliability analysis aims at determining numerical reference data, i.e., known measures of reliability of each component of the system, for the use of the same as input to the model of reliability, estimation of system reliability, under probabilistic and stochastic assumptions [12]. Quantitative methods rely on the numerical estimation of the probability of a failure event or the magnitude of the possible consequences of a risk assessment, whereby these parameters are quantified as reliability measures using statistical methods and databases. Whether a quantitative or a qualitative method will be selected depends on the availability of databases for assessing the reliability and level of analysis required to make a reliable decision. On the other hand, quantitative methods allow for more uniform analyses, but require quality data with latent results. Therefore, depending on the particular situation, it may be necessary to use a combination of qualitative and quantitative analysis, or to use methods that are qualitative but can be considered as quantitative.

An analysis of the maintenance effect according to condition on the operation of two-component systems that are interdependent was given by Phuc et al. [13]. The research focuses on two types of addiction. The first is the dependence on which the aging rate of each component depends not only on its own state (level of wear) but also on the condition of the other components of the system. The second is the dependency where the combination of maintenance activities is more favorable/cheaper than performing the individual maintenance activities separately. Selection of preventive maintenance components/groups of elements is proposed as well as the establishment of customized maintenance standards as part of custom preventative maintenance. A cost model has been developed to find the optimal value of decision parameters. Numerical indicators are introduced to illustrate the use and benefits of the maintenance optimization framework approach for two-component systems.

Rosmaini and Shahrul provide an overview of two maintenance techniques: time based maintenance (TBM) and condition based maintenance (CBM) [14]. This paper discusses how TBM and CBM techniques react in terms of making a maintenance decision. The paper analyzes and compares the implementation of each technique from a practical point of view, focusing on the issues of determining the required data, collecting, analyzing data/modeling, and decision-making. The conclusion is related to significant considerations for future research. Each of the techniques was found to have unique concepts/principles, procedures, and challenges for actual industrial practice. It can be concluded that the application of the CBM technique is more realistic and therefore more useful in the application than the TBM. The research provides useful information regarding the application of TBM and CBM techniques in maintenance, decision-making in the challenges of implementing each of these techniques from a practical perspective.

In a study [15] by Sondalini emphasizes that, based on age alone, about 80% of equipment failures are completely unpredictable and that a maintenance-by-condition strategy can cope with them. In about 20% of failures preventative and scheduled maintenance can prevent unwanted occurrences in a timely manner. But sudden system failures are not always directly linked to the nonapplication of one of the methods of TBM concept and require more complex and combined solutions. There is another important step that needs to be defined in order to have a significant impact on reducing maintenance costs, which is to recognize and avoid the condition/order for failure—failure mode. If the monitoring methods determine the preconditioning condition, it is necessary to leave in a timely manner and define the conditions for avoiding this state of the system in the future. Random errors are also a significant factor, representing a great deal of stress during work and certain events/anomalies. Until the cause of an incident is resolved, it can happen again, causing significant costs, engaging people, their time, and effort. By introducing a predictive concept of maintenance and also a case study of the cause of failure, the complete business system is qualitatively refined.

The main focus of the research [16], done by Prajapati and Ganesan, is the use of two widely applied prediction areas: statistics and neural networks. Statistical and neural networks are very powerful tools/techniques for predicting future states based on present and past states of systems and subsystems. The study is based on a case study of tire pressure monitoring as part of an evaluation of five methods to investigate the prediction of future system behavior. Statistical methods include the application of ARAR and Holt–Winters (HV) mathematical prediction methods. The ARAR algorithm is basically the process that applied memory shortening transformation and fitting the autoregressive model to the transformed data. It is used to predict the future data from existing sequence data. The algorithm was introduced by Brockwell and Davis (2000) and it consists of three phases throughout the process [3]. On the other hand, application based on time interval, linear predictor, and neural networks makes this statistical approach very complex. The paper presents detailed comparative simulation studies and demonstrates suitability and feasibility in all techniques. Sensors mounted directly on the tires were assumed to report current tire pressure for control or analysis. The control unit performs tire pressure analysis and communicates the decision of the operator or the intended group on the current pressure as well as the upcoming pressure conditions. Finally, the research concludes with the conclusion that the HV method is the best among these five approaches for predicting tire pressure and may be useful for implementing concept of maintenance for any system.

Research conducted on the reliability of robotic systems indicates that mechanical problems are the most common causes of failure of system assemblies, which relates to power transmission elements, actuators-motors, and the structure itself, as well as to bearings. On all these elements, the relative movement of different surfaces in direct or indirect contact is achieved. Some measurable outputs of the robot system are useful for diagnosing robot instability [17]. Čačko et al., using the reliability approach of technical systems as an alternative, emphasize probability through several deterministic signals [18]. The signals are analytically defined and can be collected from mechanical structures so that the probability model can capture differential equations. The equations describe the dynamic behavior of the structure, theoretical and experimental component analysis, dynamic structure analysis (modal analysis), and reliability models [18]. For the statistical model of reliability analysis, Cempel used for modeling of technical systems Weibull's distribution in order to model a group of machines (fans, robots) to calculate the cause of failures on a system [19]. Stephen emphasizes the reliability of mobile machine components to the reliability of input parameters, such as system design and functionality [20]. Robotic components whose reliability affects working conditions are a mechanical bearing. Such bearings are found on robotic motors and joints. The condition of the roller bearings is significantly influenced by operating conditions such as: temperature, speed of rpm, and loads [20]. This explains how, at a single point on the bearing, temperature and loading conditions can be extrapolated through a certain range [52]. Examining the frequency response of the function of bearing assembly presented the nonlinear response as well as the appearance of a gap in the bearing [21]. Youfang and Bin presented a manipulator-robot limitation using modal analysis, diagnostics, as well as the looseness inherent in the nature of the nonlinear structure [22]. They also introduced nonlinear systems that have modes that change with the state of the system; each difference of some manipulator has a differential inertia matrix.

In his dissertation [23], Mikić confirmed that the final research results largely depend on the applied research methodology, on the analysis of existing methods for the development of models of reliability of technical systems, on the comparative analysis of various theoretical models of reliability of technical systems, as well as on the development of reliability models for technical systems that best describe experimental data. Methods of analysis and synthesis, modeling methods using transformation matrices, and probability methods were applied. To verify the obtained transformation matrices, the Matlab software package was used, and Catia V5 software was used to model the components of the technical systems. The binary logistic regression and correlation analysis of one-factor and multifactor experiments was used as the basic method for statistical processing of experimental data. The Roc curve method was used to obtain predictions of the movement of diagnostic parameters. A method of analysis and selection of theoretical functions of binary logistic regression was used to develop the reliability model. The phases of work consisted of: systematization and classification of data, statistical processing of quantitative data, generalization and interpretation of the obtained data. Statistical data were interpreted in several ways, depending on the type and nature of the data in binary and graphic form. The analysis and the calculations for the creation of charts and tables are arranged in the software program SPSS Statistics 17.0, Microsoft Office Excel with the help of several functions, etc. Based on the results of research and application of appropriate mathematical calculations, there were adequate mathematical models of reliability of robot assemblies, pumps, and compressors, which presented the legality of the bearing behavior, the prospect of malfunction.

Advantages of using advanced technologies such as neural network systems are described by Bansal et al. in their work [24]. Condition/machine parameters can provide a wealth of useful information on system status, maintenance requirements, and productivity. So far, system parameters have been predicted based on raw motion detection and position changes of system elements. This detection requires special equipment and high maintenance costs. Particularly high requirements relate to the use of equipment for the identification and estimation of condition and motion parameters in the case of complex movements in multiple axes of motion. The need for better and cheaper techniques for estimating machine parameters motivates the development of intuitive real-time maintenance systems. This paper describes an intuitive neural system for a real-time machine assembly. The neural network approach is used to predict machine/system parameters that use the current position of system elements. Unlike many neural networks based on system state monitoring, this approach was validated as part of an off-line drilling procedure, using experimental test data on a single production machine. Comparisons of expected and actual test results showed good results. The success of the test motivated the development of an online predictive maintenance system with additional conditions/parameters in the drilling process such as friction and twisting. The paper describes the application of an online predictive maintenance system in the case of a clinical microdrilling process for human ear intervention. An online predictive maintenance system is used to evaluate the inertia of the drilling that is provided in the drilling process by the surgeon.

Advances in the collection and processing of metrics and the development of computer technology have been a major feature of experimental research in recent years. Lin and Makis analyze the concept of maintenance by condition, cost optimization, and life expectancy based on the early detection of gearbox damage [25]. The research is based on vector modeling of vibration signals, using a hidden Markov's model based on the Bayesian network. The continuous Markov's model describes three states: healthy state—0, warning of possible failure—1, and noticeable failure—2. The calculation of model parameters uses the maximum expectation algorithm. The optimization of the three-state model is based on the Bayesian diagram for a multivariate decision-making process. The chart tracks the last probability that the system is in alert state number 1 and the system stops when this probability exceeds the optimal control limit. Predicting medium remaining time uses the methodology of probability theory. Confirmation of the proposed methodologies was performed using the actual state of the gears obtained from vibration measurement data.

Henley and Kumamoto in their book [26] present the synthesis and analysis of risk and reliability methods, as well as their study of fundamental principles and applications in real industrial conditions. The full treatment of qualitative analysis methods includes: block diagram of reliability, failure tree analysis, allocation of reliability, analysis of forms and consequences of failure, etc. In addition, the quantitative analysis system is represented by Markov and Monte Carlo methods that are well described and performed.

Funk and Jackson analyzed the link between experiential diagnostics and the concept of maintenance according to state in manufacturing systems and systems using robotic components [27]. Research shows that, on average, the efficiency of production processes is about 60% and that the real expectations are about 80%. The number of failures and downtime in the production cycle is most influenced by the degree of efficiency. System maintenance is directly related to system failures so that the development and optimization of maintenance concepts directly affects the efficiency of the entire system. One way to determine the status of system components is through experiential diagnostics. Easily accessible experience and knowledge of the behavior and state of the system is a great help in engineering practice. Today, the collection and processing of experiential diagnostics data is done using computerized support and decision systems as well as artificial intelligence systems. Only experienced operators and well-educated staff can perform quality equipment data collection and archive it into an easily accessible database. This approach significantly reduces the possibility of error, shortens identification time, and allows for possible corrections in specific practical cases.

Studies conducted by Japikse et al. point to several basic causes of vibration of centrifugal pumps, without strict classification on mechanical and hydraulic causes. The cause of vibration is divided according to the ratio of the frequency of the disturbing force and the angular velocity of the pump [28].

Analyzing the vibrations of centrifugal pumps, Smith et al. and Black pointed out that most vibration problems occur due to a synchronous phenomenon such as the frequency of the number of blades multiplied by the speed of rotation, and the frequency of the speed of rotation [29,30]. Asynchronous phenomena, such as recirculation or shaft instability, can also lead to unwanted vibrations. These problems are more difficult to detect because the mechanisms of vibration excitation are not readily apparent in these cases [30,31].

Hancock emphasizes in his work that vibration measurement provides a good basis for achieving the working state of a process pump. Vibration data provide a good basis for designing a preventive/corrective maintenance program [31].

Significant contribution to the development of the theory of reliability in the world comes from Chudakov, Paraga, Calabra, Kendal, Stewart, Smirnov, Andronov, Vladimirov, Barlow, Hanter, Aronov, Zimmer, Kaplun, Birnbaum, Proschan, Fussel, Lambert, Malecev, Rushdi, Tanaka, Fan, Lai, Toguchi, Rudenko, Alefeld, Shapiro, and others [1].

In the former Yugoslavia, writing began relatively late on aspects and problems of reliability (Petrić, Todorović, Teodorović, Vukadinović, Zelenović, Stanivuković, Papić, Jovanović, Adamović, Tomić, Mesarović, Šijački-Žeravčić, Holovac, Stamenković, Soldat, Samardžić, Jugović, and others). In the period since 1995, significant investigations have been made on the application of reliability theory and application of reengineering in the field of energy and process engineering. Several scientific-professional papers have been published and several master's theses and doctoral theses have been presented (Knežević, Nahman, Papić, Macek, Boševski, Vujošević, Grujić, Bulatović, Bakić, Milanović, Vasić, Milovanović, Branković, and others) [2,3].

Thus, Papić and Milovanović provide a complex analysis of the basics of maintenance of technical systems, methodology and application of methods of improving the process of maintenance and reliability of technical systems, and the relationship of the effects of investment in the components of the production cycle according to the results of increasing efficiency (reliability, availability, and maintenance convenience) [1]. A well-chosen concept of maintenance, with proper organization, programming, and implementation of certain maintenance activities during operation, with good staff training and quality management in maintenance, also contributes to improving the economic performance of the business system. Research aimed at increasing the level of reliability and managing reliability over the life cycle of a system aims to define a set of protection measures and optimize them from the aspect of simultaneous provision of economical exploitation and implementation of complex regulations related to the protection of human environment and safety of both the business system and the environment.

In his doctoral dissertation, Kutin describes and analyzes the optimization of diagnostic techniques as the main subject of research [32]. By proving the hypothesis of the influence of thermography introduction, unconventional diagnostic method and monitoring of technical condition on increase of reliability, reduction of production costs, prolongation of exploitation life, higher environmental protection, and higher level of safety in work, the dissertation confirmed the existence of a significant and inseparable link between the effectiveness of the technical system and the application of modern diagnostic methods.

Monograph [33], by Popović and Ivanović, provides answers to the question of the role of reliability design in the design stages of technical systems, that is, before the technical system was developed. A specific example relates to motor vehicles and reliability design analysis. The importance of the research lies in the universality because the methodology applied can be used on other machine systems.

Bulatović in his book [34] presents a summarization of research and analysis of the reliability of technical systems on a concrete example of a change in operational readiness and reliability at the company Alumina Production Factory in Podgorica. Contribution to scientific research is the method of incorporating state parameters and probability of failure into the expert system. Complex maintenance problems from the point of view of determining and monitoring the status of systems in the function of predicting and preventing failures are effectively addressed by the use of expert systems.

Sćepanović et al. describe a maintenance model based on the team organization of a maintenance project with the primary goal of improving the individual in the maintenance process, extending its life, and permanently developing the maintenance process [35]. Modern tendencies in the field of development maintenance go toward the extension of the working life of the individual, the extension of the working period between the overhaul, the shortening of the overhaul period to the optimum value, with a gradual transition of maintenance to the established condition, characteristics, and planned future condition of the individual.

The problem of modeling within the development of the concept of maintenance by state is also addressed by Adamović and Radovanović [19]. Two new models of maintenance by state are presented in the paper: maintenance by state with parameter control and maintenance by state with control of confidence level. In the case of control of system parameters, the selected parameters of the technical condition (vibration, temperature, pressure, etc.) should completely define the state of the system components, which allows predicting the moment when the basic characteristics of the component parts and/or system will deviate from nominal (allowable) values. The basic approach to defining maintenance according to the condition with the control of the level of reliability is that the components of the system are used without limitation of the intermittent resource with the execution of necessary maintenance activities in eliminating the resulting failures, while the actual level of reliability is within the established (allowed) standards. If deviations occur, measures are taken to increase the level of reliability of individual components of the system, as a criterion of technical condition; in this maintenance model, the level of reliability expressed by the reliability indicators is adopted. Models of developed models and stages of development are analyzed in the paper. An algorithm for forecasting the technical state of the system has been developed.

Kostić pointed out that accelerating the movement of a robot, while preserving the given accuracy of positioning its tip and executing the desired spatial trajectory, poses eternal challenges in robotics [37]. Despite the increasingly demanding speeds and accuracy of the robot's movement, it is necessary to maintain the necessary robustness of control with respect to the uncertainties in the dynamics of the robot and the disturbances affecting the movement. He gave an experimental comparison of the managerial performance achieved by applying different management techniques. The criteria for comparison are: the complexity of the design of the control law, the accuracy of motion, and the robustness with respect to dynamic disturbances. He also made recommendations for the choice of control law design technique that is most appropriate in a given robot motion control problem [37].

Testing the reliability of roller bearings on multiple technical systems, Adamović et al. concluded that vibrations, namely its accelerations, are the best indicators of the reduced reliability of the technical systems examined [38].

According to the research of Ašonja (2012), the diagnostics of the condition of rolling bearings on machine systems shows to what extent the reliability based on the diagnostics of the condition of the bearing assemblies can influence the increase of the exploitation level of reliability on technical systems [39]. The development of modern machine technical systems and technologies used in industrial production, has conditioned the use of newer (more modern) methods for monitoring the condition of the system and its diagnostics. For the application of these methods, it is necessary that maintenance workers have a certain level of knowledge and training. It is especially important to monitor the condition of the bearings, as they are vital for all machines and technical systems, devices that house rotary parts. Bearings embedded in vital machine assemblies or heavy-duty bearings such as those in industrial production require frequent inspection by Ašonja [40]. The vibration caused by the imbalance can have a strong effect on the bearings, suspension, housing, and foundations (basis), and cause great wear and tear, that is, reduction of friction at the screw and clamp joints, and lead to the separation of components. Products that are unbalanced often have a shorter shelf life. Imbalance can significantly reduce the safety of a machine, which is a danger to humans and the machine.

Gligorić and Ašonja state that the balancing problem is most pronounced in the rotor because the value of the inertial force of the rotor, which moves uniformly circularly, is a function of the square of its angular velocity, so that very small unbalanced masses have a considerable value of inertial force [41].

Grujić [42] describes the relationship between potential risks for the operation of technical systems and the implementation of technical diagnostic measures. Risk management in engineering involves taking the necessary measures and activities in a timely manner to ensure the safety of the functioning of technical systems according to the intended purpose. Safety in the exploitation of technical systems presupposes normal operation without damage and other risks and safety for the environment especially for the human factor. The application of technical diagnostic methods is an indispensable factor in the risk management of technical systems. The reliability level is significant at all stages of the technical system life cycle. The results of technical diagnostics and reliability as indicators of the quality of technical systems can be used for revitalization, reconstruction, overhaul, etc. The study describes risk management techniques for functional diagnostics, diagnostic testing, and reliability methods.

Krunić et al. represent an information system model in industrial production with support for preventive maintenance and automatic technical diagnostics [43]. The support is implemented partly through the SCADA application and partly through the database and applications that update the state of the infrastructure.

Popović et al. represent the stages of designing reliability by applying methods of designing the reliability of motor vehicles as machine systems with the aspect of satisfying user requirements and designing with respect to reliability prior to the vehicle manufacturing process [44].

Similarly, Adamović et al. give an overview of failure considerations of hydraulic system elements, then show reliability indicators, methods of determining static characteristics, test hydraulic installations, reliability indicators are presented, reliability level increase is presented, then reliability levels are increased, and technical diagnostic elements are finally presented [45].

Defining and forecasting reliability indicators are of utmost importance in preventive maintenance of complex systems. Only on the basis of timely assessments can corrective action be taken in order to further development of failures and to prevent major damage. Milovanović proposed a modified method for estimating the optimal reliability of a condensing TPP (2000), providing a good basis for further work on refining and improving the accuracy of the estimated values, with the introduction of technical diagnostics and a modern information management system [3]. Using the benefits of modern system reengineering directions and structural or RCM (maintenance according to reliability) ways of using the best methods in reliability analysis of complex systems, a block diagram of a modified method for a 300 MW basic reference block has been formed. This method also involves calculating the level of risk and, consequently, indirect ways to maintain these systems or their components. The choice of the most appropriate maintenance task is achieved through the use of a flowchart of decision-making activities, which takes into account the technological capabilities of the proposed tasks and the verification of their realization. The activity flow diagram of this method incorporates modern knowledge regarding the application of the total productive maintenance (TPM) methodology, i.e., implementation of the concept of TPM, similar to the maintenance methodology according to the condition, with the fact that maintenance requires a special user attitude toward the system and a "total" responsibility for the quality of maintenance procedures of all employees within the TPP. In doing so, the actual maintenance costs are minimally possible for each specific situation. The dependence of the cost of generating electricity on a TPP on the level of reliability needs to be considered from two aspects: the TPP and the user. In both cases, the point of minimum cost determines the optimal reliability of both the TPP and the user.

Using exploitation research carried out in the company "Thermal Power Plants and Mines Kostolac" Ltd., Kostolac, and the company "Thermal Power Plants Nikola Tesla" Ltd., Obrenovac, in the period from 2011 to 14, Milošević within the doctoral dissertation gives an analysis of the model of providing reliability of complex plants in TPPs [46]. Technologically, the contribution of the dissertation is reflected in the application of the reliability model through the application of simulation methods for the selection of the best functioning parameters of the components and circuits, from the aspect of their reliability, which would be proved by experimental methods (using collected data and stored data from the past). The reliability analysis based on the models formed should contribute to the efficient and simple determination of parameters for making relative decisions for the reliability of complex technical systems. These parameters can define the timing of maintenance action decisions based on the required level of reliability. The simulation can thus define the optimum timing of replacement/repair of parts of a technical system before its failure or the need to act correctively. Consequently, these models, by carrying out adequate maintenance actions, serve to ensure the reliability level of complex plants in TPPs.

A review of methodologies in reliability engineering, with particular emphasis on quantitative system analysis, was provided by Dobrota (2014) [47]. It was stated that a qualitative analysis of the technical system must be carried out in order to provide answers to questions about how the system works, how it can break down, and what are the consequences of a failure. Also, a description of the methodology for qualitative system analysis is included, which includes a number of activities (familiarization with the system, functional analysis, fault classification, and component/system reliability assessment). Function division is also given, and technical methods of functional analysis of systems encountered in practice and literature are briefly described, with the purpose of identifying and describing all system functions in order to answer the question of how the system works [47].

An analysis of the effectiveness of a real production system presented in Ref. [48] by Branković indicates evident increase in the level of reliability and operational readiness and a decrease in the failure rate with a slight increase in the coefficient of operational readiness, which can be explained by the realization of major investment activities during the general overhaul of the paper plant machines in the period after implementation of technical diagnostic measures. The ability to monitor modern diagnostic systems is conditioned by an adequate level of education, training, and experience of the maintenance staff responsible for ensuring the readiness of the facility. The permanent and inseparable connection with the technological staff, who should actively participate in the program of monitoring all events during the operation of the equipment for work, is the basis of modern concepts of maintenance, such as TPM, within which technical diagnostics and maintenance principles according to the state are an integral and inseparable part. Defining the real current state of the equipment using technical diagnostic methods creates the conditions for quality repair planning. Well-planned plant shutdowns to prevent failure not only mean time savings, increased hourly output, and higher production but also lower spare parts costs, warehouse and working capital optimization, good coordination with external firms, optimal use of work equipment, rational use of human resources, and increasing the degree of safety at work. The planning of maintenance activities ensures a better fulfillment of the basic function of maintenance of production equipment, which is aimed at the main goal: achieving maximum availability and reliability of installed equipment at minimal cost, all with the aim of maximizing business system profits. Technical diagnostics of a real industrial system can give a very good result in terms of optimizing production processes and increasing the overall efficiency of the production system. By looking at the production system as part of the business system and the business system as part of the wider community, a clear link between the application of modern technical achievements and the overall advancement of the environment in which the business system "lives" and in which it develops can be established. Another dimension of the impact of modern technical solutions on the environment is the respect of environmental standards and the protection of the environment, which is given great care when selecting and applying diagnostic systems.

Previous research on the design of reliability of technical systems provides little evidence of their application, development, and modernization of models in the process of designing maintenance of machining systems in the timber industry. Numerous methods of designing reliability, while aiming to achieve the required level of reliability within the technical efficiency of a technical system, have not been created and refined on the same basis and do not always include the same parameters in the analysis of designing reliability. Therefore, there are certain limitations and difficulties in their application and the way in which the results are interpreted. Such reasons point to the need for further research, which would create universal design models for the reliability of technical systems in the maintenance design process supported by modern technological advances. Research into the process of technical diagnostics and the structure of technical systems is an important function throughout the life cycle of a technical system, starting from conception through the development of a technical system.

2. Definition of maintenance and reliability

Maintenance during the working life of a steam turbine system combines a number of supporting activities, ranging from the idea and definition of the concept, the evaluation of their cost-effectiveness, realization, exploitation, and until the system is decommissioned. The training of the maintenance system, through designing on the basis of maintenance, is conditioned by the development of the production forces of the company and aims at extending the working life, with the achievement of more optimal connections of technical, technological, and economic characteristics. The process of maintaining a steam turbine within a complex technical system of a TPP, as one of the most important parts of the overall production process in the production of electricity, has the task of preventing and eliminating system failures, primarily through streamlining and optimizing their use and increasing the productivity and economy of consumption in the process of production or exploitation. The service life begins when the idea of a new steam turbine is born, and ends when it is withdrawn from use. The main processes that help the system through the life cycle stages are: marketing (specifications), design, production, use, and finally, withdrawal from use. Work life analysis is a systematic and analytical approach to identify the resources needed to support the design, production, use, and decommissioning processes. Therefore, work life analysis is a tool for work life engineering, whose main goals are to influence work life design, to identify and quantify total resources related to work life processes, and to analytically manage work life process activities.

In other words, life cycle engineering (life cycle engineering) should enable the decision-making process to arrive at the best compromise between investing and providing the necessary resources to design, manufacture, use, and withdraw steam turbine systems from use. In doing so, this approach enables: early and continuous impact on system design from a work life (life cycle) perspective, reduction of steam turbine life cycle costs by limiting major cost generators over the life cycle, and identification of resources that accompany all processes (stages) operating life of a steam turbine system. Generally speaking, traditional (sequential) engineering is generally focused on the performance of the steam turbine system as the main goal, rather than on the development of a general integral approach. Recent knowledge and experience gained in recent decades indicate that the proper exercise of the goal function, i.e., the required degree of competitiveness of the steam turbine, cannot be ensured by investing mainly after their production and reaching the utilization phase, as is often the case. It is much more important for engineers to be sensitive to looking at the consequences of potential errors that can occur during the early stages of steam turbine design and development. This means that engineers should be able to take on responsibility for life cycle engineering (competitive, simultaneous engineering), which has been neglected most often. An indispensable part of the working life is the extension of the working life (revitalization), reconstruction, and modernization of the steam turbine, i.e., the process of extending the working life of these facilities with modernization and reconstruction, with the additional improvement of technical, technological, economic, and environmental acceptability. This procedure, by its structure, is extremely complex and is often compared to the rank of a new steam turbine. Such a systematic and comprehensive process on a part or plant of a steam turbine as a whole is an unavoidable and logical process in the working life of the facility.

The relation between the process of reengineering on the maintenance of a steam turbine with the basic and associated auxiliary equipment, with the aim of realizing the corresponding advantages and improving the reliability of the system, is given through the following characteristic elements: cost analysis related to the maintenance and availability or availability of the steam turbine system (as one of the most important characteristics of efficiency), determining the general aspects related to the motives and justification of revitalization, as well as the scope and defining the most optimal term for realization of this process. The influence of reliability and availability characteristics of the object of this system on the application of the principle of reengineering through the process of maintaining the steam turbine as a whole, or on the systematic approach to the revitalization of some of its capacities, should be emphasized. Planning, development, construction, and exploitation, while maintaining the equipment and system of the steam turbine, a large number of occurrences that can cause damage and endanger health and life, both directly involved in the main power plant (MPP) of the TPP where the steam turbine is installed, and the wider environment. In short, there is a high risk of adverse events and their consequences. For more complex technical systems, such as TPPs, which have a high interdependence of their subsystems and elements, failure of any of them can mean automatic shutdown of the whole system, or reduced power operation (or more often, operation at a technical minimum), which can result in increased operating costs of the system itself, thermal and other overloads, as well as greater damage during system outages. For these reasons, such a complex TPP needs to be reliable. The safety of steam turbines as part of this complex TPP can be considered from two aspects. The first and most important aspect is the protection of the operator (human) from injuries during system operation. Another aspect is the protection of the steam turbine itself from damage caused by external causes. Preference is given to operator safety. In doing so, the two aspects are not unconditionally complementary, and an increase in operator safety can be achieved at the expense of the safety of the steam turbine system as a whole.

Any technical system, including steam turbines, even if it performs the function of the target within the tolerances, can be damaged if it is incorrectly handled. The main causes of operator risk include: engaging body parts such as hands in the process of system operation, inattention to the operation of rotating parts of the system (especially poorly attached units), contact with sharp and abrasive surfaces, the influence of operator static on moving objects, or vice versa. The risks of steam turbines are diverse and numerous, and in the design phase, the consequences of critical types of failures must be minimized through the prediction of protective devices during the operation of the steam turbine itself. Risks in basic and auxiliary equipment of steam turbines include: shocks, vibrations, corrosion, environment, fire, and contact with high-voltage elements, then with high pressure and temperature, as well as mismanagement (overload or operation below the level of technical minimum).

Today, we are not just talking about maintenance as a discipline that is closely related to the knowledge of technical units, their working and functional abilities, the physical lawfulness of behavior toward related elements, or the environment. Maintenance today implies close integration with the complete technological principles of the production process, as it often provides answers to the source of technical problems that occur during the exploitation of system parts. From interventions that have been "waiting for cancellation," maintenance has evolved into a multidisciplinary activity with a multitude of applied scientific content in the process of fulfilling its goal function. At the 1963 OCDE Congress [1], maintenance was defined as "a function whose responsibility is the constant oversight of facilities and the carrying out of certain repairs and audits, thereby enabling continued functional capability and preservation of production and auxiliary facilities and equipment." Nowadays, maintenance is not thought of as a necessary incidental activity. Today, it is a set of segments or elements that, by acting together, ensure that the technical system is maintained in accordance with the set requirements and criteria. The process of maintaining labor resources, as one of the most important parts of the overall production process, has the task of preventing and eliminating system failures, first of all by rationalizing and optimizing their use and increasing the productivity and cost-effectiveness of the production and exploitation process itself [48].

Maintenance is today defined as a process in which all activities are carried out according to predefined goal criteria (cost, availability, effectiveness, reliability, etc.) (Fig. 8.1). Broadly speaking, the maintenance system is part of the business system and through its design/redesign should integrate: optimal organization, relevant technologies, information system as a basis for pooling available resources (material, human, finance, technology, etc.), and engineering economy.

The definition of reliability until the 1960s was: "The probability that an element will perform a given function under specified conditions for a fixed period of time." There are authors today who still prefer this definition [49]. However, most authors use the general definition of reliability given by ISO 8402: 1986 and BS 4778: 1991, which defines reliability as: "the ability of an element to perform the required function, under specified environmental conditions and operating conditions, and for a specified period of time" [50]. By this definition, the term "element" means any component, subsystem, or complex system that can be considered an object. The function required may be one or a combination of functions required to provide a particular activity. All technical elements (components, subsystems, and systems) are designed to perform one or more functions and must work satisfactorily over a period of time, with actual application for which they are intended.

Assessment of reliability as the probability of successful execution of the objective function of the observed object is performed on the basis of available test data, failures, or by observing performance under real or simulated conditions. As the results may vary, the reliability estimate may differ for certain sets of input data, even if there is no significant change in the physical characteristics of the element being evaluated. This requires that in addition to the reliability assessment, the accuracy of the security (confidence) assessment is also measured. Accordingly, Kececioglu defines reliability as follows: "Reliability is the probability with a defined level of security (confidence) that the system will perform the required functions satisfactorily and without failure, for a specified period of time, when used in a certain manner, and in environmental conditions and with the load level which is inherent to

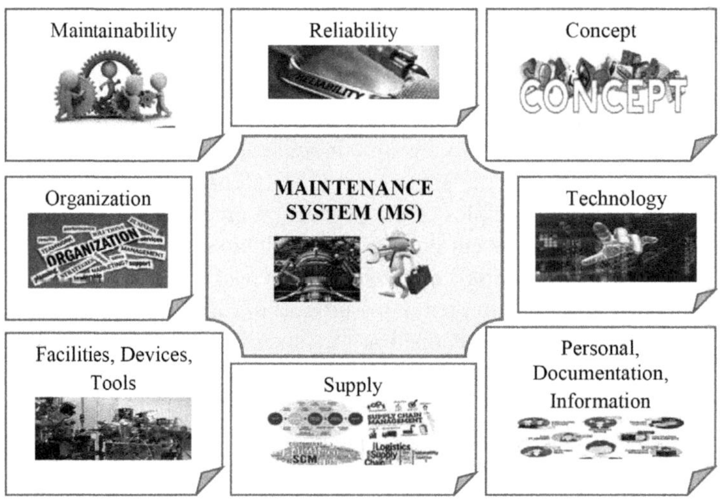

FIGURE 8.1

Systematic approach to the maintenance of technical systems.

it" [51]. The confidence level depends on the amount of observable data available and/or the results, usually presented according to some parametric distribution function. This means that the data can be represented by a mathematical expression which describes a certain statistical (discrete or continuous) probability distribution function. Most reliability and life analysis focuses on modeling the failure time distribution of an element that is modeling the properties of a continuous random variable T. The properties of a continuous random variable T can be described using the probability density function $f(t)$ or the cumulative probability density function $F(t)$, the reliability function R (t), and the fault-frequency function $\lambda(t)$.

Launching a project to adopt a single European-level maintenance terminology, the European Federation of National Maintenance Society (EFNMS) began the process of unification and first steps in 1972 in order to define a unique maintenance strategy with effects of continuity and optimization, as well as production rationality. A particular problem with the early stage of maintenance development was the study of the reliability of complex (especially energy and process) technical systems. The foundations of the Condition Based Maintenance theory were determined by the introduction of monitoring the state of certain parameters during operation (condition motoring) in the aviation industry and the military industry [52].

The motives for the study and application of the research technique in the field of complex (usually energy) plants are as follows [50,53,54]:

- complex (especially thermal power and process) plants are very specific in their complexity and the processes that take place in them, especially in cases of reduced volume of technical maintenance and impaired maximum efficiency within the superior (electricity or gas/oil) system,
- seeking to increase and extend the basic life of the plant for an additional revitalized period,
- difficulties and specifics of the process of maintaining a large number of components in terms of accessibility, possible interchangeability, or eventual repair,
- seeking to eliminate or reduce the risk of both human and material loss,
- seeking to ensure full protection of the environment and reduce the risk of emergencies,
- the desire to keep the facilities running as continuously as possible and thereby reduce the economic damage caused by system downtime or longer failures, and
- reduction of the production price of the effects of exploitation (the price of electricity or gas/oil delivered) and increase of competitiveness in the market.

Some countries have defined and adopted certain legal norms and regulations: Law on Safety and Reliability of Technical Systems in USA, 1973; 352-1987 - IEEE Guide for General Principles of Reliability Analysis of Nuclear Power Generating Station Safety Systems, 1987; GOST standards in Russia 1974; JUS standards in the field of electrical engineering and electronics in the former Yugoslavia (JUS N.N0.022/71, JUS N.N0.023/72, JUS N.N0.024/72 and JUS N.N0.025/74); Statistical terminology used in the power industry EKC/97 and other [3].

Depending on the situation observed, reliability can be quantified in various ways, such as [49]:

- mean time to failure (MTTF) for nonrepairable elements,
- mean time between failures (MTBF) for elements that are repairable and restore function after repair,
- failure rate (FR), frequency of failures, i.e., number of failures per unit of time,

- the probability that the element will not cancel in the time interval (0, t) or the probability of survival,
- the probability that the element is capable of operating at time t (availability at time).

For irreparable elements, reliability is the probability of surviving for the expected lifetime when only one failure can occur. In this case, the irreparable elements may be individual parts, such as signal lamps, electronic components, etc., or multicomponent systems (e.g., steam turbine condenser, high-pressure heaters, low-pressure heaters, etc.). During the service life of the element, the probability of the first and only failure is called the instantaneous hazard rate (IHR) and is usually referred to as λ. For elements that are repaired after a failure, the probability is that the failure will not occur during the period of interest (the period when more than one failure can occur). The reliability of repairable elements is most often quantified as the rate of occurrence of failure (ROCOF). Sometimes the elements can be seen as repairable and irreparable. Similarly, repairable elements can only be considered repairable under certain conditions. Thus, e.g., steam boiler and turbo generator can be treated as repairable only up to a certain age or cost of repair (basic working period from 25 to 30 years and extended working life after revitalization, reconstruction, and modernization lasting from 15 to 20 years) [53,54]. Repairable elements enter the repair process. So repair is a maintenance action when a failure occurs.

Like reliability, availability is the likelihood that a component or system performs the required function at a predetermined time point or beyond a specified period of time when it is operating and maintained as intended. Therefore, availability is a function of the element's inherent reliability, element maintenance convenience, and maintenance support, as shown in Fig. 8.2.

The probability distribution of failure and repair is taken into account to predict availability. Probability theory rules are applied to quantify availability. Availability can be defined as the ability of a whole to perform a required function under certain conditions at a given point in time or over a specified period of time. The assumption is that all the necessary external resources are provided. Average availability interpreted as the average availability over a long period of time in which an element is able to function is called long-run availability. If there is a limit to a long period of operation, then it is a question of steady-state availability. The forms of availability of long work or the availability of a steady state depend on the method of defining the downtime and the uptime, and accordingly, distinguish between inherent, realized, and operational availability.

According to British Standard BS 4778, maintainability is defined as: "The ability of an element to retain or restore, under specified conditions of use, to a condition in which it can perform its required function, when maintenance is performed under specified conditions and using prescribed procedures

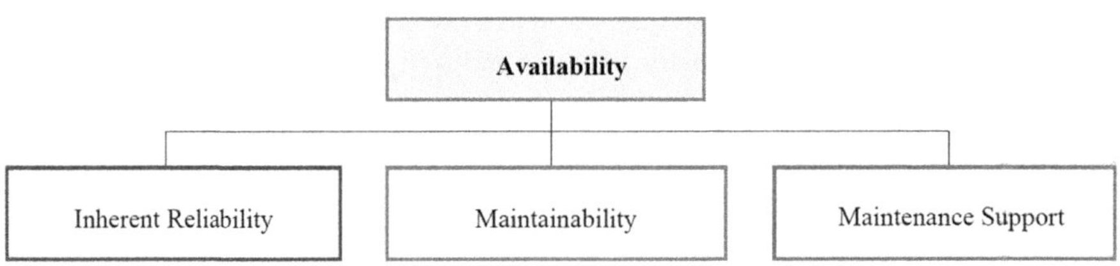

FIGURE 8.2

Availability as a function of inherent reliability, maintenance convenience, and maintenance support.

and resources" [55]. Like reliability, maintenance convenience is defined as the likelihood that a defective component or system will be restored or repaired to a specified condition within the time when maintenance was performed in accordance with the prescribed procedures. Maintenance convenience is characterized by the probability distribution of the malfunction, which must be determined. The component or system malfunction time usually refers to the sum of the factors (access time, fault diagnosis time, repair or replacement time, inspection, etc.). The length of repair of these elements depends on the ease of access, maintenance convenience, and availability of personnel, tools, and parts. For these reasons, the failure time of a particular fault is estimated based on knowledge of all these factors.

Exponential, normal, and lognormal (most often) probability distributions are usually used to estimate the malfunction time. The convenience of maintaining the element depends on design factors such as installed test equipment, modularization, standardization, disposition of parts and their accessibility, ergonomics, ease of disassembly and assembly, etc. The second group of impacts that affect maintenance suitability, and therefore availability, is maintenance support, which depends on the ability and appropriate expertise of the maintainers, their number and availability, tools, the suitability of the technical instructions under which the maintenance process is performed, the quality and availability of spare parts, and others. There are several measures of maintenance convenience. The most commonly used is mean time to repair (MTTR) [56].

Other possible measurable parameters are the median time to repair, or the probable repair time, the time at which a certain percentage of defects must be repaired, the mean repair time plus the mean stoppage time of the preventive maintenance, and the ratio of the number of maintenance hours to the hours worked. This last measure quantifies the total amount of work, while others focus on interruption times [8].

Many methods of reliability analysis, such as: cause and effect analysis of failures, failure tree analysis, and importance analysis in terms of system unit reliability, can also be successfully applied to determine the safety characteristics of a steam turbine system, such as: primary and secondary events, peak event, probability of peak event, minimum sets of cross-sections, degree of criticality of types of failure, and its whole. The causes of adverse (adverse) events are stochastic phenomena, as they depend on a number of specific but also random factors, the effects of which are often not fully understood. By preventive measures, in a way, the activities of suppression and possible response to this group of factors are planned. Ability to operate basic and auxiliary equipment of steam turbine without failure in stationary and nonstationary modes of operation, economic and technical suitability for repair of both elements and turbine as a whole, restrictions accompanying its exploitation (environment or parent system, environmental protection, financial resources and other), the possibility of using appropriate type solutions on the basis of analogy with similar facilities, standards for control and diagnostics—all these are characteristics that do not have a detailed budget and experimentally substantiated base, related to availability and reliability.

On the other hand, with the increasing complexity of technical systems, such as steam turbines, the problem of their optimal functionality arises as an accompanying problem, especially if it is known that such systems can often cause major economic losses or endanger the security of the wider macroregion and the people who serve them. Research related to increasing the degree of reliability and reliability management during the life of the turbine as a whole, aimed to define a system of protection measures and their additional optimization. Thereby, it was necessary to provide the required level of economy during exploitation, with the implementation of requirements related to regulations in the field of environmental protection and safety of both micro and macro-regions. Reliability of turbine plants in theoretical and practical considerations is the starting point for making a forecast or assessment of the correctness of the steam turbine system as a whole, its remaining working life or the life of its most critical elements and definitions seeking preventive corrective measures and assessing the feasibility, i.e., determining the likelihood that the observed part or system of the steam turbine as a whole is brought from the failure state to the operational state in the shortest possible period of time.

3. Systemic approach to reliability analysis

The requirements for controllability and reliability of modern steam turbines during their operation relate to the general conditions of operation of power systems, daily and annual schedules of energy consumption, structure of production capacities in power systems, their condition, and technical capabilities. Currently, electrical load charts of power systems are characterized by a great deal of unevenness: sharp peaks in the morning and evening, failures at night and on weekends, if necessary, to quickly increase and reduce the load. Maneuverability means the ability of a drive unit to change power during the day to cover the load curve of the power system. The periods of commissioning and decommissioning of the turbine unit are very important, taking into account its different initial thermal states: a) hot - after a preliminary interruption of work for less than 6–10 h; b) warm - after the previous switch-off from 10 to 70–90 h; c) cold - after the previous switch-off 70–90 h [52]. Consideration should also be given to the number of shutdowns/starts for the entire life cycle, the lower limit of the scope of adjustment, i.e., the lower limit of the load interval, when the power is automatically changed without changing the composition of the auxiliary equipment, and the ability to work on the load for its own load after unloading. Equipment damage statistics show that the vast majority of failures occur at the moment of implementation of transient modes, when one or the other set of parameters changes. An emergency shutdown of the turbine with or without vacuum distortion is used to avoid emergencies.

In analyzing the reliability of the technical components and systems as a whole, two different approaches are used: the physical approach and the actuarial approach [50]. The physical approach takes into account the variability of basic parameters such as mass, dimensions, coefficient of friction, strength, and stresses. All these parameters are never absolute but in practice they are variable due to changes in processes and materials, human factors, and methods of application. Some parameters change with time, and therefore all parameters are modeled as random variables. Therefore, to create a reliable product or system, as well as to solve the problem of unreliability, an understanding of the laws of probability and of the causes and consequences of their volatility is required. This approach is mainly used to analyze the reliability of structural elements. In the actuarial approach, all available data of basic parameters (e.g., stress, strength, etc.) of components and systems are described by the function of the cumulative probability of failure occurrence $F(t)$. No explicit parameter modeling is performed. Reliability properties, such as the frequency (intensity) of failure and the MTTF, are derived directly from the function $F(t)$. To model the reliability of a system of several components, different approaches can be used, and maintenance and replacement of components can be involved. When multiple components are connected to a system, then the reliability of the system as a whole is discussed.

Due to the hierarchical operation of the components of complex technological systems, a systematic approach in their design is necessary, which implies: system model representation, mathematical modeling, and system quality assessment. The essence of dynamic modeling is that the technical system as a whole or its individual modules or components are replaced by models that have the same dynamic properties as the original. The method of replacing the real system with a suitable model significantly contributes to the acceleration of the analysis of the system, making it possible to test the behavior of the mechatronic system at different combinations of parameters and different structural schemes.

Modeling can be physical and mathematical. In the physical modeling, the components of the mechatronic system have the same physical nature and are described by the same differential equations, i.e., the same physical processes take place both in the model and in the original [8]. Physical modeling, with the use of similarity theory, can best show the properties of the system under test as well as enable the identification of phenomena that cannot be determined theoretically. In this respect, physical modeling has advantages. However, it requires the creation of special models in each case, which takes time, with material costs, and yet has no universal application, so it is rarely used despite its advantages. In mathematical modeling, the corresponding elements of a real technical system and model do not have the same physical nature, but are described by the same differential equations. Therefore, multiple types of data are required to model and analyze the reliability of a component or system. Technical data are needed to understand the function and functional requirements and establish the model, and are usually provided by the manufacturer. Operational and environmental data are required to establish model components and systems. Maintenance data in the form of procedures, resources, quality, and duration of repairs are required to establish the element model and to determine the availability of components and the System as a whole. Furthermore, different types of reliability data are required to relate to fault information and fault timing. If people are active system operators, they also need information about their ability to repair failures and restore function of components or systems.

Reliability data are most often coming from generic data sources (databases, experiences of others, prior knowledge, etc.) or from one's own experience (e.g., equipment testing records), as a combined application. Using reliability models as stochastic models involves estimating unknown model parameters. Typically, the parameters are estimated based on the observed data, and the conclusions of the reliability analysis are made assuming that the parameters are equal to their estimates, i.e., fixed. This so-called a frequentist approach for estimating unknown model parameters is possible if sufficient observable data is available. However, the use of a reliability model is very often limited by the lack of adequate reliability data for parameter estimation, which is especially characteristic when introducing a new product or new technology. The problem of data gaps is not only inherent in new elements (components/systems), but also occurs in elements in operation. Therefore, when generic data or other observable data are not available or fully relevant, a reliability model can be created and validated using the collected data from similar technical components or systems (theory of similarity and use of analogues), in addition to conventional statistical methods, it is possible to use a statistical method based on the Bayesian framework of analysis [2,3]. If Bayesian analysis also includes expert knowledge of the subject area, such as information specifically related to the manufacturing process and experience of similar components and systems, then it is a powerful tool for solving the problem of parameter estimation in the case of missing or insufficiently relevant data.

In addition to estimation, data for reliability determination can be obtained by calculation and verification or naturally (unfocused), through customer experiences, own production, and other experiences, and through data from relevant service organizations engaged in maintenance work. If the object under consideration is complex (e.g., a TPP system), then the problem of determining reliability is solved if one knows the reliability of the constituent components or at least their "most critical" parts, their interconnection (structure), and operating conditions (constraints and environmental conditions) [8].

It should be emphasized that the verification of reliability, that is, testing the hypothesis in practice, is carried out at all stages of life of development, design, construction, and operation of the facility, and is mainly related to several basic limiting factors—money and time, environmental conditions, and other technical constraints. The reliability verification itself is accompanied by the corresponding mathematical apparatus, with a certain level of confidence in the parameters tested. The inadequate level of reliability during the exploitation of the complex technical system itself and the existence of irrational labor-based investments by eliminating consequences rather than causes clearly indicate the need to harmonize existing methods to achieve optimal reliability and adapt them to the system, with the prior definition and elaboration of the appropriate algorithm.

The reliability analysis of an object involves studying its measurable characteristics, expressed over the probability of performing the required objective function, under given environmental and operating conditions, and for a fixed period of time. Since reliability is defined in a probabilistic sense, the basic quantitative models are mathematical and based on probability theory. This introduces statistical methods to the analysis. In order to make the quantitative model realistic, an understanding of some of the many basic conceptions of failure is required. The future condition and behavior of complex technical systems with a large number of assemblies, subassemblies and their components is determined not only by the initial state and method of management, but also by the requirement for their economy in operation. Therefore, methods for assessing optimal reliability based on economic criteria are becoming increasingly important in the processes of designing and planning of the production, use and maintenance of such systems, as well as its parts. Also, the application of probability theory and mathematical statistics based on the history of failure data is very important for making lasting decisions in the maintenance system, which enables timely action while adequately reducing maintenance costs.

A systems approach is the basis of any reliability analysis [57]. This provides a framework for integrating the various technical, commercial, and management aspects of the analysis. Namely, the main goal of the reliability analysis must be to provide information as a basis for making decisions related to the maintenance of technical systems. This approach to solving real-world reliability problems involves a series of steps (Fig. 8.3), the successful implementation of which requires the application of various concepts and techniques from many disciplines. Reliability and maintenance estimates and predicts the likelihood of failure, whereby the methods used quantify reliability using probability and statistics to predict, measure, and analyze reliability data. Components and technical systems can involve computer programs and people in many different roles as designers, operators, or maintenance personnel. Therefore, reliability can generally be divided into three branches, namely, the reliability of technical components or systems (hardware reliability), program reliability, and human reliability. The links between the technical system, the program, and the human are very important, but they will not be the topic of this work.

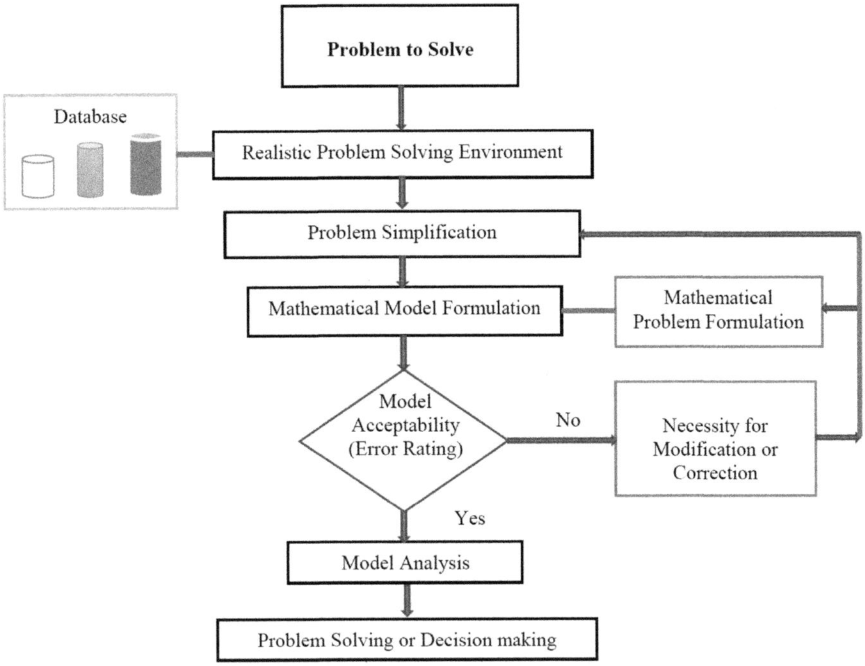

FIGURE 8.3

Systematic approach to solve the problem [50].

Analysis of complex technical systems and their facilities (such as energy-processing facilities) from the aspect of expected reliability and preventive engineering should provide the following [3]:

- assessment of reliability and load reserve of both the elements and the technical system itself, depending on the technological process and operation itself,
- analysis of the technical solution, with the detection of the so-called reliability bottlenecks, determining the operating mode and location of the technical system within a higher hierarchical level,
- special emphasis is given to the design and design process, where there are great opportunities to ensure the optimum level of reliability by optimally increasing the reliability of the operation of all elements in the structural scheme of the technical system, the choice of preventive overhaul plan, with minimal cost savings,
- display of reliability indicators of the complex of the technical system in function of the technological scheme and its tearing, with minimal reduced costs,
- display and ranking of reliability indicators of the most critical circuits, i.e., elements, depending on their parameters and characteristics, with minimal reduced costs,
- creation of unique baseline data for further research and stochastic analysis and modeling,
- defining, through an algorithm, a basic way of determining or confirming the reliability level of a complex technical system,

- accelerating the reliability assessment tests by increasing the effectiveness of the proposed model (shortened test plan for assessing the reliability of both the elements and the system as a whole), adapted to a complex technical system,
- defining the necessary activities to improve and/or optimize the reliability of a complex technical system, and
- development of a general or modified mathematical model for optimal reliability, development of reengineering processes, and defining the level of reliability and maintenance flows with the basic contours of the expert system.

The process begins with defining the real issue under consideration, setting goals, boundary conditions, and limitations of the analysis (defining a set of necessary correct, relevant, and timely inputs available to make a decision). Usually, the analytical approach to problem-solving also requires the realization of certain assumptions and other simplifications, with deviations to a greater or lesser extent than the actual environment. This is followed by the characterization of the system in such a way as to provide the necessary clarity of the essential details of the problem, that is, their elaboration and adequate modeling.

Modeling the reliability of technical systems, in addition to selecting a model, also requires parameter estimation and validation, with an assessment of model acceptability. In the initial phase of reliability analysis, the modeling process begins with a quantitative (graphical) model, usually with a block diagram showing the structure of the system (representation of parts and components of the system and their connections, which can be done at different levels of complexity and detail) (Fig. 8.4).

System reliability depends on the structure and number of elements, and also on the characteristics of each element. When the elements are integrated in different ways into the system, the task is to determine the reliability of the system by the reliability of its elements (Fig. 8.5). The mutual coupling of the system elements can be regular (serial), parallel, quasi-serial, passively parallel, etc. Furthermore, the qualitative model indicates the type and level of data, as well as other information required for analysis.

The next step in the process is to assign measurable reliability characteristics to the successful operation of the blocks in the diagram. Reliability measures are usually estimated on the basis of data available from generic data sources (Table 8.1).

If the generic data are not completely relevant, they can be adapted by expert opinions. The reliability of the system is calculated using a logical structural model (Table 8.2). The estimation of these probabilities is one of the most important aspects of reliability analyses, so the mathematical model must be simple enough to handle the available mathematical and statistical methods.

After the mathematical model is elaborated, it is necessary to check its real applicability (error rating). If the error is significant, changes are made until a sufficiently realistic representation of the real problem is reached. Only on the obtained adequate model it is possible to carry out additional analyses and, if necessary, optimization of certain parameters according to predefined criteria. This requires the use of mathematical techniques in probability theory, statistical analysis, stochastic processes, and optimization [57]. Reliability theory provides concepts and tools for this purpose. A final, very important step in the process is to interpret the results of mathematical and statistical analysis in the context of a real problem [50].

An important step in the security analysis, and therefore in the reliability of the technical systems, is the mere standardization of security, that is, the formulation of system security requirements. However, the problem of forming a minimum sufficient set of indicators characterizing the considered property of a particular system has not yet been fully resolved. Depending on the system under

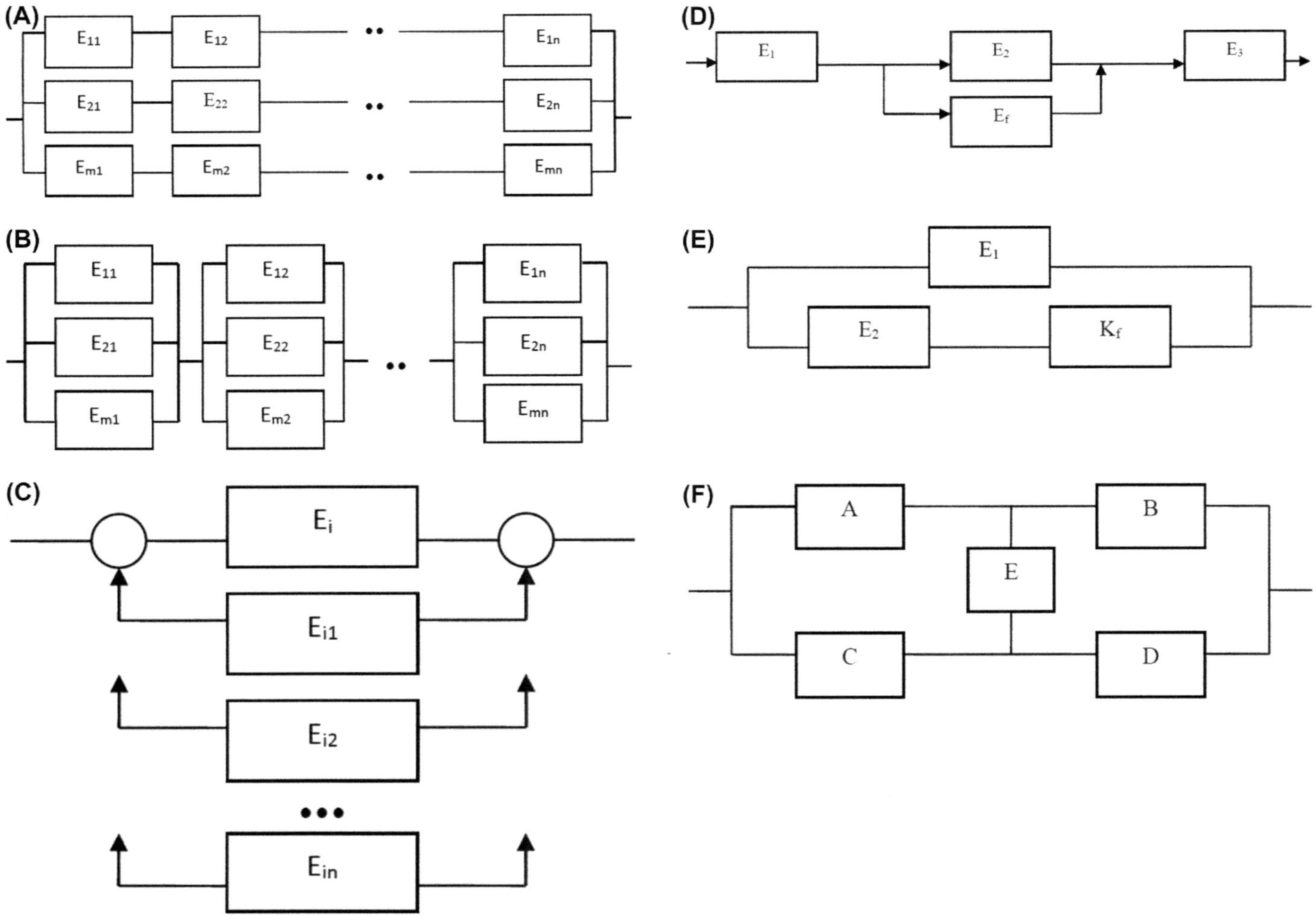

FIGURE 8.4

Depiction of different block diagrams in terms of reliability: (A) block diagram in terms of reliability for a system composed of multiple parallel connections of regularly connected elements; (B) block diagram in terms of reliability for a system with "separate" redundancy; (C) block diagram in terms of reliability for the passive-parallel connection of system elements; (D) block diagram in terms of reliability for the quasi-common connection of system elements; (E) quasi-parallel connection of system elements; (F) bridge connection of system elements.

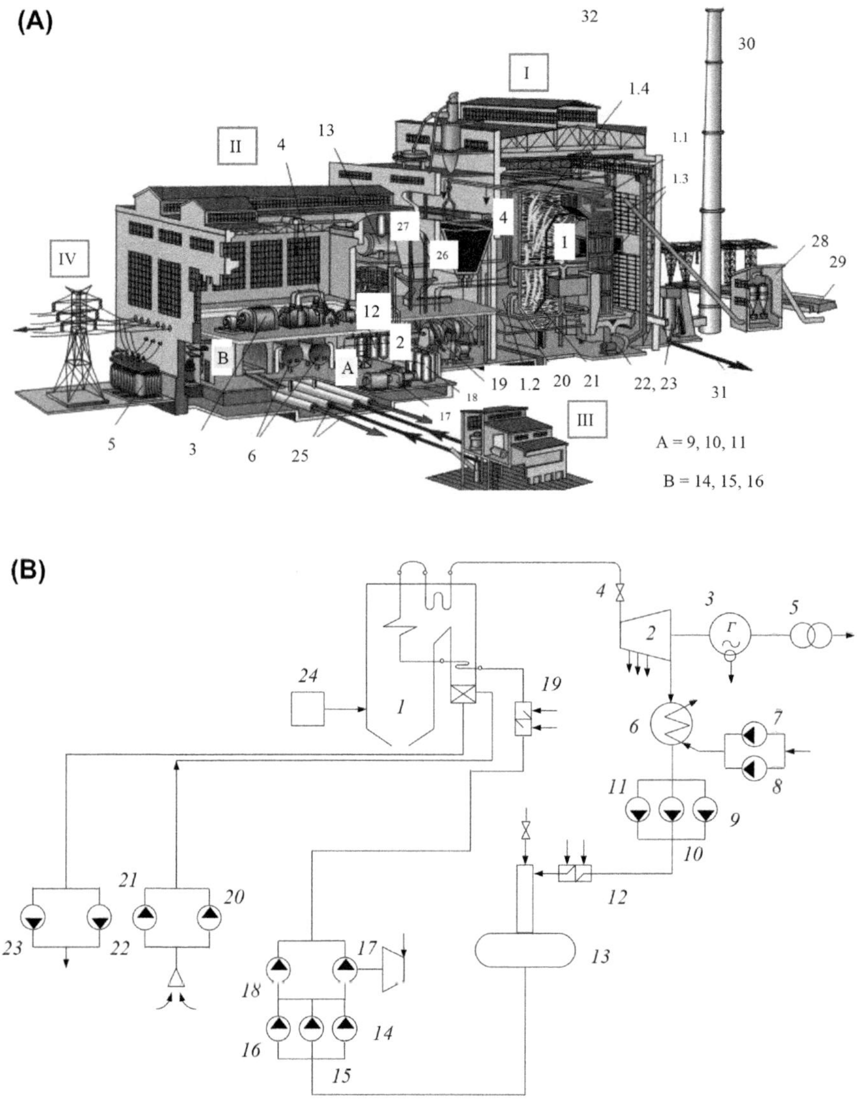

FIGURE 8.5

(A) Layout thermal scheme of the thermal power unit; (B) design thermal scheme of the thermal power unit [58].

Table 8.1 Reliability indicators of power plants [58].

Power equipment	Rated power, MW or productivity (capacity), t/h	Reliability indicators	
		ω, 1/yr.	T_V, h
Electric generator	150–165	0.55	91
	180–210	0.87	49
	250–300	0.59	66
	500	4.48	134
	800	0.89	179
Turbine plant	150–165	0.97	43
	175–210	1.45	45
	250–300	2.1	44
	500	4.22	85
	800	2.66	99
Boiler plant (steam generator)	420–480	6.14	47
	640–670	6.14	47
	950–1000	4.05	35
	1600–1800	6.59	56
	2500–2600	9.08	50
Power pumps: - Electro pump - Turbo pump	–	0.22 1.56	30 37
Heaters: - Low-pressure heaters - High-pressure heaters	–	0.024 0.22	33 21
Flue gas fan	–	0.25	27
Chimneys	–	0.27	24
Condensing plant	–	0.18	22
Regenerative air heater	–	0.28	16
Circulation pump	–	0.34	94
Condensate pump	–	0.22	37
Fuel supply (gas oil)	–	0.13	12
Degasser (deaerator, degasser)	–	0.01	33
Steam boilers	–	1.05	72
Valves, stop valves	–	0.006	27
Control armature	–	0.01	17
Technology protection	–	0.033	52
Automatic control system	–	0.01	56

Table 8.2 Baseline data and calculated reliability indicators for the thermal power unit given in Fig. 8.5 [36].

Name of the element	# in Fig. 8.5	Failure flow parameter (failure rate) ω, 1/yrs.	Generation time T_V, h	Product ωT, h/yrs.	Failure probability, P
Steam boiler (steam generator) 1.1. The steam superheater 1.2. Boiler firebox 1.3. Water heater (Economizer) 1.4. Steam boiler evaporator	1	6.69	38	254.2	$2.9 \cdot 10^{-2}$
Turbine plant	2	2.55	68	173.4	$2 \cdot 10^{-2}$
Electric generator	3	0.59	66	39	$0.4 \cdot 10^{-2}$
Steamships	4	2.1	38	79.8	$0.9 \cdot 10^{-2}$
Transformer plant	5	0.02	26	0.52	$6 \cdot 10^{-5}$
Condensing plant	6	0.18	22	4	$4 \cdot 10^{-4}$
Condensate pumps	9, 10, 11	0.22	37	8.1	$9 \cdot 10^{-4}$
Low-pressure heaters	12	0.024	33	0.8	$9 \cdot 10^{-5}$
Degasser (deaerator, degasser)	13	0.01	33	0.33	$4 \cdot 10^{-5}$
Buster pump	14, 15, 16	0.35	19,4	6.8	$8 \cdot 10^{-4}$
Power turbo pump	17	1.56	37	57.7	$6.6 \cdot 10^{-3}$
Power supply pump	18	0.22	30	6.6	$8 \cdot 10^{-4}$
High-pressure heaters	19	0.22	21	4.6	$5 \cdot 10^{-4}$
Flue gas fans	20, 21	0.25	27	6.7	$8 \cdot 10^{-4}$
Flue gas system (chimneys)	22, 23	0.27	24	6.5	$7 \cdot 10^{-4}$
Fuel supply system (gas oil)	24	0.13	12	1.6	$2 \cdot 10^{-4}$

Table 8.2 Baseline data and calculated reliability indicators for the thermal power unit given in Fig. 8.5 [36].—cont'd

Name of the element	# in Fig. 8.5	Failure flow parameter (failure rate) ω, 1/yrs.	Generation time T_V, h	Product ωT, h/yrs.	Failure probability, P
Circulation pumps	7,8	0.34	94	32	$3.6 \cdot 10^{-3}$
Chilled water pipelines	25	–	–	–	–
Coal bunker	26	–	–	–	–
Coal dust bunker	27	–	–	–	–
Crushing plant at the depot of the thermal power plant	28	–	–	–	–
Coal storage at the thermal power plant depot	29	–	–	–	–
Chimney	30	–	–	–	–
Slag and ash output	31	–	–	–	–

consideration, safety, i.e., reliability, is the result of the superposition of other more "elementary properties," such as mechanical strength, stability, fire resistance, elasticity, etc. [2].

The formation of a database and the grouping of scientifically expert methods for reliability assessment, with a critical analysis of the most commonly used ones and their adaptation to the specificities of a complex technical system, have served to form more modified methods, which as a result have a time dependence on the reliability of work and the probability of failure or failure of the technical system. At the same time, the most influential elements within the complex technical system were singled out and ranked by their importance in terms of increasing the level of reliability. The reciprocal effect and potential failures of the TPP at the level of consideration of individual units should be taken into account through the following characteristics:

- variety of constructive solutions from the aspect of principles of action and inter-subordination and coherence,
- the nature and modes of exploitation during the foreseen useful life, form of loading, liability under conditions of reliability, safety, and protection,
- the form and character of the occurrence of the failure and its effects on other parts,
- benefits for overhaul, possible forms and ways of its realization,
- seriality of details,
- the ability to perform appropriate experiments and tests under conditions that are very close to realistic and the availability to control and perform technical diagnostics.

4. Operation and maintenance of steam turbines as a condensation thermal power plant

The optimal management of complex technical systems is mainly based on quantitative assessment and complex optimization of reliability depending on the way it is provided at different intermediate stages and plant levels as a complex technical system. It should be emphasized that the optimization process represents only one link for long-term optimal management of the higher hierarchical system, which is achieved at lower hierarchical levels. The task of optimizing reliability for new plants comes down to the joint choice of the reliability indicators themselves and defining ways of their provision.

The exploitation of a steam turbine as a complex technical system primarily involves the realization of the projected objective function (production of electricity, heat, and process steam), with as little downtime as possible. Raising the level of utilization of the turbine and its auxiliary equipment to a higher level is achieved by quality and by the manufacturer of the equipment prescribed by the operation and maintenance. The basic parameter characterizing methods for maintaining them is information about their current technical condition and reliability.

Functional adaptability of technical systems is the least explored feature of technical systems, characterized as the ability of technical systems to maintain operating conditions and when changing operating conditions or changing input parameters as a result of operation of another parent technical system (e.g., operation of an energy facility within an electric power system), under which operating conditions mean environmental conditions (temperature, pressure, humidity, dust, vibration, magnetic and electromagnetic fields, dynamic influences, radiation, etc.). Sufficient reserves of material resources and a constructive solution of technical systems enable the system to remain in working order and the required economy and with acceptable degradation of the inventory of resources.

On the other hand, the functional dependence of the observed technical system within the system from the operating and wider environment also enables the functional adaptability of the system to change inputs resulting from the operation of other systems (complex serial-parallel and parallel-serial configurations of technical systems, quasi-serial and quasi-parallel configurations of technical systems, application of passive and passive-parallel configurations, etc.). The determination of the annual needs for certain forms of provision within the higher hierarchical level in the event of complete or partial plant failures as well as the amount of costs related to unscheduled maintenance and unplanned overhauls are purely incidental. Any change in the level of confidence will directly affect the change in investment required. The additional effects, which can be achieved in some forms of provisioning within the higher hierarchical level, are related to the basic effects and investments.

The choice of adequate maintenance technology has a dominant influence on the efficiency of the turbine equipment itself. The likelihood that a steam turbine will successfully enter into operation and perform the required function of the criterion within the tolerance limits for a given period of time and given environmental conditions (operating temperature, pressure, humidity, allowable vibration, noise and shocks, changes in operating mode parameters, etc.) is steam turbine efficiency.

The performance indicator is characterized by a single (unit parameter) or several performance properties (complex parameter), such as:

- reliability (the ability of the turbine to maintain continuous operating capacity within the tolerances during the calendar time period, quantified through indicators: probability of trouble-free operation, mean operating time, failure rate, and failure density),

- maintenance convenience (ability of the turbine to prevent and detect failures and damage, to restore serviceability and correctness through technical service and technical repairs, quantified through: probability of renewal for a given calendar time period, mean renewal time, and renewal intensity),
- durability (ability of the turbine to maintain operational capacity from the very beginning of its application or exploitation to the transition to limit states in which certain delays in the implementation of certain activities for maintenance, maintenance and repairs are possible, defined through indicators: medium resource, gamma-percent resource, mean lifetime, gamma-percent lifetime),
- persistence (the ability of the turbine to maintain a constant warm reserve, storage, and/or transport).

The optimal management of a complex steam turbine system must be based on the evaluation and complex optimization of the reliability indicators, depending on the means of providing them and the hierarchical level of detail as well as the current life cycle phase. For this reason, the optimization process encompasses basic structural, parametric, and structural solutions related to the steam turbine technical system itself and its associated equipment by changing its most important characteristics: efficiency (most often energy), maneuverability, reliability, and economic efficiency as a whole. The set of optimization goals is concluded in the overall choice of reliability indicators and possible ways to achieve them, given the already established rules related to the higher hierarchical level of the system (TPP, electricity system (ES)) [53,54].

An effective approach to increase efficiency is the timely and reliable identification of the condition of the equipment using technical diagnostic measures. Using adequate methods of technical diagnostics makes it easier and more reliable to diagnose the condition of individual elements of a steam turbine (rotor, blades, bearings, housing, regulating system, protection systems, foundations, etc.), on the basis of which further activities are carried out leading to the realization of the set goal, that is, maintaining the projected goal function.

The increasing trend of implementing optimization of the process of production of electricity, heat, and process steam by economic criterion aims to reduce costs and increase productivity. In order for optimization to be carried out effectively, it is necessary to define process-influencing factors as well as their possible influences on the production process itself. However, in order for the TPP and the steam turbine plant to operate optimally and most productively (when needed most), it is necessary to have reliable equipment necessary for the process to proceed.

On the other hand, operation of equipment over a period of time under conditions dictated by process optimization according to economic criteria may adversely affect the length of the basic life period foreseen for exploitation, primarily due to operation under different mechanical conditions than prescribed (overload, shortening of maintenance period and delay of certain repair planned activities, recommended by the equipment manufacturer, forced system operation, etc.). Such work very often results in increased stresses in the elements of the equipment, especially rotary machines, and leads to a faster degradation of the mechanical condition. This fact is often neglected, which in a long period of time can completely undo the positive results previously achieved by process optimization (increased maintenance and repair costs, as well as losses due to longer delays in the exploitation process). The operation of the turbine under mechanical and process unfavorable conditions generates variable stresses of the material, which leads to damage to some of its assemblies, cracks, and fractures with often catastrophic consequences. The development of microprocessor devices for complete monitoring and analysis of operation, with the possibility of determining the current mechanical state of the steam turbine, enabled a completely different approach to the maintenance of the steam turbine plant, which increasingly applies the maintenance method according to the condition (compared to the previously planned maintenance according to the constant date). In doing so, all maintenance activities are carried out only when necessary or when the condition of the turbine requires it. The deterioration of the maneuvering characteristics, in the event of partial failure, in the form of a decrease in the rate of increase or decrease of the load, leads to the appearance of deviations from the given load graph and its variability. That is why, within the higher hierarchical level, there is a need for additional high maneuverability capacities, which can be compensated operationally for these shortcomings.

On the other hand, if one looks at a complex technical system (such as a TPP) within a higher hierarchical system (such as a power system), any increase in the amount of power required to cover its own consumption additionally initiates the possibility of more failures and a gradual deterioration of the plant's consumption, not only electricity (heat) but also used fuel.

5. Steam turbine as a technical system

A steam turbine is a drive unit in which the potential energy of steam is transformed into kinetic energy, and then into mechanical work of rotating the turbine shaft. The turbine shaft is connected to the working machine (electric generator, locomotive drive wheel or some other plant, etc.) via a coupling (directly) or through a gear transmission (indirectly). It is most commonly used in power generation for electric generators, and in the industry smaller turbines are used to power large pumps, fans, compressors, and the like [52].

The steam turbine can only be used in conjunction with other energy equipment and plants that as a whole make up TPPs (Fig. 8.6). The TPP consists of [52]:

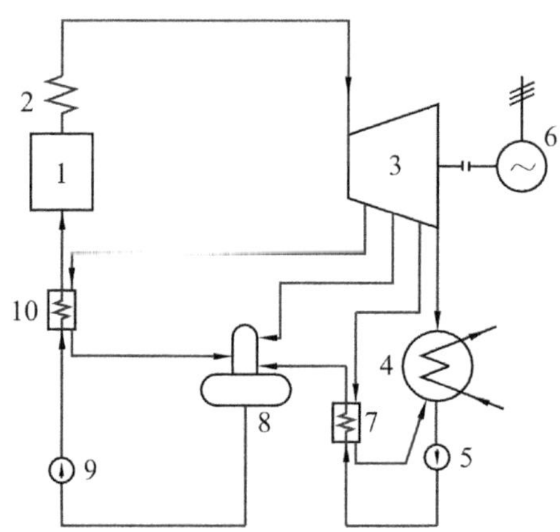

FIGURE 8.6

A simplified scheme of a steam turbine plant.

- a steam generator (steam boiler) in which the feed water is converted into cosaturated steam under certain pressure,
- steam superheater, in which the temperature (superheat) of the steam is increased to the set value,
- a steam turbine, in which the potential energy of the steam is transformed into kinetic, and this into mechanical work on the turbine shaft,
- a condenser intended for condensing the steam produced in the turbine,
- condenser pump, which returns condensate from the condenser to the system (water tank with degasser),
- a feed water tank with a degasser, in which oxygen and other gases are removed from the feed water,
- a feed pump, which supplies feed water to a steam boiler, as well as an electric generator that produces electricity.

If we apply the structural classification according to Pahl et al. to a steam turbine plant within a condensing power plant, the following basics of this system can be distinguished [59]:

(a) power system, condensing TPP, main propulsion facility, turbo generator, steam turbine plant, auxiliary equipment, machine, assembly, and components;
(b) the process of transformation of energy, the material expelled (two flows of matter: fuel-air-combustion products and feed water-fresh vapor-condensate);
(c) functional relationship (main drive: steam boiler-turbine-generator);
(d) working relationship (technological schemes, optimization models);
(e) structural interrelation (disposition and situation plan);
(f) systemic interconnection (multicountry energy community—ES, production unit—transmission system-distribution system); and
(g) objectives and constraints (energy efficiency, economy, availability, and operational readiness, as well as environmental impact).

Technical tasks are performed with the help of technical artifacts (steam boiler, steam turbine plant, and electric generator), arranged in order of complexity—plants, equipment, devices, circuits, and components. These terms may not have the same application in different areas. The device consists of assemblies and components, while the control equipment is similarly used in plants and devices, and may be composed of assemblies and components, some of the smaller devices. In addition to power generation, steam turbines can also be used as propulsion machines in various fields of industry and traffic. High operational safety, which is the result of a convenient way of converting heat into kinetic energy and then kinetic energy into mechanical work in the most favorable form of rotation of the rotor, increasing demand for electricity, is increasingly satisfied through the construction of blocks with as much unit power (above 1300 MW). High-power condensing power plants using fossil fuels are built with the highest parameters of fresh steam and the lowest end pressure (deep vacuum), thereby increasing the efficiency coefficient (EC) of the TPP. Fossil fuel−fired power plants most often use superheated steam. Today, the steam temperature in front of a turbine with classic supercritical parameters usually reaches a value of 540−560°C, at a steam pressure in front of the turbine up to 23.5 MPa. However, modern ultra-supercritical (USC) TPPs have preturbine steam temperatures of about 600°C ("600°C" turbines) and fresh steam pressures of about 30.0 MPa. Soon, turbines with USC parameters will be built with a fresh steam temperature of about 700°C and a corresponding pressure of over 35.0 MPa. Fossil fuel condensing type power plants operate per cycle without steam preheating, at initial pressure up to 8.8 MPa and overheated steam temperature at the inlet to the turbine up to 535°C, respectively, per cycle with intermediate heating having initial pressures equal to 12.7 and 23.5 MPa, where the temperature is in the range 540–560°C, at the so-called classic TPPs. Of course, thermoelectric power plants with ultracritical parameters have one or two steam superheats, with significantly higher parameter values, as previously pointed out. Under these conditions, at final pressure values of 0.0035–0.0045 MPa, the vapor humidity at the outlet of the turbine flow does not exceed 13%–14%. On the other hand, since the steam unit operates with a closed cycle of water and water vapor, without direct contact with the fuel, all types of fuel can be used. An example of a steam turbine plant for a condensing TPP is given in Fig. 8.7.

The steam turbine system involves the transformation of the potential energy obtained from the combustion of fuel (thermal energy of steam) into the kinetic energy of the ordered fluid stream by an adiabatic propagation process, and then this energy into mechanical work, which is transmitted through the turbine shaft to the working machine. For the purpose of describing the functional relationship of the steam tube as a technical system, it is necessary to determine the general function of the system with clear and easily made relationships between inputs and outputs. In doing so, the inputs and outputs of all sizes involved in the process must have known real or necessary characteristics.

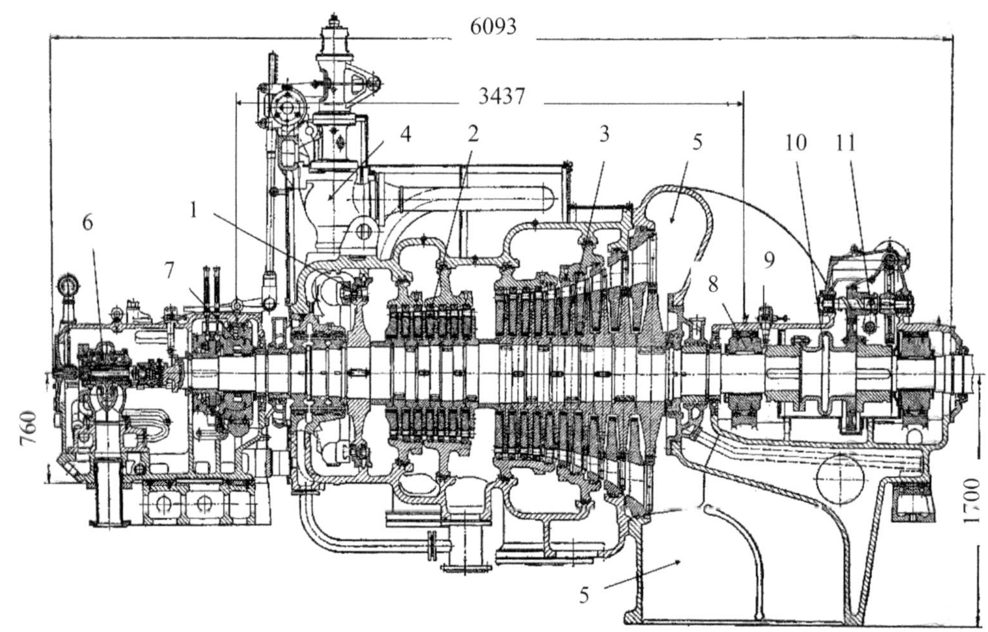

FIGURE 8.7

Longitudinal section of a 12 MW condensing multistage steam turbine [52]: 1—control degree (Curtis circuit); 2—first turbine section with seven pressure stages; 3—second (last) turbine compartment with 10 pressure stages; 4—control valve, 4 pcs.; 5—output housing according to the capacitor; 6—main oil pump; 7—front radial-axial bearing; 8—rear radial sliding bearing; 9—indicator of relative expansion of the rotor relative to the housing; 10—half-point coupling; 11—turning device.

The general function of a steam turbine plant can often be divided into identifiable subfunctions (steam turbine stage as the base unit), which correspond to the subtasks. The relationship between subfunctions and general function is very often governed by certain constraints, since some subfunctions must be satisfied before others. There are two basic parts of each turbine: a stator with fixed stator blades housed in a housing and a rotor blade impeller arranged by rim perimeter (Fig. 8.8).

All thermodynamic changes and energy transformations are realized in the channels forming the rotor and stator blades through which the working fluid flows. One or more impellers are attached to a shaft, by which torque is transmitted through the coupling to the impeller. A shaft with one or more impellers is a turbine rotor. To prevent the fluid from flowing into the environment, the workspace is enclosed in a turbine housing, which at the same time protects the rotor against foreign elements and possible damage. The rotor lies in the supporting bearings, which are assumed by the radial forces, while the axial force is taken over by the axial bearing. On the other hand, the bearing and axial bearings serve for axial and radial rotor guidance, that is, to provide axial and radial clearance when rotating. At places where the impeller passes through the housing, labyrinth-like, noncontacting seals are installed to prevent the flow of working fluid into the environment. In order to convert thermal energy into kinetic energy of a regulated current, suitable elements are needed in the form of channels with variable cross-sections, which make up a conduction apparatus or a fixed grate (recess) of a single turbine. Gas or steam from the funnel enters the impeller at an increased speed, where the reaction turbines continue the process of converting thermal energy into kinetic energy of the ordered current. On the other hand, the shape of the impeller channel must allow the kinetic energy of the fluid to be transferred to the rotor. The funnel and work blades, arranged in terms of circumference and circuit, form channels, in which appropriate thermodynamic changes and energy transfer are made. These are commonly referred to as pre-roll bars and circuit bars. In the case where the blades or nozzles in the grilles of the bell are compared on only one part of it, it is a bell or a turbine with partial charge. Very often processes in heat turbines are performed in several stages. In multistage turbine plants, fluid

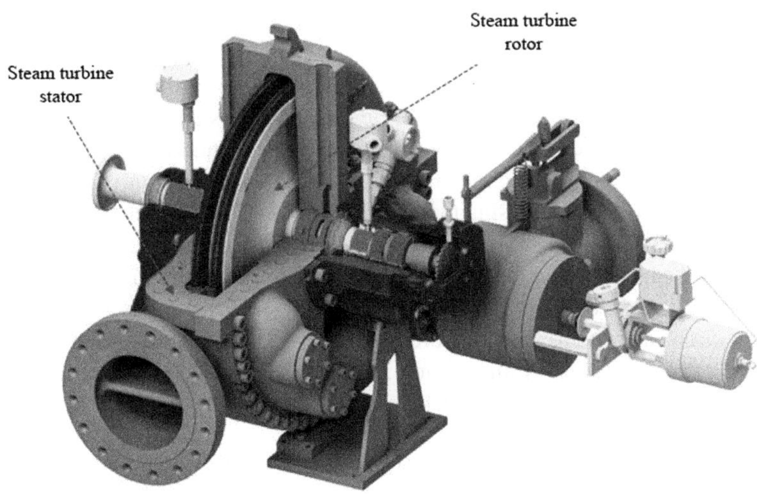

FIGURE 8.8

Single-stage heat turbines [52].

energy from the same fluid is successively transferred to the blades of successively aligned circuits, which make up the turbine stages. The process being carried out is adiabatic with friction, without bringing heat to the working fluid during the process and without, practically, greater heat removal to the environment. Often, technical solutions are used to warm the working fluid after passing through a number of stages.

The multistage steam turbines are made of action and reaction type or combined type, and are intended for use as condensation and heating plants with regulated and unregulated steam abstraction. Unlike single-stage turbines, these turbines can be designed for high unit power and high steam parameters. The number of stages in multistage turbines ranges quite wide: from 3–5 to 30 more. This number depends on the parameters of the fresh and produced steam, its flow through the turbine, the type of control stages, the required economy, and the like. Modern steam turbines are often run with action stages in the area of elevated steam pressure and with reactive stages in a low-pressure turbine. Single-stage steam turbines are characteristic for the beginning of their application, for the least unit power and for propulsion of spare or auxiliary machines, and they were used for propulsion in emergency situations. Modern steam turbines in TPPs and nuclear power plants now have available thermal drops of up to 1600–1800 kJ/kg, which cannot be used in a single stage of the turbine. In one action stage of the turbine, a heat drop of 100–120 kJ/kg can be utilized. Other limitations can be observed in the use (transformation) of energy in the steam turbine. Thus the circumferential velocity, limited by the diameter D and the turbine speed n. The speed is defined by technological requirements, such as the construction of an electric generator (number of poles) and the frequency, Hz. The diameter is also limited by the extensive speed of the working blades, i.e., by the centrifugal force, $P_{cf} \approx u^2/R$. The reason for this limitation is the strength of modern structural materials, which can withstand such high thermal and mechanical stresses. For this reason, turbine plants are built with a greater number of stages, with a gradual expansion of steam, with the extensive velocity of the blades in condensing steam turbines with several stages being 120–150 m/s, for high and medium pressure stages, respectively, 350–450 m/s, for low-pressure degrees. Basically, it should be borne in mind that due to the aforementioned limitations, the bulk velocity u is taken to be less than 600 m/s and most often up to 400 m/s. In most cases, the Mach number is less than one.

Technical artifacts or systems (Fig. 8.9) do not operate in isolation and are mainly parts of a larger system (condensing TPP). The total function of the steam turbine system includes the fulfillment of the primary function in the form of electricity production and the secondary function in the form of heat energy production and technological steam. At the same time, the working (personal) staff and their activities (input database for the condensing thermal power plant system) should not be neglected. The system responds with feedback or signals that trigger further action (stationary and nonstationary modes). In this way, people support or enable the intended operation of the technical system. In addition to the desired inputs, this technical system may also be affected by unwanted environments (ES as a higher hierarchical system, impacts of force majeure, such as floods, earthquakes, or warfare) or from neighboring systems (power transmission line failures at the energy community level). Such disturbing actions can cause side effects, such as deviations in the quality of electricity delivery, continuity of delivery, etc.

Also, there is a possibility that, with the desired operating effect, an unwanted one, such as vibrations, may appear from a single component within the entire system (steam turbine rotor). These adverse events can have a negative impact on people and the environment. Therefore, the difference between the desired action, the input action, the feedback action, the input disturbance, and the side

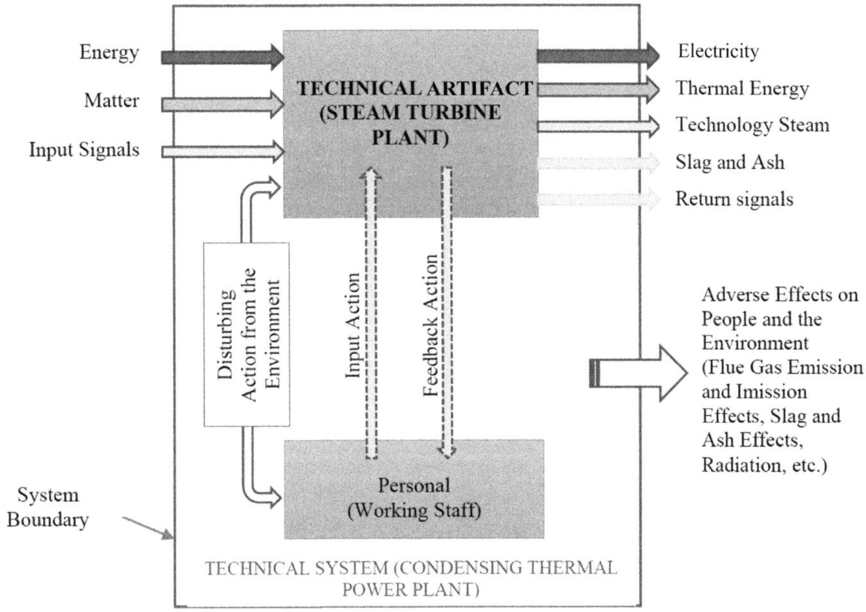

FIGURE 8.9

Interrelations in the technical system of a steam turbine including humans.

effects must be determined, and the main goals and limitations should be determined. Technical artifacts can therefore be viewed as systems related to the environment using inputs and outputs. The technical system of a steam turbine can be divided into subsystems (Figs. 8.7 and 8.8). What belongs to a particular subsystem is determined by the system boundary? One of the possible subdivisions of the subsystems for the steam turbine plant is also given in Fig. 8.10, wherein the inputs and outputs exceed the limits of the subsystems. In this way it is possible to define a particular system in every degree of its abstraction, analysis, and division.

A technical system defined by the US Army MIL-STD-882E: 2012 system standard as: "the organization of hardware, software, materials, facilities, data, and services required to perform a designed function within a specified environment with specified results," as well as a definition given by the standard ISO 9000: 2000, which defines a technical system as a set of interconnected or acting elements [60,61]. Elements of that separate unit are used together in the required work (propulsion) or supported environment to perform a specific task or accomplish a specific purpose, support, or mission requirement. The connection with the people (working staff) servicing the technical system makes it possible to perform the function of control, cleaning, lubrication, testing, and maintenance of the system or may be its users. Therefore, the technical system's commitment also depends on its connection to the environment, so it is necessary to study how these connections affect the system as a whole. An example of a technical system and its connections is shown in Fig. 8.11.

Fig. 8.11 shows the following elements [50,52]:

1. The system that is the subject of analysis usually consists of several function blocks.

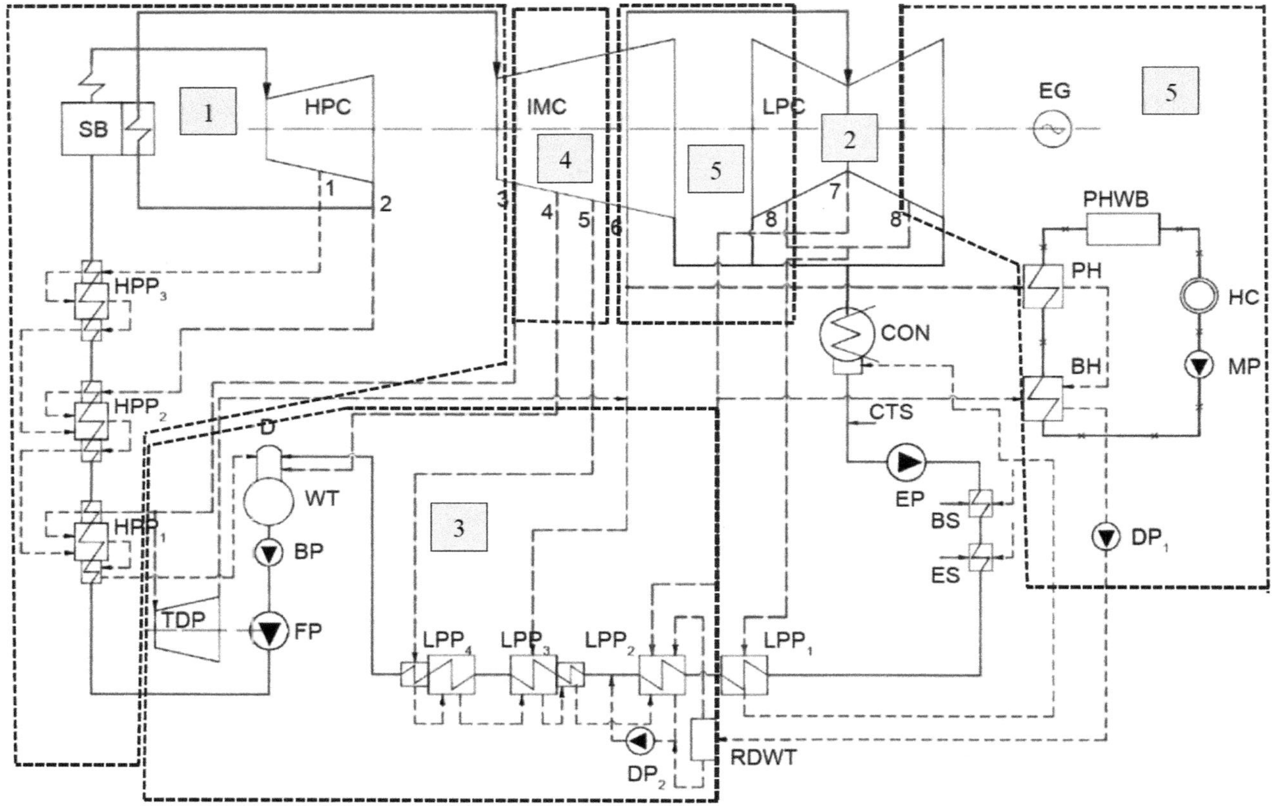

FIGURE 8.10

View of the division of the condensing thermal power plant with the K-300-240-3 LMZ steam turbine system into subsystems: 1—Unit for optimization of upper ("hot") parameters of steam power plant (parametric of fresh and heated steam, number of steam heating, high-pressure turbines and introductory part of turbine of medium pressure, boiler and steam lines between turbine and boiler, feed pump, and high-pressure heaters); 2—Unit for optimization of the lower ("cold") parameters of the steam power plant (pressure behind the last stage of the low-pressure turbine, pressure of condensation of steam in the condenser, last stage of the turbine, outlet sleeve, condenser, pumps and pipelines of cooling water and the whole cooling system, last and sometimes the penultimate regenerative low-pressure heater with parameters that define them); 3—Regenerative heating system without regenerative high-pressure heaters and the coldest low-pressure heaters with associated piping, drainage pumps, and fittings); 4—The turbine flow section, as a separate optimization unit, includes high-, medium-, and low-pressure flow parts, except for the last stage of the low-pressure turbine; 5—For a combined heat and power turbine, the subject of optimization is the removal of steam from the turbine for the needs of the heat consumer.

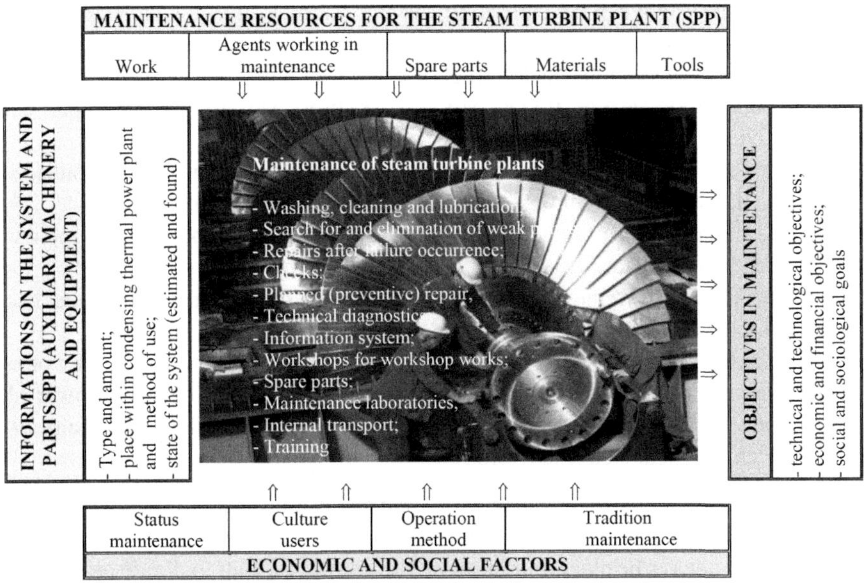

FIGURE 8.11

Technical system of thermal power plant and its connections.

2. The system boundary that determines which elements are considered part of the system and which are the elements of the extra system.
3. Outputs that can be divided into two groups:
   - Desired outputs: These are the desired results of the required function, such as materials, energy, heat, or information.
   - Unwanted outputs: Almost all systems can produce unwanted outputs. Such exits can be air, water, or earth pollution, and injuries and negative impacts on human health and the environment of the ecosystem.
4. Inputs which can also be divided into two groups:
   - Desired inputs: These are the materials and energy used by the system to perform the required function whereby the quality and quantity of the desired inputs may be subject to radiation.
   - Unwanted inputs: These are inputs associated with desired inputs that cannot be considered as normal variations of desired inputs. For example, the unwanted inlet of an impurity particle into the working fluid on the suction side of the pump to supply the TPP with the required quantities of water.
5. Boundary conditions: the operation of the system may be subject to a number of boundary conditions, such as risk acceptability and environmental criteria determined by the government (legislation related to nature protection, legal solutions related to occupational safety, and fire protection) or companies (internal regulations, regulations or decisions of governing bodies).
6. Support: The system usually requires support functions such as cleaning, lubrication, maintenance, and repair, and after the end of its basic working life and revitalization, reconstruction, and modernization for the extended service life of the technical system.

7. External threats: The system can be exposed to various external threats, which can directly affect the output parameters from the system (transmission line outage). Also, other threats can affect the input parameters of the system (lack of fuel or its poor quality, lack of cooling water, damage to the flue gas purification system, floods, earthquakes, fires, war, etc.). In principle, external threats can be divided into four groups:
   - Natural threats to the environment: These are threats to the system from the outside environment (floods, fires, earthquake, storms, war actions).
   - Infrastructure threats: These threats are caused by infrastructure shortages or interruptions, such as interruption of electricity supply or communication (transmission line).
   - Social threats: Threats from individuals and organizations such as arson, sabotage, hacking, or attack by computer viruses.
   - Threats from other technical systems: Operation of other systems located nearby or related to the technical system that is being analyzed (electricity production by hydropower plants or other production facilities with cheaper production costs per 1 kWh of electricity or 1 kJ of heat or ton technological steam delivered).

When analyzing a technical system (such as a condensing power plant or a steam turbine plant within a condensing power plant) and its connections, the distinction between unwanted inputs and an external threat need not always be clear. Furthermore, it is not important to divide different inputs, but it is important that all inputs and threats are identified and considered in the analysis. A complete description of a technical system requires the definition of: a structural model that shows the parts of the system (that perform functions), a functional model (that describes different functions) and a behavioral model (that shows how functions are performed) [62].

The way of considering the technical system of a condensing TPP/steam turbine depends on its connection with the working personnel (people), the life cycle phase (design and design, production and installation, commissioning and trial work, basic working life, extended revitalized working life, withdrawal from use, or change of purpose), and the objectives of studying the observed technical system (finding the critical assembly/component in terms of frequency of occurrence of system failures, optimization processes, maintenance tests, normative tests, etc.). Thus, in the structural description/model, the area of interest is the physical structure of the various subsystems (assemblies, subassemblies) and components and the links between them. The area of interest in the functional description is the various functions of the system (stationary and nonstationary modes of operation, normative tests, etc.). Therefore, the functional model is focused on the goals achieved by the functions and allows evaluating the consequences of loss of function. Because structural and behavioral descriptions are related to function, functional analysis is central and, therefore, paramount. Thus, in the early stages of the design process of a new system, it usually begins with defining a set of desired functions (nominal power, rated useful life, service life, safety, availability, and reliability) and the development of a system turbine capable of fulfilling those functions. This analysis is usually performed by the value analysis method, which consists of a description of the condensing power plant/steam turbine system with respect to the required functions to fulfill the user's requirements. The physical realization of the system functions is not decided at that stage. After this analysis, it is necessary to describe the technical system with the expected functions and their characteristics, which is performed by various technical methods.

In order to represent structural and functional relationships within a technical system, it is necessary to use several types of diagrams (structural or functional block diagrams), which can be quite different in terms of symbols and appearance. The function block denotes a system element that can be a component or a large subsystem. A function block diagram is a graphical representation of the operation of a system. It usually contains inputs (e.g., materials, energy sources) that enter within system boundaries, function blocks that display functions that occur within system boundaries, and outputs (e.g., by-products, electrical and thermal energy, process steam, mechanical work, signals) that leave the system boundary. In doing so, arrows are used to describe material flows, energy, signals entering and exiting systems, and function blocks. Within boundaries, each function block displays the function it must provide to the system to transform the inputs of the outputs.

Functional block diagrams are recommended by the US Army MIL-STD-1629A standard [63], which describes the procedure for performing the failure modes, effects, and criticality analysis (FMECA) method and form the basis of the maintenance-oriented method. Function block diagrams recommended by US military standards MIL-STD-1629A, [63], are describing the procedure for performing the technique failure modes, effects and criticality analysis (FMECA). These diagrams form the basis of reliability centered maintenance (RCM) method, Rausand and Høyland [49,64]. In the process industry, systems are typically represented by a process and instrumentation diagram (P&ID) and flowchart process flow diagram (PFD). These information models show the structure of the plant and the flow of mass and energy under certain conditions.

Unfolding the process on condensing steam turbines also requires monitoring the appropriate possible deviations from the predicted (nominal, design), which is the supervision of the operation of the steam turbine. The monitoring itself is based on an insight into the state of processes and equipment during stationary and nonstationary (transient) modes of operation, then effects related to the regulation and mode of manipulation of the system operator, as well as an assessment of the energy efficiency of the process. This involves the continuous measurement of certain process sizes, the conversion of the signal obtained from the measurement and its processing, with the printing of certain parameters important for the decision of the operator or system owner. The monitoring involves monitoring the parameters (pressure, temperature, flow, chemical composition, etc.) of the working fluid (steam, condensate, cooling water, lubricating oil) on the appropriate equipment (steam turbine, condenser, heaters, pumps, etc.), with the realization of certain measurements characterizing the mechanical (number of revolutions, elongation or displacement, gaps, vibrations, eccentricity) and thermal (temperature of the rotor, housing, bolts, flange, bearings) condition.

The development of computer equipment has enabled the development of modern systems for condensing steam turbine automation in order to accelerate and increase the accuracy of the measurement data obtained, as well as methods for long-term monitoring and editing of databases, with the aim of conducting certain analyses (first of all, analysis of accidents, accidents, and major disturbances) in the system. As a result of the use of computer technology, the technical diagnostics of individual elements of the system or the turbine as a whole is improved (especially important when introducing maintenance according to the condition), better overall economy and greater availability and safety in the operation of the turbine. The development of computer technology has also enabled the accompanying development of expert systems of different purposes, with special participation in energy and process plants. To use them, knowledge bases are needed, encompassing theoretical and experiential knowledge, whereby these bases (as the core of the system) allow for a certain dialogue, then processing of knowledge, as well as for solving and interpreting results. The first expert systems used at turbine plants were diagnostic expert systems, the structure of which was built up of several units (database, analytical problem-solving, data processing and diagnosis, considering possible deficiencies, and establishing the facts).

The possibility of using expert systems at condensing power plants, combined cycle power plants, and industrial power plants goes toward introducing them into the control of technical processes, as an aid to the process regulation and control system and the operator for making corrective decisions. In principle, they cover three areas: current-thermal processes and degree of usefulness, vibrations of turbines and other rotating devices and installations, as well as the condition of the materials of individual elements of the turbine (especially high-load ones) parts. They can also be applied to predicting the duration of thermal and electrical load (consumption), consumption of process steam, as well as ancillary tool in making strategic decisions, such as managing and optimizing the process of electricity and heat production, maintaining and scheduling reconstruction, modernization and extension of work turbine systems lifetime.

The application of expert systems from the point of view of the higher hierarchical system (Power Company, Corporation or Holding company) should be reflected in solving complex optimization tasks related to the operation of individual objects from the system, with special attention paid to increasing their reliability during exploitation as well as to increasing safety and quality in supplying consumers with some energy.

6. Qualitative analysis of a steam tube plant

A technical system usually consists of a number of subsystems and components connected in such a way that the system is capable of performing a number of required functions. Very often, the subsystems and components that make up the system are of different technologies, which increase the complexity of the system. In this chapter, the term functional block will be used to refer to a system element, whether it is a component or a subsystem.

A major interest in reliability engineering is to identify potential failures and prevent these failures from occurring [4]. According to British Standard BS 4778, a malfunction of a functional block is defined as: "Interruption of its ability to perform the required function" [55]. Therefore, in the reliability engineering, it is necessary to identify all the significant functions and performance criteria that relate to each function of the technical system.

The methodological presentation of the scientific prediction of the state or behavior of the technical system is based on deterministic, stochastic, or, most commonly, combined deterministic–stochastic prediction methods [3].

The development of technical diagnostics is made possible by the application of scientific knowledge and technical advances in the fields of measuring technology, cybernetics, theory of scientific prediction, and theory of reliability of technical systems. Functions to be fulfilled by the system of technical diagnostics are given in the form of certain checks of the technical state of the system, checks of working ability, checks of functionality, location of failures at the lowest possible hierarchical level, as well as giving estimates of the remaining time of use or the trend of failure. Undoubtedly, the use of technical diagnostics has opened new possibilities for plant management, which created all the preconditions for a significant reduction of corrective and preventive maintenance activities, while maintaining the same or achieving an even higher level of reliability of the plant as a whole. This is of particular importance for energy-processing plants operating within a higher hierarchical system (power system, oil industry, petrochemical plant, etc.) and for which a high level of reliability is required during exploitation and increased protection for both personnel and personnel of the environment.

Reliability analysis for the purpose of providing reliability estimates plays an important role throughout the life cycle of a technical system, and the ultimate goal of reliability analysis is to model and calculate the reliability of a technical system (Fig. 8.12). Qualitative analysis of the technical system involves a number of activities, such as familiarization with the system, functional analysis, fault classification, and reliability analysis, in order to provide answers to questions regarding the system, the mode of operation of the observed technical system, the mode of occurrence and the assessment of the criticality of the fault on the technical system, as well as the analysis of all consequences of the occurrence of failures, i.e., the appearance of a condition in the failure of the technical system. Functional analysis of the system gives an answer on the way of its operation, with identification and description of all functions of the technical system, taking into account its position within the hierarchical "higher" or superior complex (steam turbine plant within the condensing TPP, condensing thermoelectric power plant within the electric power system, electric power system within energy communities, etc.). Familiarization with the system, which is the first step of qualitative analysis of the system in order to build knowledge about how the system works, must first be done. In this step, it is necessary to determine the modes of operation of the technical system. The system may have the following conditions: normal operation, test operation, operation in unforeseen circumstances caused by failure, malfunction, non-stationary operating modes (commissioning, planned and forced

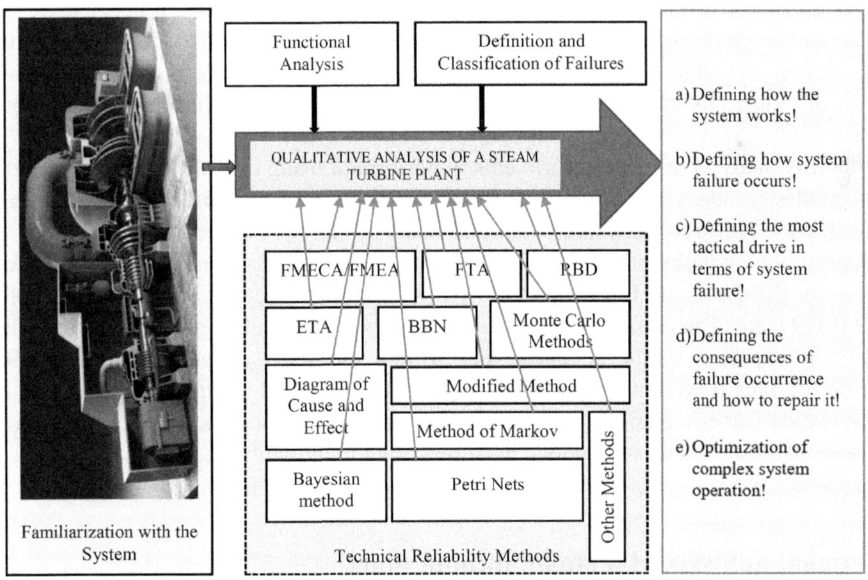

FIGURE 8.12

Quantitative analysis of the technical system of a steam turbine.

shutdown). Familiarization is then continued with a review of the hardware and software documentation (technical drawings, functional diagrams, operating instructions, maintenance instructions, testing procedures, accident procedures, etc.). If test equipment is installed, testing of hardware and integrated software of the system is performed in its working phase, using information on the experiences of similar systems and the actions of the environment and work staff (employees). The purpose of this test is to detect hidden faults. Also, it is possible to get a confirmation of the validity of the realized modifications of the hardware and software of the system.

The optimal management of a complex technical system must be based on the assessment and complex optimization of the reliability indicators, depending on the means of providing them and the hierarchical level of detail of the system as a whole, as well as the current stages of the system life cycle. For these reasons, the optimization process includes basic structural, parametric, and constructive solutions related to the system itself by changing its most important characteristics: energy efficiency, maneuverability, reliability, and cost-effectiveness. The set of optimization goals is concluded in the overall choice of reliability indicators and possible ways to achieve them, given the already established rules related to a possibly higher hierarchical level, design, installation, and exploitation, as well as maintenance of complex technical systems. On the other hand, the realized level of reliability has an impact on the level of failure and its consequences (material, human, and environmental disasters), on the availability of technical systems and on the costs that depend on the level of reliability in the continuous delivery of the product or the realization of the intended service.

Reliability analysis provides answers to the questions: "How can a system be broken?" and "What are the consequences of a failure?" The answers require a previous description of the malfunction, i.e., failure modes that describe how failure is viewed [50]. Since system functions are usually subdivided into subfunctions, the classification of failures by cause and effect is important because it is necessary to link the failure modes with the lower-level functions to the higher level, that is, the main function of the system. Causes of failure are essential information that prevents failure or recurrence of failure. The division of failures by consequences is important because some failure modes are more critical than others.

The reliability analysis of technical systems is performed using one of the technical methods [3]: analysis of modes (modes), effect (consequences) and criticality of failure (FMECA), analysis of failure tree (FTA—fault tree analysis), event tree analysis (ETA), consequently cause and effect diagram, reliability block diagram (RBD), Bayesian method (BM), Bayesian Belief Networks (BBN), Method Markov (MM), Monte Carlo Method (MCM), modified method for estimating optimal reliability (MMEOR), etc. The choice of technique depends on the available reliability data, the type of system, and the expertise of the personnel performing the analysis. Very often, some standards, depending on the type of system, suggest the techniques used, and several other methods are of industrial application (other methods), if they could be applied in process and energy conditions under certain conditions. Many of these methods have been integrated with other analysis tools to improve their applicability [65].

6.1 Functional analysis of a steam turbine plant

In order to be able to identify all potential failures, it is necessary to thoroughly understand the various functions in each function block and the execution criteria related to different functions. A function is a predicted effect of a function block and must be defined so that each function has one specific intended purpose. Therefore, functional analysis is an important step in reliability analysis, which aims to: determine all functions in the system, determine all the functions required for different operating modes (modes) of the system, give a hierarchical decomposition of system functions, describe how each function is realized, determine the connection between function, and determine points of contact with other systems and environments. It is desirable that the name of the function be expressed in such a way that it has a declarative structure in the form of a verb plus nouns by saying what to do, as shown in the example in Figs. 8.10 and 8.11. Function execution criteria are specified by function requirements. If, for example, diesel engine functions to provide torque, the functional requirement may be that the torque on the engine flywheel, at maximum speed is between 345 and 350 Nm. In some cases, some of the functions may have several functional requirements. In order to ensure the required reliability in the operation of steam turbines and to avoid possible downtime due to the failure of some of its components, it is necessary to carry out certain maintenance tasks (audits and overhauls) and to monitor the behavior of the material or to monitor the operation of the turbine. The designed service life of the turbines, which was previously used in the calculations, was 100,000 h, while today it is calculated with the value of 250,000 h (improved materials, more accurate calculation procedures for the structural parts of the turbine, improved and new test methods, as well as the introduction of new criteria and procedures to evaluate embedded material). Extending the service life requires additional assessments due to the need to change the operating mode from basic to peak loads and because of increased requirements regarding the degree of utilization (replacement of worn parts, change of operating mode, setting new intervals in the maintenance process, targeted material testing, etc.). The complexity of the assessment of the remaining service life is often illustrated by providing an overview of the causes of the destruction and breakage of steam turbine structural parts (Table 8.3).

In addition to that, the basic approach to budgeting does not change. The permissible voltage is determined on the basis of the relevant material characteristics and the degree of safety. The introduction of computer data processing provides better and more detailed determination of voltage concentration (voltage state), which is a prerequisite for the design and optimization of structural solutions of vital parts of the turbine plant (construction of multishielded housings, construction of welded rotors with reduced voltage at the same diameter, for better behavior at nonstationary thermal loads, determination of blade loads—model of corrosion fatigue life, etc.). A large number of influential factors that need to be taken into account when assessing the situation relate to the conditions of exploitation, the material, and the calculation procedure selected. It is of particular importance to have data for the commissioning period or a recorded "zero" initial state. From the conditions of exploitation, it is possible to obtain data on all impacts to which the material was exposed during the exploitation, especially from the aspect of realized turbine modes, material weather characteristics and stress distribution across the component, unexpected material behavior in exploitation, and occurrence of failures (hidden material errors, residual voltages, poor maintenance, etc.). It should be noted that newer turbines have built-in special systems that prevent uncontrolled start and unwanted power changes. Also, in the design phase, the expected number of starts has already been evaluated, with the projection of the material behavior in the first 100,000 and 250,000 h of operation, respectively. On the other hand, procedures for testing nondestructive methods have been developed, which very reliably determine the existence of faults even before the turbine plant is commissioned (ultrasound, radiography, magnetic particles, acoustic emission, holography, endoscopy, etc.).

Table 8.3 Overview of the root causes of damage and breakage of structural parts of turbine plants [52].

Destruction as a result of processing errors	Destruction as a result of incorrect structural application or inappropriate material	Destruction resulting from deterioration of exploitation conditions
Errors as a result of incorrect composition: inclusions, "brittle impurities," unsuitable material	Tough fracture (excessive elastic or plastic deformation, cavity collection or shear fracture)	Overloads or unforeseen loading conditions
Processing errors: folds, seams, hot cracks, increased local plastic deformation	Fatigue fracture: cyclic loading, cyclic deformation, thermal loads, corrosion fatigue, fretting fatigue, low-cyclic fatigue	Corrosion: chemical, stress, corrosion fatigue, graphitization cast iron, atmospheric contamination
Irregularities and errors in machining, grinding, or pressing: grooves, burrs, cavities, edges, cracks, brittleness	High-temperature destruction: creep, oxidation, local melting, relaxation	Lack of maintenance or poor maintenance or poor repairs: welding, grinding, inaccurate hole punching, cold reinforcement
Welding faults: porosity, reminiscence, residual stresses, nonwelding, insufficient penetration	Delayed static fractures: hydrogen brittleness, caustic brittleness, slight crack growth stimulated by an aggressive environment	Disintegration due to: chemical agents, liquid metal, elevated drag
Irregularities in heat treatment: overheating, cracking of grains, grain growth, excess austenite residual, decarburization, deposition	A pronounced stress concentration inherent in the design	Radiation injury (sometimes it is necessary to perform decontamination due to testing, which can destroy healthy material and cause destruction)
Surface hardening errors: carbide separation by grain boundaries, soft core, wrong thermal cycle	Incorrect stress analysis or inability to calculate stress in a complex part	Accidental (unexpected) circumstances: abnormal operating temperature, strong vibrations, thermal shocks
Errors due to surface treatment: cleaning, coating, chemical diffusion, hydrogen brittleness		

6.1.1 Function division

A steam turbine plant within a hierarchically "higher" energy system of a condensing TPP, as a complex technical system, can have a large number of required functions. However, not all functions are equally important and therefore must be classified for recognition and analysis [4,49]. One of the possible ways to divide a function for this technical system is within Table 8.4. It should be pointed out that the displayed classification of functions within Table 8.4 is not necessarily excluded, which means that individual functions can be classified into more than one type.

Functions are divided into two basic categories: primary (have the same purpose as essential functions) and secondary functions (have the purpose of additional functions), Moubray [66]. In doing so, the function must be described by word, means, and a certain measure of performance. They are subsequently subdivided into various subtypes. Secondary (ancillary) functions can be divided into several categories: security/structural integrity, environmental integrity, management, storage, comfort, visibility, protection, economy/efficiency, unnecessary function.

In many cases, it is important to distinguish between visible and hidden faults, and accordingly the functions can be divided into active (online) and inactive (off-line) functions [4] or visible and hidden [66]. The first are those functions that are in operation continuously or very often those for which the operator knows their condition. The termination of an active function is called evident failure. Inactive features are those features that are used occasionally or infrequently, so the availability of these features is not known without special checks or tests. Many protective functions are of this type. The termination of the ability to perform an inactive function is called a hidden defect. In the same context, visible and hidden functions are considered.

A steam turbine system and its functional blocks (auxiliary equipment) may have several modes (operating modes) and several functions per operating mode. Operating modes may include normal (drive, nominal), test, and transient modes, and alternative modes caused by a malfunction or failure of a system or operator on a technical system. Identifying the different modes of the system is important because it also identifies other functions that could be predicted because of too much focus on the essential function. Furthermore, identifying different system modes provides a structural basis for identifying fault modes that are fully coupled and thus dependent on the default operating mode [66].

The control system of a steam turbine within a condensing TPP should ensure the normal operation of the turbine, provided that all its elements are satisfactory. However, malfunctions of different origin may occur in the control system itself and in the turbine as a whole. Some of these failures or malfunctions can be very serious and dangerous, so the need for immediate shutdown of the turbine is indicated by closing the steam supply to the turbine. On the other hand, the turbine works in a complex with very complex plants or aggregates such as: steam boiler or nuclear reactor, condenser, regenerative heaters, and through the electric generator the turbine is connected to the electrical grid. The complexity of these facilities also does not exclude the possibility of their failures and disturbance of their modes of operation, which can create a danger for turbines. Therefore, the turbine is equipped with protection system which automatically ensures its operation from disturbances of different origin, either by itself or by other equipment connected to it in the power plant. Thus, the possibility of urgent closing of the steam supply to the turbine is provided in the following cases: increase of the turbine speed above the permitted value, impermissible axial displacement of the rotor relative to the turbine housing, unauthorized increase of the condenser pressure, as well as unauthorized reduction of pressure in the lubrication and control system from these disturbances it has special protection, which registers the deviation of certain parameters in the range of impermissible values, and each of them

Table 8.4 One possible way of dividing the function for the technical system of a steam turbine plant.

Function name	Short description	Moubray classification (1997) [36]		Application to a steam turbine plant
Essential functions	Required functions to fulfill the purpose of the function block	Primary functions (have the same purpose as essential functions)		Steam expansion, production of mechanical work on the turbine rotor shaft
Auxiliary functions	Those functions that are required to support basic functions. In many cases, these functions can be as important as basic functions, and failure can be much more critical in terms of safety than basic functions	Secondary (ancillary) functions are intended for additional functions	Safety/structural integrity, protection	Steam reduction for regeneration system
Protective functions	Those functions designed to protect people and the environment from harm and injury. They can be classified into safety, hygiene, and environmental protection functions, environmental integrity, protection		Environmental integrity, protection	Steam turbine protection systems
Information functions	Those functions that include condition monitoring, various measuring instruments, and alarms (signaling and alarm)		Controls, visibility, economy/efficiency	Systems for lubricating and regulating a turbine plant; Steam turbine protection systems: - Increase of turbine speed above the permitted value; - Impermissible axial displacement
Interface functions	Functions that enable interaction between elements of a function block and other function blocks. They can be active and passive.		Storage, comfort, economy/efficiency	

Table 8.4 One possible way of dividing the function for the technical system of a steam turbine plant.—cont'd

Function name	Short description	Moubray classification (1997) [36]		Application to a steam turbine plant
	For example, passive interaction is present when a function block is a support or basis for another function block.			of the rotor relative to the turbine housing; - Impermissible increase of pressure in the condenser; - Impermissible pressure reduction in the lubrication and control system
Superfluous function	Those functions that the function block never uses or may not necessarily need, unnecessary functions		Unnecessary functions	

acts on the stop and control valves for the purpose of urgent interruption of steam supply to the turbine. During the operation of some of the turbine protections, the turbine cannot be restarted automatically, regardless of the fact that the causal parameter has assumed nominal values. The turbine can be restarted when the cause of the protection and assurance of safe recommissioning has been fully understood.

The danger of exceeding the turbine speed above the permitted limit requires additional precautions to be taken. Unauthorized turbo charging may occur for two main reasons: immediate disconnection of the generator from the mains or breakage of the coupling connection between individual turbine rotors, including the generator. The turbine's turnaround time is about 0.30–0.35 s. During this time, when the turbine is unloaded, the turbine rotor speed changes by $\Delta\omega_{max}$. It should be borne in mind that there is a relatively large amount of steam left in the space between the valves and the first row of the nozzles, which additionally increases the number of revolutions despite the current closing of the steam inlet to the turbine. It is precisely such possible phenomena that limit the higher unit power of the turbine, as it is difficult to provide such strength of blades, especially the last stages of the turbine that would withstand large centrifugal forces when the turbine is fired. Therefore, it is necessary to provide a very fast operation of the control system in order to prevent the speed increase above 10%–11% of the rated speed. Turbine rotor starter protection should be very reliable and independent of the control system. In other words, this protection system should have: independent turbine rotor speed measuring device, independent impulse transfer to the actuator, one or more independent stop or stop valves to prevent steam from entering the turbine at the command of the safety automat.

The turbine shutoff switch can be connected to the control valves and cause them to close, together with the shutoff of the valve in the event of a turbine overspeed. The turbine rotor impulse and encoder is called a safety automatic. There are several types of safety automata, but they are all mechanical designs, based on the effect of centrifugal force on the corresponding eccentrically positioned element of the automaton (roller of certain dimensions, eccentric ring).

The axial displacement of the turbine rotor may result from major damage to the axial bearing of the turbine, for example, the melting of white metal on segments (hammers) of that bearing. Such occurrence is a dangerous turbine accident. Inadmissible (excessive) axial displacement of the rotor causes contact between the rotor and the fixed parts of the turbine, which can cause severe damage, which will exclude the turbine from exploitation for a long time.

The contacts of rotor and stator labyrinth seals, rotor blades with diaphragms, and the like cause heating and thermal deformation of the elements in contact. It can also lead to impeller unbalance, increase in turbine vibration, as well as a progressive increase in "turbulence" to complete destruction of the turbine. For the aforementioned reasons, a higher protection is applied to the turbines, which prevents steam from entering the turbine in the event of axial displacement of the rotor. If this protection is applied, the generator must also be disconnected from the mains in order to prevent it from entering motor mode and rotating the turbine rotor, regardless of the steam supply being switched off. Axial displacement protection is performed hydraulically and electromagnetically.

The number and type of automatic turbine protection devices are not precisely defined. There are general technical standards that are taken into account by turbine manufacturers, and their competence is to choose the number and types of protection, with the approval of the turbine buyer. Here are some of the protections that apply to turbines.

Pressure boosting in the condenser is the third most important for the turbine. A rapid drop in the condenser vacuum can result from interruption or a significant decrease in the flow of cooling water, as well as certain failures (malfunctions) on the vacuum maintenance system. The deterioration of the vacuum leads to an increase in the temperature in the outlet part of the turbine casing, and as a consequence of this, curvature of the casing and damage to the bearings can occur, causing increased turbine vibration, increased blade stress, and even their fracture in two ways—two-step. A special vacuum relay gives impulse to the electromagnetic ejector of the turbine, which causes the protection of the turbine to work. In some turbines, the pressure limit in the condenser is set at 70 kPa. The second-tier vacuum protection is made of thin sheet safety membranes mounted on the turbine outlet housing. In normal operation of the turbine, the strength and sealing of these membranes is sufficient to prevent the suction of air into the condenser, and with increased pressure in the outlet housing, the membranes burst, with the release of steam into the engine room. Such cases rarely occur. Regulated steam extraction or pressure turbines may cause impermissible pressure drops in the exhaust chambers or at the outlet of the pressure turbine. In this case, the last diaphragm may be overloaded, its greater bending and the rotating elements of the rotor may be impinged. The large pressure drop in the chamber of the reactive turbines can cause the last row of working blades to break in front of the extraction chamber. In order to prevent such an emergency, automatic protections are applied, which exclude the turbine from operation in the event of an unacceptably large pressure drop in the last stage of the turbine. Turbine protective devices are coupled with sound and light signals, which alert the turbine operator of abnormal modes. With the further growth of signaled hazards, automata shut down the turbine and the entire power block from operation.

6.1.2 Technical methods of functional analysis

Functional analysis of a steam turbine technical system can be performed using several technical methods (functional tree—FT, structured analysis and design technique—SADT, functional analysis system technique—FAST, multilevel flow modeling—MFM, goal tree–success tree—GTST) (Table 8.5).

Unlike the FAST model, the SADT model can consist of several diagrams, which allows only one diagram to perform a functional breakdown of the system. There are several types of functions in the MFM model:

- Mass and energy flow functions (functions of source, sink, transfer, barrier, storage, equilibrium, where these functions can also be used to describe flow information);
- Information flow functions (functions of observation, decision-making, and execution);
- Support and management functions (referring to organizational functions), which include network functions for grouping the flow structure by purpose and management functions, which describe control and management systems including personnel (working personnel) as operators.

An understanding of the basic concept of the MFM method with a detailed description of its formalism can be seen in the works of Linda [67] and Larsson [69,71,72]. In addition to the recommended application as a tool to represent the goals and functions of a complex industrial plant, this method has been proven to be an effective inference tool for plant failures, control strategy, diagnostics, and alarm analysis [71,73–75]. The MFM method has also been used in studies of the safety of nuclear facilities and the monitoring and diagnosis of patients after surgery [72].

The GTST model, in relation to knowledge recognition and knowledge representation, has in principle the same properties as the MFM model. The main advantage of the GTST method is that it provides a different representation of aspects of the system, modeled in textual form using natural language (better understanding of the system). The disadvantage of using this method is that in the case of a more complex structure of the system, this method produces a complex model, thus losing the overall view of the system as a whole. In this case, the dynamics of the observed system cannot be easily depicted, so hybrid methods have been developed that combine the advantages of the MFM and GTST methods (hybrid MFM-GSTS or HMG method) [70].

6.2 Basic concept of fault analysis

Starting from the most accepted definition of reliability as the ability of a functional block to perform the required function, under given environmental and operating (exploitation) conditions and for a fixed period of time or number of cycles [50,82], it is possible to determine the ability of a functional object using probability theories, or it can be set deterministically. The deterministic approach is based on understanding how and why a malfunction occurs on a functional block, and how it must be designed and tested to prevent, i.e., minimized the occurrence and recurrence of failures. These include certain analyses (deterministic analyses and review of fault reports, understanding of fault physics, role and degree of testing and testing, performing redesign or reconfiguration, transient purposes). In practice, this is an important aspect in reliability analysis.

A technical system failure is defined as the cessation of the ability of an element or system as a whole to perform the functions for which it was designed. Reduction or loss of working capacity of the technical system during exploitation is due to the effect of various factors (built-in, accidental, or

Table 8.5 Technical methods for functional analysis of the technical design of a steam turbine.

Name	Short description	Method and rating of applicability to a steam turbine plant
FAST method	In complex systems theory, the functional tree method is a diagram showing the dependency between the functions of the system. This method breaks down functions by hierarchy. A functional tree is created by asking how an already established function is executed. This activity is repeated until the lowest level functions are achieved. A diagram can also be developed in the opposite direction by asking why a function is needed.	When analyzing an existing system, such as a steam turbine plant within a condensing thermal power plant, it is more appropriate to use a physical breakdown of the technical system rather than a functional breakdown. The breakdown by physical structure is the same as a function tree, but each box shows a physical element instead of a function. A physical element can be a technical element, an operator, and even a process. When each function is performed with only one element, the two approaches produce similar results. In case the system has redundancy, the tree is different. In the functional tree, redundancy is represented as a single function, while in the breakdown by physical structure it is represented as two elements.
SADT method	A widely used functional modeling approach in various technical fields is the SADT method. This method was introduced by Ross [68], which was originally intended for the field of computer engineering (computer software development), but quickly found application in other technical fields (aviation, energy and process engineering, mechatronics, automation, control and regulation, etc.). The graphical representation of the SADT model is given by a diagram consisting of boxes (function, i.e., activity described by a verb) interconnected by arrows (enters entered by arrows on the left side of the box, exits on the right side, control from above pages and bottom-up mechanisms that are usually defined by data). The activity adds extra size to the input data by transforming the inputs into outputs by means of mechanisms or supporting data. Management data are the conditions that guide or adjust an activity. The basic difference between input and management data is that management data are never changed by activities. Management data include limitations and influence on the transformation of input data. The mechanism defines the resources that carry out the activities (usually expressed in the form of "how" and "who" questions). The output of a functional block may be an input to another functional block or it may act as a control to another functional block. In this way, the functional blocks are connected and thus form a diagram.	According to Lind [67], a system representation can be made at different levels of abstraction with respect to two major axes: whole-part (corresponding to the hierarchical breakdown of the steam turbine system up and down into several subsystems of less complexity down to the basic level) and means—purpose (meets, for one level of system breakdown, requirements for goal attainment). Such a system breakdown is an advantage of the SADT method. Therefore, this method is based on a hierarchical and modular description of the system with respect to functions. Each function block according to the same structure is modeled with five main elements: function (definition of the function being supplied), inputs (energy, materials and information required function), control system (controls and other elements that limit or control how the function is performed), implementation mechanism (people, systems, plants or equipment required to perform the function), outputs (the result of a function that can be divided into desired results). The SADT method, in addition to the described activity model (actigram), can use another type of diagram in a steam turbine system to display a data model (datagram). In this case, the boxes represent the data and the activity arrows. This type of diagram is mainly used in the data processing process. When designing the SADT model, an up and down approach is used. The model is organized in a hierarchical and modular way. At the highest level, the required function of the system begins (transformation of potential energy into kinetic energy of the fluid stream, and then into the mechanical operation of turning the turbine rotor shaft to the coupling with the generator shaft, etc.). The functions required to perform the system function form the SADT diagram at the next lower level (subtraction for the regenerative heaters system, condensate outlet from the low-pressure turbine and inlet to the steam turbine condenser, etc.). Then, each function at this level is broken down into a lower level of functions (high-pressure regenerative heaters, low-pressure regenerative heaters, network heaters, etc.) until the required and required level is reached. The diagram shows serial, parallel, and back links. The hierarchy is maintained by numbering the system.

FAST method	The system functional analysis technique, FAST, conceived by Bithenay in the late 1960s [50], belongs to the functional tree family and comes from a value analysis method aimed at designing a new system. Also used to study existing systems [62]. The FAST method consists of a graphical representation of the connections between the functions of a technical system. A number of system functions need to be obtained using the following questions: "How this function is realized", "why", and "when". These issues determine the ascending and descending functions. The diagram between two transverse dashed lines determines the area of functional analysis. Outputs are defined to the left of a certain area and inputs to the right.	The formalism of the FAST method is simple. The diagram is drawn from left to right. First, the main function of the system must be determined, and then a box containing the system function described by the verb and noun is drawn. The question is then asked "how" that function is performed (identification of the function at the first level, which is then entered into the diagram). The question "how" is asked until the desired level of detail is reached along the horizontal axis of the FAST diagram. The diagram can be obtained in the opposite direction by asking the question "why" a function is needed, thereby identifying a higher level function than the initial one (the hierarchical structure of the FAST method). Along the vertical axis of the FAST diagram, functions are determined by the question "when". This question is obtained by an answer that begins "if, at the same time" and thus answers the necessary conditions, that is, functions that must be performed at the same time. The lower level functions can be connected by logical operators—"I (AND)" or "OR (OR)", which displays redundancy and parallel operation.
MFM method	The technique of multilevel flow modeling, that is, the MFM method, was developed by Linda in 1990 [67], for the purpose of demonstrating the goals and functions of a complex process plant. This method also breaks down the system into two major axes: whole-part and means-purpose. Such a breakdown provides an overall view of the system and a detailed view of its parts [10]. The function of a system can generally be represented by the following definition: "Function is the role that the system plays in achieving the goal" [69]. In the MFM method, a function is always associated with a goal and accordingly goals are always associated with functions. Functions may or may not be attached to physical components in the system, where components may also be attached to multiple functions. Without knowing the exact goals (purpose, safety goals, economic goals) it is not possible to create a good MFM model (the purpose of the system is to meet the defined goals).	The MFM method is commonly used for systems that can be characterized by energy, material, and information flows, such as a steam turbine system within a condensing power plant. According to this method, the axis of the means/purpose can be broken down into three levels, namely: the level of goals (purpose or purpose required of the steam turbine system: "keep the turbine rotor shaft speed within acceptable limits," "generate electricity," "cool the pump"), function (means to achieve the goal, related functions in the MFM model show the capabilities of the system such as "energy transfer," "power supply," or prevent "heat transfer"), and parts (physical components that build the system, which may be the elemental stage of the steam turbine, the lubrication and regulation system of the steam turbine, the condenser plant with the system for maintaining the vacuum in the condenser, the protection systems of the steam turbine, etc.). Components are not typically displayed in the MFM model. In addition to goals that describe the purpose of the system, goals can also be security (used for reasons of safe system operation, which means in practice that some process variables need to be kept within the security area) or economically (used to demonstrate the optimal process of the technical system, and most often through a complex function that depends on the operating constraints and the economic efficiency of the system as a whole and of the system within a higher hierarchical system).

Continued

Table 8.5 Technical methods for functional analysis of the technical design of a steam turbine.—cont'd

Name	Short description	Method and rating of applicability to a steam turbine plant
GTST method	The objective tree—success tree functional analysis method—The GTST method was developed by Kim and Modaress in 1987 [50], for the purpose of demonstrating a hierarchical in-depth knowledge of various aspects (characteristics) of a technical system [70]. The difference between the success tree and the goal tree is only that the success paths must be satisfied to reach the subgoal. The performance tree can be displayed as a tree where the individual components are connected to the lowest level of functions they accomplish. Higher levels of hierarchy continue downward. Each block of the GTST model may have conditions (used to show how different aspects of the system change according to the new conditions) and properties, i.e., attributes (allow for further description of each block such as its priority or the order in which it must be considered).	The model consists of two parts, the first part is the goal tree (showing knowledge of system level), while the second part is modeled as the success tree (showing knowledge of system components). The goal tree is created by parsing the overall system goal into a series of necessary and sufficient subgoals, subgoals into functions, and finally merging functions into components, as needed, using computer programs, or with the physical structure required to meet the subgoal. At this point, the formation of a performance tree for a particular subgoal begins. In order to create a success tree, all the different paths that the subgoal can be reached must be displayed. The contradiction between blocks and sequences can be expressed by adding conditions and attributes, the inclusion of which for each individual target or successful element in the GTST model of the system allows to show changes of some information regarding the system as a whole, and the order and importance of conditions and attributes that reflect the dynamics of the observed system. When it is necessary to express explicitly the dependency of each lower level of function on the components, the Master Plan Logic Diagram (MPLD) method can be used to draw the performance tree.

time), which change the initial parameters of the system, causing different levels of damage. A failure is an event that occurs in a certain time that may or may not be spotted. Failure is a basic concept in any reliability analysis, by which stochastic relationships of functional block behavior are observed. Namely, due to the nature of many factors that may be involved, the time of occurrence of a failure is unpredictable, i.e., by chance, and therefore predicting their occurrence is a probabilistic problem. It also follows that reliability analysis is a probabilistic process that assesses and predicts the probability of failure, whereby the methods used quantify reliability using the probability and statistics to predict, measure, and analyze reliability data.

According to Rausand and Høyland [4], the quality of the reliability analysis depends strictly on the analyst's ability to identify all required functions and any failures of the functional block that is the subject of the analysis. In order to be able to consider problems related to the identification and classification of failures, it is necessary to clarify the terms and definitions used in the failure analysis, which relate to the function of the functional block, failure, fault, error, and mode of failure, cause of failure, effect of failure, and the consequence of failure. Specifically, these terms are often used in many ways, leading to confusion and ambiguity. Using these terms, functional block failure analysis is based on the concept of "What, How, and Why".

The development of the occurrence of failures can be observed through the case of a change in correctness (monitoring the change of condition in the time interval) and through the case of the occurrence of failures (monitoring the number of these failures in a given time). In doing so, developed methods are used to monitor the occurrence with distribution modeling, all with the aim of obtaining a mathematical model of changing the state of the technical system, thus creating the necessary assumptions for the possibility of making diagnostic decisions. The determination of mathematical models is performed by deterministic, stochastic, or commonly combined deterministic–stochastic prediction methods. The modeling of system behavior is mostly based on specific operational research (for real conditions, with treatment of system functionality only), application of mathematical statistics methods (defining and selecting distributions, estimating observed parameters, test hypotheses, defining ranges, and evaluating characteristics), and probability theory method (various mathematical models) and stochastic method. In addition, the built-in causes of failure may include, but are not limited to: faulty design, reliability design, occurrence of design voltages, manufacturing errors, assembly errors, errors due to adjustment, errors due to incorrect material selection, thermal processing and residual thermal stresses, failures of technical controls, etc.

On the other hand, accidental causes of failure include: environmental conditions, overload, poor handling and maintenance, unstable (especially nonstationary) operating modes, inadequate control, load gradient, etc. Temporal causes of failure may include: exploitation mode, maintenance mode, lubrication, fatigue of material, heating, erosion and corrosion, embedded parts, working medium, contamination, etc. The sooner the failure research begins, the greater the opportunity to determine the real reason for the failure, with the basic rules of conduct being applied [1]:

- Do not destroy broken items. In particular, do not touch the surfaces where the cracking occurred, or break pieces.
- Look at elements that have been damaged and which are to be inspected only after receiving all the documents (reports, explanations, photographs, etc.), ensuring that all dismantled parts can be correctly identified, reassembled, undamaged, etc.

- Research should not only concentrate on the location of the fracture, but also the state of the environment. Because of the possibility of a chain failure occurring, it is necessary to consider whether the failure occurred is merely a consequence of something else or is it caused entirely by other causes.
- Do not easily form a conclusion about the occurrence of failure. Collect all the information, then discard the insignificant ones, and form a definitive conclusion only when it is clear that all other possibilities have been eliminated.
- In addition to the material evidence and reports from other experts, when preparing the cancellation documentation, do not take anything without verification ("on trust"), especially your own opinion, since conclusions, estimates, and decisions can be subjective (in terms of making decisions based on your own experiences).

The term failure is sometimes often confused with the terms fault and error [1,3,81]. In any failure analysis, it is important to determine the difference between a failure or a malfunction and an error because this difference describes the boundary between what is failure and what is not (Fig. 8.13).

The curve shows the observed level of performance variable of a functional block over time. Initially, the observed performance meets the target value, after which it starts to deviate gradually (a nonfaulting error occurs because it is the same within acceptable limits). A failure is defined as an event in which the observed performance exceeded the acceptable tolerance limit, after which the functional block is in a malfunctioning state (failure state). After repair and restoration to a functional state, for which a certain amount of time is required, the observed variability of the functional block performance returns to acceptable limits. Therefore, in order to determine whether a failure has occurred, performance parameters as well as their target levels and acceptable limits must be defined for all functions.

The performance of a functional block can be described by four elements [50]: the competency or ability of a functional block to satisfy functional requirements, efficiency, or the ability of a functional block to efficiently and easily achieve, reliability, or the ability of a functional block to start and operate continuously, and the availability or ability of a functional block to quickly become operational after malfunction.

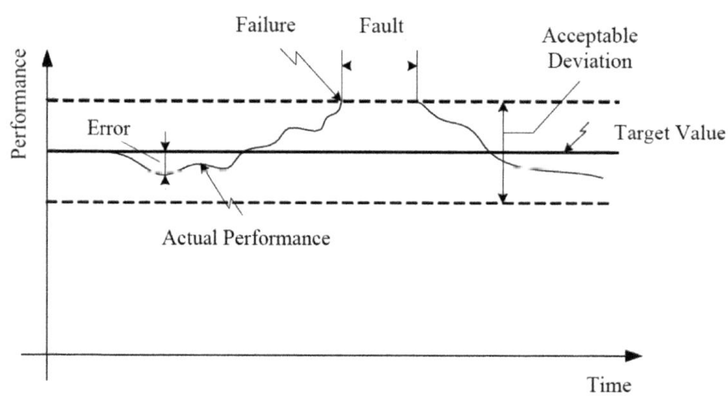

FIGURE 8.13

Displays of the difference between failure, malfunction, and fault.

The basic requirement that a TPP has to fulfill is the reliability of operation, that is, the continuity of electricity and/or heat production in accordance with the needs of customers and the dispatching load diagram. Power units must have the necessary maneuverability to accept rapid load changes (ability to quickly increase or lower loads, quickly start downtime, quickly shutdown drives, etc.). This should certainly not be to the detriment of the safety, reliability, and durability (lifetime) of the plant.

The design and manufacture (production) of functional blocks within the condensing TPP also affect the ability and efficiency of this facility within the power system. It is the task of designers to provide the required level of maneuverability of a functional block determined by the design requirement to achieve its functional requirements. It is important to note that the stage of designing and construction of new TPPs is characterized by conditions of incompleteness and unavailability of input information. Complete information on the safety and reliability of the TPP is obtained by observing much of the service life, the period of winning and designing technical and economic solutions, with complete stabilization of the reliability indicators. However, such long reliability tests require high costs and the involvement of a large number of experts, and more often resort to setting up a group of models for incomplete conditions of information on the operation of analogue plants and their approximate solution. Therefore, the decisions on ensuring the reliability and safety of the TPP are largely defined at the stage of development and design. This, on the other hand, results in poor adaptability of the TPP to changing conditions and demands of exploitation.

The appearance of serial parts, subsystems, and systems increases the applicability of statistical and similarity methods under analogous operating conditions, which also determines certain specificities related to the choice and definition of reliability indicators of a TPP, and consequently the steam turbine system, the appropriate methods of their calculation, and the most efficient ways of providing them. The concept of structural solutions of a TPP is based on providing [53,54]: a general sequence of flows in one thermodynamic and production process, clear flows of realization of the process and functioning of individual parts and equipment, as well as a hierarchical tree of details and their complexity into subsystems, that is, systems and the TPP itself as a whole. Fig. 8.14 shows the hierarchical structure of a stationary TPP. As the main systems, aggregates, and elements of the technological scheme, the minimum necessary groups are specified to fully complete the transformation of energy from the primary resource (coal) into the final form of energy, while all other plants are classified as auxiliary or reserve. The very structure of decomposition and aggregation of parts of an object is a consequence of the results of the stage of solving the tasks of reliability, security, availability of information, available methods of calculation, and means of calculation. The interplay and detrimental effects of TPP details at the level of consideration of individual details, as well as at the level of the physically technical systems formed by them, should be taken into account through the following characteristics:

- variety of constructive solutions from the aspect of principles of action and mutual subordination and coherence;
- the nature and modes of exploitation during the expected service life, the form of loading, liability under conditions of reliability, safety, and protection;
- the form and character of the occurrence of the failure and its effects on other parts;
- overhaul benefits, possible forms, and ways of its realization;
- seriality of details;
- the ability to perform appropriate experiments and tests under conditions that are very close to real and the availability to control and perform diagnostics.

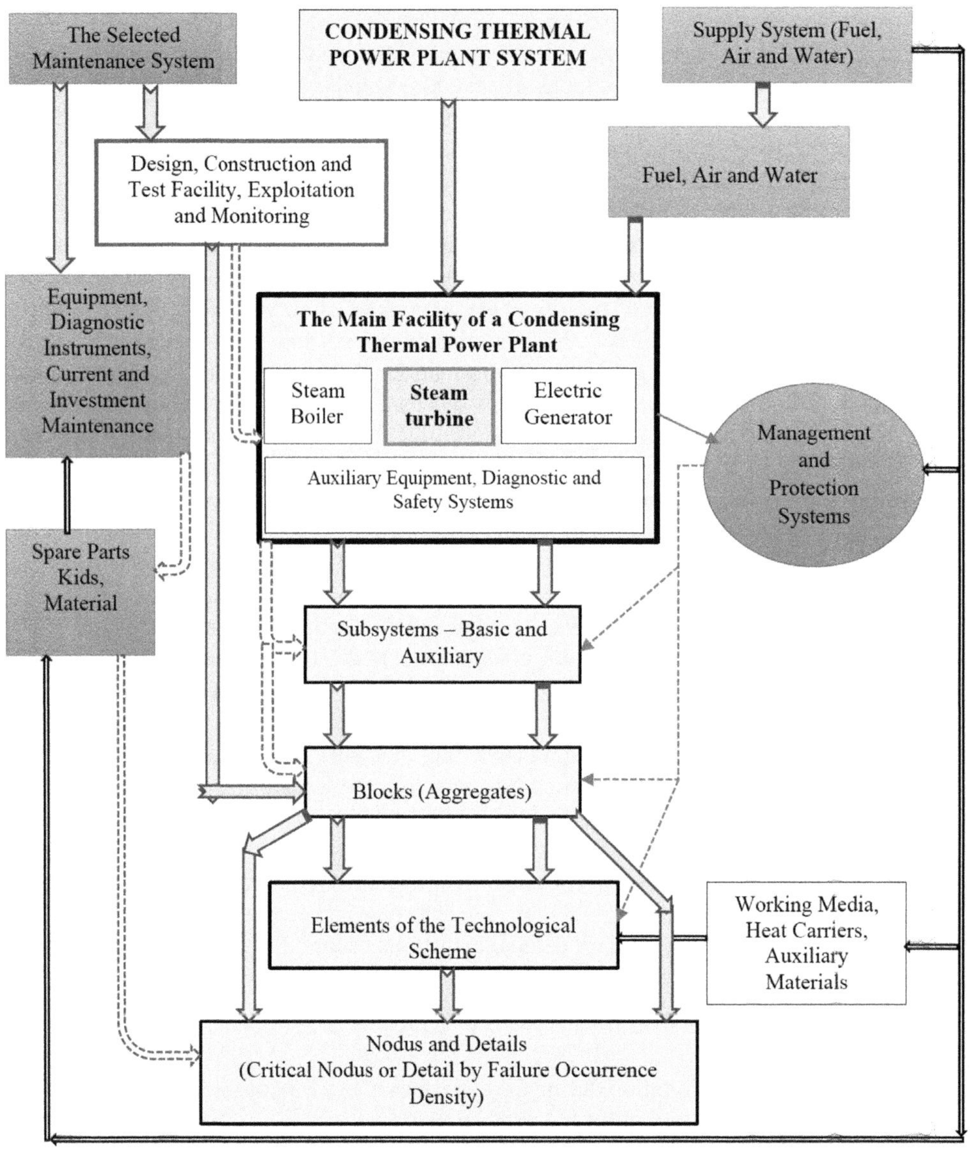

FIGURE 8.14

Outline of the structure of a complex thermal power plant system [53,54].

Reliability, on the other hand, is a problem related to the operability of a functional block and which is influenced by its potential to remain in operation. For a repairable functional block, the ease with which the functional block is maintained, repaired, and returned to service is measured by its maintainability. Based on the above definitions, it is possible that the functional block has high reliability but cannot achieve high performance, i.e., its specific design goals. Therefore, the high quality of design and workmanship results in a low incidence of failures, effective maintenance and repair, and ultimately high functional block performance. According to Moubray, performance can be described in two ways: desired performance, that is, what the user wants to block functionally (defined by the nominal operating mode), and performance embedded in functional block capabilities (defined by normative tests of the steam turbine system), respectively, what a functional block can do [66].

In doing so, performance measures can be divided into: multiple, qualitative, quantitative, absolute, performance measures of variables, and upper and lower bounds. Of these four elements that define the performance of a function, the area of interest of this paper is reliability only. Clearly, reliability is an important element in achieving high performance of a functional block because it directly and significantly affects its performance and ultimately its life cycle cost. Specifically, poor reliability directly increases costs within the warranty period, liability, and repair (maintenance during exploitation time).

Several different definitions of failure are available in the engineering literature, defined by international and national standards, professional association standards, or defined by the authors themselves. Thus, the standards of the International Electrotechnical Commission IEC 60050 (191): 1990 [76], the International Organization for Standardization (abbreviation ISO), ISO 14224: 2006 [77], and the British Standard BS 4778: 1991 give the definition of failure as: object to perform the "required function". By IEC 60050 (191), the term "required function" is defined as a function or combination of functions of an object that is deemed necessary to provide a given service. The term "service" refers to a series of functions that are given to a user or organization. A similar definition is given by Smith as: "the inconsistency of some specific performance measures" [78]. According to these definitions, a failure is an event when the required function of a functional block is interrupted, or when it has crossed acceptable limits. Therefore, these defect definitions focus on the actual apparent performance of the functional block and the loss of functionality. The failure concept defined by these standards does not apply to functional blocks consisting solely of software. "Inability of systems or components to perform their required function within specified performance requirements" is the definition of a failure according to the Institute of Electrical and Electronics Engineers (IEEE) glossary.

For better transparency, definitions of the broader context of parameters that define the state of operation and the transition to failure state are often given [3]. Correctness is the property of a product or system to fully meet the requirements of its specification. A defect is a deviation of quality characteristics that causes the system to fail to meet the specified and foreseeable requirements of use. Faults can be classified into four classes according to the degree of impact. Malfunctioning deviation of product quality or system output compared to the set and specified requirements of the criterion function. Reconstruction is an activity that changes the characteristics of the system most often to improve performance. Maintenance (repair) is return of system elements or the system itself to a state of readiness or a specified work activity. Supervision is a specific maintenance activity that provides an insight into the state of the system by means of constant or periodic inspections. It denies termination of the system's ability to work, or inability to perform projected activities or to perform projected activities function. The term cancellation is essentially related to the notion of the criterion function, since defining the criterion function can also change the qualification of the cancellation. The failure associated with the criterion function is a broader term, which contains notions of downtime, errors, malfunctions, and even system readiness. Failure intensity is the ratio of the function of the density of occurrence of a condition to failure and the cumulative density of occurrence of a condition in operation. In simple terms, it is the ratio between the number of elements that failed and the total number of elements during system operation over a period of time. Criterion function is a function that determines the acceptable output of a system. Most often, this is the limit that defines the lowest acceptable level of output realized by the operation of the system, but can also be defined by the field of acceptable outputs, whether static or dynamic. Fix ability is the property of an element or system as a whole that determines the ability of the system to perform specified work activities again defined by the criterion function after the maintenance procedure, i.e., durability is the property of an element or system to maintain the projected and function-specified work activity during exploitation to the limit state. During the shelf life, the element or system may be repeatedly switched on and off and subjected to certain maintenance operations.

There are also definitions of failures by other authors who advocate a conceptual revision of the definition, such as the following definition of a failure: “A failure occurs when an object interrupts its required function” [79]. An alternative approach to defect definition is inspired by the concept of product life cycle, which introduces a clearer concept of failure and allows for the classification of failures to take into account the roles of stakeholders involved in the various stages of the product life cycle [80]: “an engineering process, product, service, or system to meet the design goals of the team for which it was developed”, is the definition of a product failure and is different from the traditional definition of a failure concept (exploitation) and maintenance. A failure is an event that occurs at a certain time that may or may not be observed [64]. When a fault is present, the functional block has a malfunction. Therefore, a malfunction is a condition resulting from a failure as an event. It is similar to IEC 60050 (191) and ISO 14224, which defines a malfunction as: “a condition of an object characterized by an inability to perform the required function, excluding incapacity during preventative maintenance or other planned activities or lack of external resources.”

According to Rausand and Høyland [4], failure is always associated with the function of a functional block and occurs when a functional block is unable for a long time to perform its function according to certain performances. A malfunction may be such that the functional block performs its function when not required or when required, without interrupting when the required function is performed. Therefore, the notion of failure covers functioning when required as well as when not required by Rausand [64]. The importance of recognizing all functional block functions together with certain performance measures for the purpose of a reliability-centered maintenance process (abbreviated as RCM), also cited by Moubray [66,67]. Assuming that a function block can have more than one (often several) different functions, it follows that the function block can have a number of different malfunction states. This means that there may be a situation where the functional block is in a state of malfunction with respect to the primary function (steam turbine system within the condensing TPP in failure and no power generation), but not with respect to the secondary function (when steam boiler is in proper condition, fresh steam is produced, which is further transformed through the cooling-reducing stations into a pair of parameters necessary to achieve heating during the winter period of operation). Therefore, for certain systems, in addition to the failure of the functional block as a whole, it is important to determine the failure in relation to the loss of a particular function. In doing so, the line between satisfactory performance and the failure of a particular function is determined by the measure of performance. A single function defect is defined as a functional failure, that is, "the inability of a functional block to fulfill a function by user-friendly performance measures" [50]. The IEC 60050 (191) and ISO 14224 standards define an error as: "deviation between calculated, observed or measured size or condition and actual, specified or theoretically correct size or condition." The error is not yet a failure because it is within acceptable limits. In some literature, the error sometimes refers to the initial failure.

The practical importance of accurately defining terms related to the operation of a technical system is based on monitoring and prediction of the state of elements and systems. If the relevant definitions of damage are taken: "Damage is the condition of an item or equipment in which the item or equipment still has the capacity to perform the required function, but may develop a failure" and the definition of a failure is: failure is an event of termination of an element ability or equipment by performing the required function, then it can be concluded that the sequence of events illustrated in Fig. 8.15 [81].

With more complex technical systems (such as condensing TPP), which have a great interdependence of their subsystems and elements, failure of any which can mean automatic shutdown of the entire system, or reduced power operation (or more often, operation at the technical minimum), which can result in increased operating costs, thermal, and other overloads, as well as greater damage at system outages. For this reason, it is necessary for such a complex system to be reliable in operation. The elements or equipment is initially in proper working order. After some time, there is a gradual or sudden degradation of the element or equipment when a condition occurs when one element or equipment is damaged, but others still perform the required function (with the same or less intensity). The moment, i.e., an event when elemental equipment comes to such a state that it can no longer perform a given function is called a moment (event) of failure, and after the failure of a critical element

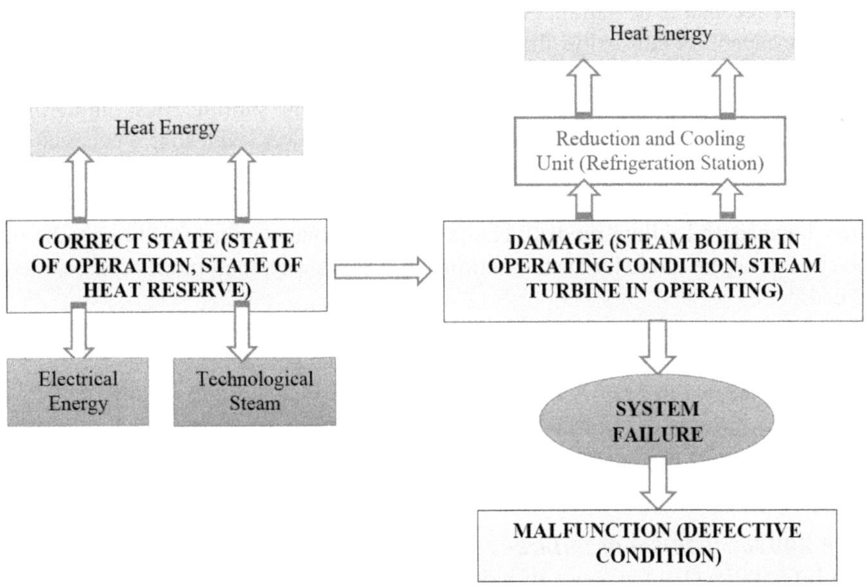

FIGURE 8.15

Sequence of events on the condensing thermal power plant system.

(steam boiler within the condensing TPP), the condition of the condensation TPP as a whole occurs, the term "deadlock" with the same meaning is also used, the deadlock being also caused by the external connection elements of the observed system (portable transmission line system) within the higher hierarchical power system. Depending on a number of factors, the process of transitioning from a correct state to a faulty state can be gradual or rapid. The safety of technical systems can be considered from two aspects. The first and most important aspect is the protection of the operator (human) from injuries during system operation. Another aspect is the protection of the system from damage caused by external causes. Preference is given to operator safety. In doing so, these two aspects are not unconditionally complementary, and an increase in operator safety can be achieved at the expense of system security.

6.2.1 Failure mode

The failure mode as a malfunction description is a way to observe that the functional block is malfunctioning. Sometimes the term malfunction mode is used instead of malfunction mode, as recommended by IEC 60050 (191) [76]. Specifically, according to this standard, fault mode is defined as one of the possible states of a faulty function block, for a given required function, and the term fault mode is therefore discarded. However, in the literature in the field, the term failure mode is mostly used. Defective mode is defined according to ISO-14224 and BS 5760-5: 1991 (replaced by BS EN 60812: 2006) as: "the effect with which the defect is observed on defective object." The failure mode, according to Moubray [66], represents any event that causes a functional failure. For the purpose of determining the failure modes, the outputs of different functions of the functional block must be studied. Some functions may have several outputs. Some outputs can be very well defined, so it is easy to determine whether an output request is met or not. In the second case, the outputs can be specified as a target value with acceptable deviation. Accordingly, failure modes can be classified into one of the triggers: when the capability falls below the required performance, when the desired performance rises above the initial capability, when the agent (functional block) is not capable of doing the initial failure.

In functional system analysis, functions are usually divided into subfunctions, failure modes at one hierarchical level are often caused by failure modes at the next lower level, so it is necessary to link the fault modes at the lower levels with the response of the highest level. Identification of all possible modes of function block failure can be more difficult than recognizing its functions (each of its functions may have several failure modes). Faults that cause one or more functions to be interrupted are described by the failure mode for each function, e.g., in this way, fault modes can be traced back to their initial cause.

Examples of failure modes for electrical components are, e.g., short circuit, deflection, etc., and with mechanical components it breaks, breaks, creeps, fatigue, etc.

In practice, there appear to be no formal procedures that can be used to identify and classify fault modes [4]. The main sources of information on malfunction modes are: manufacturers and sellers of equipment, generic lists of malfunctioning modes, other users of the same equipment, technical records of usage history, and people operating and maintaining it [66].

6.2.2 Cause and mechanism of failure

According to IEC 60050 (191), a cause of a defect is defined as: "design, construction and use circumstances that lead to a defect," while ISO-14224 supplements this definition with circumstances during installation (montage) and maintenance. Identification of the potential causes of a functional block failure is important in order to perform its proper repair in order to avoid failure or recurrence of failure. The failure mechanisms according to IEC 60050 (191) and ISO-14224 are defined as: "physical, chemical or other processes leading to malfunction." This term is usually understood to mean processes that cause wear, hardening, tearing, and oxidation. However, this level of description of the cause of the failure is not sufficient to assess possible repair. For example, wear and tear can be the result of misspecification of the material, use beyond certain limits, insufficient maintenance, inadequate lubrication, etc. These root causes are sometimes referred to as root causes, which may require a decision to take corrective action. Modarres et al. [82] define root causes as most of the root causes that can be reasonably identified by experts, and can be corrected to minimize their occurrence. The process of identifying root causes is generally performed by a group of experts. The root cause analysis (RCA) itself involves three steps: determining events and causal factors, marking and documenting root causes, and generating recommendations (corrective actions).

Failure mechanisms are physical processes whose occurrence either results in stress or is caused by stress, and can also reduce the ability of a functional block to resist the action of these stresses, e.g., reduction of strength or durability of functional block [82]. Since the failure mechanisms for mechanical and electronic/electrical systems are sometimes different, they divide them into mechanical and electrical ones. Mechanical mechanisms of failure are divided into induced stresses (e.g., elastic deformation, bending, etc.), mechanisms of strength reduction (wear, corrosion, scraping, etc.), and mechanisms of stress increase (e.g., fatigue, radiation, thermal shock, shock, etc.). For electrical failure mechanisms, they state that they tend to be much more complicated than mechanical systems because they are caused by the complexity of the electrical components themselves. They are divided into three types: electrical voltages, internal and external failure mechanisms. Malfunctions caused by electrical stress are usually due to the action of higher voltage levels or current. Internal failure mechanisms are the result of poor workmanship or design. External failure mechanisms result from the packaging or coupling of electrical components.

Of particular importance is the root cause analysis (RCA) as an essential element of reliability based maintenance (RBM) and asset integrity management (AIM) [50]. RCA failure analysis is a structural process that uncovers the physical, human, and other latent (hidden) causes of any undesirable workplace events. RCA is a required part of proactive reliability-focused maintenance that utilizes advanced maintenance techniques to apply corrective actions, eliminate prior failures and extend equipment life, and minimize maintenance costs. Realization of the set requirements for high reliability of complex technical systems, such as the thermal power plant system, results in a large consumption of materials, time and resources for continuous monitoring and testing of the condition with diagnosing the remaining service life of the system [52]. Intensive research results in the creation of certain small costs, so the practical application of procedures for accelerated testing and enhancing the effectiveness of the system as a whole is of great importance. The accelerated testing procedure itself does not usually cover all the elements, that is, the technological system completely. Only those units or elements that have the least durability, or are the most unreliable or most often canceled, are evaluated. Of course, the question remains how to identify these elements and how to evaluate and rank their impact.

Complete or partial failures of TPP systems operating within the power system have several characteristic forms, differing in their levels of performance and efficiency. The criteria with which they are to be described are determined on the one hand by the quantitative characteristics of the deterioration of the working capacity of the condition and the determined efficiency of use of the TPP as a whole within the electric power system (reduction of labor power, degree of efficiency, or deviation of operating parameters of the equipment and parameters of the given electricity and/or thermal energy), and on the other hand, by the ratio of the level of efficiency of the TPP for a certain regime and the working capacity of the power system itself. In doing so, the following forms of partial failure criteria exist, adapted to the specifics of the operation of the TPP system within the power system [2]:

(a) Exceeding the required productivity (power) for the given regime $W(t)$ over the operating productivity of the TPP system $N(t)$ at the observed moment of time t

$$0 < W_{\min}(t) < N(t) < W(t) \leq N_{nom}, \tag{8.1}$$

where: N_{nom}—nominal productivity (power); $W_{\min}(t)$—allowed technical minimum load of the TPP system.

(b) Exceeding of the exploitation indicator of the TPP system—the composite cost of fuel energy cost in the second (peak) zone of energy characteristic $C_{g.2}(t)$ over analogous size for incompletely loaded energy facilities within the power system, with the participation of network losses and losses related to the commissioning of the TPP system $C_g^{m}(t)$ due to partial or the complete failure of any component of the TPP

$$C_{g.2}(t) > C_g^{m}(t) > C_{g.1}(t), \tag{8.2}$$

where $C_{g.1}(t)$ is the composite cost of fuel energy cost in the first (lower) zone of the energy characteristic of the TPP system at its default state for the moment of exploitation. In accordance with the requirements of the optimal load distribution within the power system, the load of the TPP is reduced within the limits of the first (lower) zone, i.e.,

$$W_{\min}(t) \pi W(t) \leq N_1(t) \pi N_{nom}, \tag{8.3}$$

where $N_1(t)$ is the higher boundary of that zone. In this case, the optimal load distribution between the power plants is achieved in the case of the load of all TPPs, which should be put into operation at the observed time, such that the total cost of fuel in them is as minimal as possible.

(c) Exceeding the output values of one of the set energy parameters or controlled technical parameters of the TPP system $x(t)$ with respect to certain lower $x_{lower}(t)$ or higher $x^{higher}(t)$ limits, corresponding to the set operating modes and operating conditions at the moment t. This means that the operation of a TPP is stopped when the system of a TPP with or with some productivity is, at which it is

$W_x(t)$ has $x(t) < x_{\min}(t)$ or it is $x(t) > x^{\max}(t)$, then it is $W_{\min}(t) \leq W_x(t) \leq W(T)$,

(d) A combination of the conditions set out in points (a) to (c) with all the above justifications.

Similarly, the criteria for complete failures of a TPP system within a power system can be defined:

(a) Complete failure to provide the required productivity (power) for the default mode, where the workforce $N(t)$ is lower than the permitted technical minimum for the specified exploitation moment t, that is, $0 \leq N(t) \leq W_{\min}(t)$.

(b) Relative time inefficiency of the utilization of the TPP system within the power system, due to deterioration of the operating state, which may be given by the following inequalities

$C_{g.1}(t) > C_g^{m}(t)$ or $C(t) > C^m(t)$, where with $C(t)$, $C^m(t)$ the corresponding cost prices are given for the energy extracted from the power plant system and from insufficiently loaded separate power plant installations, with participation network losses and start-up related losses.

(c) Exceeding the output quantities beyond the permissible limits of one or more default energy parameters or controlled technical parameters of the TPP system, which may be represented by the following inequality $x_i(t) < x_{i.min}(t)$ or $x_i(t) > x_i^{max}(t)$, $i \in |1, I|$.

The failure and damage criteria for details, subsystems, and systems as a whole are given by the parametric nonequivalence of components:

$$X(t) \geq X^{\max}(t); X(t) \leq X_{\min}(t), \tag{8.4}$$

where $X = (x_1, x_2, \ldots, x_i)$ is the vector of components.

Here, in addition to the sizes that are controlled in the operation of the TPP system, there are also the sizes that are controlled in the stand and overhaul diagnostics of the TPP, maintenance, as well as exploitation tests.

6.2.3 Effect and consequence of failure

The failure effect is a description of what happens when a failure mode occurs [66]. This term is sometimes confused with the term fault consequence in the literature. Also, many standards define and interpret these concepts differently. It should be emphasized that the effect of the failure answers the question "what is happening?", while the consequences of the failure answer the question "what is the impact of the failure?" According to US Army Standard MIL-STD-1629A (recalled in 1998), which describes the procedure for performing the FMECA technique, a failure effect is defined as: "The effect of a failure mode on the operation, function, or condition of an object. The failure effects are classified as a local effect, next higher level and end effect" [63]. The local effect refers to the consequences of a failure on a specific component of a functional block, while the effect of the next higher level refers to the consequences of a failure of the next functional block having a higher level of complexity. The end effect relates to the highest levels of complexity. Each failure mode can be caused by several causes leading to several different failure effects. A consequence of a failure is the way (or more of them) in which the failure mode or the fault multiplication is affected by the dependence on the functional context of the functional block, the performance standards applicable to each function, and the physical effects of each failure mode [2]. Any failure with associated consequences, except in an operational context, can have consequences for human and environmental safety. The concept of failure consequences, except in the RCM approach, is also established in the discipline of risk management (financial loss, injury to people, and other unintended consequences).

Tam and Gordon, defining the terminology of plant asset management failures, refer to the consequences of a failure as: "the effect that a failure event has on a plant failure in an operational context, including business, people and environment" [83]. Basically, the consequence of a failure refers to the effect on the outside world, while the effect of a failure refers to a failure of the equipment function. Failure symptoms occur prior to a failure event, and may be visible or measured as a warning sign of a failure for deterioration of the physical conditions of the plant or if the plant is exposed to harmful effects. A failure event is the occurrence of an interruption of a plant's ability to perform the required function, which results from a link between the cause and effect of the failure or deterioration of the physical operating conditions of the equipment and operation actions, or good physical operating conditions of the equipment and inadequate operation actions. Deterioration of working conditions is observed by subjective methods (human senses) or by objective methods (verified by measuring ion instruments). In this case, the plant may send a signal or symptom, i.e., warning sign before actual failure. After a failure event, symptoms can be observed or recognized as the effects of the failure [84]. The effect of the failure, as a direct consequence of the failure event on the operation, function, or condition of the installation, requires that the associated consequences of the failure be analyzed in relation to the operation of the installation. If it is necessary to rank or scale the severity of a consequence, then the term "consequence ranking" (used for the concept of ranking magnitude of consequences) is often used, that is, the term "risk ranking" (the case where the ranking of consequences is required to identify hazards and assess risks to individuals, property, and environment). The failure effects can be sorted and logged throughout to allow a historical account of the plant's behavior (source of information for operation history and/or statistical analysis) (Table 8.6 and Fig. 8.16).

When investigating a plant failure, the warnings and effects of the failure are taken into account in order to identify the failure mechanisms (how, i.e., how the failure occurred) and the cause of the failure (why the failure occurred). Based on the research findings, a strategy can be developed that can consist of preventive actions (referring to goals designed to maintain the plant and its environment within the desired specifications as well as preventing improper use) and proactive actions (related to selection) and the installation of instruments capable of detecting and acting upon available fault warnings. This requires that measurable parameters (termed "health indicators"), which represent physical performance (work performance), be determined in order to identify differences between normal (nominal or favorable) and abnormal (unfavorable) conditions. Steam turbine propulsion instructions, supplied by the manufacturer together with the turbine, accurately define the steady-state and nonsteady-state conditions of the steam turbines, whereby the start-up and shutdown of the steam turbine, as nonstationary modes, are particularly sensitive operating states. The rapid change in temperature inside the turbine and the accompanying increase in relative stretching, as well as greater thermal and mechanical inertia, have certain limitations in the form of permissible maximum pressure parameters and temperatures in terms of the materials used. Long-term and trouble-free operation of the turbines is possible only in the case of their careful and professional technical service, both during nonstationary and stationary modes of operation during operation, as well as quality execution of the planned overhaul activities.

Commissioning and shutting down the steam turbine are the most responsible operations during the operation of the steam turbine plant. These operations are associated with significant changes in the mechanical and thermal state of the turbine and steam line elements.

Proper commissioning and shutoff significantly affects the operational reliability and life of the turbine. Long-term experience in the operation of high-power turbines has shown that much of the damage and damage to the turbine equipment resulted from the start-up of the turbine due to improperly controlled heating mode, malfunctioning turbine plant, as well as due to certain structural defects at the plant. It should be borne in mind that poor turbine start-up and shutdown does not have to immediately lead to failure and damage to the plant, but will be reflected during future turbine operation, housing of turbines, valves, and steam lines, then with bending of horizontal flange joints, change of structural state of metal, increased wear of bearings, and other unintended consequences. The commissioning of a turbine is a particularly complex case from the point of view of a transient thermomechanical condition, since in such processes thermal and mechanical stresses in the elements of the turbine are summed up. In addition, when the turbine is released from the warm state, there are additional difficulties that do not occur during the cold start. Likewise, different problems occur when releasing blocks with critical parameters versus blocks with supercritical steam parameters [52].

Table 8.6 Display of condition parameters in failure of TPP Ugljevik condensing thermal power plant system and rated power of 300 MW.

General information		Boiler plant, downtime			Other equipment, downtime			Boiler plant, downtime			Other equipment, downtime		
Year	Failure number	Failure group I	Failure group II	Failure group III	Turbine plant	Electrical equipment	Other (overhaul + net suppression)	Failure group I	Failure group II	Failure group III	Turbine plant	Electrical equipment	Other (overhaul + net suppression)
1985	22	**3**	5	5	0	8	1	536.7	479.62	843.03	0	131.71	580.13
1986	34	**4**	**10**	**9**	9	1	1	230.85	554.9	91.18	363.16	0.08	1088.43
1987	19	3	11	1	0	3	1	255.6	714.86	2.13	0	205.39	2056.22
1988	13	5	1	3	1	2	1	207.38	67.82	169.88	171.05	171.83	967.05
1989	19	10	3	1	0	4	1	421.21	265.17	4.53	0	43.88	1603.35
1990	11	5	2	2	0	1	1	308.57	259.31	66.95	0	12.51	988.3
1991	17	10	1	1	1	2	2	779.05	33.03	2.97	35.55	49.53	1824.8
1992	3	1	0	1	0	0	1	34.38	0	108.2	0	0	6060.2
1993	1	0	0	0	0	0	1	0	0	0	0	0	8760
1994	1	0	0	0	0	0	1	0	0	0	0	0	8760
1995	7	1	1	1	4	0	0	70.5	90.37	115.93	49.92	0	7863.27
1996	18	2	0	6	1	6	3	750.04	0	324.5	17.32	189.57	2530.2
1997	16	4	0	4	1	6	1	404.86	0	133.2	477.42	100.53	2778.97
1998	18	7	4	1	0	5	1	533.92	281.58	8.45	–	113.52	1683
1999	24	1	7	4	1	10	1	60.78	506.83	54.35	47.35	172.88	1375.2
2000	28	10	2	1	1	11	2	993.82	86.8	15.8	53.33	479.43	1431
2001	17	7	5	0	0	3	2	510.58	196.48	0	0	211.88	1749.4
2002	26	13	3	1	0	8	1	1074.95	366.73	8.12	0	116.49	1492
2003	22	15	4	1	0	1	1	895.15	490.07	175.38	0	8.88	1358.65
2004	23	10	2	5	0	5	1	713.68	262.42	337.64	0	370.15	2037.4
2005	24	14	1	2	0	5	2	885.65	132.3	7.88	0	1056.96	2307.05
2006	27	19	2	1	0	3	2	1346.71	152.43	22.58	0	118.33	1133.96
2007	22	12	4	2	0	3	1	833.59	322.77	6.02	0	28.73	1124.2
2008	16	9	3	3	0	1	1	603.65	312.9	40.24	0	4.75	1142.3
2009	24	14	1	3	0	3	3	709.11	68.08	19.19	0	101.3	1156.23
2010	16	4	2	1	0	7	1	156.93	147.67	7.5	0	227.3	2545.2
2011	21	6	2	4	3	5	1	189.13	91.45	85.77	354.85	121.56	707.93
2012	11	5	2	1	2	2	1	200.27	112.55	25.22	103.15	21.77	887.33
Total:	500	194	78	64	24	105	36	13,707.06	5996.14	2676.64	1673.1	4058.96	67,991.77

Failure group I: Firing of the pipe system on the boiler P64. Failure group II: Drossing of the firing screens, slashing of the slit, failure of the deflector, backfill with flying ash of the inductive part of boiler P64, clogging of bright openings of convective surfaces, and the like. Failure group III: Failures on other elements of the boiler plant and auxiliary boiler equipment.

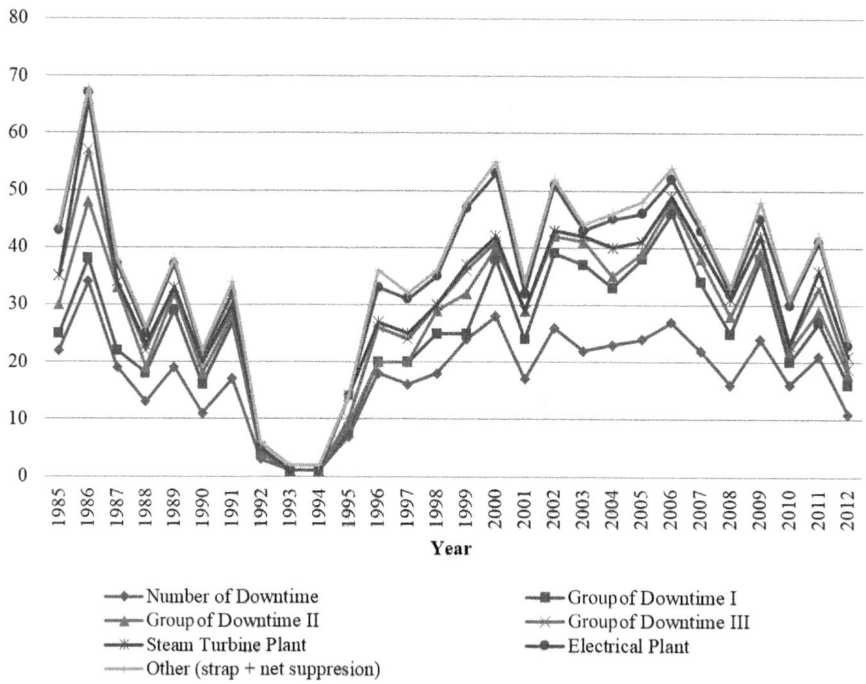

FIGURE 8.16

Trend of "number of failures" parameters in condensing thermal power plant (TPP) Ugljevik nominal power of 300 MW for data from Table 8.6.

During start-up and loading of the turbocharger, certain changes occur in the mechanical condition of the turbine elements and associated equipment, i.e., individual elements of the turbine are exposed to different complex stresses [56]:

- stresses in the steam, turbine, and valve housings occur under internal pressure;
- flexural stresses occur in diaphragms, rotor discs, routing blades, and working blades;
- stresses from centrifugal forces occur in the blades, discs, and other rotating elements of the turbine;
- tangential stresses occur on the turbine shaft due to the transfer of torque to the electric generator;
- alternating voltages occur due to vibrations in the blades, shaft, and other elements of the turbine plant;
- axial forces occur that act on the axial bearing of the turbine.

Due to the nonstationary thermal state of the turbine, the following phenomena occur:

- thermal stresses in the walls and flanges of the turbine body, steam lines, stop and control valves;
- additional tensile stresses for the bolts that maintain the upper lower housing of the turbine, as well as the bolts of the flange connections of the valve and the steam line;

- thermal stresses in the turbine rotor;
- distortion of the housing due to impermissible temperature difference of the upper and lower parts of the housing;
- linear stretching of the turbine rotor and stator;
- change of axial clearances in the flow part of the turbine due to different elongation of the rotor and housing;
- change of radial clearances in the flow part of the turbine.

These phenomena complicate the start-up of the turbine, increase its start-up time, and can cause accidents when deviating from the prescribed heating mode. Various safety devices (safety valves, centrifugal regulators, and safety ejectors) protect the turbine from mechanical overloads, while the turbines are not protected against unauthorized thermal stresses. In such a case, the safety of the turbine depends entirely on the properly chosen release method, as well as the qualification and training of the turbine plant operator. Although the control system monitors the operation of the steam turbine, mandatory direct supervision is required, carried out by the turbine operators (steam turbine engineers, turbine engineers). Control of the turbine operation includes activities on the maintenance and control of the parameters of the steam turbine and the plant as a whole, taking measures in case of unexpected operational events, in order to prevent damage or failure of the turbine from operation, and then keeping protocols (recording the operating parameters read at equal time intervals, recording of characteristic operating events and objections related to the operation of the plant, relevant to the turbine maintenance service, etc.), as well as periodic testing of the functionality of individual turbine components and keeping protocols of these tests, in accordance with the technical instructions. The tasks of the operator also include cooperation with the maintenance services in the elimination of failures on components and periodic inspections and overhauls of the turbine plant. The most important duty of turbine operators is to maintain the plant's absolute cleanliness (preventing oil, condensation, steam, etc.) from leaking. In the event of any noise that deviates from normal occurrence, metal shocks, impingement, and large vibrations, it is necessary to stop the machine as soon as possible and to take measures to determine the cause of its occurrence, in order to eliminate it effectively and quickly. Also, continuous monitoring on the measuring and measuring instruments (KMP) is necessary, as well as their display, with regular recording of the obtained measurement, results during the operating time and according to defined protocols.

The scope of activities for individual segments of the turbine plant during regular operation includes [52,56]:

(a) Maintenance of the system and protection and control of the steam turbine during normal operation:
- continuous daily inspections and controls for oil leaks that can cause a fire;
- monitoring of the condition of the nuts, manifolds, stop bodies of the steam distribution (especially if they operate under vibration conditions and disturbed operating modes);
- monitoring of vibrations of regulatory bodies, which may cause their housings and supports to burst;
- monitoring of changes in pressure and pulsation in oil pipelines (any change in these parameters indicates poor regulation, poor lubrication, then oil system dirt, increased air content, etc.).

(b) Checking the tightness of the steam turbine regulators and stops:
- continuous monitoring and control of the leakage of stop and control valves, with the aim of preventing the turbine from "run-in";
- monitoring of valve leakage (the leakage must not be such as to cause the turbine rotor to start while the turbine is still in standstill);
- monitoring and control of the seals on the nonreturn valves on the steam extraction from the turbine.

(c) Valve design (in order to prevent the control valves from collapsing, salt deposition, or temperature deformation, the regulating valves must be worked out daily by forcibly adjusting the position of one of the valves).

(d) Steam check of steam turbine protection (all modern turbines have a device for checking the speed protection operation at nominal speed, idle, and load, whereby the protection check period is defined on the basis of the recommendations given in the instruction for each turbine).

(e) Tests of the control system in accordance with the operating instructions (to check the control system it is necessary to record the static characteristic of the control once a year, with recording being made before and after major overhaul, as well as after all major interventions on the control system).

(f) Maintenance of oil and steam turbine lubrication systems:
- ensuring the functionality of the oil system, especially the oil tank;
- ensuring long-term oil quality, removal of air, impurities, and water, by monitoring the oil level in the tank, as well as signaling the levels in the dirty and clean part, i.e., on sieves (sieve cleaning is required when increasing the level difference);
- during operation of the unit, it is necessary to check at least once a month the activation of the backup and emergency oil pumps, as well as at each start and stop of the turbine;
- the water pressure before and after the oil coolers must also be monitored in order to assess the purity of the water on the water side (the maximum permissible oil temperature is usually around 41°C and the bearings 60–65°C;
- monitoring of oil quality in chemical laboratories, through regular oil analysis.

(g) Monitoring and control of the operation of the turbine, through the primary monitoring of parameters whose exceeding of the limit values endangers the operation of the turbine. These are, first of all, the relative elongation of the rotor and the offset. For each turbine in the operating instructions, the limit values of these parameters are defined, whose disruption leads to impingement of the rotor and stator parts and the failure of the unit. These parameters should be monitored during the operation of the unit when the load changes abruptly. Vibration monitoring makes it possible to prevent turbine parts from being impeded in a timely manner. Steam parameters in front of the turbine are very important for reliable operation of the unit. They must be maintained within the permitted limits. Economical operation of the turbine is only possible if the condensing plant is operating properly. The deterioration of the condensing plant is usually associated with pipe contamination, lack of cooling water, nasal sealing of the vacuum part of the turbine, etc. It is also important to monitor the operation of the feed water regeneration system in low- and high-pressure heaters, whereby the efficiency of the heater operation is judged by the heat exertion of the heater, i.e., the difference in saturation temperatures, which corresponds to the vapor pressure of the extraction vessel and the condensate. This difference should be 2–3°C for low-pressure supply water heaters (LPP) and 1–2°C for high-pressure supply water heaters

(HPP). An increase in temperature stress for the LPP is an indicator of pipe contamination or a deterioration of the vacuum, and for the HPP a sign of a decrease in water flow.

The operation of the turbine plant must be monitored within the energy block as a whole. During start-up of the turbine, as well as under reduced load, due to the increase in losses and the return of heat in degrees, the steam temperature at the outlet of the turbine rises, which can again cause a change in the relative stretching and an increase in the rear turbine bearing temperature. These operating modes are very disadvantageous and are often time-limited by manufacturers. Often, during the operation of the turbine (self-induced oscillation) or its commissioning (improper clearances in the turbine bearings), unexpected oscillations of the rotor occur. The cause of the oscillations of the rotor may be the transverse deformation of the rotor, as well as inadequate turbine oil quality or bearing oil temperature. Prior to making any decision on the further operation or shutdown of the turbine plant, these oscillations must be measured and their detailed analysis performed. One way to make a decision on further continuation of the turbine plant operation is to compare the values obtained and measured with the values given in the diagram in Fig. 8.17.

It should be noted that steam turbines of more modern construction have a device for continuous monitoring of rotor oscillations, as an integral part of the turbine protection system.

In case of impermissibly high oscillations, the steam turbine is automatically shut off. With the addition of vibration-diagnostic monitoring of the state of the steam turbine, in conjunction with the existing analytical methods, the dynamic behavior of the machine can be significantly better investigated. The monitoring thus constructed allows to determine when and where the damage occurred (technical diagnostics), then to evaluate how the damage will continue to develop over time, and finally to estimate the time until definitive failure (technical prognostics), with a preliminary identification of the cause of failure (technical genetics).

For reliable and safe operation of machines, it is necessary to know the effects of the process on the machine and vice versa and at all times to know the state of the steam turbine, in order to manage it in the process of electricity production in the optimal and economically most favorable way.

Recently, there is a growing trend of optimization of the electricity production process with the aim of reducing costs and increasing productivity, with the focus on process-influencing factors, with a view to getting to know them more, and considering their effects on the production process itself. However, in order for the thermal power unit as a whole, and therefore the steam turbine plant, to work optimally and most productively when needed, it is necessary to have reliable equipment necessary for the process to take place. It should be noted that operation of equipment over a period of time under conditions dictated by process optimization (usually only by economic criteria) can be adversely affected because by removing certain process limits, the turbine operates under different mechanical conditions, which often result in increased stresses in the equipment elements, which in rotary machines can lead to faster degradation of the mechanical condition of the equipment. With increasing maintenance and repair costs, as well as unplanned losses due to production failure, the positive results achieved by previous process optimization are often completely nullified. On the other hand, the operation of machines under mechanical and process unfavorable conditions generates variable stresses of the material, which can lead to damage to the machine assemblies, cracks, and breakage, with often catastrophic consequences. The development of microprocessor devices for complete monitoring and analysis of work with possible determination of the current mechanical condition of machines, as well as other diagnostic equipment for monitoring the condition of the turbine plant,

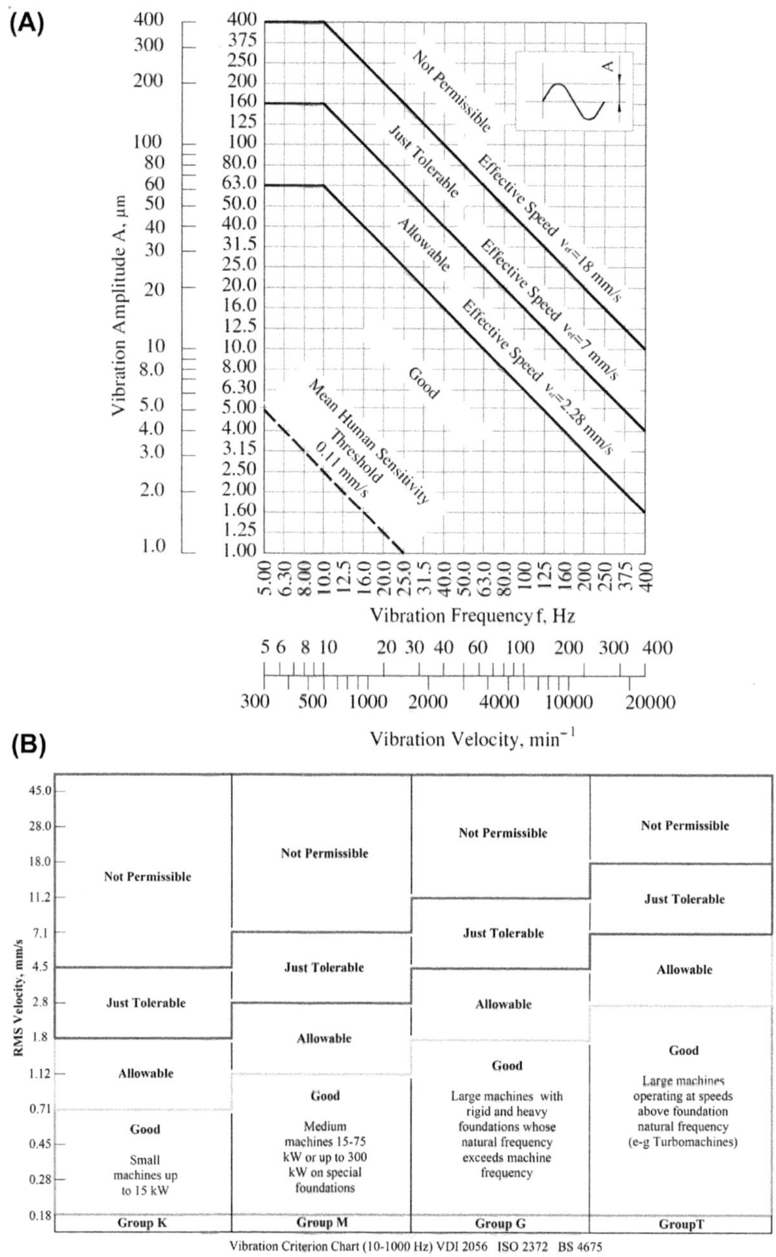

FIGURE 8.17

Vibration criteria for (A) rotary machines and (B) VDI 2056—group T [85].

enabled a completely different approach to plant maintenance. The result is an intensification of Condition Based Maintenance applications. Thereby, maintenance activities are performed only when necessary and when required by the condition of the turbine.

Steam turbine shutdown procedures, i.e., its stopping, must also be such as to allow the temperatures of the turbine parts (especially the casing), as well as the relative stretching of the casing and rotor, during the period required for the operation of the steam turbine shutdown operation to remain within the permitted limits. In doing so, the stopping procedure differs from the so-called normal exploitation, from the procedure at the so-called "accidental" or "accidental" stops.

6.3 Classification of failures in the general case

In general, failures can be classified in different ways. Their division is mainly based on the basic concept of fault analysis (Table 8.7). Malfunctions are often classified into fault modes, which represent malfunction descriptions, that is, ways in which it can be observed that the functional block is malfunctioning. Fault modes are identified by studying the functions and performance of the functional block. British Standard BS 5760-5 defines general failure modes for: time-out failures, failures to work in due time, failures to stop work in due time, and overtime. When analyzing functional block failure modes, it is usually useful to investigate their cause. Defects according to the causes of their occurrence, according to IEC 60050 (191), can be divided according to the life cycle of the functional block [4]. Different causes of failures are not necessarily excluded, such as obviously overlapping defects of defects, design, and construction. In some cases, it is useful to divide failures into primary, secondary, and control command failures. IEC 61508 standard distinguishes two major types of failures by their cause: accidental hardware failure and systematic failure. In most cases, it is usually assumed that all elements of a functional block are independent, which is not always the case in practice. Security systems often have a high degree of redundancy, and therefore the reliability of the system is strictly dependent on potential common cause failure (CCF). Therefore, it is very important to identify CCFs and take the necessary safety measures to prevent these failures [4].

There are several definitions of CCF events. Thus, ISO 14224 defines CCFs as: "failures of different objects resulting from the same direct cause and occurring within a relatively short time in which these failures are not due to other failures." Similarly, IEC 60050 (191) defines the same as: "failures of different objects resulting from one event and where these failures are not due to each other" [76]. The US Nuclear Regulatory Commission, in its database standard for CCFs and system analysis NUREG/CR-6268: 2007, links CCFs closely to understanding the nature and significance of dependent events, and defines the event of this failure as: "a dependent failure in which two or more components exist at the same time in a faulty state or within a short time interval and which are a direct result of a common cause" [86,87]. Modarres et al. define dependent failures as events in which the probability of occurrence of each failure is dependent on the occurrence of other failures [82]. According to Rausand and Høyland, CCFs can be divided into two types—multiple failures occurring at the same time due to a common cause and multiple failures occurring due to a common cause, but not necessarily at the same time [4]. There are many causes for the occurrence of CCFs, of which the following are distinguished [88]:

- Shared outdoor environment (includes temperature, lighting, humidity, vibration, dust, and environmental pollution);

- Operation and maintenance errors (includes factors such as: negligence, improper handling, improper maintenance, incorrect adjustment, or calibration);
- Defects in equipment design (the result of a failure during the design phase);
- External catastrophic events (occurrence of natural events such as earthquakes, fires, floods, etc., the occurrence of which can cause the simultaneous failure of a redundant device);
- Functional defect (may be the result of poorly designed safeguards or inadequate instruments that monitor changes in system performance);
- Shared external power source (a redundant unit powered from the same power source, directly or indirectly, may fail at the same time in the event of a power failure);
- Common external transmission system (unit used for transfer/transport of end products to end users);
- Joint manufacturer (a redundant unit purchased from the same manufacturer may have the same factory or other defect).

CCFs are very important in reliability analysis and should be given appropriate treatment to minimize overestimation of reliability. Also, a number of procedures and checklists have been developed to assist engineers in designing a system that is resilient to CCFs. The MIL-STD-882E standard defines "severity" as: "the magnitude (magnitude) of the potential consequences of accidents including: death, injuries, occupational diseases, damage or loss of equipment or property, environmental damage or monetary loss" [60]. In reliability analyses, situations may be considered in practice when system components may schedule independently of one another or that their malfunction may not be independent of one another (dependent failures). There are two types of dependencies: positive (implied if failure of one component leads to an increase in the tendency of failure or scheduling of the other component) and negative (if failure of one component leads to a decrease of the tendency of failure or scheduling of the other component.

6.4 Classification of failures for the steam turbine system

The failure of a steam turbine system is defined as the interruption of the ability of the unit to perform its intended function. Reduction or loss of working capacity of the technical system during exploitation is due to the effect of various factors (built-in, accidental, or temporal), which change the initial parameters of the system, thus causing different levels of damage. There are several categorizations of failures according to the selected criterion (Table 8.7). The development of the occurrence of failures can be observed through the case of failure (monitoring the change in the state in the time interval) and through the case of the failure (monitoring the number of these failures in time). In doing so, developed methods are used for monitoring the phenomena with modeling of distribution, all with the aim of obtaining a mathematical model of changing the state of the technical system, thus creating the necessary assumptions for the possibility of making diagnostic decisions. The determination of mathematical models is carried out by a deterministic and stochastic method.

Causes of steam turbine failures and failures are most commonly divided into three groups: systemic (built-in), random and monotonous, or temporal causes of failure. In the initial period of operation of a steam turbine, systemic or built-in system failures usually occur, which, among other things, can be: incorrect design (poor construction with defects in the structure, errors as a result of wrong material selection, etc.), design reliability, occurrence of design voltages, manufacturing errors,

Table 8.7 Classification of failures in the general case based on the basic concept of fault analysis.

Criterion	Classification of failures according to the selected criterion				Author/Standard
Classification of failures according to fault modes	Intermittent failures: Failures of a functional block that result in the loss of some of its function in a very short period of time (after this failure, the functional block returns to its full standard operation).				According to Rausand and Høyland [4]
	Extended failures: Failures that result in the loss of a function of a functional block and lasts until a part of it is repaired or replaced.	Complete failures: Failures that cause a complete loss of the required. Partial failures: Failures that result in the loss of a function of a functional block, but not the complete loss of its required function.	Sudden failures: Failures that cannot be predicted without first testing or testing. Gradual failures: Failures that can be predicted by testing or testing (the procedure may indicate a gradual deviation from a specified range of functional block performance values, the recognition of which requires comparison of actual performance with the specified functional block performance).	Catastrophic failures: a failure that is both sudden and complete. Degraded failure: A Failure that is both partial and gradual.	

Continued

Table 8.7 Classification of failures in the general case based on the basic concept of fault analysis.—cont'd

Criterion	Classification of failures according to the selected criterion		Author/Standard
Classification of failures by cause of failures	Design failures: Failures that occur due to inadequate functional unit design. Weakness failures: Failures due to defects in the functional block itself when exposed to loads within the indicated capabilities of the functional block (defects may be inherent or induced in the functional block so that it cannot sustain the load in a normal environment). Manufacturing failures: Malfunctions due to manufacturing mismatch due to the design of a functional block or to specific manufacturing processes (all failures caused by a manufacturing process error, inadequate manufacturing, or a manufacturing control error during the design, testing, or repair period). Aging failure: Failures whose probability of occurrence increases over time as a result of an inherent process in a functional block, i.e., of its use and aging. Misuse failure: Usage failures due to the application of loads exceeding the indicated functional block capabilities. Mishandling failure: Malfunction caused by improper handling of a functional block or lack of care by the person concerned.		In accordance with IEC 60050 (191) [76]
	Primary malfunction: A malfunction caused by the natural aging of a functional block that occurs under conditions within the designed block of a functional block. Secondary malfunction: Malfunction caused by excessive stress outside the designed function block framework (such stresses can be shocks caused by the action of thermal, mechanical, electrical, chemical, or radioactive energy sources, and stresses may be caused by nearby components, environment, and operators of the system or plant personnel; in the event of a primary or secondary failure, a repair action must be performed to return the functional block). Control command Malfunction: A malfunction caused by an incorrect control signal or interference (it is not usually necessary to perform a repair action to restore the functional block to functional state; these failures sometimes relate to transient failures).		According to Rausand [50]
	Accidental hardware failures: Failures resulting from the natural mechanisms of hardware failure.	Aging failures: Defects that occur within a designed system framework. Fault failures: Failures occurring under the action of excessive stress, i.e., stresses above the designed frame.	According to IEC 61508: 2000 concerning the functional safety of electrical, electronic, and programmable electronic security systems
	Systematic malfunction: A malfunction having the definition of: "malfunctioning in a deterministic manner by a specific cause, which can only be remedied by modification of the design or manufacturing process, work procedures and other relevant factors" (failures of a nonphysical nature and manifested by a deviation of performance from required, without any physical degradation of the functional block).	Failures initiated by design, fabrication, or assembly: Failures that may be latent (hidden) from the first day of operation (e.g., hidden software failures, sensors (wrong readings), or if any piece of equipment is installed in the wrong place). Maintenance-related failures: Operating and assembly failures initiated by human errors during system operation, maintenance, or testing.	

Sorting of failures according to the effect and consequences of the failure	Dangerous failure: Failure that has potential to put safety connected system into dangerous state or state of functional malfunction. Safe malfunction: A malfunction does not have the potential to put a securely connected system in a dangerous or functional malfunction state.	Detected failure: A defect that is detected immediately when it occurs, e.g., by automatic self-testing. Undetected malfunction: A malfunction that can only be observed by functional testing or when a system is required.	According to standard IEC 61508
	Failures with security consequences: Failures that cause significant consequences for the safety of the technical system. Environmental failures: Failures that cause significant irreversible environmental impacts. Failures with operational consequences: Failures that cause significant consequences for the operational readiness of the system. Failures with nonoperational consequences: Failures that do not affect the operational readiness of the system.		According to the RCM approach with respect to consequences [66]
	Catastrophic malfunctions: Malfunctions that cause death or complete loss of the system, and a significant irreversible impact on the environment or monetary loss. Critical failures: Failures that cause serious injury or damage to the system, and significant environmental impact and monetary loss. Marginal malfunctions: Malfunctions that cause minor injuries or damage to the system, and moderate environmental or monetary loss. Negligible failures: Failures that result in very minor injuries or damage to the system, and minimal environmental impact or monetary loss.		Concerning the severity of the failure mode
	Critical failures: Failures that cause an immediate and overall loss of the ability of a functional block to provide output within specified limits. Gradual failures: Failures that are not critical but that prevent the functional block from coming out (they can usually, but not necessarily, develop into critical failures). Initial failures: Failures that do not immediately cause the functional block to lose the required outputs (if these failures are not remedied in the near future, it may result in a gradual or critical failure).		According to ISO 14224 [3,50]
Dependent failures	Common cause failures: Failures that simultaneously affect two or more identical redundant components and have the same root cause (these are causes that represent residual, sometimes difficult to grasp, interdependence among components whose mechanisms are not explicitly incorporated into logical plant models). Cascade failures: Multiple failures initiated by a single component failure in a system that result in a chain reaction or domino effect. Negative dependency failures: Single failures that reduce the likelihood of failures on other components.		According Rausand and Høyland [4]

installation errors, adjusting errors, heat treatment and residual thermal stresses, technical control failures (insufficient control and testing), etc. On the other hand, accidental causes of failure include: unstable environmental conditions, overload or unstable (especially unsteady) operating modes (instability of technological parameters), poor handling and maintenance, inadequate control, instability of structural parameters (load gradient, etc.).

Monotonous or weathering causes may include: exploitation mode, maintenance mode, lubrication, fatigue of materials (fatigue processes and changes in material properties), warming, erosion and corrosion of parts, wear of embedded parts, contamination of working medium, regulation and concentration of parts, etc. The sooner the failure research begins, the greater the opportunity to determine the real reason for the failure, with the basic rules of conduct being applied [1]:

- Do not destroy broken items. In particular, do not touch the surfaces where the cracking occurred, or break pieces.
- Elements that have been damaged are to be inspected only after receiving all the documents (reports, explanations, photographs, etc.), ensuring that all dismantled parts can be correctly identified, reassembled, undamaged, etc.
- The research should not only concentrate on the location of the fracture, but also the state of the environment. Because of the possibility of a failure chain occurring, it is necessary to consider whether the failure occurred is merely a consequence of someone else or is caused entirely by other causes.
- Do not easily form a conclusion on the occurrence of dismissal. Collect all the information, then discard the insignificant ones, and form a definitive conclusion only when it is clear that all other possibilities have been eliminated.
- When forming documentation on dismissal, together with material evidence, also take reports from other experts, without taking anything without verification ("on trust"), especially your own opinion, since conclusions, estimates, and decisions can be subjective (in terms of making decisions based on your own experiences).

The results of numerous investigations and analyses of damage to the turbines in operation have shown the susceptibility to more frequent damage and failure (with a greater likelihood) of some parts of the turbine, while for other elements this probability is very low. Therefore, the classification of damage to the turbine parts is often made according to the categories of probability of their occurrence: damage with a high probability of occurrence, then damage with a medium probability of occurrence, and damage with a low probability of occurrence.

The parts with high probability of damage during operation include: labyrinth seals, bearings, first and last stage rotor blades, first stage nozzles, spindles and plates of control valves, condenser tubes, etc. The average likelihood of damage to the parts most often includes: rotor blades of other stages, turbine shafts, partition walls with stator blades, valve housings, and the like. Elements of a steam turbine with a low probability of failure include all remaining components of the turbine, which are not exposed to high temperatures and operate at approximately constant pressures (rotor discs, turbine housing, turbine condenser jacket, ejector, oil cooler, etc.).

The previous categorization is of a conditional type and can only serve as a preliminary estimate for the procurement of new spare parts and a prerequisite for defining the positions that need to be paid more attention to when working. Applying technical diagnostics in the course of performing activities for maintenance of the turbine and introducing systematic monitoring of the condition of parameters and damage for specific conditions of exploitation can determine the categorization of damage of individual parts for a particular turbine. In this way, it is also possible to further reduce the scope of research into the causes of failure of the most critical element (drive) on the turbine, which will result in an adequate reduction of the costs necessary for the research (preliminary or preliminary assessment of the situation). Also, the results obtained in this way can serve for further reconstruction, revitalization, and modernization of individual parts of the structure, that is, the steam turbine system as a whole.

During the operation of the turbine, its individual parts operate under different operating conditions, which simultaneously determine the most loaded and most endangered components and parts of the steam turbine according to the critical influence. Similar to the boiler plant, the critical parts are determined in the turbines, with special attention being paid to the parts where large defects of design may occur, as well as loads that may cause material degradation (Fig. 8.18).

In the transformation of the kinetic energy of a fluid, gas, or vapor into mechanical energy, rotating blades which are turning represent components of a turbine exposed to different loading conditions. Thus, turbine blades are subjected to changes in pressure, temperature, corrosion, and may be particularly sensitive to changes in dynamic conditions.

The most common damage to steam turbines during operation is due to improper operation of the plant and poor construction and materials. The most common causes of accidents that do not affect equipment damage are: protection devices, vacuum damage, jamming of control valves, and direct operator errors, when changing modes with a sudden reduction in load.

Steam in the critical and supercritical pressure turbines has an inlet temperature of 500–650°C and a pressure of 8.8–30.0 MPa. These conditions during operation result in material degradation. Fig. 8.18 shows the occurrence of damage to several high-pressure turbine rotors of the K-500-240 HTZ (manufactured by Kharkov Turbo Institute, Russia) type after 70–100 thousand hours of effective operation in the area of the first joint blades, operating in the highest temperature ranges.

6.4.1 Propulsion damage due to erosion and corrosion

Wet steam turbine stages often have erosion of the surface of the rotor blades, which results in the removal of the metal so that the surface becomes uneven and rough with certain cavities (Fig. 8.19). Erosion can capture a significant part of the profile and lead to the separation of a portion of the blade. Even slight erosion changes the vibratory characteristics of the blade, its firmness, and lowers the degree of its beneficial effect. Corrosion, which occurs on elements made of weakly alloyed steels, is caused by the action of salts and acids, which enter the turbine together with steam.

Salt deposition, which may result from poor chemical preparation of feed water, as well as the composition of sediment by turbine stages, is different. It depends on the temperature, pressure, and velocity of the vapor in degree, the condition of the surface of the blade, and the chemical composition of the salt. Salts are usually deposited under the bandage. During a longer period of work, they can salt the blade along its entire height. Salt deposition on the blades decreases the turbine flow surfaces, and thus reduces the power of the turbine, which can cause accidents, as the stresses in the blade and the bandage increase due to the centrifugal force of the deposited salts, the axial force acting on the bearing increases due to the increased reactivity in the turbine stage to reduce the cross-sectional area of the interchannel ducts, as well as to unbalance the rotor, due to the unevenness of the sediment, which causes vibrations. Examples of corrosion damage are given in Figs. 8.20–8.23.

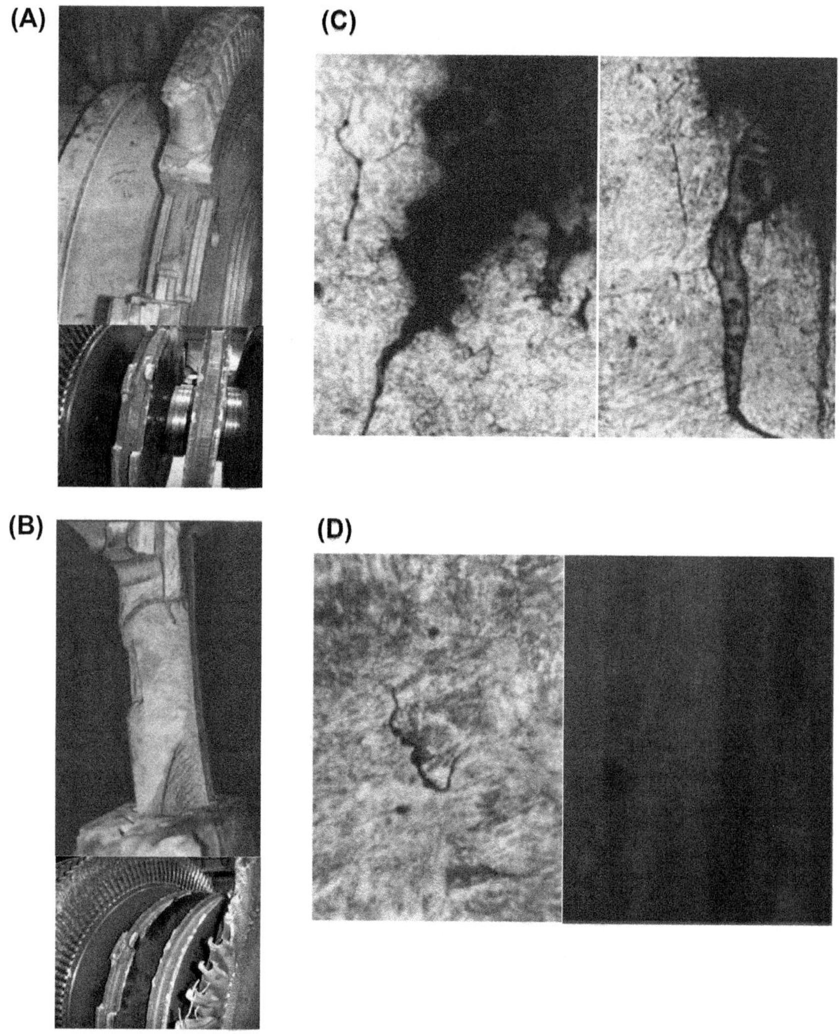

FIGURE 8.18

Demonstration of rim deflection of high-pressure rotor of turbine K-500-240 HTZ [52]: (A) general appearance; (B) damage surface; (C) secondary cracks in the perimeter; (D) micropores in the metal.

The use of stainless steel blades is the most effective measure of corrosion control. In parallel and independently, it is necessary to ensure the tightness of all valves connecting the turbine to the steam pipes, where the high-pressure steam pipes between two consecutively installed valves must be drained for drainage of steam (condensate), which eventually flows through the first valve.

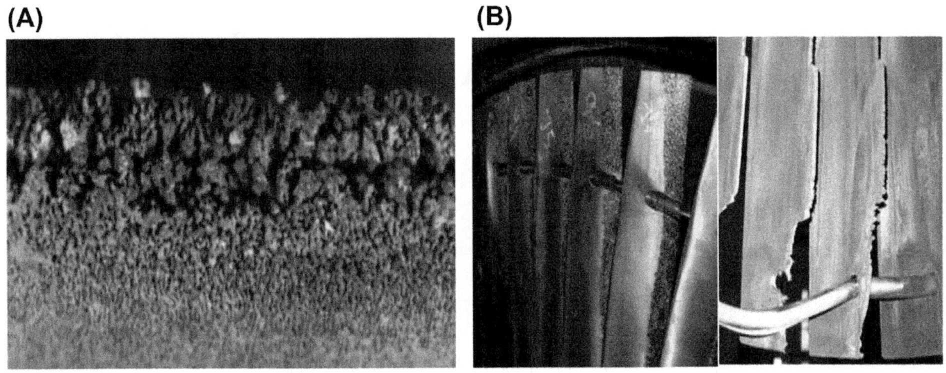

FIGURE 8.19

Layout of damaged surface of steam turbine blade caused by erosion by dropping water into low-pressure turbines.

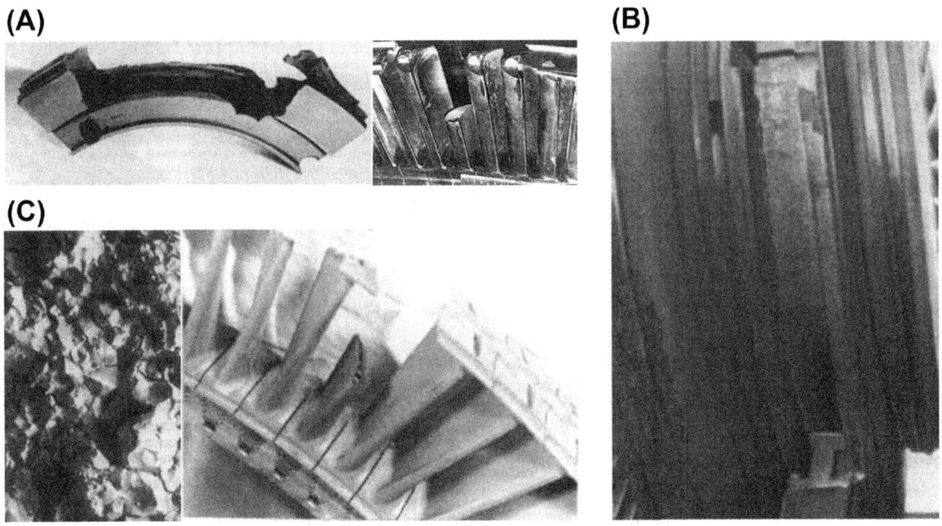

FIGURE 8.20

Damage to the rim of the 12th stage of the high-pressure cylinder (HPC) turbine K-800-240 LMZ due to thermal corrosion [52]: (A) segment of rim damage with secondary cracks; (B) general appearance of the fracture; (C) appearance of the rim fracture surface.

On the other hand, certain types of Cr—Ni stainless steel austenitic structures have a tendency, at high temperatures (more than 500°C), to intercrystalline corrosion, hardly noticeable on the surface of the blades, but which can cause the blades to break even under by the action of less force. By adding a stabilizer to the steel, this effect can be substantially neutralized.

FIGURE 8.21

Damage to individual turbine elements as a result of erosion and corrosion [52].

Erosion measurement of the turbine blades must be performed on the back rows of the low-pressure rotor, the inlet edge of the rotor blades, and the outlet edge of the rotor blades. Erosion of the inlet edge of the rotor blades occurs on the low-pressure rotor blades. These degrees of blades are subject to greater erosion due to higher moisture content and high centrifugal force. The magnitude of erosion of the inlet edge of the rotor blades depends on the structural design of the housing, the technology of fabrication and assembly of the blades, the quality of the protective stellate plate, and then on the thermodynamic and machine parameters. The size of the blade damage can be up to 75% of the length of the blade from top to bottom. If the inlet side of the blade is found to have erosion damage of 60% of the length from top to bottom, or 20%–32% of the tendon of the profile, the entire row of blades must be changed. On the entrance side, if only the stellate or protective stellar plate is damaged, it is sufficient to remove the old tiles and install new tiles.

(A) (B) (C) (D)

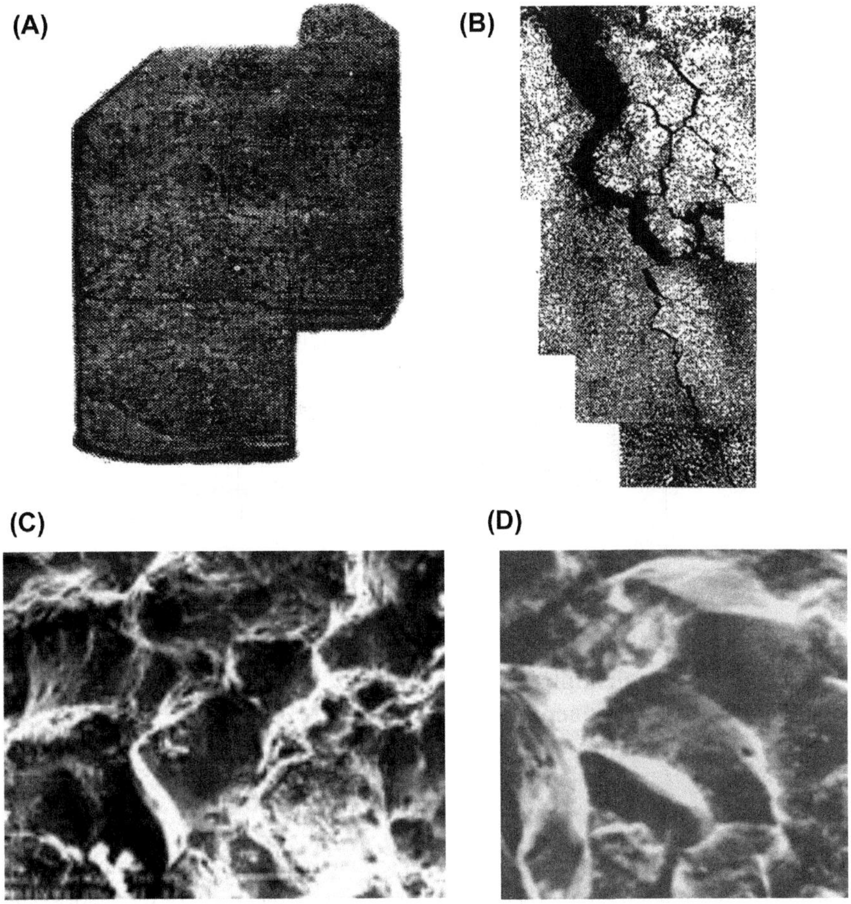

FIGURE 8.22

Corrosion damage of the rim of a 20th-degree turbine VK-100-90-6 LMZ [52]: (A) general appearance of cracks; (B) cracks enlarged 100×; (C) appearance of fracture surfaces at magnification; (D) appearance of fracture surfaces at magnification 500×.

Any damage to the blade changes the vibration characteristics and strength of the blade, which means that the blade may break and crash. Because of this, it is important to control the wear of the blade and to properly record values in each overhaul.

The size of the largest eroded area is measured for each package. Measurement of wear is performed periodically by the same blades in the package. Damage to the output edges of the impeller blades may also occur, which occurs when the turbine is operated for a long period of time with reduced load or frequent start-up, when eddy currents and back currents are applied which cause damp steam to the output edge of the impeller. The size of blade outlet edge damage is usually 25%–64% of the length of the rotor blade and 10–18 mm in width. This size and any larger than this one require replacement of the blades. It is also necessary to keep a record of the extent of damage from overhaul to overhaul. Erosion damage to the outlet edge of the blade extends from the bottom-up.

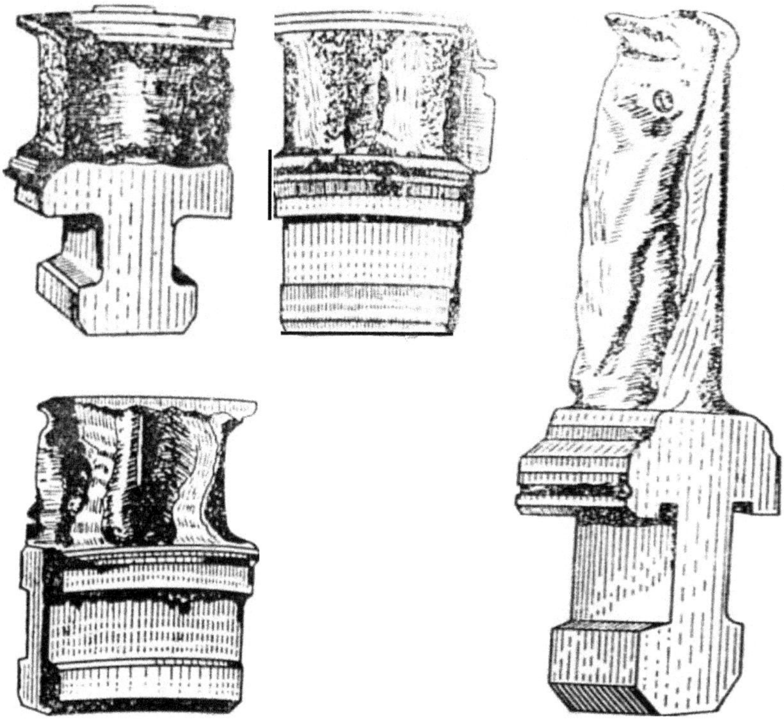

FIGURE 8.23

Abrasive removal on the blades of the first stages of a steam turbine as a form of erosion damage [52].

In the process of exploitation, salt deposits on the blades do not infrequently occur, causing the blades to corrode faster, reducing the efficiency of the turbine, the safety of the blades, and increasing the axial force on the axial bearing, rotor imbalance, and increasing the force of the blades. It is therefore necessary to clean the blades. Cleaning can be done by sandblasting, warm condensate, then heating, as well as mechanically. To clean the blades by sandblasting, it is necessary to provide: sand with a granulation of 0.3 mm, sandblasting machine, space for sandblasting (outside the engine room), good protection of the surfaces of the sleeves and rotor couplings, compressed air of 0.5 MPa, as well as personal protective equipment (protective suit with spacesuit and other protective equipment). Sandblasting is done until the metallic shine of the blades is obtained. After blasting is completed, the impeller is purged with air. It should be noted that sandblasting has a negative effect on the surface of the blades (they lose smoothness, thus increasing the possibility of erosion). Cleaning the blades and blade wheel with a hot condensate temperature of about 100°C and a pressure of 0.15–0.2 MPa is most advantageous because in this way they do not cause mechanical damage, so the blades retain the required smoothness of the surface. For faster drying of the blades, the impeller is carefully purged with warm air. Salt deposits can also be cleaned by an autogenous apparatus, with the blades heated from 100 to 110°C, with heating being very careful to prevent metal overheating. Mechanically, they are cleaned using a metal preparation, and bent exactly to the shape of the blades.

6.4.2 Water damage due to water shocks

Water surges occur when water or moist water vapor enters the working part of the turbine. Water from the steam generator usually enters the turbine due to illicit overloading of the steam generator, then raising the supply water temperature in the generator above the maximum allowed, as well as increasing the water level in the boiler drum or foaming the supply water due to its poor composition. The effects of water shocks are usually very large. Reaching more water leads to the complete destruction of the flow part of the turbine. The cause of water entering the turbines may be a malfunction of the steam boiler drainage system, as well as the bursting of the superheater pipe. Most often, a water stroke occurs at the start of the turbine, when the vapor overheating relative to the metal temperature is not large. Water shocks can also occur when the boiler load is suddenly reduced. As a result of a water shock, the following side effects occur [52]:

- a sudden decrease in the temperature of fresh steam;
- a sudden decrease in the temperature of the metal on the part of the HPC (high-pressure cylinder) or IPC (intermediate pressure cylinder) steam release;
- hydraulic shocks in fresh steam or intermediate steam lines;
- metallic sound and shocks inside the casing, due to thermal expansion and deformation;
- increase in axial displacement of the rotor, with increase in temperature of the axial bearing;
- the appearance of whitish steam (due to high humidity) on the stop and control valves, as well as on the seals and turbine couplings.

Due to the fact that the hydraulic shock sometimes results in the complete destruction of the electric structure of the shovel apparatus, with considerable damage to the rotor, diaphragm, labyrinth seals, and other turbine elements, it is necessary to shut off the turbine in the event of a water shock by the procedure for emergency shutdown. In order to prevent water shocks, all auxiliary turbine equipment, heaters below the angle of the turbine, subtraction with the inclination from the turbine to the customer, with regular control of nonreturn valves, and regular control and checking of the protection system, need to be properly installed. Thanks to these measures, as well as the good training of the operating personnel for the operation of the power unit as a whole, as well as the accompanying array of automation, damage due to hydraulic shock is very rare.

6.4.3 Diaphragm deflections

Diaphragm deflections are a very dangerous occurrence, especially if the rotating parts of the turrets are stationary, with the inevitable breakage of the blades and with the possible destruction of the turbine housing. A characteristic sign of strangulation is the appearance of unusual noise that can occur when reducing load. The diaphragm can bend both in the direction of steam and vice versa. The diaphragm deflection in the direction of steam can occur abruptly (at turbine overload or water stroke) and slow (due to sediment in the flow section). Diaphragm sagging can also occur gradually due to material creep as a result of high temperatures, especially at the inlet diaphragms HPC and IPC. Deflection of the diaphragm opposite to the direction of the vapor occurs as a result of the blockage at the flange or clamp. As a result, the front has a higher temperature than the rear. The blockage can also be due to insufficient mounting gaps, a sudden increase in the vapor temperature, with the diaphragm expanding faster than the housing, or due to the growth of the casting.

6.4.4 Control and maintenance of bearing operation

Modern maintenance approaches rely on recording the condition of steam turbine bearings to further determine the extent and type of maintenance based on subjective or objective measurements of specific sizes (temperature, noise, vibration, etc.). The absolute vibrations of the bearing represent the movement of the cover and bearing housing against a fixed "zero" point in space. They are measured on the bearing housing in a horizontal, vertical, and, where necessary, axial direction, with electrodynamic or piezoelectric transducers, which give the effective value of vibration velocity in mm/s or in/s.

Steam turbine bearings basically receive all static and dynamic loads. For safe and quiet operation of the turbine, the radial bearings must provide [52]:

- **(a)** good position of the rotor in the bearing, where the fit must be the entire length of the bearing and the width of the bearing is 30°C;
- **(b)** good heat dissipation caused by friction;
- **(c)** there must be an oil pin between the sleeves and the white metal;
- **(d)** all bearings must have lateral and upper clearances, in accordance with the standards of the turbine manufacturer;
- **(e)** the contact surfaces must be clean;
- **(f)** the bearing slippers must fit 90% on the housing.

Disassembly and mounting of bearings is done in a certain order, using the prescribed standard and special tools. After dismantling the upper half of the bearing, the side clearances and the upper clearance must be measured. The lateral clearances are measured with measuring slips. The depth of entry of the dipstick when measuring lateral clearances is given by the turbine manufacturer. The upper bearing clearance is measured with a lead imprint.

After completing the revision, repair, or assembly of the bearing, the position of the rotor in the bearing must be measured using the templates and the value and number of the templates measured in the form entered. When the bearing is reopened, the measurement with the template should be repeated in order to measure the wear of the white metal (Fig. 8.24).

It is necessary to check the bearing load in cases where the rotor, usually high pressure, has no two bearings but one, so that part of its weight is loaded by the second, which can automatically relieve the third bearing, which means that all the steam turbine bearings will not be equally loaded.

In order to achieve equal load on all the bearings, the couplings should be separated between the HPC coupling and the IPC coupling on the lower side. The value of the openings of the couplings on the underside is specified by the turbine manufacturer. If this information is not available, the magnitude of the separation of the couplings can be determined by calculation and by a dynamometer. For the K-200-130 turbine, the LMZ (manufactured by Leningradsky Metallichesky Zavod, Russia) gives the separation size on the underside of the coupling 0.25−0.42 mm. The calculated value of the separation of the couplings requires information on the weight of the rotor and the surface of the rotor in cm^2.

6.4.5 Damage to the blades

Fractures of stator and rotor blades of steam turbines are due to water shocks and foreign objects reaching the flowing part of the steam turbine, insufficient static strength of blades, bandages and connections, then from material creep due to variable loads due to vibrations, as well as corrosion

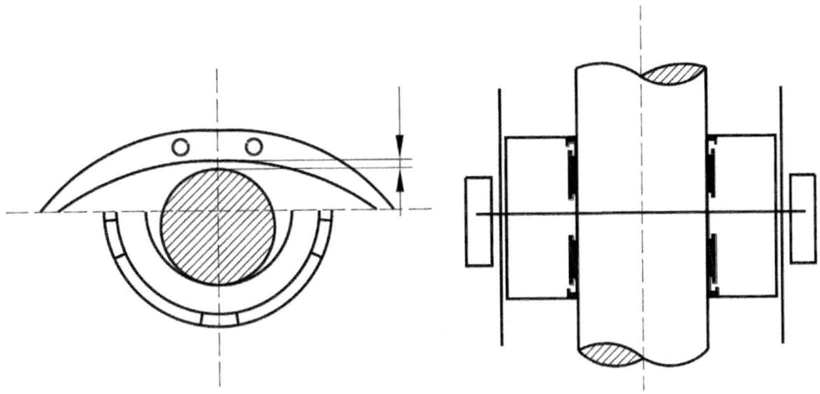

FIGURE 8.24

White metal wear control using templates.

deformations and erosion of metal surface blades. Bursting of the blades is also due to gross disruption of the technology of electricity production and exploitation of the energy block as a whole. Very rarely, turbine blades burst due to a significant increase in RPM when the load is reduced or when the turbine is "run over" due to the insensitivity of regulation or the tightness of the regulators. The most dangerous is the bursting of the blades of the last stages of the steam turbine, due to their large mass and centrifugal force, which results in the appearance of large vibrations and the accidental stoppage of the turbine. Bursting of blades is usually preceded by cracks. Cracks are caused by plastic deformation, corrosion processes, and poor repair. As a rule, a crack increases its dimensions to reach a critical size, when the blade bursts. The critical size of the crack depends on the dimensions of the blade itself. The larger the scale and the greater the force at the root of the blade, the smaller the critical size of the crack is required to break the blade itself.

6.4.6 Malfunctions of the condensation plant

The most serious impairment of the condenser operating mode is the increase of pressure in the condenser and the deterioration of its water tightness. A sudden increase in the pressure in the condenser is caused mainly by a break or a substantial decrease in the flow of cooling water. In this case, it is necessary to relieve the turbine immediately, in order to avoid overheating of the outlet section and its centering. When the pressure in the condenser increases to the protection level, the turbine shuts off accidentally. A characteristic of the rapid increase in pressure in the condenser is the rapid increase in the level of condensate in the condenser. A slow increase in pressure in the condenser occurs very often. When analyzing the cause of the pressure increase, it must first be ensured that it is not due to an increase in the cooling water temperature. An increase in the pressure in the condenser is most often associated with a gradual decrease in the flow of cooling water or a deterioration of the heat transfer coefficient. Thus, a 10% flow reduction reduces the condenser vacuum by 0.4% in summer and 0.2% in winter. The decrease in the flow of cooling water through the condenser can be due to two reasons: due to the increased hydraulic resistance of the circulatory tract and due to the decrease in the water level in the cooling tower pool. The clogging of the condenser tube causes not only an increase in the resistance of the tract, but also a deterioration of the heat transfers at the same time. In such cases, the condenser must be cleaned. The method of cleaning is determined by the nature of the precipitate. Cleanings can be carried out chemically, by thermal method, as well as mechanically. Another reason for the gradual increase in pressure in the condenser is due to the decrease in the heat transfer coefficient. The main reasons for this are the decrease in the velocity of water in the pipe due to the decrease in flow, the increase of the resistance of the pipe to increase the sediment, and the deterioration of heat transfer due to the increase of the air content in the water. An increase in the air content of the water is due to the malfunction of the sealing system and the ejector. In the process of exploitation, leakage occurs on the corrugated joints, as well as mechanical and corrosion damage of the condenser tube.

6.4.7 Steam turbine rotor control and centering

The most important condition for the smooth operation of the steam turbine is the achievement of rotor centricity and the alignment of the rotor axes with the axles of the bearings and the housing. To ensure this, it is necessary to ensure that the rotors, couplings, and bearings are in good condition. For this reason, due to the centering of the rotor, it is necessary to control the radial stroke of the couplings, to control the radial throw of the rotor, then the axial stroke of the forehead of the couplings, as well as the position of the rotor toward the housing and the tilt of the rotor sleeve. If the rotor is centered incorrectly, it causes a number of problems in operation, such as: turbocharger vibration, improper coupling operation, damage to the bearing, bearing wear, increased lubricating oil temperature, etc. The rotor centering control is performed through the couplings, which provide such mutual position of the rotor in the operation of the turbine, so that the axis of one rotor is in extension of the axis of the other rotor, so that the axes of the connected rotors as a whole represent one continuous elastic line. Each time a turbine is opened, overhauled, or inspected, a check of centering over the couplings is necessary.

During operation, due to bearing wear, deformation of individual turbine assemblies, foundation settling, and thermal deformation of the casing, the rotor is displaced, with loss of ability or loss of rotor centricity. Since the rotor is rotated at 3000 min^{-1} during operation, if the lower intensity of the rotor touches any of the stator parts, local heating of the rotor will result, resulting in its final distortion. Due to high friction, the part of the rotor that touches the stator part curves the rotor in the opposite direction from the point of contact after cooling the rotor. The rotor is to blame the other way for shrinking material. Causes of touching can be varied: touching blades of labyrinth seals and intermediate seals due to small gaps, if the housing does not expand freely, if there are no thermal gaps, improperly mounted discs, plugs and other mounting details, their inclination, uneven cooling of the turbine body, maximum difference in temperature of the turbine body upper and lower housings during turbine operation, incorrect start-up of the turbine, water shocks inside the turbine, disappearance of white metal on the bearings, etc.

The distortion of the rotor must not exceed the prescribed size. Any major distortion allows contact with the stator parts and thus the possibility of turbocharger failure. Therefore, it is necessary to determine the magnitude of the distortion, the location of the distortion, the depth of the spur, the depth of the damage, the structure of the material at the distortion site, and to compare whether the rotor had a distortion, the magnitude of the distortion, and the location in the previous overhaul. The rotor distortion is measured by rotating the rotor in the bearings with a steel rope using a crane, with the comparators mounted sideways on the housing. In order to be sure of the "throwing" of the rotor, the oval sleeve oval check must be checked, which must not exceed the defined size (e.g., 0.2 mm for 200 MW condensing turbines). If the curvature is more than 0.10 mm, the rotor should be leveled.

To be able to start centering the rotor, if it is an overhaul, it is necessary to perform the first centricity check after opening the bearing blocks with the condenser filled with water, with the upper halves of the housing and the shields mounted on the generator. The second centricity check is performed after the housing and condenser are opened without water and used to determine the effect of the capacitor on the centricity. The effect of the capacitor on the centricity should be taken into account when centering the rotor.

Measurement of centricity is done by a comparator or measuring paper, depending on the preparation used for the measurement. The notes are guided by a circle so that the axial values are written inside the circle and the radial ones outside. To calculate the rotor line repairs, the dimensions of the couplings, distances, the method of calculating the repairs, and the values for the rotor lines provided by the turbine manufacturer must be prepared. In measuring the centricity of the couplings, it is mandatory to enter on which half coupler the preparation (subset) for the measurement is attached. The measurement is made at four points: 0, 90, 180 and 270°. One radial data and three axial data are taken at each point: up, left, and right.

During the overhaul, it is necessary to perform certain tests on individual parts of the rotor, using penetrating colors, magnetic methods, then ultrasonic methods, as well as to perform certain dynamic tests. If rotor blasting, changing blades, or any other major operation has been performed during the overhaul, static and dynamic balancing of the rotor must be performed. Penetrating dyes are used to test the stellar plates of the low-pressure impeller rotor at definite degrees by the manufacturer of the turbine. The blades, rivets, and bandages of the HP, IP, and LP rotors, as well as the front surfaces of the half-couplings on all rotors, should be examined by the magnetic method. Using the ultrasonic method, it is necessary to examine the bearing sleeves on all rotors, the front surfaces of the half-couplings of all rotors, and the rivets of the end blades on the HP and IP rotors.

Dynamic testing requires the control of the natural frequencies of the blades of each stage on the low-pressure rotor. All tests must be recorded and only certified companies or persons authorized for this type of activity may test. Once the rotor is centered through the couplings, the level of the rotor must be checked with a special level (Fig. 8.25). The level control is measured along the longitudinal direction of the sleeve. One notch on the nonius level indicates a slope of 0.1 mm at 1 m.

The high-pressure rotor is connected by a rigid coupling to the medium-pressure rotor. Due to the uneven tightening of the coupling bolts or a slight tilting of the half-coupling face, a noncentricity occurs, which is manifested by "throwing" the front end of the high-pressure rotor, which further causes the entire front end of the turbine to vibrate. In order to prevent the front end of the turbine from vibrating, after the centering is completed and the coupling is tightened, it is necessary to perform a "throttling" control of the front end of the high-pressure rotor. After tightening the coupling, the high-pressure rotor should be lifted by a 0.30–0.40 mm crane and the lower half of the bearing removed.

Before lifting the rotor, two comparators should be installed, which will allow the rotor to return to its original position after removing the lower half of the bearing. To prepare the front end of the high-pressure rotor, one of the preparations must be enabled so that the rotor can be easily moved by hand and that the rotor returns to its original position after each movement. Usually the rotor is separated by a crane with a special rope (Fig. 8.26).

Due to the mass of the rotor, the steel rope is stretched, so that the rotor occasionally needs to be lifted by a bend so that the needle on the upper comparator takes the position "0". Rotation of the rotor

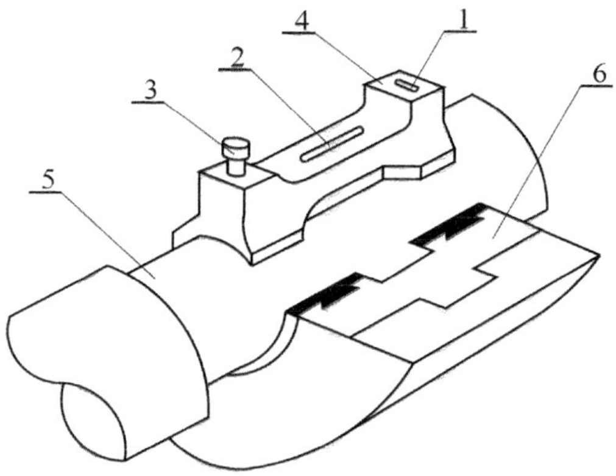

FIGURE 8.25

Position control of the rotor sleeve with a special level [52]: 1—transverse alignment adjustment lever relative to rotor sleeve; 2—nonius leveling lever; 3—leveling adjustment screw; 4—special rotor sleeve positioning control; 5—rotor sleeve; 6—bearing segment.

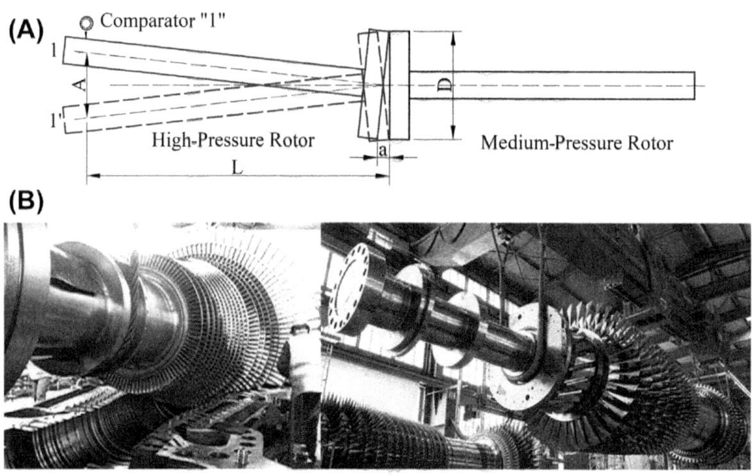

FIGURE 8.26

View of the "throwing" control of the front end of the rotor and the use of crane with special ropes: (A) control of the throwing at front end of the HP rotor; (B) use of crane with special ropes.

is done with a steel rope, which is tied to the crane. The front end of the rotor should be marked with chalk numbers, which correspond exactly to the amount and position of the holes on the high-pressure rotor coupling. The rotor must be rotated with a stop at each mark and read values read from the side comparator.

The maximum permissible "throw" of the front end of the rotor is 0.16 mm [52]. If the deviation size is up to 0.30 mm, retightening the second row coupling and tightening force can eliminate the "throwing" of the front end of the high-pressure rotor. A larger throw of 0.30 mm indicates poor machining of the contact of the touching half-couplings, i.e., indicates that there is an inclination that needs to be eliminated.

Curving should be eliminated by showering the head of the high-pressure rotor half-coupling. The size of the shower is calculated using the formula [52]:

$$a = \frac{A \cdot D}{2 \cdot L}, \tag{8.5}$$

where: a is the maximum amount of material removed; L is distance from the head of the rotor coupling to the position of the side comparator; D is high-pressure rotor half-diameter; A is maximum impact.

After removing the rotor and putting it into the stands, the rotor sleeve must be checked. The rotor arm must be examined by ultrasound to detect any cracks or porosity. Visually inspect the sleeve for mechanical damage or leaks. Measure the diameter of the sleeve with a grip micrometer to control taper and oval shape, which must not exceed the prescribed value (for the K-200-130 LMZ turbine of 0.02 mm). Smaller abrasions should be sanded using fine sandpaper soaked in oil, with the aim of making the sandpaper wider to cover a larger area at a time. If the oval shape or taper is greater than the prescribed value and if the sleeve has deeper leaks or mechanical damage, the sleeve should be equalized. The balancing of the sleeve is carried out at the construction site or at the factory, depending on the equipment available to the repair team. When inserting the impeller into the stands, strict care must be taken to ensure that the sleeve is protected by a cloth and nylon.

The rotors are interconnected by couplings. The number of screws on one coupling and their diameter is calculated accurately. Each bolt must be loaded with equal force. In order to achieve this, all the screws must be measured before the tightening with a gauge micrometer, record the values, and calculate their elongation.

The bolt elongation value is calculated by the formula [52]:

$$L = (L + D) \cdot K, \tag{8.6}$$

where: L is the working length of the screw; D is hole diameter; K is screw elongation coefficient (for LMZ turbines, $K = 0.0012$ [52]).

The bolts must be tightened in a certain order so that their tightening reaches a relative elongation value of $\Delta L \pm 0.03$ mm.

6.4.8 Turbine oil quality control in the function of maintaining the steam turbine system

During the exploitation of energy oils, reliable operation of technological systems with oil-filled equipment must be ensured, then the continuous maintenance of the exploitative properties of the oils, as well as the collection and regeneration of the produced oils in order to reuse them. All energy oils must have an adequate quality certificate, and periodically undergo laboratory analysis to determine their characteristics according to the standard. Oils that do not meet the requirements of the standards by which they are manufactured must not be used. Turbine oil for steam turbines, electric and turbo power pumps must meet the following standards:

(a) Petroleum oil:

- acid number—below 0.3 mg KOH per 1 g of oil, water, dirt;

- free of mechanical impurities and dissolved precipitate (to be determined at an acid number of oils 0.1 mg KOH per 1 g of oil and more);
- thermo acid stability for oil Tp—22 S (acid number—not exceeding 0.8 mg KOH per 1 g of oil);
- oil oxidation conditions: test temperature 120 ± 5°C, time 14 h, oxygen delivery rate 2000 m^3/min;
- thermo acid stability of oil, determined once a year before the start of the autumn-winter maximum for oil (or mixtures thereof) with an acid number of 0.1 mg KOH per 1 g of oil and more (for oils from oil systems of electric and turbo power pumps this indicator does not specify);

(b) Synthetic regulatory oil:
- acid number—not exceeding 1 mg KOH per 1 g of oil;
- water-soluble oxide content—not exceeding 0.4 mg KOH per 1 g of oil;
- mechanical impurities by weight—not exceeding 0.01%;
- a change in viscosity not exceeding 10% of the initial value for heavy oils;
- content of dissolved precipitate (according to BTI procedures)—change in optical transparency—not exceeding 25% (determined at an acid number of oil 0.7 mg KOH per 1 g of oil and more).

Regulatory synthetic oils, which have reached the limit of performance of an acid number, must be shipped to the manufacturer for quality regeneration. Operation of refractory oils must be carried out in accordance with the requirements of the special instruction.

During storage and operation, turbine oil must be periodically inspected by visual inspection and summary analysis. An abbreviated analysis of petroleum oil includes: determination of acid number, determination of presence of mechanical impurities, dirt, and water. An abridged analysis of refractory oil includes: determination of the acid number, determination of the content of water-soluble oxides, then the presence of water, and determination of the quantitative content of mechanical impurities by any of the rapid methods with satisfactory accuracy. Visual inspection of the oil involves checking the appearance of the oil for the presence of water, dirt, and mechanical impurities, in order to make a decision on its purification.

The periodicity of the turbine oil summary analysis is for oils Tn-22 S (TU 38.101.821-83) [89] not less than 1 month after being poured into the oil system and still during operation at least once every 2 months at acid number 0.1 mg KOH per 1 g of oil and at least once a month for an acid number exceeding 0.1 mg KOH per 1 g of oil, while for refractory oils not later than 1 week after commencement of operation, at least once every 2 months at an acid number of $\leq$0.5 mg KOH per 1 g of oil and at least once every 3 weeks at an acid number > 0.5 mg KOH per 1 g of oil. The visual control of the oil used for steam turbines and turbo pumps should be performed once a day [52,89].

A constant reserve of petroleum turbine oil must be stored in the power plant in an amount equal to (or greater than) the volume of the oil system of the aggregates and a reserve for refueling of 45 days. This turbine oil reserve must be greater than or equal to the annual refueling consumption of a single turbine unit.

The quality control of fresh and exploitative energy oils, the provision of data on the use of oils, as well as the formation of graphs of their control, as well as the technical management of processing technology, must be carried out by a chemical-technological laboratory.

6.4.8.1 Water getting into turbine oil

Water ingress into oil is a widespread phenomenon when operating steam turbines. This usually occurs when the sealing mode of the end seals is disturbed. When the vapor pressure of the sealing gas is high, it enters the atmosphere and enters the oil through the oil seals. Steam can also get into the oil due to impaired sealing ejector operation. In addition, the oil has some hygroscopicity, that is, it has the ability to absorb water vapor from the surrounding environment. The steam, when it enters the oil, condenses and drains it. When the water seals of the power pumps are disrupted, water enters the shaft and enters the oil. If water is not mixed with the oil, it falls to the bottom of the oil reservoir, from which it can be separated through a drainage pipeline (drainage basin).

During sedation, permanent oil—water emulsions are formed in the oil. As the water content of the oil increases, the viscosity of the oil increases and the conditions of oil circulation in the pipelines decrease, while reducing the reliability of the bearing operation. Water in the emulsion results in the formation of oxides in the oil, and the corrosion of the lubricated parts and the entire oil system. The emulsion is a good carrier of abrasives. It follows that the oil must have the ability to deemulsify. An indicator of the deemulsifying capacity of oil is the time of deemulsification, which is determined as follows: 20 mL of water and 100 mL of oil are poured into a 250 mL graduated cylinder and then boiled for 10 min. The condition of the hydrocarbon emulsion in the cylinder is monitored. The time at which water is separated from oil is called the time of deemulsification. The time of oil deemulsification has a major impact on the structural performance of the oil system. The volume of the oil tank is determined so that the oil, with a given flow rate of the oil pumps, will be stored in the tank long enough to separate the water from it, that is, at least 8 min. Pure, unmixed oil always has better deemulsifying properties than the same oil if it is dirty with oxides.

In order to prevent water from entering the oil, the following activities should be carried out continuously by the maintenance and operating staff of the steam turbine [52]:

- monitor the operation of the sealing of the main turbine and the drive turbine of the feed pump, preventing the steam from escaping from the sealing;
- control the pressure of the steam that is fed into the manifold and at each seal and pressure of the working mixture in front of the ejector from the seal suction;
- control the operation of the water seals of the power pumps, preventing water leakage;
- monitor the condensate pressure applied to the seal and also monitor the operation of the pipeline to remove leakage from the seal;
- do not allow the cooling water pressure in the oil coolers to be greater than the oil pressure.

If evaporation and leakage through seals cannot be prevented by changing the mode, then it is expedient to install steam or water repellant "paronite" shields on the shaft between oil and steam or water seals to prevent the direct flow of steam or water into the oil-repellent rings. With a significant increase in the emulsion, it is necessary to add up to 0.01% of the special deemulsifying admixture (e.g., diproxamine-157) in the oil. The best effect is achieved when the admixtures are added to fresh oil.

6.5 Reliability and initial database analysis

The primary objective of reliability analysis is to analyze failures, their cause, and effect. Usually, reliability measures are incorporated into the reliability analyses, and to verify that the reliability of the functional block is below a certain level, the analysis of the reliability of the functional block is focused on assigning measurable characteristics expressed by the likelihood that it will perform the required function under the given environmental and operational conditions for a fixed period of time. This involves the creation of a model, which must be presented logically and mathematically, and which is based on the distributions of component failures of which the functional block is composed.

Planning, development, construction, and exploitation, while maintaining energy facilities and systems in the energy sector, bring with it a large number of phenomena that can cause damage and endanger the lives of people directly engaged in the facility itself and the wider environment. The system of a TPP is considered as a complex system, with a great interdependence of the constituent components, in which failure of any of them can also mean automatic interruption of the whole system or reduced operation. This can result in an increase in the cost of operating the system, thermal and other overloads, as well as the possible occurrence of major accidents. Condition of working capacity of TPP system with stable fault-free operation, which due to static structure and dynamic influence of many factors from operational and wider environment, often turns into unstable condition in failure, can be through scientific approach of competitive engineering (life cycle engineering), with previous formation databases and use of classical methods, kept under control.

Depending on the purpose and stage of the calculation, availability and degree of accuracy, and applicability of the starting information, methodological approaches differ. In this case, the possible budget targets can be classified into the following subgroups [3]:

(a) assessment and research of the most critical facility, that is, the most critical details of that facility, by setting influential basic and supplementary research criteria and determining the rank of critical plants and summarizing them (determining and comparing them using the ranking method);

(b) optimizing the means of ensuring reliability, with an analysis of their internal and external links;

(c) analysis of the interrelation of the requirements for the reliability of the parts, that is, of the system as a whole and the total cost of providing them;

(d) forecasting the optimal reliability of the TPP as a component of the power system and linking it with the solutions of reliability optimization tasks at that level, considering the functional relationship between the demand curve and the cost curve necessary for their realization.

Today's stage of development and exploitation of a TPP is characterized by an increased degree of complexity of both the technological scheme itself and the construction of individual assemblies and elements of equipment, which results in a number of issues related to ensuring and improving the reliability of the system as a whole. The ultimate working connection and ability of the block as a whole depends on the work and behavior in exploitation of each of the elements in the scheme. Conducting analyses of the reliability and availability of the plant is also a requirement for conventional fossil fuel power plants. On the other hand, any disturbance of the regime of steady operation of a TPP within the ES results in a decrease in the amount of electricity produced, that is, an increase in the power reserve in the power system, which will increase additional investments and production costs.

Starting from the applicable methods for these analyses and single failure criteria, it is possible to determine the reliability and availability of a complex TPP system by breaking it down into constituent elements, determining the appropriate reliability and availability parameters using statistical analysis and identifying and defining the interconnections or the influence of individual elements on the system as a whole. Analysis of the reliability of a steam turbine system implies a concept for several model-based methods that allow quantification of reliability measures. Since reliability refers to probability, it is necessary to define the relevant aspects of probability at the very beginning of the reliability analysis. Subsequently, a mathematical description of the characteristics (measures) that quantify reliability and the main methods used in reliability analyses is provided. It should be noted that structural reliability analyses exist, which include the study of the effects of static and dynamic stresses, corrosion, material fatigue, dirt, and the like, on individual elements (components) of a complex TPP system, and based on that they draw conclusions about the reliability of the analyzed component. The system of the TPP requires, among other things, a high level of automation that is maintaining the parameters at a given level (pressure, temperature, and power). In order to achieve this, in addition to a good knowledge of the dynamic characteristics of processes and facilities, it is also necessary to know the control systems of a TPP [1,2].

Combining the formation of a mathematical model based on known physical laws with experimental identification, which decides the dynamics of the process, is possible based on the information collected on the process itself (most commonly measured). Identification process means knowing a priori information about the object being observed, which means knowing the structure of the object and the class of model to which the object belongs [2,3]. In this way, the task of identification is to evaluate unknown parameters or condition of objects, as well as to give forecasts for the next period in the form of short- and long-term forecasts. The flow of identification in TPP systems is most often monitored through the processes of transmission and transformation of thermal energy into electricity. The system of partial differential equations describing such a system can be simplified by adopting justifiable assumptions that simplify the set equations and translate them into linear differential equations. The basic approaches to identification can be: identification of system parameters, identification of the state of the system, a combination of the previous two identifications, and estimation.

The optimal management of the TPP system must be based on the evaluation and complex optimization of the reliability indicators, depending on the way of providing them and the hierarchical level of detail of the system as a whole, as well as the current stages of the plant life cycle. For these reasons, the optimization process includes basic structural, parametric, and constructive solutions related to the TPP system by changing its most important characteristics: energy efficiency, maneuverability, reliability, and economic efficiency as a whole. The set of optimization goals is concluded in the overall choice of reliability indicators and possible ways to achieve them, given the already established rules related to the higher hierarchical level of the power system.

The level of reliability required has direct and indirect impacts on the costs associated with the life cycle of an energy plant. Management and cost as two important aspects of reliability are of particular importance in the TPP system. The main role of management is to establish the organization and control all activities related to the reliability of TPP operation and continuous supply of electricity of appropriate quality (voltage, frequency, etc.). The reliability organization itself ensures the participation of all structures from the top (top management) to the lowest executor of the lowest level, with the aim of realizing and establishing reliability management, such as providing a high level of reliability and product safety, reducing development and production costs, and increasing customer satisfaction and uplift reputation of the company. To carry out the activities related to reliability requires the training and experience of specialists with knowledge of the power plant system (functionality, production processes, technological scheme), reliability methodology, as well as methods of statistical and reliability analysis. It is especially important to know the failure mechanisms, the prehistory of system elements, and the procedures related to testing and testing.

Meeting the general requirement of realizing a high level of reliability is closely linked to the increase in costs and constraints present in the processes of development and design, installation and operation, as well as the maintenance of a complex TPP system. On the other hand, the realized level of reliability has an impact on the level of failure and its consequences (material and environmental catastrophes), on the availability of the TPP system and on the costs that depend on the level of reliability in the continuous supply of electricity.

The goal of TPP drive balancing is the requirement to control the cost-effectiveness of electricity generation in TPPs by monitoring the operation of the components of the units, comparing all (subjective and objective) causes of change in specific heat consumption in relation to its nominal (base) value. Deviation of the actual specific consumption from the base (determined on the basis of the supplier of equipment and conditions for optimal operation) consists of the part of the deviation independent of the exploitation and the part which is in the function of exploitation.

The first group covers those deviations that cannot be influenced by the power plant personnel and which are due to external causes, such as [54]:

(a) deviations due to load plan, e.g., adopted plan, deadlock, planned start-up, etc.;
(b) deviations due to atmospheric conditions, e.g., cooling water temperature, outside air temperature, humidity, etc.;
(c) deviations due to nonfulfillment of the guarantees (differences in the results of hand-over tests of individual equipment at TPP, most often the inability to achieve installed power);
(d) deviations due to special causes, such as operation of the auxiliary diesel group, operation of turbo charging pumps, etc.;
(e) deviations due to the so-called force majeure (earthquakes, floods, storms, tsunamis, etc.)

The second group represents deviations that can be more or less directly influenced by the personnel at the power plant, and includes:

(a) deviations caused by the actual condition of the individual equipment at the power plant (dirty condensation, dirty heating surfaces of the steam boiler, dirty turbine blades, failure of regenerative heaters, or shutdown of high-pressure heaters during operation, e.g., TPP Gacko, etc.);
(b) deviations due to the regulation of the combustion process, e.g., impact of losses due to mechanical incompleteness of combustion and losses in exhaust gases from steam boilers, etc.;
(c) deviation due to changes in the fresh steam parameters at the inlet of the turbine or the parameters of the subsequently superheated steam;
(d) deviations due to changes in own consumption of TPP;
(e) deviations due to increased losses of condensate, water, and steam at the power plant;
(f) deviations due to special causes, such as heating of auxiliary fuel (fuel oil), room heating, defrosting of primary fuel (coal), soot blowing, unplanned downtime, etc.

Specific heat consumption deviations at TPP include steam boiler deviations, turbine plant deviations, deviations due to changes in actual own consumption, blockage deviations, and other deviations (Table 8.8). Steam boiler deviations are based on the assumption that these are relatively small deviations; otherwise necessary is to accept that in the areas in question the loss curves are approximately linear depending on the cause of the change.

Table 8.8 Display of deviation of specific heat consumption on TPP [54].

Group	Subgroup	Cause of change
Steam boiler deviations	Deviation of losses due to mechanical incompleteness of combustion	- Lower thermal power of the fuel - The content of the unburnt part in slag and ash - The ash content of the crude fuel
	Deviation of losses in the exhaust gases from the steam boiler (steam generator)	- Lower thermal power of the fuel - The content of the unburnt part in slag and ash - The ash content of the crude fuel - Content of C in crude fuel - CO_2 or O content in the dry exhaust gases - Water vapor content in flue gases
	Deviation of other losses due to change in the lower thermal power of the fuel	- Loss due to chemical incompleteness of combustion q_3 - Loss due to external cooling q_5 - Loss of physical heat of slag q_6
	Deviation due to change of steam boiler load	- Order by order of dispatcher - Operation with reduced fuel supply (due to flow-through) - Technical condition of individual elements and plants at TPP (work with lower loads)
	Deviation due to nonfulfillment of the given guarantees	- Deviations due to deviation of the actual efficiency coefficient (EC) steam boiler from the guaranteed value - Other deviations for other plants and equipment at TPP
Turbine plant deviations	Deviations due to change in active load	- Change in power at generator terminals - Change by order of the dispatcher due to restrictions on the transmission network
	Deviation due to changes in parameters of fresh/superheated steam	- Temperature of fresh/fresh superheated steam - Temperature of the postsuperheated steam - Pressure of fresh/fresh superheated steam - Pressure drop in the system of subsequent superheating of steam

Continued

Table 8.8 Display of deviation of specific heat consumption on TPP [54].—cont'd

Group	Subgroup	Cause of change
	Condensation plant deviations	- Change in condenser pressure (change in cooling water temperature, change in cooling water flow, uneven distribution of cooling water flow to the left and right half of the condenser, change in ambient air pressure, soiling of the heat exchange surfaces in the condenser, etc.) - Change in cooling water temperature
	Deviations of electric generator and main switchgear	- Change of reactive load (at given active load for generator and main transformer) - Change in the active load of the main transformer due to the change in the degree of utility of the main transformer
	Deviations due to the injection of water into the subsequent superheater (high, medium, and low pressure)	- A tour of the HPC by steam generated from the injected water (injecting water into the line of subsequently superheated steam due to the regulation of its temperature results in an increase in specific heat consumption) - Changed steam flows at turbine outlet points
	Deviations due to heat losses due to leaks and leaks in pipelines, poor insulation on pipelines, etc.	- Steam leaks in the block steam system - Leakage of water and condensate in the block piping system - Poor insulation on steam lines
	Deviations due to change in nominal (theoretical) own consumption	- Change of EC transformer of own consumption and changes due to other causes
Deviations due to change in actual own consumption	The share of theoretical own consumption in power at generator connections Other deviations due to changes in actual heat consumption	
Deviations due to blockages	Deviations due to own electricity consumption during blockages	- Planned delays (at the request of the dispatching service)
	Deviation due to heat consumption for starting after block deadlock	- Unplanned delays (accident or plant failure at TPP)
	Other deviations due to blockage	- Organization of maintenance at TPP

Table 8.8 Display of deviation of specific heat consumption on TPP [54].—cont'd

Group	Subgroup	Cause of change
Other deviations	Deviations due to the consumption of coal defrosting heat	- Extremely low outside temperature
	Deviation due to heat consumption for soot blowers	- Quality of coal and combustion process
	Deviation due to actual losses of steam, water, and condensate	- Tightness and sealing on installations and reinforcement within the TPP
	Deviation due to heat consumption for heating workspace at TPP	- Extremely low outside temperatures
	Deviation due to heat consumption for heating the nitrogen	
	Other deviations not covered by previous classification	- Specific causes not previously covered

The results of the propulsion balancing are presented in the form of reports, which can be daily, decadal, and monthly, and their recapitulation is given in the final monthly, semiannual, and annual reports. It is necessary to form a database of continuously measured quantities, such as: production of a steam boiler (steam generator), dry flue gas content (especially CO_2 or oxygen), flue gas outlet temperature, microclimatic characteristics of the environment in which the TPP is located and operating (especially the pressure and temperature of the surrounding air), the amount of coal and other supplies, the continuous results of the technical analysis of the coal (lower thermal power, moisture content, ash content, etc.), the continuous results of the elemental analysis (determining the composition and content of combustible in slag and ash), data related to the nominal (budget mode), and data related to the performance during the warranty tests (especially steam boiler, steam turbine, and generator with main transformer and transformer for own consumption).

Concerning downtime tolerances, the data required to calculate specific heat consumption deviations include the duration, description of the cause of downtime, and own consumption during downtime and during start-up after downtime (this information is taken from the equipment supplier and subsequently checked through special tests). Also, data related to other discrepancies are usually determined decadal (similar to discrepancies due to blockages), and only the data required for their determination (calculation) are entered in the daily report: steam flows and conditions for coal defrosting, soot blowers and heating oil, losses due to steam leaks and water and condensate leaks, heat consumption for heating, etc.

The total heat consumption of a block is the sum of the heat input by the base fuel (usually coal) and the heat input by the subbase and support fuel (fuel oil). The value of net energy produced for the observed real time is determined on the basis of readings on the appropriate instruments. Deviations related to specific heat consumption for the realized net energy production are most often grouped and divided into two categories: deviations independent of the exploitation of the block and deviations dependent on the exploitation of the block. According to the decadal reports, a monthly report on the balance sheet of the block is made, whereby, on the basis of the decadal and monthly reports, the final decadal and final monthly reports are made on the same principles. Final reports usually have a more transparent structure, on the basis of which management can monitor the cost-effectiveness of the operation of individual parts of the block and the block as a whole. After calculating all values of monthly deviations, a recapitulative monthly report is prepared.

The balancing method is very suitable for the use of computers, with manual input of the necessary data needed to calculate the calculation of individual deviations and provide a feedback effect in order to reduce deviations or increase the efficiency of the block.

6.5.1 Basic mathematical definitions related to the reliability of technical systems

System reliability is defined as the probability that the system will perform the intended function of the target without failure at the predicted time and under certain conditions [3]. The above definition of reliability can be mathematically represented as:

$$R(t) = P(T > t), \tag{8.7}$$

where: R—reliability, P—probability, T—running time, t—specific time moment.

As it is very important in practice to analyze the time period in which a component or system will function (life cycle analysis or failure time analysis), it is necessary to include a positive continuous random variable analysis, which requires different statistical models than successful/unsuccessful data analysis.

The operating time spent by the facility in drive t can be measured according to different scales: calendar time, time spent in operation, number of operating cycles, number of miles driven, number of switching "on/off", number of revolutions at the bearing, etc. In these examples, the running time t and also the failure time T can often be a nonnegative integer value, i.e., a discrete variable, which can, however, be approximated by a continuous variable, i.e., by a real number.

Most reliability analyses focus on modeling the distribution of element failure time, that is, modeling the properties of a continuous random variable T. The properties of a continuous random variable T can be described using the probability density function $f(t)$ or the cumulative probability density function $F(t)$, the reliability function $R(t)$, the failure frequency function $\lambda(t)$, and mean time to failure (MTTF). There are multiple probability distributions that model failures. Some of the most used in reliability are: Beta, Exponential, Gamma, Log-normal, Normal, Rayleigh, and Weibull [3]. Their features depend on the area and width of application. So, e.g., normal distribution is used mainly to model mechanical failures resulting from fatigue or wear [90]. On the other hand, exponential distribution is widely used in reliability because it is used to model completely random failures of irreversible functional elements (constant failure rate model), and often is a real situation.

A derivative of the time reliability function represents the value of the failure distribution density function:

$$\frac{dR}{dt} = -f(t). \tag{8.8}$$

Integration of expression (8.8) gives:

$$\int_1^R dR = -\int_0^t f(t)dt. \tag{8.9}$$

The reliability function can be expressed over the density of failure distributions as:

$$R(t) = 1 - \int_0^t f(t)dt. \tag{8.10}$$

A very significant and widely used reliability feature, now called failure rate λ (t), can now be introduced:

$$\lambda(t) = \frac{f(t)}{R(t)} = -\frac{R'(t)}{R(t)}. \tag{8.11}$$

Size λ (t) represents the mean number of failures in a unit of time, which at a given moment comes to the unit of those elements that did not fail. From expression (8.11) we obtain:

$$\lambda(t) = \frac{f(t)}{R(t)} = -\frac{R'(t)}{R(t)}, \tag{8.12}$$

or

$$R(t) = e^{-\int_0^t \lambda(\tau)d\tau}. \tag{8.13}$$

Reliability is expressed as probability (0—1), where value 1 refers to absolutely correct, while value 0 refers to absolutely unreliable system. Accordingly, reliability can be represented by the unreliability of the system Q (t):

$$R(t) = 1 - Q(t), \tag{8.14}$$

so it can be concluded that:

$$Q(t) = \int_0^t f(t)dt. \tag{8.15}$$

From expression (8.13), the mathematical expression for the probability of failure-free operation over a time interval follows (t_1, t_2):

$$R(t_1, t_2) = e^{-\int_{t_1}^{t_2} \lambda(\tau)d\tau}. \tag{8.16}$$

For practical analysis, it is important to consider the expected time to failure or the MTTF. The expected time until failure occurs, as a function of the density of failure distribution $f(t)$, is mathematically expressed as [97]:

$$E(t) = \int_0^\infty t \cdot f(t)dt. \tag{8.17}$$

The MTTF is:

$$m = \bar{t} = E(t) = \int_0^{\infty} t \cdot dt. \tag{8.18}$$

Solving the previous integral gives the expected time to failure, as a function of reliability:

$$m = \int_0^{\infty} R(t)dt. \tag{8.19}$$

In practical analyses, there are two cases related to the MTTF occurrence m:

- for systems or devices that are being repaired or maintained for the expected time m, the MTBF is used,
- MTTF is used for nonrepairable systems.

For engineering applications, it is necessary to know the methods of calculating, testing, and installing the reliability of complex systems, which include software (so-called software systems) in the development phase, on the one hand, but it is also necessary to know the methods of increasing the reliability of such systems during the period of use, on the other side. System reliability built into the development phase is called inherent reliability. If the system is well designed, thoroughly tested, well maintained, a high level of reliability is expected with proper use [3]. However, the reliability of the system is significantly affected by the environment. Reliability is measured and has a practical interpretation. The exact value of reliability is never known, but its numerical rating can be obtained close to the true value. Such an estimate can be obtained by stochastic methods based on the data obtained by measurement in a particular set. Reliability estimation on the basis of a certain number of data sets is expressed by the level of confidence, which represents mathematical probability and relates the estimated and real (but unknown) value of reliability. The confidence level represents the probability that a certain value of reliability will be between the lower and upper bounds. The reliability of a complex system depends on the reliability level of each of its entirety. There are mathematical connections that show the dependence of the reliability of a complex system on the reliability of the whole.

Mathematical descriptions and features of commonly used fault probability distributions and reliability measures can be found in various sources (Meeker and Escobar (1998), [3,4,54,82,88,90–95]).

Functional block reliability function R (t), is defined as the probability that the failurc of the functional block will not occur at interval (0, t) or in other words that it will survive the established time interval (0, t) and will still function in time t. If T is a random variable indicating the time of failure occurrence, then the reliability of the functional block can be expressed as:

$$R(t) = P(T > \mathrm{t}) \ za \ t > 0. \tag{8.20}$$

A possible representation of the distribution of failures as a continuous random variable T is also called the hazard rate function, also known as the failure rate function. The failure frequency function is very important in reliability analysis because it determines the measure of the aging of the functional block.

If it is necessary to determine in the reliability analysis the probability that an element that has successfully worked for some time t will cancel in the next time interval $[t, t + \Delta t]$, then this problem is solved by the conditional reliability function $R(\Delta t \mid t)$. This feature is often used to provide guarantees, to evaluate the success of the next period of work after a cycle or after periodic inspections, that is, the functionality of an element in a time period Δt that began after successful operation at a time point t.

The failure frequency function is important in reliability analysis since it displays changes in the probability of failure throughout the life of the functional block. In practice, $\lambda(t)$ often expresses the shape of a bathtub, and is therefore referred to as the tub bathtub curve [82]. Furthermore, [11,92] also states that the concept of a bathtub curve is a well-known concept used to represent the failure behavior of various engineering functional blocks because their failure frequency is a function of time, i.e., changes with time (Fig. 8.27).

In the beginning (burn-in), early (initial) failures occur, which are caused by: poor production process, poor quality control, human error, substandard materials, etc. The number of early failures during the run-in period declines sharply, so the frequency of failures (DFR) decreases during this period. After the run-in period, the so-called period begins useful life, in which only random failures

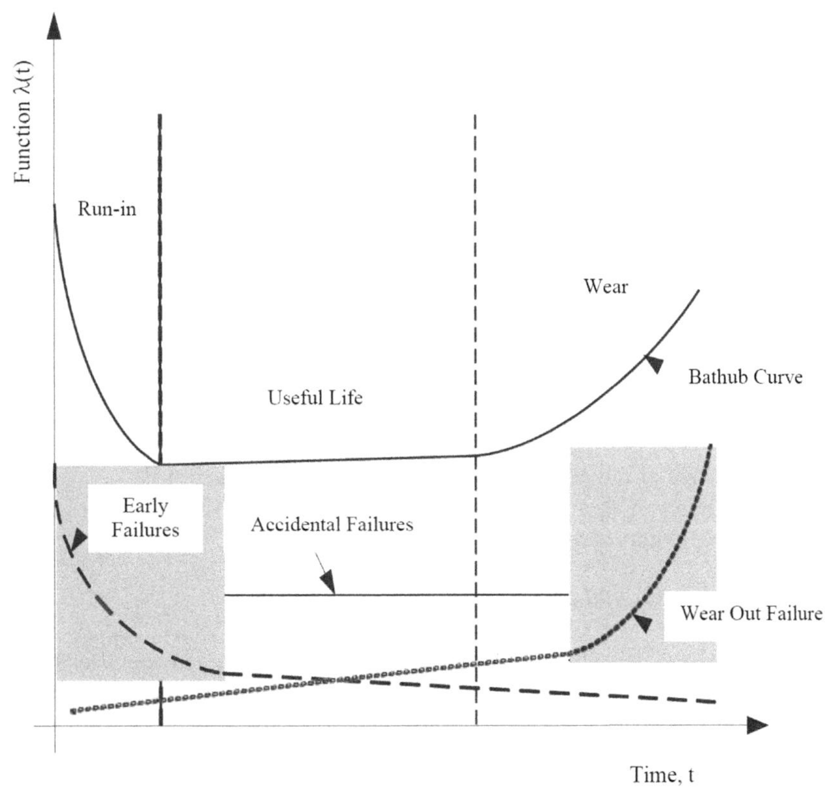

FIGURE 8.27

Bathtub-shaped failure rate function.

occur, whose frequency is constant (CFR, $\lambda = 1/m$). The causes of these failures may be higher accidental stresses than expected, human error, abuse, imperceptible defects, etc. After this period of time, the wear out period begins, and from that moment onward accidental failures are associated with wear and tear, which causes the frequency of failures to increase (*IFR*), which can be caused by wear and tear, poor maintenance, improper repair (corrosion), and corrosion, short designed life span, aging wear, and more. Mathematically, this curve can be obtained by combining several fault distributions or as a function of the parts of linear and constant fault frequencies [50].

Usually, in reliability analyses of interest, the expected time to failure is referred to as the *MTTF*. The MTTF of a functional block is the expected time to occur (first) failure. If T is a random variable representing the time of occurrence of a failure, then the MTTF is equal to the mathematical expectation of a random variable T, i.e., $MTTF = E\ (T)$. Assuming that the reliability function of the functional block is given by $R\ (t)$, $MTTF$ can be calculated as:

$$MTTF = E(t) = \int_0^\infty t \cdot f(t) \cdot dt = \int_0^\infty R(t) \cdot dt. \tag{8.21}$$

If it takes time to repair or replace a defective one very briefly when comparing it to MTTF, MTTF also displays MTBF. If repair time cannot be ignored, MTBF also includes MTTR. MTTF is just one of several measures of distribution location. The second location measure is the median T_{med}, defined as:

$$R(T_{med}) = 0.5. \tag{8.22}$$

The median divides the division into two halves.

The third metro location distributes the T_{med} mode, which is very similar to the failure time. This is the time where the probability density function $f\ (t)$ reaches its maximum:

$$f(T_{med}) = \max_{0 \leq t_P \infty} f(t). \tag{8.23}$$

Mod represents the time with the highest expectation of failure.

Another parameter used in reliability analysis is the design life of an element, t_R, defined as the instant of time in which an element has a confidence level equal to:

$$Rt(t_R) = 100 - X\%, \tag{8.24}$$

where X is the percentage of the population of the elements they scheduled.

Thus, e.g., a reliability of $R = 0.98$ means 2% of the failures in the element population, and is referred to as "B_2 life", denoting $t_{0.98}$.

6.5.2 Reliability methods and techniques

The reliability analysis of the technical system most often starts with a functional block model, composed of its corresponding functional elements (components). Each functional element is then assigned an estimated individual failure distribution, and then the reliability (current or future) of the functional block is evaluated using the appropriate methods or techniques available. Many methods of reliability analysis have been developed and put into practice to evaluate the reliability of a technical system in the last 50 years, and their effectiveness, advantages, and disadvantages can vary greatly depending on the technical system and the environment in which it operates. There is no unified rule for selecting the best method applicable to assess the reliability of a particular functional block, and their use for particular applications depends on various factors, including specific requirements and needs, type of functional block, and preferences of the persons involved in the analysis (ease of use and requirements for certain analyst experience). On the other hand, given the available data, international and national standards and professional association standards provide some guidance to support engineers in choosing the appropriate technical method used during the earlier, later, or throughout the life cycle phases of the functional block (Table 8.9).

The divisions of the reliability analysis methods reported in the literature are with respect to the goal of reliability analysis on qualitative and quantitative, and inductive and deductive (the difference between inductive and deductive methods is in the way of investigating the relationship between cause and effect of failure), as well as depending on the starting point analysis (bottom-up methods and top-down methods).

Depending on the nature of the elements of the functional block (system), the methods of reliability analysis are divided into methods based on state space models and nonstate models. This division is based on the ability of the reliability model to show the dependence of multiple states that a functional block, such as a condensing TPP, may have after a failure (the system operates in low mode) or a regular dependence of the functional block failures (the system will schedule if event A precedes event B); the basic methods of reliability analysis in terms of their application over the life cycle are given in Table 8.10.

All of the processes listed in Tables 8.8 and 8.9 are based on the assumption that repairable functional blocks (components and systems) can be in one of two states: the functional state and the malfunction state. In the case where it is necessary to model a repairable functional block that has multiple states, e.g., state of operation, failure status, and readiness for operation (cold state and state of warm reserve of a TPP), then a special type of stochastic process, called Markov chains, is used. The failure state (failure state) and the ready state are then used by a special type of stochastic process, called Markov chains repair of individual components to the state of the system. Markov chains can be

Table 8.9 Basic methods of reliability analysis by various standards [50].

Methods	Standards
Failure mode and effect analysis (FMEA)	SAE ARP 5580:2001; SAE J1739:2002; IMO-HSC Code:1994; SEMATEC:1992; Marine Contractors Association-IMCA:2002; IEC-60812:2006, DNV-RP-D102:2012
Failure mode, effect, and criticality analysis (FMECA)	MIL-STD-1629A:1980; ANSI/IEEE-STD-352:1987; IEC 60812: 2006; Reliability Analysis Center:1993; The International; ABS-RCM:2004
Failure tree analysis (FTA)	NUREG-751014 (App.2):1975; NUREG-0492:1981; ANSI/IEEE-STD-352:1987; SAE ARP 4761:1996; NASA:2002; EN 61025:2006
Markov analysis	ANSI/IEEE-STD-352:1987; IEC-61165:2006
Event tree analysis (ETA)	NUREG-75/014 (App.1):1975; IEC-62502:2010
Petri nets (PN) analysis	ISO/IEC-15909-1:2004
Reliability charts (RBD) block	ANSI/IEEE-STD-352:1987; ISO 17359:2003; IEC 61078:2006
Bayesian method (BA)	NUREG/CR-6823:1999

Table 8.10 Characteristics of basic reliability analysis methods [50].

Method	Application period within life cycle (LC)	Aim	Bottom-up or Top-down	Relation research Cause–effect	Dependence modeling
Failure mode and effect analysis (FMEA)	Throughout LC	Qualitative	Bottom-up	Inductive	No
Failure mode, effect and criticality analysis (FMECA)		Quantitative			
Failure tree analysis (FTA)		Qualitative/ quantitative	Bottom-up	Deductive	Partial
Markov analysis	Later stages LC	Quantitative		-	Yes
Event tree analysis (ETA)	Throughout LC	Qualitative/ quantitative	Top-down	Inductive	
Petri nets (PN) analysis	Later stages LC		Top-down		
Reliability charts (RBD) block	Throughout LC			-	Partial
Bayesian method (BA)	Later stages LC		-		-
Stochastic processes		Quantitative			Yes

defined for a discrete-time series, where time takes on values $\{0, 1, 2, \ldots\}$ or for a continuous series of time, where time takes negative real values. When time is continuous, then it is a Markov chain of continuous time, and in discrete time it is a Markov chain of discrete time. A particularly suitable probabilistic model for reliability analysis is the Markov continuous-time chain (also called the Markov Process), which is essentially a Markov discrete-time chain at any given time. Therefore, Markov's process for analyzing the reliability of engineering systems is based on probabilistic models characterized by the system (can be fully described at any time by determining its state at that time) and by time (as an exponentially distributed random variable, while changing states from one to other). The Markov process is characterized by terms such as system state (determined by the state of its components) and transitions (occurs when a component and/or system changes states). System states indicate whether the entire system under consideration is in an operational (working) state or in a malfunctioning state. Because the system is made up of components, system and component transitions occur because of a change in the state of the components. Transition of component state can be caused by component failure or repair. In Markov models, transitions between different states are characterized by constant frequency of transitions, which need not be constant in practice. A stochastic process that records the state of a component and/or system at a time point, called a half-Markov process, while a process that counts the number of times each state of a component or system is reached, is called a Markov renewable process (Ross, 1972, [96]). Reliability analysis by the Petri net (PN) method uses an algorithm defined in 1962. Using qualitative transition and status diagrams, qualitative system information can also be collected.

Reliability analysis using the Petri net (PN) method uses a formalism defined as far back as 1962. The Petri net is a graphical and mathematical tool applicable to the modeling of technical systems characterized as competitive, parallel, asynchronous, distributed, and nondeterministic and/or stochastic. As a graphical tool, Petri net can be used as an aid for visual communication, similar to flowcharts, block diagrams, networks. As a mathematical tool, it can, if possible, produce state equations, algebraic equations, and other mathematical models that describe the behavior of a system. In the general case, the Petri net is described as a bipartite graph made up of places that are graphically represented by circles, transitions crossings to represented rectangles, and direct arcs connecting places and transitions. In fact, the Petri net is a network of places and transitions in which places have the meaning of conditions, and transitions of events in terms of the concept of conditions and events. Transitions have a number of entry and exit points that represent the pre- and postevent state. Places may contain tags, marked with black dots. The presence of a tag interprets that component or resource data are available. PN behavior is guided by the activated transition and ignition rules. Transition is active for default tags when all of its entry points contain at least one tag. Each network ignition removes one tag from each of its inputs and adds one tag to its output locations, causing the status to change. Reliability analyses using the PN model allow quantitative analysis and qualitative analysis. From the aspect of quantitative analysis, this refers to the quantification of reliability in the transient (MTTF) and stable ($\lambda(t)$) states. This method has a very wide application in reliability, availability, and maintenance (RAM parameters) analysis of information systems, in particular. In the literature, in reliability analyses and probabilistic risk assessment of engineering systems, various types of PN models are used in combination with other classical combinatorial methods (e.g., FTA, RBD) to model the multiple states in which a system may reside. However, its application in this area has remained largely on the authors' hypotheses, which means that it is not possible to assess the reliability of their results.

When the reliability model is limited by the lack of adequate data to estimate the parameters of a functional block, then a statistical method based on the Bayesian framework can be used in addition to conventional statistical methods. In reliability engineering, the BM for modeling the reliability of a functional block gives the opportunity to take into account prior knowledge (e.g., databases) and the real available reliability data about the functional block obtained during its operation. Based on this information, the appropriate uncertainty of the component's reliability is calculated, which can then be modified using expert opinion. In this way, the development of a functional block reliability model becomes a learning process, and knowledge is continually upgraded as more information becomes available [97].

6.5.2.1 Qualitative and quantitative methods

The division into qualitative and quantitative is based on how the probability of failure is determined. Four methods are predominantly used to determine the likelihood of a failure event, depending on the available data: statistical method (include the direct handling of relevant fault data and the calculation of the probability of failure event), extrapolation method (to be used when available fault data are not available and appropriate, and includes the use of prediction models, consideration of similarities with other similar systems and components, and Bayesian concept, partly using limited expert judgment to estimate unknown quantities as inputs to the extrapolation model), expert estimation method (includes direct likelihood estimation by subject matter experts), as Hybrid methods, combined on the basis of the three methods described above (can be used together to estimate the probability of an event as efficiently and reasonably as possible) (Fig. 8.28).

The main objective of a qualitative reliability analysis is to identify the potential failures of the functional block under consideration, their consequences at the same level, and the cause and effect relations of the failures, as well as identify possible repair strategies. For qualitative methods (e.g., FFA and FMEA), the magnitudes of the probability of failure occurrence are estimated by subjective judgment of the analyst or expert opinion. Thus, quantitative methods aim at generating a list of potential failures, with the failure data relying heavily on previous experience or expert opinion.

Quantitative reliability analysis aims at determining numerical reference data, i.e., known reliability measures for each component of the system (e.g., *MTTF*, $\lambda(t)$), to use the same as input to the reliability model, to evaluate the system reliability, under probabilistic and stochastic assumptions. Thus, quantitative methods (e.g., FMECA, ETA, RBD, FTA, PN, and Markov methods) rely on numerical estimation of the likelihood of occurrence of quasi magnitudes of possible consequences in risk assessment, whereby these parameters, as measures of reliability, are quantified by statistical methods and databases.

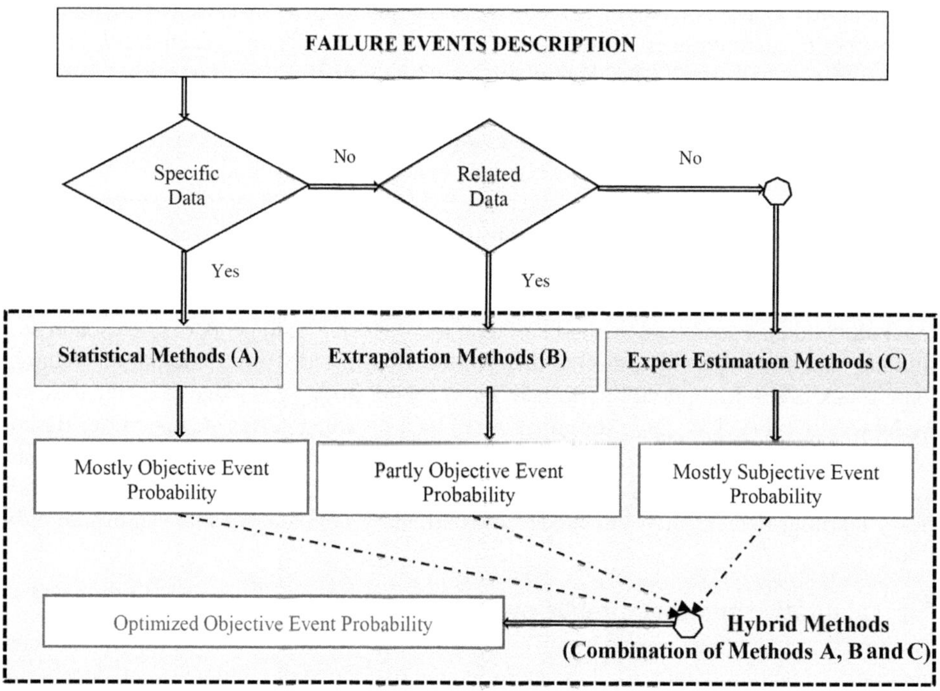

FIGURE 8.28

Methods for determining the probability of a failure event depending on the available data.

Which method (quantitative or qualitative method) will be selected depends on the availability of data to assess the reliability and the level of analysis required to make a reliable decision. It is important to note that qualitative methods provide analyses without requiring detailed data (intuitive and subjective processes used in these methods may lead to deviations in similar analyses), while quantitative methods allow for more uniform analyses (requiring quality data for accurate results). The combination of qualitative and quantitative analysis is also often used, or methods that are qualitative but can be considered as quantitative (e.g., ETA, RBD, FTA, PN, etc.) are used.

Unlike quantitative methods, qualitative methods do not require a lot of intensive work, so it is logical to favor them in many industries and standards. What is important is that qualitative methods are very useful as a starting point in the analysis of system reliability (identification of all potential failures, their modes, and consequences). In this case, it is important in all cases to choose the most appropriate technique, as support in making a decision with sufficient confidence (accuracy). This requires a clear specification of the scope and objectives of each reliability analysis, in order to select the most appropriate method, both from the aspect of the problem under consideration and from the aspect of resource efficiency with which it is available.

6.5.2.2 Inductive and deductive methods

As the relation between cause and effect of failure is investigated, reliability methods can be divided into analytical inductive methods (e.g., ETA, FMEA, FMECA, where the analysis process starts from known causes to predict unknown effects) and deductive methods (e.g., FTA, FFA, where the analysis process starts from known effects to unknown causes). In the inductive approach, one must first assume the specific failures or states of the functional block (the technical system and its components) and then try to determine what are the appropriate effects of each failure or condition on the operation of the entire functional block. Inductive approaches are also called bottom-up approaches because they start from the bottom (from the beginning of the failure) and then continue upward to determine the effects of an individual failure on the system. For deductive methods, the basic starting point is that the functional block itself is at a particular mode that has a malfunction and then one tries to find which behavior modes of the functional block (technical systems and its components) contribute to the malfunction that causes its malfunction. Inductive methods are applied to determine the possible failure states of a functional block, while deductive methods are applied to determine the occurrence of default failure states of a functional block.

6.5.2.3 Dependency modeling

In terms of reliability, it is possible that a component or subsystem has some influence over other subsystems, which characterizes the dynamics of system reliability [98]. Basic examples of such dependent relationships are: power sharing, cascading work, standby redundancy, interference, on-demand work, sequentially dependent failure events, CCFs, etc. Furthermore, the configuration of the system in terms of reliability (and availability) may vary (a faulty component or subsystem, in accordance with the maintenance policy, can be repaired with varying efficiency). Other examples are the application of a reliability growth model, which is mostly related to repairable systems or that the system works in phase missions. The ability to model dependencies is a very important criterion to consider when choosing the reliability analysis method for the technical system under consideration.

6.5.3 Reliability sources

Modeling and analyzing the reliability of components or systems as a whole requires several types of data, such as: technical and operating data, environmental data, maintenance data, and various types of reliability data. The reliability data refer to the fault data/fault modes and fault distribution times. The availability and recognition of relevant failure data is the most important part of any qualitative reliability analysis. Thus, for reliability analysis, the basic requirement is related to the availability of such data. These data should be: current, specific, revised, large (large sample with many recorded failures), applicable to the environment, as well as suitable with respect to the life cycle stage. Starting from the principle technological scheme and equipment composition, and the basic flows of substances involved in the technological process, it is possible to adequately view the technical maintenance system as a system for managing the reliability and safety of technical maintenance strategies. Given the high complexity of the classification, the following factors are included: complexity of the technical system, mechanisms of occurrence and possible consequences of failures, level of training of staff to carry out maintenance and performing overhaul activities, formation and possession of information about possible failures of elements of the technical system, possession of control systems and diagnostics, ownership of infrastructure that monitors the exploitation of the system (engaging specialist firms in the maintenance of technical systems), evaluation of the existing load structure and their frequency, as well as the associated exploitative impacts and conditions.

For these reasons, it is necessary to select the criteria for defining the reliability indicators. Budget criteria for failure of TPP elements are criteria for reaching the limit states of strength and service life, but they cannot always be used (different life and strength levels of individual components, difficult to reach and diagnose elements, presence of associated harmful effects and their accumulation, various technological defects, initial malfunctions, residual stresses, etc.). For this reason, suitable correction coefficients, corresponding general relationships in the probabilistic interpretation of processes occurring in equipment, etc., are encountered in the literature, which results in a certain generation of errors in the process of reliability estimation and its optimization according to the chosen, most often economic criterion. In further studies, it is necessary to optimize the choice of reliability criteria by increasing the level of accuracy of the correction coefficients and minimizing the resulting errors and reducing them to an acceptable level. In order to analyze the importance of the elements of a complex TPP, in terms of operational certainty, starting from the structural thermal scheme of the TPP, it is necessary to know the type of connection in the reliability scheme, certain data on failures and their intensities (literary recommendations or data from exploitation), the results of calculation of the stationary coefficient of certainty individual components as well as drive groups.

There are many sources for collecting functional block life cycle data (warranty claims, previous experiences of similar or identical functional blocks, testing, records generated during the development phase, functional block user failure reporting system, manufacturer quality control records or users). The "lower" stages of the reliability forecast of energy plants are characterized by a lack of information on their overhaul convenience, as well as in dynamic time budgeting. At the stage where it is possible to carry out appropriate experiments, tests of certain characteristics, and possible refinement of samples (prototypes), it is possible to give a preliminary estimate of the time between the planned overhauls, as well as the parameter of the failure flow, using statistical processing of existing information on the essential characteristics and forms of equipment. In the case of the stage of development and design of serial equipment, where a certain development of technology and organization of variant solutions of the system of maintenance and technical maintenance is possible, a combination of engineering calculations with research and experimental results is performed, in order to obtain basic parameters.

If the problems of optimizing reliability and standardizing certain reliability indicators are addressed, it is necessary to define in detail the aspects of frequency, degree of depth, and duration of failure. At the stage of production of standard equipment, it is necessary to determine the guaranteed values of the reliability indicators of the TPP in relation to the norms prescribed in the relevant framework power system. It is similar to the assembly and test stage. At the exploitation stage, it is necessary to increase the level of accuracy of the previous data, while solving complex optimization tasks, studying the causes of phenomena, distribution patterns and ways of predicting the failure state of the system. In further case-by-case studies, it is necessary to adjust the selection of both criteria and reliability indicators, with the aim of optimizing and accurately assessing the reliability of the TPP system in the short or long term.

The available reliability data collected from various sources mainly come from three types of databases: the quartile event database, the accident or incident database, as well as the component reliability database. Data quality may vary due to incompleteness and level of detail, and for one or more reasons such as the method of collection, the specification of boundaries by the company, the description of the effects of the failure, the subjectivity and experience of the data collector, and the time since the failure.

Today, when modern and profitable production rests on market demands, optimal technology, and optimal technological process, from the aspect of technical and economic life of using a complex technical system, the need for intensive revitalization (extension of basic working life), reconstruction, and modernization of existing technological processes is of particular importance, and related equipment. This requires further dissemination and implementation of modeling, simulation, and optimization methods based on the use of information technologies. Changes in the maintenance system should be approached in the order of importance of the factors that influence the change of performance indicators of the complex technical system, as well as the possibilities of application from the aspect of the required investments. The aim is to achieve such a sequence of development steps that provide the greatest effects with the least investment (economic criterion for optimization). The specific tasks of maintenance technology are to provide the process of maintenance optimization and to refine the directions for achieving higher quality, reliability, and economy of the complex technical system and its production itself. The decision on the type and activities of maintenance can also be made based on the costs of the company related to maintenance and operation and the methods of maintenance chosen. The order of development steps to be implemented affects the efficiency and effectiveness of the maintenance system.

A commitment to a maintenance strategy or technology will certainly influence the character, scope, and frequency of maintenance work to be performed in a specific technical system (implementation of the integrated logistics support (ILS) concept). In doing so, the greatest attention is paid to reducing administrative time through the implementation of the necessary changes in the type and form of management and organizational structure and the application of the concept of computerized maintenance management system (CMMS). Another factor in importance is the shortening of logistic times by determining the most optimal levels, methods of managing and allocating spare parts by levels, as well as accelerating material and information flows related to them. Merely improving the quality of work execution requires more changes in the behavior of the equipment repairer and maintainer than the material investment itself. In addition to the introduction of modern overhaul equipment, the use and application of diagnostic analysis methods can further improve the performance indicators of the maintenance technology applied.

Choosing the right maintenance organization defines the basic elements of a work-sharing structure that will allow all required maintenance tasks to be performed from the aspect of the selected maintenance technology. On the basis of all the defined characteristics of the maintenance process of a particular technical system, it is not difficult to conclude that the characteristics of the chosen maintenance technology depend on the characteristics of the technical system itself, that is, they must also be considered within the characteristics and content of the "technical factor" [1].

Maintenance technology jobs, activities, and operations include, in the most common terms, the following works: inspections (condition checks, inspections), tests and measurements, audits and controls, cleaning, washing and lubrication, corrosion protection (deconservation, conservation, and reeventual conservation), external monitoring, servicing, technical diagnostics, safety measurements, occupational safety, servicing and finishing, assembly and disassembly, overhaul (small, medium, and large, general audit), reconstruction, revitalization, and modernization (restoration, corrective and preventive programs, etc.). Starting from the data on embedded materials, their structural dimensions, and design and working parameters of work surfaces, failure analyses are performed, with a detailed review of all previous tests performed. On the other hand, knowing the mechanisms of influence on the material under consideration for which there are certain possibilities (identification of pressure and associated temperature, possibility of testing the structure onsite by nondestructive methods, possibility of measuring dimensions, wall thickness or ovality, computational possibilities of checking the saturation of the material, etc.), it is possible to effectively define the scope and dynamics of testing embedded materials. The very scope of testing depends primarily on the knowledge of the main mechanisms of damage (failure records, experientially or on the basis of analogs), such as erosion, corrosion, fatigue, damage, combined mechanisms, etc.

Systematic procedures for determining the causes, types, and consequences of failures that may occur, it is necessary to define and specify activities to minimize the catastrophic consequences of failures, especially those related to the means and the environment (preventive engineering). Managing the remaining life of complex technical systems (complex units of TPP [53,54]: steam boiler, steam turbine, electric generator, etc.), with the inevitable analysis and specification of its "weaknesses", is today a multidisciplinary task for a team of experts, whose implementation requires new methods and concepts, as well as the corresponding algorithms for working methods. The main tendency in the development of these methods is efficiency, speed, and cost, i.e., obtaining certain numerical values on the basis of which an appropriate and timely decision can be made in the maintenance process (decision optimization). In addition to estimation, reliability determination data can be obtained through calculation and verification or naturally (unforced), through customer experiences, manufacturing and other experiences, and through data from relevant service organizations engaged in the maintenance of the TPP system as a whole. If the observed object is complex (as a TEP system), then the problem of determining reliability is solved if the following data are known: the reliability of its constituent components (steam boiler, steam turbine, electric generator, auxiliary systems, auxiliary equipment, etc.) or at least its "the most critical" elements (steam boiler, steam turbine, electric generator), their interconnection (connection structure) and operating conditions (limitations and environmental conditions).

It should be emphasized that the verification of reliability, that is, testing the hypothesis in practice, is performed in all life stages of development, design, construction, and operation of the facility, and is mainly related to several basic limiting factors—money and time, environmental conditions, and other technical constraints. The reliability verification itself is accompanied by the corresponding mathematical apparatus, with a certain level of confidence in the parameters tested. The inadequate level of reliability during the exploitation of the complex technical system itself and the existence of irrational labor-based investments by eliminating consequences rather than causes clearly indicate the need to harmonize existing methods to achieve optimal reliability and adapt them to the system, with the prior definition and elaboration of the appropriate algorithm. The primary goal of the system user is to keep the system in a working state for as long as possible. To achieve this, it is necessary for the system to "assist" by performing certain maintenance tasks. Important decisions about the responsibilities, content, and timing of individual maintenance tasks define the methodology or philosophy of maintenance [52].

In terms of philosophy or methodological approach to maintenance, there are two schools (lines) that have been attracting the most attention lately [2]: RCM and TPM. The maintenance-centered reliability methodology also includes failure analysis in the maintenance decision-making process. In most different industries, such as nuclear power plants, TPPs, and the petrochemical industry, businesses are now required to assess the likelihood of risks, capturing the growing risks of the system as a whole. These systems will have a large number of protection devices (fire alarms, gas detectors, critical switches, pressure relief valves or safety valves, overload protection, safety switches, and related equipment) designed to prevent malfunctions and damage to the main TPP circuits as a whole. Most of these devices do not fail in safe conditions, and regular checks are required to confirm that the overall protection system remains operational. As these systems become extremely complex with the increase in the amount of security devices and alarm systems, it is very difficult to estimate the risk resulting from an increase in the probability of failure.

6.5.3.1 Functional block failure event database

Many companies use computer systems and adequate programs to manage their maintenance, with records of failures of various components, preventative and corrective actions performed continuously, through programs. In this way, a component failure event database and maintenance actions are carried out. The data is used to plan maintenance and optimize the cost of maintaining and purchasing spare parts. Some companies have a failure reporting analysis and corrective action system (FRACAS) implemented, where failures are formally analyzed and classified before failure reports are stored in the database [50].

6.5.3.2 Accident or incident database

Accident and incident databases contain information about them within specific categories and are managed by various organizations, consulting companies, and official bodies. Some of the databases are very detailed, while others contain only a brief description of the accident or incident. Component failure data can sometimes be read from the description of accidents or incidents. In energy and process engineering, one of the main concerns is the safety of an energy plant within a hierarchically higher system (power system). Namely, due to heavy penalties in case of loss of life, pollution and destruction of the environment, as well as loss of electricity and heat production, as well as technological steam, special attention is given to activities in the design and production (with installation) of the energy or process plant, operations during its exploitation (stationary and nonstationary modes of operation), and the education and training of all persons involved in plant management in both stationary and nonstationary modes of operation [50,52].

6.5.3.3 Database reliability components

The component reliability databases provide an estimate of the failure rate for individual elements, some of which describe the effect of failures and repair times. The commercially available component reliability databases are based on generic data that is the result of analysis based on actual fault and maintenance event data. Generic databases are databases in which components are grouped without manufacturer information, production, or a detailed description of component characteristics. These databases typically contain data on the incidence of failures, which can be based on recorded failure events, expert opinion, and laboratory testing. There are numerous generic databases in different industries, most importantly shown in Table 8.11. It is important to note that generic databases are always being developed and supplemented because of the increasing number of components, new methods and techniques, and the application of computers.

6.5.3.4 Data analysis and data quality

In the reliability analysis, it is usually necessary to estimate the failure frequency of each component and the confidence interval for that assessment. A simple approach that does not take into account the variability of the samples is to combine all the data into one sample and estimate the confidence interval according to the general terms for the exponential/Poisson's model. Apart from the possible uncertainty that accompanies this approach, estimation of the failure rate is usually biased, and the estimated confidence interval is unrealistically narrow. As the quality of the data presented on the databases depends on the way they were collected and analyzed, in addition to activities related to the implementation of the best possible process of data collection and processing, several guidelines and standards have been issued with the aim of achieving high-quality analysis of the collected data. Most commercially available reliability databases only provide a constant incidence of failures, even in the case of mechanical equipment subject to damage due to erosion, corrosion, and fatigue breakage. With the knowledge of the mechanism of decay due to these damages, the frequency of failures must increase. For example, the US Army Military Manual for the Evaluation of the Reliability of MIL-HDBK 217F: 1996 Electronic Components [113] uses a method based on a detailed stress analysis that results from environmental performance, quality applications, temperature, complexity, construction, etc. Therefore, the frequency of failures according to this method, called the part stress analysis prediction technique, is estimated by:

$$\lambda = \lambda_B \cdot \pi_Q \cdot \pi_E \cdot \pi_A \cdot \ldots\ldots \tag{8.25}$$

where λ is the basic failure rate, which is estimated according to the reliability test performed under standard ambient conditions, i.e., standard stresses and temperature conditions.

Parameters π_Q, π_E, π affect factors that take into account part quality, environmental impact on equipment, application stresses, etc. The MIL-HDBK217F manual describes a special method for evaluating a system where all its components are in operation, which implies the serial configuration of the system. Therefore, the failure rate of the system λ consisting of n components is calculated:

$$\lambda_S = \sum_{i=1}^{n} \lambda_i \quad t > 0. \tag{8.26}$$

Table 8.11 Examples of generic databases [50].

Database	Equipment	Source
OREDA Handbooks [112]	Process equipment (approval industry)	Det Norske veritas, Norway
MIL-HDBK-217F-Reliability Prediction of Electronic Equipment	Electronic components	US Military Handbook
EPRD—Electronic Parts Reliability Data	Electronic components	Reliability Analysis Center (RAC), USA
NPRD—Nonelectronic Parts Reliability Data	Mechanical and electromechanical components	
NOOP—Nonoperating Parts Reliability Data book	Electronic, mechanical, and electromechanical components that are not in operation or are stored	
FMD—Failure Mode/Mechanism Distributions	Electronic, mechanical, and electromechanical components	
PDS Data Handbook	Sensors, detectors, valves, and control logistics	SINTEF, Norway
FARADIP—Failure Rate Data In Perspective	Electronic, mechanical and pneumatic equipment	TECHNIS
Handbook of Reliability Prediction Procedures for Mechanical Equipment	Mechanical equipment (hydraulic and pneumatic)	Naval Surface Warfare Center Carderock Division
IEEE 493-1997-Institute of Electrical and Electronics Engineers	Production and distribution of power energy	ISBN1-55937-066-1
STF18 A83002, Reliability of Surface Controlled Subsurface Safety Valves	Surface control below surface safety valve	Exprosoft, Norway
STF75 A89054, Subsea BOP (Blowout Preventer) Systems, Reliability and Testing, Phase V	Underwater BOP devices	
STF75 A92026, Reliability of Surface Blowout Preventers (BOPs)	Underwater BOP devices	
STF38 A99426, Reliability of Subsea BOP Systems for Deepwater Application, Phase II DW	Underwater BOP devices for deep sea submarines	
Subsea Master and Well Master	Components in oil sources	
EIREDA Database—European Industry Reliability Data Handbook, Electrical Power Plants	Valves, sensors, and control logic (nuclear power plant data)	EUROSTAT, France

In the OREDA manual [112], the failure rate λ is assumed to be a random variable in order to receive different values for different samples. In order to combine the nonhomogeneous data into one multisample estimated with average failure rate (AFR), a special estimation method called "OREDA - Estimator" was used, which is based on the Bayesian point of view of probabilities and in fact represents one semi-Bayesian access. Using the "OREDA - Estimator", an estimate of the mean value of λ

was calculated for each component failure mode with the corresponding lower and upper confidence intervals for λ. In addition, an estimate of the standard deviation of the distribution λ was calculated, whose high value indicates inhomogeneity. OREDA has established a comprehensive reliability and maintenance database for offshore oil and gas exploration and production equipment for a variety of geographical areas, installations, equipment types, and operating conditions. The data is stored in databases, and a specialized computer program has been developed to collect, retrieve, and analyze information. Access to the electronic database is restricted to OREDA participants only. Commercially available reliability databases are published in the OREDA Reliability Data Handbook [112]. An important feature of the OREDA manual is that it contains the characteristics and drawings of the physical boundaries of components/systems. The data in the OREDA manual are primarily reliability data used in the analysis of availability. However, in the initial phase of the maintenance system application, historical data may be insufficient, in which case OREDA provides the best available data source for determining MTTF failure modes.

The program for raising the competitiveness of energy and process plants is highly complex and multidisciplinary. In addition, each plant has its own specifics that must be taken in the right way, otherwise the output results in a sudden loss of value (if a positive result is achieved at all). Therefore, great importance must be attached to the architecture and hierarchy of activities leading to the final results. After that, the level of computer analysis of the input data needs to be agreed with the plant personnel, as it can go from simple data entry and grouping, which would then be "manually" analyzed, to high-tech computer so-called "expert" systems, which "independently" make engineering decisions based on the software tools contained therein. In addition to the basic plant information, the input bank must contain relevant exploitation and maintenance information. The most important feature of this data is the truthfulness, since their subsequent statistical processing leads to basic indicators, which determine the further strategy. Inaccuracy of the input data leads in the wrong direction and reduces the level of raising the competitiveness of the plant. In the event that the relevant information is not known, the method of estimating or assuming these values must be determined in agreement with the plant personnel. This controls the probability level of accuracy of the output data.

7. Costs as indicator of economic efficiency of securing reliability

The complex nature of the tasks of this type is most often due to the following features of the system under consideration: the essential connection of basic reliability solutions with multiple intersystem factors of the facility and its application, as well as the multivariate different and interconnected ways of possible reservation and overhaul of technical maintenance of the plant, with the analysis of hierarchical connection, its components, and constituent accessories. On the other hand, there are a number of sharp technical limitations related to the provision of reliability characteristics, from basic research, certain experiments, and model tests, through the development of simulation models to check the design and budget analysis and statistical information analysis of the existing exploitation of existing plants, construction of diagnostic systems and technical and technological control, its automation, as well as material and technical security and overhaul and technical maintenance of certain equipment.

The optimization solution obtained from the budget through the modified method model is based on the differences of the starting information obtained from various sources about the exploitation of the reference block so far, while providing the conditions for sufficiently accurate choice in the conditions of incomplete determination of the starting information. Starting from its form, scope and accuracy, input and output information, as the end result of the implementation of such budget-optimization calculations, obtain optimal plans of experiments, controls, and diagnostics. In this case, the system of the TPP cannot be viewed separately from the ES, as well as the realized savings obtained on the basis of the obtained results of the application of reliability methods and defining ways for their provision.

There are two ways to implement the process of optimizing the reliability indicator. The direct way of ensuring optimization of the reliability of an object is to maximize the level of its reliability at a given maximum allowed level of losses (costs) at the object, while the indirect way is based on the minimization of losses (costs) at a given minimum allowed level of reliability. Optimization by the modified method algorithm is based on the indirect method and interval predetermined range of variation of certain reliability characteristics [3]. This method allows for certain simplifications of most of the indefinite influences and conditions of exploitation, that is, the accuracy of this method is limited by the accuracy of optimal results, for example, with the given baseline data, with the possibility of its continuous correction at the stage of exploitation of the object. Methods and programs for dealing with the reliability of individual parts of a technological scheme or system as a whole, with adequate planning of programs, contents, and duration of planned overhauls, are based mainly on statistical analysis and the use of results obtained on the basis of analogues.

Models and methods for solving optimization problems can be classified into several subgroups:

(a) consolidation of the influence of the basic technological scheme of the plant on the mono and double block structure without its detailing and analysis of losses by equipment structure, through analysis on the basis of nominal power;
(b) variant estimates of the plant scheme through the application of the FMEA/FMECA analysis and the failure tree of the most critical facility (obtained by the ranking method), combined with the analysis of semi-Markov or Markov failure processes by the criterion of no-fault or certainty;
(c) variant assessments of a complex set of reliability indicators for the relevant design scheme and structural reservation of the plant based on the previously conducted higher hierarchical reservation level of the ES, with simplified or detailed analyses (depending on the prescribed accuracy).

The formulation of tasks related to optimizing the reliability of a TPP system in the general case can be defined as minimizing the losses on the construction and application of a serial unified power plant, consisting of losses related to the plant itself and losses due to its ES dependence, depending on the reliability indicators and possible ways of providing them for the given conditions at each stage of the life cycle, as well as the system parameters taken and the known minimum necessary functional structure of the plant. Solving this problem, with all the given limitations and overall analysis of interconnections and overlaps of certain expressions, is quite complex from the aspect of the totality of the system of equations, and therefore for this reason considerable simplifications are made for the level of estimation of the reliability indicators.

Since the total period of construction and use of a complex TPP system takes several decades, most often due to the indeterminacy of information, a uniform block analysis of certain groups of 150–800 MW is used. In this case, the general term for defining all types of costs takes the form [3]:

$$Z = E \cdot \sum_{\tau=1}^{T} (KU_\tau + U_\tau) \cdot (1 + E_{nor})^{T_s - \tau} + U_{nor.ers}; T \leq T_s + T_{sl}, \tag{8.27}$$

where: T_s—the average time value from the start of exploitation for a given series of TPPs; E—normative coefficient of effectiveness; KU_τ—capital investments for the year τ; U_τ—investment for ongoing maintenance (without the participation of the modernization and reconstruction part); E_{nor}—normative coefficient of taking different time losses (costs); $U_{nor.ers}$—investments related to the maintenance of normal operation with stable reliability indicators and operating modes.

For these reasons, an overview of all the equations related to the valuation of capital total investments is given, followed by investments for routine maintenance and investments for the introduction of diagnostics as an integral component of preventive maintenance. Cost structures related to providing the predicted levels of reliability indicators at stages are also given: development and design, installation and trial commissioning, operation of the plant, as well as during the extended life after reconstruction and modernization. The analysis also included some additional effects related to site specificity and the role of TPP installations within the ES. Starting from the general scheme of reliability calculations, with the choice of optimization criteria in the form of realization of the minimum required investments for the selected variant solutions of certain levels of reliability, as possibilities and ways of their provision, it is possible to define the costs per plant of the TPP.

At the stage of development and design, in the absence of modernization and reconstruction, the amount of capital investment covers all costs in the period before the start of exploitation, while maintenance costs have their place in the exploitation stage. They may also include any costs related to modernization and reconstruction incurred after the trial commissioning and during the operation itself. In this case, based on expression (8.27), it follows [3]:

$$Z = E \cdot \left[\sum_{\tau=1}^{T_s} KU_\tau \cdot (1 + E_{nom})^{T_s - \tau} + \sum_{\tau=T_s+1}^{T} U_\tau \cdot (1 + E_{nor})^{T_s - \tau} \right]. \tag{8.28}$$

The preliminary costs can be divided into the costs for the development and design of the batch plant, their general reservation and overhaul maintenance within the power system, and the costs related to the production and installation of each of the components in the TPP system. In this case, KU_τ the size consists of the amount of capital investment in the plant for the year τ (including the part for winning and developing the new series) and the costs of production and installation with control and trial commissioning, as well as the costs related to capital investment in the ES (relative capital investments for planned and unplanned general provisions within the ES), as well as for the overhaul and technical maintenance of the power plant within the ES (initial costs related to the overhaul base, the material, and technical maintenance base for the development and conquest processes) and the costs related to the lost power ES.

The costs of ongoing maintenance, due to the irregularity of their release and the different con ditions of their use, are presented with size U_τ, varying for individual units in the system. They consist of two components: the first, related to the conditional annual cost for normal functioning without reliability, then the cost for the individual participation of components related to overhaul and technical maintenance (in the event of complete or partial failures) and the ongoing redistribution of fuel due to deterioration of its quality and unplanned push-shutdown modes and others related to annual maintenance in the ES (including unplanned plant reservations for partial or complete cancellation of various forms, compensation for the plant's own consumption, etc.).

Sometimes, costs related to the hierarchical connection of the TPP system with the environment and measures for the implementation of its supplementary protection, as well as the appearance of possible restrictions, are also included in the optimization process. The next step is to group the above costs by facilities, life cycle stages, and asset allocation. One way of possible grouping is given in Refs. [2,3], with a concept adapted to electronic data processing. This system should be supplemented with the costs associated with providing reliability for each of the stages of the life cycle of the TPP system.

7.1 Some aspects of cost estimation related to providing reliability at the stage of development, design and conquest of a thermal power plant

Investments related to reliability assurance at this stage are an integral part of the costs related to the development and conquest of the production of standard thermal power equipment, which consist of a part of the reliability costs for the development of the new plant and other costs related to the development of the supporting production base necessary for its realization. Parts of this stage contain two conditional cost categories: theoretical budget documents with the development of project documentation, development of a prototype unit and its testing, with the providing of a complex related to energy efficiency. The lack of information on all these costs is most often due to ignorance of certain limitations of information related to the scope of individual tests and their duration. Based on the shortened scope of tests and the time required, using recommendations [2,3], the following estimate is proposed:

- estimate the first group of costs over the following quantities: costs related to laboratory tests on the reliability of parts of the heat diagram obtained from the determination of the criticality, costs related to testing of other elements of the scheme based on existing analogues, and costs related to possible testing after refinement within the manufacturer's factory (industrial testing);
- to estimate the second group of costs on the basis of experience in the production of similar thermo-energy parts or on the basis of analogues (production and installation, technological process and control, required labor force, coefficient of capital investment duration at this stage, etc.).

In doing so, the following facts should also be borne in mind [2,3]:

- The basic factors of changing the total costs to ensure a certain level of reliability in the design and engineering of the plant are: the design price of its components, the number of tested parts and their possible finishing, duration and scope of effective tests, design characteristics of load reserves according to nominal power, and additional characteristics related to the elaboration process in the form of certain correction coefficients;
- the use of statistics for the above estimates is the subject of specific research, which can be obtained from the use of data from analogue plants already in production (starting from the structure of the component eligible costs compared to the corresponding conditions).

7.2 Some aspects of estimating costs related to ensuring reliability at the plant design and installation phase

The structure of these investments consists of costs related to the process of preparation, production, and quality control of the plant itself, and costs related to the process of their installation at the power plant. These costs also include the corresponding start-up costs related to: structural, time, functional, information, and load provisioning; costs related to entry, intermediate, and final quality control; provision of conditions for maintenance and repair, system of diagnostics, protection, and management; as well as for material and technical provision of resources within the power plant (spare parts, etc.). The system of providing accelerated assessment of reliability indicators at this stage through the selection of the optimal plan of abbreviated tests and automation of "on-line" procedures for reliability assessment, is based on simplifying the dependence of equipment element prices and its operational characteristics. They are used as orientation relations between the appropriate quantities for the considered variants and the corresponding analogues with known data. This is especially important for heat exchangers, pipelines, steam turbines, compressors, pumps, fans, electric motors, etc. The accuracy of such an estimation will depend primarily on the quantity and quality of the starting information, construction conditions, and similarity of the construction with the analogue under consideration which are included in the total system of TPPs, as a rule, lasts for 1–2 years, their total breakdown by years can be simplified by allocating the costs of production to the first, i.e., the cost of installation in the second year of the total process [2].

7.3 Determining the amount of capital investment for provisions and the process of switching on plants within the electricity system

This type of capital investment is based on the size and shape of the structure of the ES itself, that is, on specific indicators of the operation of its components in the observed Armenian period (usually 1 year). There are different ways of simplified project forecasting of the size of the emergency reserve of a TPP, considering its power, type (seriality), and the level of required reliability during exploitation. In this case, the size of the regime portion of the provisioning can also be estimated on the basis of the size of the emergency reserve with a coefficient $k_{res} < 1$, which is determined on the basis of tests of analogue plants and normative values adopted within the ES itself. An example of such a system of equations is given in Ref. [2] and is related to the research of the ES of the former USSR for the period up to 1980.

In addition to the model outlined above, attention should also be paid to possible forms of system reserve and coverage of loss of power that were not previously covered, relating to [2]:

- loss of manpower due to reduced EC for a given load, with possible analysis of the justification of its full or partial replacement with other plants with better EC, or anticipation of additional costs in the case of using it;
- the frequent occurrence of a gradual deterioration of the EC system in the framework of interannual planned overhaul activities, without conducting an efficiency analysis and replacing them with other plants, which also entails additional costs;
- loss of plant manpower due to the appearance of reduced working capacity due to the aging of the facility itself and an increase in the number of hours worked, which may also be a consequence of previously inadequate repairs, etc.;
- occurrence of design modification of the heat scheme and type equipment within the TPP in terms of providing reliability, which can lead to a change in total own consumption and change in the value of the utilization coefficient;

- the required period of winning production after the trial run and reaching the design nominal parameters of the plant, related to the lost power;
- the need to operate the TPP at the technical minimum due to network constraints and the appearance of "force" elements;
- increase of the necessary labor for own consumption due to the gradual deterioration of the hydraulic, aerodynamic, thermal-mechanical, and electrical parts of the plant within each interrepair period.

Most often, the participation of the preceding elements is taken through the corresponding corrective individual coefficients, which reflect the very structure and efficiency of the ES, or generally over the highest value of one of them.

7.4 Reliability limitations due to force majeure

Natural phenomena (atmospheric discharges or thunder, earthquakes, volcanoes), then various meteorological events (floods, droughts, high winds), as well as catastrophes caused by higher level human factor involvement (state of imminent danger of war, state of war, fires), may occasionally be of particular importance to the reliability of the ES as a whole. The effect of natural phenomena can be characterized by the corresponding intensity of outages λ_f^{np}, while the probability of a fall in TPP due to the above effects of "occurrence of force majeure" can be represented by Poisson's distribution in the following form [2]:

$$P_f(t) = \frac{\lambda_f^{np} \cdot t \cdot e^{-\lambda_f^{np} \cdot t}}{z!}, \tag{8.29}$$

where z is the number of repetitions of one occurrence in the observed period of time over 1000 h. Therefore, the reliability with respect to the effect of these phenomena is

$$R_f(t) = e^{-\lambda_f^{np} \cdot t}. \tag{8.30}$$

When the types of causes of failure related to the reliability of the operator and due to the effect of force majeure act independently, then the reliability of the TPP system is determined as a product of reliability with respect to the influence of individual groups of causes.

It should be noted that the Poisson's process is a special case of the Markov's process, in which the number of events over time is observed, with events of a discrete type, while time is of a continuous type. In this case, the following assumptions apply [2]:

(a) the probability of a transition from n to $n+1$ state over time, where Δt is $\lambda^* \cdot \Delta$ where λ^* is a constant value representing the events in the unit of time (cancellation/hour);
(b) each event is independent from the other;
(c) the events are irreversible, which means that the number of events increases as a function of time, while in the case of repairable systems, transitions are allowed in both directions;
(d) the probability of occurrence of two or more events in the interval Δt is negligible.

7.5 Basic aspects of project predicting cost estimation related to ensuring reliability at the exploitation phase

The level of anticipated reliability and the manner of anticipated current maintenance are a function of the lifetime service, which includes the performed overhaul of planned and unplanned works, replacement, reconstruction, and modernization of certain parts of the TPP system, as well as possible repairs due to accidental failures of some of its components. This type of cost is determined and given for each of the anticipated years of operation, and includes all current maintenance per plant, as well as certain additional costs associated most often with the first year of operation due to the need to increase the level of efficiency of the plant. Conducted in-box analyzes and more knowledge of this problem systematized, into favor the choice of a combined two-step calculation method [1,3,52]. The first stage determines the direct (normative) annual costs related to capital investments, current overhaul, and planned technical maintenance, including salaries for overhaul personnel, depending on the level of maintenance, the chosen variant of overhaul, and the use of technical diagnostics for maintenance according to the condition. This also allows for some generalizations related to the optimization of these costs according to their unified and typified classification in the form of interval averaged values (fuzzy technique). In the second stage, they are improved to the level of accuracy required by decomposing and detailing their individual components in the form of giving a more accurate estimate for ongoing maintenance. This also includes the redistribution of costs related to fuel consumption ΔB by the TPP and within the power plant, in the event of complete or partial failures and in relation to normal operation, of its EC, is to calculate and compare the change in the specific consumption of fuel by the electricity produced in relation to the normative-budget value and the reduction of the plant's workforce compared to the given standard value. Often, this type of analysis comes down to the analysis of critical elements in a technology scheme, with their consequent or parallel relations by reliability indicators (failure tree analysis, FMEA/FMECA). This type of analysis by plant and its individual parts does not include external system losses, which need to be further considered.

7.6 Supplementary effects related to the analysis of costs and forms of reservation of the system of thermal power plants within the electricity system

Determining the annual needs for certain forms of provisioning within the higher hierarchical level of ES in case of complete or partial plant failure as well as the amount of costs related to unscheduled maintenance and unplanned overhauls are of a purely random (stochastic) character. Any change in the level of confidence will directly affect the change in the required investment. The additional effects that can be achieved in some forms of provision within the ES are related to the main effects and investments. The deterioration of the maneuvering characteristics in the event of partial failures in the form of a decrease in the rate of increase or decrease of the load leads to the appearance of deviations from the given load graph and its variability. Therefore, the availability of additional high maneuverability capacities is required within the ES of each country, which can operationally make up for these shortcomings.

On the other hand, any increase in the amount of power needed to cover its own consumption additionally initiates the possibility of more failures and a gradual deterioration in the consumption of the plant not only electricity (heat) but also the fuel used. There are certain software packages designed for quick calculations of the heat scheme in the absence of some of its elements (e.g., high-pressure heaters) or to change certain parameters and characteristics of their parts or components. The following are examples of applications from Thermoflow, Inc., USA: STEAM PRO, STEAM MASTER, RE-MASTER, RECIPRO, THERMOFLEX, QT-PRO2. The number and duration of such events are significantly related to the failure flow parameters and the occurrence of additional costs to remedy them. For these reasons, simplification of solutions with determined intermediate intervals of the reliability indicators of the type equipment as well as the technical and economic indicators of the plant as a whole or parts thereof are often used. Starting from the initial stage of elaboration, designing, and winning production of a certain type of thermal power equipment in order to fulfill all the requirements without limitations, arising from its purpose, the designer is given a multivariate choice, with the need to optimize most often according to a certain algorithm. The aim is to design such a plant that has a satisfactory structure in terms of reliability indicators, with minimal maintenance costs over the expected working life.

Defining the plant that best meets the requirements for reliability and the process of operation and maintenance of the plant itself must be the result of the optimization process (in this case, based on the selected minimum investment criterion).

7.7 Reliability optimization based on minimum cost criteria at a hierarchically higher level (thermal power plant—electricity system)

Reliability optimization based on the economic criterion of minimum costs is performed according to a previously adopted algorithm, where finding the optimum represents the final activity in the optimization process. Based on the selected objective function in accordance with the selected criterion, the interval estimation of the reliability indicators in the form of expression (min, max) is determined. In the case of optimization when the constraint conditions do not exist, the procedure is reduced to calculating and equalizing with zero partial derivatives of the objective function, operating conditions, or conditions of exploitation [2]. For these reasons, it is necessary to determine the most optimal solution for each considered set of such conditions or for a shorter regime of operation of the TPP. Because this is associated with many calculations, it is necessary to use computers and classic software packages. Based on the set of solutions thus obtained, it is also possible to properly standardize the parameters, that is, to define certain technical solutions for the interval to give a set of reliability characteristics. In doing so, the methodology for defining and optimizing certain reliability indicators can be used to design expert systems for a TPP, with a given mathematical optimization model representing only one part of them. Efficient design and design based on reliability of TPP systems require the use of powerful computer-aided design (CAD), with the improvement of neural network based microprocessor systems and fuzzy logic, which is the main modern direction in this field. Defining the plant that best meets the requirements for reliability and the process of operation and maintenance of the plant itself must be the result of the optimization process (in this case, based on the selected minimum investment criterion).

8. Repair activities under the steam turbine system and the impact on reliability

The overhaul activities carried out within the steam turbine system are based on the instructions of their manufacturer, using the accompanying modern diagnostic and control equipment, as well as personnel specially trained to perform complex operations during the disassembly and assembly of

FIGURE 8.29

Outline of the overhaul of a 235 MW turbo generator unit in the TPP Kakanj.

individual turbine assemblies and their parts. The main activities in preventive maintenance for a certain period of time are performed during the overhaul of the steam turbine (Fig. 8.29). Their planning and implementation should ensure reliable and economical operation of the turbine until the next overhaul period. The most commonly scheduled period for a regular annual overhaul is 45–60 days, while the time for a capital overhaul is slightly higher (60–90 days).

The basic works, which are realized during the overhaul of the steam turbine, can be classified into several groups [52]:

(a) works prior to stopping the steam turbine (measuring the vibration of the bearings, measuring the basic thermodynamic quantities, measuring the characteristics of regulation, measuring the clearance on the benchmarks, measuring the thermal expansion, etc.);

(b) works after stopping the steam turbine (dismantling of turbine instrumentation, dismantling of oil pipelines and insulation, control of mounting the bearing blocks, control of height of bearing sleeves, etc.);

(c) steam turbine dismantling works (dismantling of upper half, high-, medium-, and low-pressure housings, dismantling of upper bearing housings, control of radial and axial clearances of the turbine flow, centering of rotor couplings, removal of HPC, IPC, and LPC rotor, removal of bearings, removal of radial walls and labyrinth seals (sheet metal), dismantling of servomotors, dismantling of regulating and quick-closing valves, dismantling of other control and signaling equipment, etc.);

(d) works after disassembly of the steam turbine (blasting of casing, partition walls and rotors, control of dimensions of individual parts, testing of parts by methods with and without destruction of materials, control of centricity and balancing of rotors, replacement of worn out and repair of damaged parts, etc.);

(e) steam turbine mounting works (mounting and centering of partition walls, installation of intermediate and outer labyrinth plates (seals), installation of internal housings in outer housings, mounting of rotor bearings, mounting and centering, control of axial and radial clearances of the flowing part, check of the backing and bearings clearance, mounting of upper inner and outer housings, control of centricity of couplings and lines of turbine and generator rotor, final assembly and insulation installation, fine adjustments, etc.);

(f) works after the start of the steam turbine (static and dynamic testing of the regulation and protection of the turbine, vibration measurement of the bearing housing, measurement of basic thermodynamic quantities, etc.).

In doing so, activities undertaken as a result of a capital overhaul should result in the restoration of the steam turbine plant to the condition that characterizes the efficiency that was when the turbine was installed and put to the test. The revitalization process (extension of service life) is an extension of the service life, which is mainly accompanied by the modernization and reconstruction of the steam turbine, while improving its technical and environmental acceptability. Such a systematic and comprehensive process for steam turbines is an indispensable and logical process in its basic working life.

The connection of the process of reengineering with the maintenance of the steam turbine system within the TPP system is aimed at realizing the corresponding advantages and increasing the reliability related to the following characteristic elements: analysis of costs related to the maintenance and availability or availability of the system, then determining the general aspects related to motives and justification of the revitalization, as well as determining the scope and defining the optimal term for the implementation of this process. In addition to the boiler plant, as the most critical part of the plant within the TPP, the characteristics of steam turbine reliability and availability have a great influence on the overall application of reengineering principles through the system maintenance process, i.e., the systematic approach to revitalization and modernization of individual thermal capacities.

The data systematization and the sequence of works and activities during the overhaul indicate the need to carry out certain preparatory (usually control) works necessary for the overhaul before and after the shutdown of the steam turbine. During dismantling of the turbine, it is necessary to check the alignment of the existing surfaces of the housing, as well as to control the centricity of the turbine rotor joints. Fig. 8.30 shows the high- and low-pressure turbine rotors of the installed capacity of 235 MW, prepared for performing the overhaul activities in the TPP Kakanj. Previously, before dimensional

FIGURE 8.30

High- and low-pressure turbine rotor of 235 MW in TPP Kakanj.

inspection and nondestructive testing (NDT methods), all parts of the turbine should be cleaned of corrosion and salt deposits (usually by sandblasting).

The main part of the activities and works in the overhaul of the steam turbine are performed after the dismantling of all its parts, and include works on defecting the disassembled parts, then works on dimensional control of the turbine parts, as well as works related to testing the turbine parts by nondestructive methods. If appropriate reconstructions and improvements are made to individual elements of the steam turbine (modernization), such as replacement of blades with more modern 3D blades, changes to the cooling and sealing system of the turbine, etc., then the effects of the activities carried out can give better parameters of steam turbine operation. Table 8.12 gives an overview of the improvements achieved for the reconstruction and modernization of a 200 MW steam turbine LMZ turbine.

Within the capital repairs, in accordance with the manufacturer's instructions, among other activities, inspections and defects of the turbine housings are carried out depending on the type of turbine, nozzles, clamps, diaphragms, seals, heating systems for flanges and screws, working blades and bandages, shafts, bearings, supports, oil seals of couplings, etc., with the elimination of the observed defects.

Of the major works on HPC, IPC, and LPC rotors, dimensional control and ultrasonic testing of the rotor sleeves and couplings, control of the surface of the shaft, discs, and rotor blades and its rivets (ferroflux) are performed, as well as the centricity and balancing of the rotor. On the partition walls with the stator blades, the control is performed on the existing surfaces, then the dimensional control of their outer and inner diameters, and the examination of the surfaces of welds and stator blades with ferroflux. The following works are carried out: overhaul of the housing, with material control and replacement of diaphragms as needed, scaling of horizontal joints, centering of the flow part, ensuring clearance in accordance with standards, as well as inspection, defecting, and repair of the rotor, with its centering, overhaul of bearings, overhaul of couplings, and the overhaul and adjustment of the control system and oil lubrication system.

Significant work on the internal and external housings of the turbine is to distinguish the control of the current surfaces, the testing of the transition radii inside and outside, as well as the testing of the existing surfaces by ferroflux in 100% range. The turbine bearings are subjected to dimensional control, ultrasound white metal deposition testing, and penetrant edge white metal deposition. As for

Table 8.12 Performance improvement of LMZ's 200 MW turbine after reconstruction and modernization [52].

Turbine housing	Utility level, %			Improvement measure	
	Designed	Real	After modernization	%	*MW*
High-pressure cylinder (HPC)	82.0	78.0	87.0	5–9	2.7–4.9
Intermediate pressure cylinder (IPC)	91.0	89.0	92.5	1.5–2.5	1.5–3.0
Low-pressure cylinder (LPC)	78.0	65.0	88.0	10.0–23.0	7.0–10.0

the work on the turbine screws, in addition to the dimensional control, a ferroflux test of 100% and a test of their hardness are performed. Dimensional control of the spindle and guide guns, but also the control of the tray position on the seat, the testing of the sealing (sealing) layer on the tray and the penetrant seat, as well as the testing of the transient radii on the outside of the ferroflux case are important on fresh steam valves and control and stop valves.

On the basis of the obtained results, after the examinations (defects), controls, and tests, the worn parts are replaced as well as the damaged ones are repaired. Following the realization of this activity, the process of assembling the parts follows, with particular attention to the installation and centering of the partition walls and segments of the labyrinth sheets (seals) in the inner housings. This is followed by the realization of the assembly and centering activities of the internal and external turbine housings and the mounting of the bearings into the bearing blocks. Depending on the installation and alignment of the rotor, with the control of the clearance and the bearing sleeves fit into the bearing, as well as on the axial and radial centering in the flow turbine section, the centering of the couplings and the control of the rotor line will depend on the operation of the turbine in the future (quiet and reliable operation of the turbine).

After completing all rotor centering and inspection, the upper half of the inner and outer housings is assembled, as well as the heating and tightening of the screws on the existing surfaces of the housings. A more significant final overhaul activity is the realization of control of the turbine rotor and generator rotor line and the connection of their coupling. Installation of the insulation on the high- and medium-pressure parts of the turbine is done after completion of all assembly work. This is followed by the oil circulation, the static adjustment of the turbine regulation, and the operation of its safety protections. After that, the steam turbine is put into operation and dynamic adjustment of both the control system and its protection system is performed. The vibrations of the turbine unit bearing housing are tested and dynamic regulation adjustment and protection are performed and thermodynamic quantities are measured and compared with the values they had before the repair. On the basis of these analyses, the final conclusion is reached on improving the dynamic state and increasing the utility of the turbine plant as a consequence of the realized overhaul activities.

If work is required in a high-risk facility, certain fire safety measures must be observed. Diagnostic tests are performed primarily to detect failure, but in areas threatened by explosive atmospheres and high fire risk, not only to detect failure, as a parameter to reduce the working condition of the equipment, but also to detect and monitor failure as the initial cause of ignition, and to detect the possible presence of explosive atmospheres. Special diagnostic methods and diagnostic equipment are used for normal maintenance work, that is, when diagnosing an explosive atmosphere in a plant or potential ignition sources for electrical and mechanical equipment. At that time, it is obligatory for all engaged personnel to control the process and take organizational and technical measures during gas welding, electric welding, soldering, as well as melting of bitumen and resins, performing vulcanization and other fire-risk works. Diagnostic testing of the presence of gas in plants generally provides very important information on the prevention of explosive atmospheres. Preventive maintenance with vibration and infrared thermography diagnostics is one of the best tools for detecting failures while still in the initial stages of occurrence, whereby information obtained through diagnostic methods through the maintenance process provides primarily the facilities with the necessary data to increase the level of safety in the operation of these facilities. All places for welding and other fire-risk work, which are related to the use of open flames and heating of parts up to the ignition temperature of the material and structure, are divided into [52,54]:

- permanent places of execution of fire-risk works, which are designated in the part of the plant, workshop, or on open surfaces and are intended for these purposes;
- temporary places for welding works to be determined directly on the premises and on the equipment (unless there is no possibility to remove the part and bring it to the place intended for permanent execution of the welding works).

Permanent places for the execution of welding and other fire-risk works must fully comply with the requirements of the "Instruction on fire safety measures when performing welding work on energy facilities." Execution of welding and other hazardous works may be performed by personnel who have completed training and assessment of fire safety knowledge and instructions when performing fire-risk work, as well as existing rules and other branch normative documents, previously aligned with the requirements for the professional preparation of personnel performing these works. During the execution of welding and other hazardous works, the personnel must be obliged to possess the certificate of the competent electric power organization, then a valid certificate of fire protection technique, as well as a permit for temporary execution of works at a specific place.

When performing welding and other hazardous work at height (from a ladder or scaffold), measures must be taken to limit the dispersion and fall of molten material to combustible structures, equipment, or materials. If necessary, especially in places where there are combustible materials and where people are passing, the downward angles must be protected and the people to be monitored must be identified, with appropriate warning signs visible beforehand.

In the case of temporary fire-risk works in buildings, structures, and equipment, these places must be equipped with fire extinguishers. If there is a fire hydrant in the immediate vicinity, a fire hose with a nozzle must be installed on it and a line drawn to the place where the fire-risk work is performed. In hazardous areas, fire-hazardous work may only be carried out if they cannot be performed at permanent welding locations or in environments which are not hazardous in fire conditions.

It is forbidden to carry out fire-risk works unless all fire-fighting measures have been fulfilled, no fire-extinguishing agents have been prepared, and a permit for temporary execution of these works has not been obtained, then if the equipment is defective, if there are freshly painted paint on the equipment and in the vicinity of machine structures (up to 20 m), surface or surface protection is performed. Also, work is prohibited when work clothes and gloves are soaked with flammable liquids and grease, when welding cables are damaged, poorly insulated, or not insulated at the joints, and also if their cross-section does not provide the permitted nominal current, and if the intersections of welding electric cables with pipelines (especially with flammable gases and liquids) at the intersection points, no additional insulation or height hang of those cables was made.

In the case of an accident, temporary welding and other fire-risk work must be carried out under the direct supervision of the shift manager or other responsible engineer of the facility designated by him. Obtaining a special permit in this case is not necessary because of the urgency, but all fire-fighting measures must be complied with in order to avoid fires, with continuous supervision by the appropriate supervisor.

Prohibition is the use of gas cylinders and other apparatus for welding and gas cutting, directly in explosive environments, then installations with cables, as well as their exposure to longer direct sunlight and also heating from other sources of heat. The following are also prohibited:

- the use of open flames to heat frozen pipelines, installations within a building and structure located more than 3 m away from flammable structures or fire-hazardous equipment;

- performing welding as well as other fire-hazardous works in buildings extracted from metal-coated structures with combustible polymer insulation on these structures;
- performing welding work on panels of metal-coated structures with combustible polymer insulation to dismantle or attach these panels or to cut openings for installation of other parts or appliances.

9. Result discussion

Assessing the current state and forecasting the behavior of a complex system, such as a TPP, in a form suitable for application to the selection of a maintenance system is most relevant by analyzing the failure flow as a function of exploitation time. The next step is to rank the cause of unforeseen setbacks and analyze the costs. The point of ranking according to the Pareto statistical method is to allocate 20% of causative agents, which account for 80% and more of plant failures. The statistics of the exploitation and maintenance parameter can define the causes of the greatest number of unforeseen delays (failures) from the domain of the power plant subsystem. A further step is to implement the results of the control (geometry, microstructure, etc.) of the most vital components of the plant, to determine the mechanism causing the failures, and to quantify the "age" of the plant by quantifying the parameters *PoF* (Probability of Failure)/*CoF* (Consequence of Failure), without which the trend of changes in the exploitation characteristics of the plant as a whole cannot be estimated. The *PoF*/*CoF* parameters allow the technical aspect of the competitiveness problem to be reduced to the narrowest extent of the causative agents, which have the highest risk of failure. This introduces the optimum types of maintenance for risk-based maintenance (RBM) and RCM facilities [1,50].

Optimization is primarily based on cost optimization, with maximum reliability (availability) by managing maintenance against the riskiest components. The reliability assessment is based on the calculation of the probability that the risk component will reach a given moment in time (under exploitation conditions), without failure. It is an analogous estimate of reaching the expected exploitation life on the basis of probability, only at shorter time intervals (up to the first predicted downtime due to overhaul, for example).

The complexity of the forecasting problem is that the law of change of the monitored parameter must be defined, which is adequate not only to the prehistory, but also to keep the adequacy at the forecast interval. This means that the mathematical apparatus for defining a successful model must treat the physical essence of the process, which governs the change in the tracked parameter. Otherwise, the prediction itself is meaningless. Once the reliability of the most critical components is determined, then, based on the scheme of the plant, the reliability is calculated on a given interval of the entire plant using a mathematical apparatus of reliability theory, based on statistics and probability theory. Then, analogous to the analysis carried out in the first approximation, a recalculation in the second approximation is performed that uses the results obtained by applying module activity for all components that cause a minimum of 80% of unforeseen downtime. In this way, all relevant data related to the current state of the plant's competitiveness is prepared and analyzed. In order to determine the way to raise the level of competitiveness, it is necessary to make a comparison with best practice, then to determine the optimal level of competitiveness in the technical, economic, and business sense (to be aspired) by comparison, with the accompanying development of programs for raising the level of competitiveness and control of relevant parameters during exploitation.

The optimization of competitiveness indicators is done precisely because of the changing business philosophy by introducing deregulation in the electricity market (in the field of the process industry the market conditions have always been deregulated): earlier the "*price = cost + profit*" was in force, while now the equality of "*profit = market price - cost*" is in force [1,3,52].

In cases where cascade failures and negative dependencies may have to be taken into account in reliability analyses based on reasonable assumptions about independent failures, then these dependencies can be clearly modeled using FTA, RBD, and the Markov model. For each reliability analysis, FMECA also provides a framework for identifying and investigating dependent failures. IEEE Std. 352 provides two additions to FMECA, which describe how to capture potential interdependencies between system components in the analysis, for common cause and cascade failures. In most cases, analysis based on the assumption of fault independence leads to unrealistic results and limited value for practical purposes, so in the last 40 years more suitable models have been developed that take into account various types of dependencies: the square root method, the β-factor model, and the binomial failure frequency model. All of these models are mainly used to model CCFs, with the β-factor model being the most widely used [50].

Reliability models based on the FMEA, FMECA, FFA, ETA, FTA, and RBD methods are based on the assumption that components or systems can be in one of two possible states: the functional (correct) state and the fault state. Since these models are quite static, they are not suitable enough for the analysis of repairable systems, so counting processes are used to model these problems. State-based methods, such as Markov or PN analysis, are more comprehensive and allow for clear modeling of complex relationships in a complex system, and hence modeling of system reliability dynamics. The literature also provides a breakdown of analytical stochastic models used for qualitative assessment into two main categories, namely, combinatorial models (which cannot fully describe the dynamic dependence between system components) and space state models (mainly based on Markov chains—Markov discrete-time chain, Markov continuous-time chain, Markov regenerative process, Markov gain process, semi-Markov process, which overcome all constraints of combinatorial models but can become too big and difficult to deal with) [99,100,101]. There is also a division of methodologies for reliability dynamics into three categories: state transitions or Markov models, continuous dynamic event trees, and direct system simulations (e.g., Monte Carlo simulation, discrete event simulations, etc.) [101]. It is also pointed out that the purpose of these methods is to complement the classical methodology when the dynamic behavior of complex systems needs to be taken into account.

Stochastic processes and simulation methods are powerful tools for reliability analysis, but also a rather complicated application in the analysis of the complexity of complex technical systems. Therefore, on the formalism of the FTA and RBD methods, which give a closer look at their structural diagrams, several researchers have proposed FTA and RBD methods for solving the reliability dynamics of complex systems (Vaurio [102] describes an FTA method for modeling systems consisting of repairable and irreparable components and working in phase missions during which system logic changes; Hurdle et al. [103] recommend the FTA method for identifying potential system component failures that takes into account its dynamics).

An alternative approach to analyzing a fault tree of independent repairable components based on binary decision diagrams (BDDs) is described by Rauzy [104], Sinnamon and Andrews [105], and Dutuit and Rauzy [106]. This method is an alternative to conventional FTA and ETA techniques to perform qualitative and quantitative analysis, with this method not directly analyzing the fault tree but converting the tree to a BDD [107]. The BDD approach involves working directly with logical expressions, instead of working with minimally reduced sets. Thus, the BBD method does not directly analyze the fault tree, but converts it to a diagram in the form of a graphical representation of the Boolean algebraic expression for the main event in a bottom-up approach. Also, the BBD algorithm requires a large amount of memory, so it is not applicable to large fault trees (possible application of the method to reduce the error in the procedure of truncation of minimal reduced sets with a low probability of occurrence, as errors in approximating the quantification of rejected minimum reduced sets).

Reliability-oriented maintenance also alters the reliability of the system in a general sense, thereby creating the assumptions for applying a fault-frequency calculation method, called the condition-based fault tree analysis (CBFTA) [108]. This method is explained as an extension of the use of the FTA method after the system design phase, which makes it useable during the remaining stages of the system life cycle. The CBFTA method combines breakdown statistics used during design and condition monitoring data, with the resulting data being further used to upgrade the fault intensity values of sensitive components.

In recent years, several new methods have been proposed for reliability and risk assessment that combine the good properties of combinatorial and state space methods (dynamic reliability block diagram (DRBD); dynamic fault tree (DFT), Markov process driven by Boolean logic, Boolean logic driven Markov process (BDMP abbreviation), and Stochastic Petri net (SPN)). Based on DFT and DET approaches, Distefano and Puliafito [109,110] and Distefano [111] propose a new methodology for dependency modeling in complex systems reliability analysis based on block reliability diagrams. The method is called dynamic block diagrams of reliability or DRBD. The main advantage of this model is the ability to model dependencies between subsystems and components with respect to their impact on reliability. Quantitative analysis of the DRBD model can be performed with existing state space techniques (e.g., Markov chains and Petri nets, and simulation methods).

Finally, it should be noted that failures are an integral part of the normal operation of the turbine plant and its auxiliary equipment and as such must be further analyzed. Most equipment failures develop gradually, and the time until the cause of the failure occurs depends on a number of causes (turbine operating parameters, operating conditions, lubrication, material, type of load, stress, etc.). The analysis of all data is intended to make a decision on the process of carrying out certain activities and actions, as well as improving the maintenance, and ultimately how to reduce the number of failures in the equipment and the plant downtime, which is directly related to plant safety, maintenance costs, and overall business. Attention should also be paid to the importance of training staff to monitor and diagnose turbine plants and associated equipment. It is necessary for the management of the plants to understand the importance and the need to train operational personnel for performing professional activities in operation, primarily repair, maintenance, and servicing of the related equipment, in accordance with the applicable legal regulations and instructions given by the equipment manufacturer [98].

Modern concepts of steam turbine maintenance, implemented with the use of modern equipment for technical diagnostics and monitoring, in accordance with the instructions from the normative-technical documentation and instructions prescribed by the manufacturer, as well as the adopted regulations by the users of the equipment, ensure the safe and reliable operation of the turbine plant and its accessories. The number, then the period of realization, as well as the scope of activities that must be performed during the steam turbine overhaul are closely related to the location and role of the associated power plant within the higher hierarchical ES.

10. Conclusion

Reliability, as the likelihood that a complex technical system will fulfill the required function over a period of time and under certain conditions, has four important factors: probability, required function, time periods, and operating conditions. Since the entire system of tasks and direct paths for achieving optimum reliability of a particular plant is still not formulated and elaborated in certain fields of technology (especially in the energy-processing industry), in practice it is commonly used. The consequence principle is through the elimination or enhancement of the performance of "weaknesses" at all stages of the life cycle of the facility itself. In doing so, the results obtained from the qualitative and quantitative analysis, that is, the experience gained in achieving reliability at all stages of the life cycle of a technical (usually energy or process) plant, are used as a basis. Knowing the basic reliability characteristics, on the basis of which the occurrence of failures is constantly predicted in one go, the forecasts of the future states of the system are made, on the basis of which decisions are made about the necessary preventive maintenance procedures and timing of their implementation, in order to prevent the accumulation of damage and sudden occurrence, cancellations, that is, unplanned downtime, additional costs, or major accidents. The optimal management of complex technical systems must be based on the evaluation and complex optimization of the reliability indicators, depending on the means of providing them and the hierarchical level of detail as well as the current life cycle phase. For this reason, the optimization process includes basic structural, parametric, and constructive solutions related to the technical system itself by changing its most important characteristics: efficiency (most often energy), maneuverability, reliability, and economic efficiency as a whole. The set of optimization goals is concluded in the overall choice of reliability indicators and possible ways to achieve them, given the already established rules related to the higher hierarchical level of the system. Only on the basis of timely assessments can corrective action be taken in order to further develop failures and prevent major damage. In doing so, the actual maintenance costs are minimally possible for each specific situation. The dependence of the cost of generating electricity on a TPP on the level of reliability needs to be considered from two aspects: the TPP and the user. In both cases, the point of minimum cost determines the optimal reliability of both the TPP and the user. An analysis of the available literature that treats dynamic reliability models has shown that they represent powerful tools primarily for information systems and computer technology. However, their possible application in the analysis of the reliability and risk of technical systems in various industries for the time being remains on their own hypotheses, since the results of their research are not compared with the results of the classical methods recommended standards, and therefore it is very difficult to assess their reliability.

References

[1] L. Papić, Z. Milovanović, Maintenance and reliability of technical systems, in: DQM Monograph Library Quality and Reliability in Practice, Book 3, Prijevor, 2007, p. 501 (in Serbian).

[2] Z. Milovanović, Optimization of Power Plant Reliability, University of Banja Luka, Faculty of Mechanical Engineering Banja Luka, Banja Luka, 2003, p. 300 (in Serbian).

[3] Z. Milovanović, Modified Method for Reliability Evaluation of Condensation Thermal Electric Power Plant (Ph.D. thesis), University of Banja Luka, Faculty of Mechanical Engineering Banja Luka, Banja Luka, 2000 (in Serbian).

[4] M. Rausand, A. Høyland, System Reliability Theory: Models, Statistical Methods and Applications, second ed., Wiley, Marvin, 2004.

[5] E.V. Lewis, Principles of Naval Architecture, vol. 3, Motions in Waves and Controllability, Jersey City, NJ, 1989. SNAME.

[6] T. Hashimoto, K. Murayama, K. Hiejima, Aspect and Consideration on Occurring and Detecting of Maintenance Works Belonged to Marine Engine Department, 1997. ICMES.

[7] Reliability Centered Maintenance, 1990. IEC Draft 56 (Sec.), 317.

[8] Z. Milovanović, L. Papić, S. Dumonjić-Milovanović, A. Milašinović, D. Knežević, Sustainable Energy Planning: Technologies and Energy Efficiency, DQM Monograph Library Quality and Reliability in Practice, Book 9, Prijevor, 2017 (in Serbian).

[9] B. Blanchard, W. Fabrysky, System Engineering and Analysis, Prentice Hall Inc., New Jersey, 1981.

[10] V. Benard, L. Cauffriez, D. Renaux, The safe-SADT method for aiding designers to choose and improve dependable architectures for complex automated systems, Reliability Engineering and System Safety, Elsevier Ltd. 93 (2) (2008) 179−196.

[11] B.S. Dhillon, Maintainability, Maintenance, and Reliability for Engineers, CRC Press LLC, Boca Raton, 2006.

[12] S. Benardi, Dependability Analysis Techniques, Book Model-Driven Dependability Assessment of Software Systems, Springer-Verlag, Berlin, 2013 (Chapter 6).

[13] D. Phuc, S. Phil, I. Benoit, Condition-based maintenance for a two-component system with dependencies, 9th IFAC symposium on fault detection, supervision and safety for technical processes, Safe Process (2015) 946−951, https://doi.org/10.1016/j.ifacol.2015.09.648. Paris.

[14] A. Rosmaini, K. Shahrul, An overview of time-based and condition-based maintenance in industrial application, J. Comput. Ind. Eng. 63 (1) (2012) 135−149.

[15] M. Sondalini, How to Use Condition Based Maintenance Strategy for Equipment Failure Prevention, Lifetime Reliability Solutions HQ, 2009. http://www.lifetime-reliability.com.

[16] A. Prajapati, S. Ganesan, Application of statistical techniques and neural networks in condition-based maintenance, J. Qual. Reliabil. Eng. 29 (3) (2013) 439−461.

[17] H. Asada, J.-J.E. Slotine, Robot Analysis and Control, John Wiley and Sons, Toronto, 1986.

[18] J. Čačko, M. Bily, J. Bukoveczky, Random Processes: Measurement, Analysis and Simulation, Elsevier, New York, 1986.

[19] C. Cempel, Passive diagnostics and reliability experiment: application in machine condition monitoring, ASME J. Vib. Acoust. Stress Reliab. Des. 111 (1) (1989) 82−87. https://doi.org/10.1115/1.3269828.

[20] S.B. Stancliff, Planning to Fail: Incorporating Reliability into Design and Mission Planning for Mobile Robots (Ph.D. Thesis), Robotics Institute, Carnegie Mellon University, Pittsburgh, Pennsylvania, 2009. Tech. Report, CMU-RI-TR-09-38.

[21] M. Xiaojiang, Y. Jingxia, L. Peide, Dynamical tests of a bearing assembly and nonlinear analysis, in: Proceedings of 8th International Modal Analysis Conference, Kissimmee, FL, 1990, pp. 685−689.

[22] L. Youfang, W. Bin, Modal analysis problems in the non-uniform inertial field, in: Proceedings of 8th International Modal Analysis Conference, Kissimmee, FL, 1990, pp. 784−787.

[23] D. Mikić, Modeling of mechanical technical systems by using matrix of transformation (Ph.D. Thesis), University of Novi Sad Technical Faculty, Mihajlo Pupin, Zrenjanin, 2016, p. 224 (in Serbian).

[24] D. Bansal, D. Evans, B. Jones, A real-time predictive maintenance system for machine systems - an alternative to expensive motion sensing technology, in: Proceedings of the ISA/IEEE 2005, Sensors for Industry Conference, Houston, Texas, USA, February 8−10, 2005.

[25] C. Lin, V. Makis, Optimal Bayesian maintenance policy and early fault detection for a gearbox operating under varying load, J. Vib. Contr. 22 (15) (2014) 3312−3325.

[26] E. Henley, H. Kumamoto, Reliability Engineering and Risk Assessment, Prentice-Hall, New York, USA, 1981.

[27] P. Funk, M. Jackson, Experience based diagnostic and condition based maintenance within production systems, in: M. David (Ed.), The 18th International Congress and Exhibition on Condition Monitoring and Diagnostic Engineering Management, COMADEM, August 2005. United Kingdom.

[28] D. Japikse, W.D. Marscher, R.B. Furst, Centrifugal Pump Designand and Performance, Concepts ETI, Inc., Wilder, VT, 1997.

[29] D. Smith, S.M. Price, F.K. Kunz, Centrifugal pump vibration caused by supersynchronous shaft instabuility, use of pumpout vanes to increase pump stability, in: 13th International Pump Users Symposium, Tamu, Texas, 1996.

[30] H. Black, Effects of fluid-filled clearance spaces on centrifugal pump and submerged motor vibrations, in: Proceedings of the Eight International Turbomachinery Symposium, TAMU, Texas, 1979, pp. 29–38.

[31] W. Hancock, How to control pump vibration, Hydrocarb. Process. 53 (3) (1974) 107–113.

[32] M. Kutin, Optimization of the Application of Diagnostic Techniques and Their Influence on the Reliability of Technical Systems (Ph.D. thesis), Mihajlo Pupin Faculty, University of Novi Sad, Zrenjanin, 2010 (in Serbian).

[33] P. Popović, G. Ivanović, Reliability Design of Machine Systems, Monograph, Vinča Institute of Nuclear Sciences, Beograd, 2005 (in Serbian).

[34] M. Bulatović, Maintenance and Efectiveness of Technical Systems, University of Montenegro, Faculty of Mechanical Engineering Podgorica, Podgorica, 2008 (in Serbian).

[35] S. Šćepanović, L. Vujović, J. Vujović, Developmental maintenance of the individual, in: Proceeding of the OMO 2014, Beograd, 2014 (in Serbian).

[36] Ž. Adamović, L. Radovanović, Models of Maintenance Based on Technical Diagnostics, Mihajlo Pupin Faculty of Engineering, Zrenjanin, 2008 (in Serbian).

[37] D. Kostić, Data-Driven Robot Motion Control Design, Technische Universiteit Eindhoven, Proefschrift, Eindhoven, Netherlands, 2004.

[38] Ž. Adamović, M. Gavrić, Ž. Grbavac, Functional Safety of Technical Systems, Academy of Maintenance Engineering, Beograd, 2009 (in Serbian).

[39] A. Ašonja, D. Mikić, E. Desnica, Ž. Adamović, Roller Bearing Damage, Proceedings of the 36th conference may meeting of maintainers of Serbia “measurement of maintenance performance indicators”, in: Society for Technical Diagnostics of Serbia, No. 4, pp. 29–37, Vrnjačka Banja, May 2013 (in Serbian).

[40] A. Ašonja, Use of Software Systems for Determining the Service Life of Rolling Bearings During Use (Master’s thesis), Faculty of Agriculture Novi Sad, Novi Sad, 2006 (in Serbian).

[41] R. Gligorić, A. Ašonja, THEDIS, Smederevo, Balancing and Vibration Problems of Mechanisms, Machine Maintenance, 5, 2005, pp. 52–56 (in Serbian).

[42] N. Grujić, M. Simonović, D. Grujić, Risk management using technical diagnostics and reliability methods, Tech. Diagn. 12 (2) (2013) 7–13 (in Serbian).

[43] V. Krunić, M. Krunić, N. Četić, Information systems in industrial production with support for preventive maintenance and technical diagnostics, Tech. Diagn. 12 (2) (2013) 19–25 (in Serbian).

[44] P. Popović, G. Ivanović, R. Mitrović, A. Subić, Design for reliability of a vehicle transmission system, ISSN 0954-4070, Proc. Inst. Mech. Eng. D J. Automob. Eng. (2011) 194–209, https://doi.org/10.1177/0954407011416175. SCI-M23, Suffolk, United Kingdom.

[45] Ž. Adamović, L. Paunović, K. Paunović, Reliability of Hydraulic Systems, Academy of Maintenance Engineering, Beograd, 2007.

[46] D. Milošević, Reliability Ensuring Models of Complex Facilities in Thermal Power Plants (Ph.D. thesis), Mihajlo Pupin Technical faculty, Zrenjanin, 2015 (in Serbian).

[47] Đ. Dobrota, Modeling the Distribution of Prior in Failure Analysis of Marine Hydraulic Equipment (Ph.D. thesis), University of Split, Faculty of Electrical Engineering, Mechanical Engineering and Naval Architecture, Split, 2018 (in Croatian).
[48] D. Branković, Reliability Optimization of the Production System for Hygiene Paper Manufacturing by Using the Concept of Condition Based Maintenance (Ph.D. thesis), University of Banja Luka, Faculty of Mechanical Engineering, Banja Luka, 2018 (in Serbian).
[49] M. Rausand, A. Høyland, System Reliability Theory, Models, Statistical Methods, and Applications, second ed., John Wiley & Sons, Inc, New Jersey, 2004.
[50] Đ. Dobrota, Qualitative Analysis in the Reliability Assessment of Ancillary Marine Systems, Qualifying Doctoral Exam, University of Split, Faculty of Electrical Engineering, Mechanical Engineering and Naval Architecture, Split, 2014 (in Croatian).
[51] D. Kececioglu, Reliability Engineering Handbook, vol. 1, DESteh Publications, Inc., Lankaster, Pennsylvania, USA, 2002.
[52] D. Miličić, Z. Milovanović, Monograph of Energy Machines - Steam Turbines., University of Banja Luka, Faculty of Mechanical Engineering Banja Luka, Banja Luka, 2010 (in Serbian).
[53] Z. Milovanović, Monographs: Energy and Process Plants, Tom 1: Thermal Power Plants - Theoretical Foundations, University of Banja Luka, Faculty of Mechanical Engineering Banja Luka, Banja Luka, 2011 (in Serbian).
[54] Z. Milovanović, Monographs: Energy and Process Plants, Tom 2: Thermal Power Plants - Technological Systems, Design and Construction, Exploitation and Maintenance, University of Banja Luka, Faculty of Mechanical Engineering Banja Luka, Banja Luka, 2011 (in Serbian).
[55] BS 4778-3.1:1991, Quality Vocabulary. Availability, reliability and maintainability terms. Guide to Concepts and Related Definitions, 1991, pp. 1−7.
[56] Z. Milovanović, D. Miličić, Steam Turbines for Cogeneration Energy Production, Library of Monographs, Energy Machines 3, University of Banja Luka, Faculty of Mechanical Engineering Banja Luka, Banja Luka, 2012 (in Serbian).
[57] W.R. Blische, D.N. Prabhakar Murty, Case Studies in Reliability and Maintenance, John Wiley & Sons, Inc., New Jersey, 2003.
[58] G.V. Nozdrenko, V.G. Tomlov, O.K. Grigoryeva, Reliability of Thermal Power Plants, NSTU Publishing House, Novosibirsk, 2009 (in Russian).
[59] G. Pahl, W. Beitz, J. Feldhusen, K.H. Grote, Engineering Design - A Systematic Approach, Springer-Verlag Ltd., London, 2007.
[60] MIL-STD-882E, System Safety Standard Practice, Department of Defense, U.S., 2012.
[61] BS EN ISO 9000:2000, Quality Management Systems - Fundamentals and Vocabulary, European Committee for Standardization (CEN), 2000.
[62] M. Lambert, B. Riera, G. Martel, Application of functional analysis techniques to supervisory systems, Reliab. Eng. Syst. Saf. 64 (2) (1999) 209−224. Elsevier Ltd.
[63] MIL-STD-1629A, Procedures for Performning a Failure Mode Effects and Criticaly Analysis, Department of Defense, U.S., 1980.
[64] M. Rausand, Risk Assessment Theory, Methods, and Applications, John Wiley & Sons, Inc., New Jersey, 2011.
[65] A. Pillay, J. Wang, Technology and Safety of Marine Systems, Elsevier Ltd., Oxford, 2003.
[66] J. Moubray, Reliability Centred Maintenance, Butterworth-Heinemann, Oxford, 1997.
[67] M. Lind, Modeling goals and functions of complex industrial plants, J. Appl. Artif. Intell. 8 (2) (1994) 259−283.
[68] S.M. Ross, Introduction to Probability Models, ninth ed., Elsevier Inc., Oxford, 2007.
[69] J.E. Larsson, Diagnosis based on explicit means-end models, Artif. Intell. 80 (1) (1996) 28−93. Elsevier Ltd.

[70] A. Jalashgar, Identification of Hidden Failures in Process Control Systems through Function Oriented System Analysis (Ph.D. thesis), Risø-R 936(EN), Risø National Laboratory, Roskilde, Denmark, 1997.

[71] J.E. Larsson, Knowledge engineering using Multilevel flow models, in: Proceedings of the 2nd International Symposium on Engineering of Intelligent Systems, Paisley, Scotland, 2000.

[72] J.E. Larsson, Support tools for situation assessment, in: Proceedings of the 3rd Seminar on Alarm Systems, Safety & Nuclear Division, IBC Global Conferences, London, 2002.

[73] B. Öhman, in: Proceedings of the 5th European Control Conference, Karlsruhe, Germany, 1999.

[74] M. Lind, Modeling Goals and Functions of Control and Safety Systems - Theoretical Foundations and Extensions of MFM, Nordic Nuclear Safety Research, 2005.

[75] M. Lind, Control functions in MFM: basic principles, Nucl. Saf. Simulat. 2 (2) (2011) 22−32.

[76] IEC 60050(191), International Electrotechnical Vocabulary (IEV) - Part 191: Dependability and Quality of Service, 1990.

[77] ISO 14224 (E), Petroleum, Petrochemical and Natural Gas Industries Collection and Exchange of Reliability and Maintenance Data for Equipment, second ed., 2006.

[78] D.J. Smith, Reliability, Maintainability and Risk-Practical Methods for Engineers, sixth ed., Butterworth-Heinemann, Oxford, 2001.

[79] A. Birolini, Reliability Engineering Theory and Practice, fifth ed., Springer - Verlag, Berlin, 2007.

[80] L.D. Frate, Failure of engineering artifacts: a life cycle approach, Sci. Eng. Ethics 19 (3) (2013) 913−944.

[81] A. Halep, Defect, failure and breakdown, in: Proceedings of the Maintenance 2016, Zenica, 2016, pp. 11−14 (in Bosnian).

[82] M. Modarres, M. Kaminskiy, V. Krivtsov, Reliability Engineering and Risk Analysis - A Practical Guide, Marcel Dekker Inc., New York, 1999.

[83] S.B. Tam, I. Gordon, Clarification of failure terminology by examining a generic failure development process, Int. J. Eng. Bus. Manag. 1 (1) (2009) 33−36.

[84] J.E. Larsson, On-line root cause analysis for nuclear power plant control rooms, in: Proceedings of the International Symposium on Symbiotic Nuclear Power Systems for the 21st Century, ISSNP, Tsuruga, Fukui, Japan, 2007.

[85] VDI 2056 Evaluation of Mechanical Vibrations of Rotating Machinery (with Drawn); Replacement: ISO 10816.

[86] NUREG/CR-6268, Rev 1, INL/EXT-07-12969, Common-Cause Failure Database and Analysis System: Event Data Collection, Classification, and Coding, Idaho National Laboratory, U.S. Nuclear Regulatory Commission Office of Nuclear Regulatory Research Washington, USA, 2007.

[87] Reactor Safety Study: Accident Definition and Use of Event Trees, U.S. Nuclear Regulatory Commission, 1971, pp. 5−210.

[88] B.S. Dhillon, Design Reliability: Fundamentals and Applications, CRC Press LLC, Boca Raton, 1999.

[89] TU 38.101.821-83, Turbine Oil Ti-22C. Technical Conditions TU 38.401.5848-5892. (in Russian).

[90] E. Ebling, An Introduction to Reliability and Maintainability Engineering, McGraw-Hill Companies Inc., USA, 1997.

[91] W.Q. Meeker, L.A. Escobar, Statistical Methods for Reliability Data, John Wiley&Sons Inc., New York, 1999.

[92] B.S. Dhillon, Engineering Maintenance: A Modern Approach, CRC Press, 2002.

[93] E. Zio, An Introduction to the Basics of Reliability and Risk Analysis, World Scientific Publishing Co. Re. Ltd., Danvers, 2007.

[94] M.S. Hamada, A.G. Wilson, C.S. Reese, H.F. Martz, Bayesian Reliability, Springer Science, New York, 2008.

[95] M.P. Kaminskiy, Reliability Models for Engineers and Scientists, CRC Press LLC, Boca Raton, 2013.

[96] S.M. Ross, Introduction to Probability Models, ninth ed., Elsevier Inc., Oxford, 2007.

[97] J. Barle, D. Ban, Maritime Component Reliability Assessment and Maintenance Using Bayesian Framework and Generic Data, Advanced Ship Design for Pollution Prevention, Taylor & Francis Group, London, 2010, pp. 181–188.

[98] S. Distefano, A. Puliafito, Reliability and availability analysis of dependent - dynamic systems with DRBDs, Reliab. Eng. Syst. Saf. 94 (9) (2009) 1381–1393. Elsevier Ltd.

[99] F. Chiacchio, L. Compagno, D. D'Urso, G. Manno, N. Trapani, An open-source application to model and solve dynamic fault tree of real industrial systems, in: IEEE Proceedings of the 5th International Conference on Software, Knowledge Information, Industrial Management and Application (SKIMA), Benevento, September 2011, pp. 8–11. ISBN: 9781467302487.

[100] F. Chiacchio, L. Compagno, D. D'Urso, G. Manno, N. Trapani, Dynamic fault trees resolution: a conscious trade-off between analytical and simulative approaches, Reliab. Eng. Syst. Saf. 96 (11) (2011) 1515–1526. Elsevier Ltd.

[101] M. Marseguerra, E. Zio, J. Devooght, P.E. Labeau, A concept paper on dynamic reliability via Monte Carlo simulation, Math. Comput. Simulat. 47 (2) (1998) 371–382. Elsevier Ltd.

[102] J.K. Vaurio, fault tree analysis of phased mission systems with repairable and non-repairable components, Reliab. Eng. Syst. Saf. 74 (2) (2001) 169–180, https://doi.org/10.1016/s0951-8320(01)00075-8. Elsevier Ltd.

[103] E.E. Hurdle, L.M. Bartlett, J.D. Andrews, Fault diagnostics of dynamic system operation using a fault tree based method, Reliab. Eng. Syst. Saf. 94 (9) (2009) 1371–1380. Elsevier Ltd.

[104] A. Rauzy, New algorithms for fault trees analysis, Reliab. Eng. Syst. Saf. 40 (3) (1993) 203–211. Elsevier Ltd.

[105] R.M. Sinnamon, J.D. Andrews, New approaches to evaluating fault trees, Reliab. Eng. Syst. Saf. 58 (2) (1997) 89–96. Elsevier Ltd.

[106] Y. Dutuit, A. Rauzy, Approximate estimation of system reliability via fault trees, Reliab. Eng. Syst. Saf. 87 (2) (2005) 163–172. Elsevier Ltd.

[107] K.A. Reay, J.D. Andrews, a fault tree analysis strategy using binary decision diagrams, Reliab. Eng. Syst. Saf. 78 (1) (2002) 45–56. Elsevier Ltd.

[108] D.M. Shaleva, J. Tiran, Condition-based fault tree analysis (CBFTA): a new method for improved fault tree analysis (FTA), reliability and safety calculations, Reliab. Eng. Syst. Saf. 92 (9) (2007) 1231–1241. Elsevier Ltd.

[109] S. Distefano, A. Puliafito, Dynamic Reliability Block Diagrams: Overview of a Methodology, Safety and Reliability Conference (ESREL07), Stavanger, Norway, June 2007, pp. 1059–1068. ISBN 978-0-415-44786-7.

[110] S. Distefano, A. Puliafito, Dependability evaluation with dynamic reliability block diagrams and dynamic fault trees, IEEE Trans. Dependable Secure Comput. 6 (1) (2009) 4–17.

[111] S. Distefano, Dependability of complex, large, dynamic systems, in: Proceedings of the 8th International Conference on Reliability, Maintainability and Safety (ICRMS 2009), IEEE Chengdu, China, July 2009, pp. 27–31, https://doi.org/10.1109/ICRMS.2009.5270247.

[112] OREDA, Offshore Reliability - Data Handbook, fourth ed., Sintef Industrial Management, Trondheim, Norway, 2002.

[113] MIL-HDBK 217F:1996, Reliability Prediction of Electronic Equipment, Department of Defense, Washington, DC, 1996.

CHAPTER

9 Methods of risk modeling in a thermal power plant

Zdravko N. Milovanović[1], Ljubiša R. Papić[2], Snježana Z. Milovanović[3], Valentina Z. Janičić Milovanović[4], Svetlana R. Dumonjić-Milovanović[5], Dejan Lj. Branković[1]

[1]*Department of Hydro and Thermal Engineering, University of Banja Luka, Faculty of Mechanical Engineering, Banja Luka, Republic of Srpska, Bosnia and Herzegovina;* [2]*DQM Research Center, Čačak, Serbia;* [3]*Department of Materials and Structures, University of Banja Luka, Faculty of Architecture, Civil Engineering and Geodesy, Banja Luka, Republic of Srpska, Bosnia and Herzegovina;* [4]*Routing Ltd., Banja Luka, Republic of Srpska, Bosnia and Herzegovina;* [5]*Partner Engineering Ltd., Banja Luka, Republic of Srpska, Bosnia and Herzegovina*

1. Introduction

The safety of technical systems can be considered from two aspects: the protection of the operator (person) from injuries during the operation of the system and the protection of the system from damage caused by external causes. Preference is given to operator safety. In doing so, these two aspects are not unconditionally complementary, and an increase in operator safety can be achieved at the expense of system security. The need to reduce the cost of exploitation of the technical system, while achieving the required level of reliability, also requires the continuous development of reliability engineering. The risk involved should be distinguished, which implies the continuation of the activity, regardless of the possibility of the hazard occurring. Any technical system, even if it performs the function of the target within the tolerance limits, can be damaged if it is incorrectly handled. The main causes of operator risk are: engaging body parts such as hands in the process of system operation; carelessness in the operation of the rotational parts of the system (especially the weakly attached units); contact with sharp and abrasive surfaces; the effect of operator static on moving objects or vice versa; ejection of waste material (especially in production) in the form of sawdust, shavings, sparks or molten metal, and the like. The causes of system risks are diverse and numerous, and in the design phase, the consequences of critical types of failures must be minimized through the prediction of protective devices during system operation. The risks of technical systems include: shocks, vibrations, corrosion, environment, fire, mismanagement, etc. Their occurrence is related to all projects, as well as to all processes and decisions made within the project life cycle.

Scientific modeling is the process of generating the physical, conceptual, or mathematical phenomena that are actual, which are hard to observe directly [1]. They are used as explanations for prediction of the behavior of a real object or system. They are often shown as approximations of the objects and systems they represent, although they are not their exact replica, so additional engagement

The Handbook of Reliability, Maintenance, and System Safety through Mathematical Modeling.
https://doi.org/10.1016/B978-0-12-819582-6.00009-5

is required to further enhance them. Scientific modeling, depending on the stage of their development or complexity of the system which is the object of modeling, can result in an established physical, abstract, conceptual, graphical, or mathematical model. Mathematical models, as a segment of the scientific model, are mathematical picture of reality and represent a description of the object or system by applying particular mathematical concepts and certain characters. The process of forming mathematical models is the process of mathematical modeling. Although there are many definitions of mathematical modeling, the most commonly used is the following definition, "Mathematical modeling represents the implementation of mathematics in solving unstructured problems in real situations" [2]. Problems under consideration at the technical system are conceptually transformed into a mathematical problem and are solved using mathematical techniques [3]. It is important to note that mathematical models are often both a tool for optimization and the basis of control mechanisms of objects or systems. Numerical models are a segment of mathematical models. They use a specific numerical time-sampling procedure to obtain a picture of the behavior of the technical system over time. The paradox is that the numerical procedure of model creation can be carried out without knowing the basic properties of the system that is the object of modeling (structure, number of elements, and their interconnections). Application of tool for numeric modeling in predicting the behavior of the lesser known system may cause some errors in comparison with the real measurements based on the experiment (the rule is that the more the system that want to model is known, the better the results that the final model will outcome).

Each description of the system can be considered as a model of the system, which still allows a control of the effects of time and space, feature extraction, and therefore simplification, as the retention of only those details that are relevant for a given problem [4]. Most importantly, the need of the model reduces the need for real experiments and allows many different goals to be achieved with the reduced cost, risk, and time required for analysis. The model can be a mathematical or statistical description of a particular aspect of the process. The model, on the other hand, may take the form of a qualitative description of the process behavior.

From the point of view of previous research and more general classification, models can generally be classified into two modeling classes: the first class of symbolic, in most cases numerical, models and the second class of real (physical, material) models [5].

The development of models of different technological processes for optimization has resulted in the development of tools for optimization of technical-technological, economic and, more recently, environmental aspects in the considered process. Also, a large number of modeling applications deal with the analysis of the greenhouse effect and climate change, as well as the consequences of the action of process technologies. This is especially important in complex thermal power plants [6]. One of the initial works with the problems of modeling and computer simulation is the work "Modeling and Simulation" [7]. This paper provides an overview of various aspects and the problem of modeling and simulation techniques applied in the development of microprocessors, which also influenced the expansion in the field of research in this area. The various aspects show the relationships between the technological parameters of the process, the physical effects, and the performance of the final products themselves. They also describe the applicability of different simulation models from the perspective of different levels of model complexity (the level of computer technology development at the time limited the models to include a smaller number of influential parameters in the final models).

In the early 21st century, areas of interest became renewable and alternative sources of energy, and works involving modeling in this area appeared. Mathematical modeling of biomass fuel formation processes is addressed by Gaska and Wandrasz [8]. The analysis of the biomass fuel extraction process based on a detailed technical, environmental, and economic analysis, with the requirements defined to optimize the blending process of the fuel components, required the formation of a suitable mathematical model of the process. The model formed, in the form of structured databases, uniquely identifies the real process and converts given data into an algorithm based on a linear programming problem. This further enables the optimization of the parameters of the fuel formation process, using a modified simplex algorithm with polynomial time.

Modern computational techniques have been used to develop numerical models of technological processes in the last 10 years. Modeling and analysis of process value chains for the development of new technologies are also addressed by Großmann et al. [9]. They analyze the cause–effect relationship between material properties and production conditions, in terms of the required quality of the final products. Finite element method (FEM) was used to analyze and define the basics for process modeling. Based on the experimentally obtained results from the process of mechanical expansion of thin-walled tubes, a numerical model was formed to identify material properties using an inverse method and then to study technological of the tube expansion process using a simplified description sufficiently accurate to the practical model applicable in practice [10]. In Serbia, as part of the consideration of the operation of TPPs using lignite for fuel, using computational fluid dynamics (CFD) model, sensitivity of kinetic factors and possibilities of their use for numerical modeling of the process of removal of volatile components of Serbian lignite were analyzed [11]. Also, there is an increasing presence in the literature of developed modeling methodologies in the field of ecology and sustainable development. Thus, methods of selecting adequate models in ecology are addressed by Johnson and Omland [12].

The wide applicability of different modeling tools in different fields of modern science and technology is evident, with the presence and diversification of potential modeling tools and techniques applied by the authors to the various technological processes that are the object of their study. Selection of particular models for specific technical systems and fields of application is usually made by simply selecting one of the modeling methods used by other researchers to model similar systems.

2. Basic concepts, definitions, and risk sharing

All the definitions of risk used in different disciplines today are based on the concept that risk depends on uncertainty about possible events and their effects, whereby to view the risk the approach may be one-sided or viewed in a broader context. In most definitions, risk is reduced to the likelihood of loss or negative scenarios. Engineering definition sees the risk as the product of the likelihood of an adverse event occurring and the estimated expected damage should an event occur [13]. Definitions of risk from an organizational perspective treat risk as a threat to an organization that reduces the likelihood of achieving one or more of its goals [14]. As risk represents the possibility of occurrence of positive or negative deviations from the expected results, the situation when there is a probability of a negative deviation (where the probability of positive deviations is equal to zero) is still marked as risky. In the case where there is only a probability that a positive deviation will occur while the probability of a negative deviation is zero, then we cannot speak of risk [15]. Any approach that focuses solely on minimizing risk exposure at the same time also reduces the possibility of offsetting the positive effects of risk. Risk-taking capacity is the only way to improve entrepreneurial performance [16]. Risk elimination stifles the sources of value creation and growth potential of Knight and Petty [17]. The definitions of risk in the last few years have come from organizations that work toward standardizing terminology and risk management processes.

An aspect is an element, activity, product, or service of a company that has or can have an impact on the environment and the work environment.

Accident, in this case, is defined as an unplanned event, leading to environmental and occupational pollution or work-related illnesses and injuries.

Risk, as an indicator of determining the degree of danger, is a combination of the likelihood of an accident (in this case) in the environment and the work environment and the consequences of an accident in that environment. The risk has two different aspects: quantitative (calculated on the basis of the known probability of occurrence of events and consequences) and qualitative (related to human perception, i.e., depends on the emotional state of the person). Risk can be characterized as a complex magnitude that simultaneously describes the likelihood of adverse events occurring and the expected magnitude of the consequences of such events in a rounded system and over a fixed length of time interval, or during a particular process [18]. Very important risk concepts are risk perception and risk preference. This applies to people and groups of people who are willing to take risk in collective decision-making. Many authors describe the term risk as a possibility that a certain random event will have consequences on. The concept of risk can be subdivided into various elements, which makes it possible to adjust the general definition to specific risk areas, including the logistics systems of Kaplan and Garrick [19]. They view risk with respect to the three dimensions that they generate, three essential questions: What can happen? How often will it happen? And if it does, what will be the consequences? The answer to these questions is called a triplet.

The term natural disaster refers to an event caused by the operation of natural forces, which generally result in big number of individual losses, including a large number of insurance policies. The magnitude of damage (as a result of disasters) depends not only on the strength of natural forces but also on other factors (construction, design, or effectiveness of disaster control in that region) and leads to financial, environmental, and human losses. In this case, the natural disasters are divided into specific categories, such as: floods, storms, earthquakes, droughts/forest fires/heat waves, cold waves/ frost, hail, tsunami, etc.

Management is a set of actions (planning, organizing, providing, managing, and controlling) that manage the processes of the business system in order to achieve the set goals, while using management rules and certain procedures to identify, conduct analysis, provide risk assessment or estimation, as well as its treatment and monitoring or reporting, it all represent risk management. Of particular importance is the application of risk management to the project, with all factors involved (technical, market, political, economic, legal, etc.).

There are different types of risks, and one of the general divisions is:

- sources of risk: technological risk, human-made risk, which is the source of human economic activity (industrial incidents, nuclear incidents, transport incidents, cyberattacks, terrorist attacks, public riots, contamination of food, water and land, atomic biological and chemical attacks, refugees, etc.), and the natural risk associated with the impact of natural phenomena (floods, bad weather, pandemics/epidemics, forest fires, and earthquakes);
- the level of risk impact: local and global;

- frequency of exposure: constant risk (risk of exposure is constant), periodic (risk from time to time), and one-off (risk arising from an unusual situation);
- people's perception: voluntary risk (for staff working in a hazardous manufacturing facility) and mandatory (for the population living near a hazardous manufacturing facility);
- in relation to the spheres of human activity: commercial, social, political, technological, and environmental risks;
- the nature of the damage caused: economic, environmental, and social risks;
- level in acceptability: negligible, acceptable, maximum permissible, excessive;
- the maximal part of the risk that influences concrete company (according to the management mode): financial, operational, and strategic.

There are specific risks in the business system, so one of the divisions of business risks is:

- strategic risks (most often related to the strategy of the company, but also personal, which can have serious consequences, such as the collapse of the company or the personal bankruptcy);
- business—financial risks (influence the work and finance, for example, market and credit);
- program—project risks (poor defined purpose, scope of work, terms, costs, etc., and too many projects, subcontractors, poor communication, lack of support from senior management, etc.);
- operational risks (fault management, computer error, human error, a mistake in the process and procedures, etc.);
- technical risks (new products, new technologies, design error, material, etc.);
- external risks (political, local community, shareholder influence);
- risks of environmental protection (permits, changes in environmental conditions, lack of experts on an issue of protection, local community);
- organizational risks (loss of significant staff, additional staffing needs, pending approval, lack of time for quality planning, change of priorities, incomprehensibility in procedures or responsibilities).

One of the divisions of risk in the business world is the division of risks given in Table 9.1.

Insurers and global corporations have an interest in better studying macro-catastrophic risks in order to effectively diversify and manage risks. The following macro-catastrophic risks can be identified:

(a) Rising risks—cause potentially extreme losses that become more apparent or significant than before, either because the threat itself is growing or because society is increasing its vulnerability to that risk (cyber risks, climate change, pandemics originating from laboratory conditions);

(b) Cascading or associated risks—one type of hazard leads to events of another type, which causes more extreme events (a large earthquake triggers a tsunami which then leads to a nuclear incident);

(c) Networked risks—an event that causes losses in several lines of business or causes losses in unexpected places or in multiple geographic markets due to the interconnectedness of business connections (floods in Republic of Srpska caused losses through disconnection in the B&H market);

(d) Systemic risk (exogenous and endogenous risks)—systemic risk term is often used in the management of financial risks and significance of the event which may cause consequent effect throughout the financial system (real estate price bubble);

Table 9.1 One of the divisions of risk in the business world.

Division criterion	Classification and brief description	
Concerning the country (country risks)	Country risks include, for example, political, social, economic, local transport, infrastructure, administration, import of equipment, local suppliers and production, insurance, and the like	
Technical risks	Technical risks include: Borders and scope of work, innovation, design defects, the condition of the existing equipment, technical specifications, technical interfaces, climatic conditions, building site (access, ground), environmental protection, installation, commissioning, service life of element, assumed characteristic estimates (parameters) of availability (wear) through time, the beginning of the warranty period (commercial operation date—COD), acceptance conditions (the acceptance certificate (Provisional Acceptance Certificate—PAC)), etc.	Risks: - which directly affect the ability of the product to corresponds to declared in the form of feasibility, operational reliability and performance guaranty, - that are relevant to the shape and definition of the product, - in which the essential is WHAT we do, - offering something that cannot be achieved, - where consciously (intentionally) product, performance, or service is "decorated" (misrepresentation).
Industrial risks	Industrial risks include manufacturing process, materials (quality and availability), capacities (bottlenecks), subcontracting, subcontractors, technology transfer, ability to purchase spare parts over the life cycle, etc. These are risks that affect parts of tasks and internal processes, especially production, but which are not relevant to the shape and definition of the product. These risks do not matter what is done, but HOW it works	
Planning risks	Planned risks are the risks of delivery time, project planning, available resources, critical dates, testing and commissioning, documentation and customer acceptance, delay of partner/subcontractor/customer, and other risks related to the implementation of the plan	
Concerning pricing (pricing risks)	Risks related to pricing are risks related to exchange rate risk, inflation, cost estimation accuracy, pricing and validity of the offer (contract), supplier's offer, and the like	
Financial risks	Financial risks are risks of payment terms, customer credibility, billing insurance, financing, etc.	
Fiscal and taxation risks	Tax risks are import taxes and duties, double taxation, the stability of the tax rules (laws) giving on salaries, etc. It is necessary to look at the overall problem on this point: our part, the subcontractor's part, the customer's obligations, the material, and services divisions	
Contractual/legal risks	Contractual legal risks cover the area of duties and responsibilities, penalties, termination of contract, indirect and/or consequential damages, acceptance and takeover, warranty period, applicable law, arbitration, and other possible legally consequential risks	
Customer/subsupplier risk	Buyer—subsupplier risks cover the ownership structure of the buyer, new (unknown) buyer, insolvency, bankruptcy, new subcontractors, "old" known problematic subcontractors, etc.	
Risk management	Oversight and management risks cover the handover phase from the tender phase to the execution phase, the project management organization, external partners, consultants, internal organization, international organizations, and the like	
Acts of God	Force majeure is the risk of war, civil unrest, terrorism, atomic explosions, thunderstorms, earthquakes, floods, and similar unusual situations	

(e) Black Swans (and identified/unidentified risks)—unlikely and strategically surprising events beyond regular expectations, characterized by difficulties in predicting occurrences due to certain preconditions (breakup of Yugoslavia, USSR);

(f) "Dragon King" event—an event that is achieved on a larger scale than anticipated (a terrorist attack 9/11 in the United States);

(g) Risks that cannot be modeled (the insurance industry uses this term for risks that are underresearched and below the margin of interest of the insurance industry but can cause large losses—risks with low probability but large consequences, which are predictable and amenable to analysis risks—volcanic eruptions with dust clouds, meteors falling).

Many authors, explaining the essence of risk and the way to determine the definition of risk, represent different approaches and concepts of the notion of risk, present to the framework of the logistics to the process [20,21]:

- Risk as a deviation from the goal is an approach that involves making the wrong decision that led to the deviation from the goal, which involves the interaction of these two factors, because the decision cannot be judged to be wrong without analyzing the goal. Deviation from the goal contains the intensity and probability of realization of a negative event.
- Risk as a possibility of a wrong decision is an approach that is part of the goal deviation approach. It is very difficult to measure the risk of a particular decision because the risk assessment is made after the event, analyzing the scenario of the event and how certain decisions affected the realization of the risk. This approach also includes correlation between the set goals, for the reason that a particular decision cannot be judged wrong without analyzing the goal.
- Risk as information deficit is an approach that involves a lack of information in situations where a particular decision has to be made. The lack of information causes people or business systems to be unaware or insufficiently aware of the risks involved. This approach involves situations where there are restrictions on the availability of information under the conditions of making an appropriate management decision.
- Risk as a combination of information deficits and possible goal deviations is an approach based on the combination of the goal deviation concept and the information deficit concept, on the basis of which the risk can be divided into two components: a description of the risk based on the likelihood assessment and possible goal deviations for symmetric and asymmetric risks.

According to the International Association of Actuaries, risks can be broadly divided into four groups: hazardous, financial, operational, and strategic risks (Table 9.2).

3. Planned working life cycle of the thermal power plant

The life of the plant elements is influenced by static loads and dynamic loads. The effect of static loads is taken into account over the exposure time of the elements to the operating pressure and temperature. Dynamic loads are taken into account over the number of load changes with a given temperature gradient during the element's current life. Based on the analysis of the operation of the existing TPPs, it can be taken as initial information that their basic working life is 20–25 years (depending on the quality of the maintenance process), and that through the revitalization process (with mandatory reconstruction with accompanying modernization) this period can be extended for an additional 15 years. This procedure, by

Table 9.2 Classification of risk, Casualty Actuarial Society, in 2002 [22].

Hazardous risks	Financial risks	Operational risks	Strategic risks
- Fire and other property destructive risks; - Storm and other risks of incident disasters; - Theft and other criminal, personal injury; - Business disorders; - Illness and disability (including inability to work); - Damage resulting from personal liability	- Price; - Liquidity; - Inflation/purchasing power; - Hedging (exchange rate risks, changes in interest rates, changes in the value of shares, etc.)	- Business operations (product development, human resources, supply chain management, etc.)	- Reputation problem; - Competition; - Customer requirements; - Demographic and sociocultural trends; - Technological innovations; - Availability of capital; - Regulatory and political trends

its structure, is extremely complex and is often compared to the level of realization of a new technical facility. The very process of planning and implementing the process of revitalization and exploitation of the plant within the system under consideration is realized with the aim of achieving a high level of operational safety, which implies defining and discovering possible sources of unreliability. Measures to eliminate and mitigate their effects must be defined, and the economic criterion is most often used as a criterion.

The revitalization process is an extension of working life, which is mainly accompanied by modernization and reconstruction and improvement of its environmental friendliness. Such a systematic and comprehensive process at a technical facility or plant is an indispensable and logical process in the working life of the facility. The connection of the process of reengineering to the maintenance of technical systems, with the aim of realizing the corresponding advantages and improving the reliability of the system, is given through the following characteristic elements: analysis of costs related to maintenance and readiness, or availability of the system (as one of the most important characteristics of reliability), then, determining the general aspects related to motives and justification for revitalization, as well as the scope and definition of the most optimal term for the realization of this process. In particular, the influence of the reliability and availability characteristics of the system object on the application of reengineering principles through the process of system maintenance, that is, on the systematic approach of revitalization of some of its capacities, should be singled out.

At all stages of the TPP life cycle, there are certain risks that may have a negative impact on their construction or future operation. Knowledge of the possible risks allows timely action to be taken to minimize them. This primarily refers to good design, good selection of components, good installation, and subsequent proper management of the unit's block in exploitation, with the ultimate goal of achieving the maximum possible reliability and availability of the unit, all with the aim of maximizing electricity production. In the last 15–20 years, a large number of new steels have been developed for

TPPs. The improved characteristics of these steels are based on advanced technology. Initially, the development was based on 9%–12% Cr martensitic steels with increased creep resistance for thick-walled boiler components and rotor turbines, and more recently for boiler components such as evaporator and superheater. On the other hand, the supply of coal from its own mines gives additional security for the safe and reliable operation of TPP, among other things, due to the available coal reserves during the foreseeable future, the possibility of its easy storage in power plants, whereby the production of coal-based electricity is not time dependent. Also, no special transport routes (conveyor transport) are required for the use of coal. No additional protection is required in the case of coal conveyor shipping.

3.1 The requirements concerning thermal power plants' useful (service) life

The plant will be designed for a minimum useful (service) life of 25 years, with possibility of extension up to 40 years taking into account the following conditions relating to the challenges and delays, as well as the intensity of load changes during the life cycle [23,24]:

- cold start (75 starts): required time 4½ h (after a delay of more than 120 h);
- warm start (50 starts): required time 3 h (after a delay of more than 12–120 h);
- hot start (100 starts): required time 35 min. (after a delay of up to 12 h).

Also, it is expected to be about 2000 load changes between 50% and 100% of rated power. The power plant will be able to operate with a load change gradient of at least 5% of rated power per minute. The stated requirements should be calculated with about 25% of the total estimated lifetime of the power plant, in order to provide sufficient safety limits (boilers, steam lines, steam turbines) according to the characteristics of coal-fired boilers, such as load changes and temperature overruns.

3.2 Maintenance requirements

On average, 2–3 weeks of scheduled downtime are scheduled for regular maintenance. The existing block of these TPPs is known to have scheduled maintenance at a predetermined date of 30–45 days per year (regular maintenance) and 60–90 days every fourth year (major overhaul). By installing diagnostic equipment by condition, it is possible to achieve extended operation of these units without regular annual shutdowns to perform regular overhauls.

3.3 Determination of block guarantee points

The calculation of net specific heat consumption will be based on 100% of nominal power according to the following formula:

$$\mathrm{NSHC} = \mathrm{HI}/(\mathrm{GO} - \mathrm{LT} - \mathrm{LAT} - \mathrm{OC}), \tag{9.1}$$

where: NSHC is net specific heat consumption, kJ/kWh; HI is gross heat input to the boiler, kJ/h; GO is generator output, kW (at generator terminals after subtraction of excitation power); LT is losses of the block transformer, kW; LAT is losses of auxiliary transformer, kW; OC is own consumption, kW (including coal and ash feed and storage system, filters, and flue gas purification system).

4. Risks in the design of thermal power plant

The process of construction of a thermal power facility includes a number of activities, ranging from the preparatory and exploration phase related to the location, analysis of location conditions, reserves and quality of coal, etc., to the design, construction and installation, commissioning, and commercial exploitation phases. In all these phases, there are certain risks that may have a negative impact on the construction or future operation of the TPP. Knowledge of the possible risks allows timely action to be taken to minimize them. This primarily refers to good design, good choice of components, good installation, and subsequent proper operation of the unit in operation, with the ultimate goal of achieving the maximum possible reliability and availability of the unit, all with the aim of maximizing the production of electricity and heat (or technological steam). The basic assumptions made in the research of complex thermal power engineering systems are that the state of their working ability with stable fault-free operation, which due to the static structure and dynamic influence of a large number of factors from the operational and wider environment, often turns into an unstable state in failure, can be controled by scientific approach of competitive engineering (life cycle engineering). Previous research has shown that the scientific approach (scientific prevention through design, scientific recognition, and scientific application) can, in the most favorable way, through reliability management, lead to an optimal level of reliability according to the criterion of life cycle cost, that is, to the prediction of the moment of necessary implementation of reengineering. This refers above all to good design, good selection of components, good installation, and subsequent proper management of the block drive, first in trial and later in continuous operation, with the ultimate goal of achieving the maximum possible reliability and availability of the block. The ultimate goal is to create the conditions for maximum electricity production. Potential risks that may arise during the implementation of this project, with a given likelihood of their occurrence, as well as the impact they have on the project, with proposed preventive measures to prevent their occurrence or possible minimization are shown in Table 9.3. This assessment does not address social risks, such as strikes or adverse political situations, which may affect the investor's willingness to fund the project, as well as the so-called industrial risks including fires, floods, thefts, and failures of machinery that can be insured. The likelihood of a turnout and the impact on a project of a particular event may qualify as high (H), medium (M), and low (L).

5. Risks in exploitation of thermal power plant

Systematic procedures for determining the causes, types, and consequences of failures that may occur, it is necessary to define and specify activities to minimize the catastrophic consequences of failures, especially those related to the medium and the environment (preventive engineering). Managing the remaining life of a TPP, with the inevitable analysis and specification of its "weaknesses," is today a multidisciplinary task for a team of experts, whose implementation requires new methods and concepts, as well as appropriate algorithms for working methods. The main tendency in the development of these methods is efficiency, speed, and cost, i.e., obtaining certain numerical values on the basis of which an appropriate and timely decision can be made in the maintenance process (decision optimization). In addition to estimation, data for reliability determination can be obtained by calculation and verification or naturally (unfocused), through customer experiences, own production, and other experiences, and through data from relevant service organizations engaged in maintenance work. If the

Table 9.3 Project risks and preventive measures to eliminate them.

Risk description	Probability of taking place	Impact on project	Preventive measures
Obtaining agreements, permits, and approvals	L	M	Term clearly define with realization plan and determine steps in the process of obtaining permits and their time duration. Provide continuous control over the course of project activities
None or partial fulfillment of the contractual obligations of the contractors	M	H	Provide payment in stages, depending on the degree of completion of the work. The contract defines compensation for the delay in the execution of works. Contract requires performance guarantees. Carry out rigorous control over the performance of suppliers or contractors. Conduct regular monitoring
Inadequate design solution	L	M	Perform all necessary site testing when designing the main and the execution project
Low productivity in construction	L	L	Adequate organization of site activities to maximize construction productivity. Exercise constant control over the execution of planned activities. Climatic conditions should be taken into account in the project implementation plan
Location security	M	M	Take all necessary measures to ensure that the construction and operation of the plant complies with safety regulations or tests. Adopt an adequate security control program based on properly selected security standards and procedures. Prepare a safety assurance plan during project implementation and conduct appropriate staff training on a regular basis
Improper project management	M	H	Hire a competent and experienced project management team. Clearly define the workload of each team member. Hold regular meetings with responsible staff
Inadequate implementation of the program security or management quality on the project	M	H	Develop a detailed program of security and whether the quality management of the project and adopt adequate procedures. Conduct regular controls on the implementation of quality management programs, as well as conduct regular training of staff in the field of quality management
Damage to existing objects	M	M	Develop a detailed plant construction plan at location: Plan for construction of facilities, plan for excavation, transportation (entrance/exit from the site), as well as handling of emergency situations. Define personnel and systems that will be engaged during project implementation
Counterparty disputes and claims regarding payment terms	M	M	Define the terms of payment with the contract

Continued

Table 9.3 Project risks and preventive measures to eliminate them.—cont'd

Risk description	Probability of taking place	Impact on project	Preventive measures
The lack of skilled labor force	L	M	Define the conditions of employment and selection of new staff. Organize and conduct regular training of existing and future staff
Exchange rate change	L	L	The financial analysis includes the costs of hedging against changes in the exchange rate over the period of loan utilization and repayment
Change in equipment price	L	M	The contract defines in detail the scope of delivery and characteristics of equipment and works. Agreement to define the conditions and manner of correcting prices

Note: These preventative measures are only given as recommendations and guidelines for conducting more detailed analyses. Most of these risks, primarily those that may have an impact on the cost of operating facility or investment, are included in the feasibility analysis by changing key parameters in the sensitivity analysis.

object under consideration is complex (e.g., a TPP system), then the problem of determining reliability is solved if one knows the reliability of the constituent components or at least their "most critical" parts, their interconnection (structure), and operating conditions (constraints and environmental conditions). It should be emphasized that the verification of reliability, that is, testing the hypothesis in practice, is carried out at all stages of life of development, design, construction, and operation of the facility, and is mainly related to several basic limiting factors—money and time, environmental conditions, and other technical constraints. The reliability verification itself is accompanied by the corresponding mathematical apparatus, with a certain level of confidence in the parameters tested. The inadequate level of reliability during the exploitation of the technically complex system itself and the existence of irrational labor-based investments by eliminating consequences rather than causes clearly indicate the need to harmonize existing methods to achieve optimal reliability and adapt them to the system, with the prior definition and elaboration of an appropriate algorithm. In the stochastic behavior of complex technical systems with a large number of circuits, subassemblies, and their components, the state of the future is not determined solely by the initial state and mode of control, which is why methods for estimating optimal reliability on the basis of economic criterion play a role in the design and planning processes, use, and maintenance of the system as well as its parts. Also the application of probability theory and statistical methods based on the history of failure data is very important for making long-lasting decisions in the maintenance system, which enables timely actions with adequate maintenance cost reduction.

6. Methods for risk assessment of thermal power plant

Risk, as a potential threat to people and material goods, can be viewed as a function of the likelihood of an adverse event occurring and negative consequences (human and animal health and environmental hazards, environmental and material goods, risk of monetary loss in business, etc.).

$$R = f(P, C), \tag{9.2}$$

where R is risk, P is the probability of occurrence of adverse events, and C is the negative consequences that may cause an undesired event.

The risk can be reduced by reducing the likelihood of an adverse event P or reducing the negative effects that an adverse event C can cause or by reducing both.

The methodology of conducting the risk assessment procedure defines the algorithm, tools, and method of carrying out the risk assessment procedure, and the procedure for conducting the risk assessment process defines a standardized set of steps that ensures the implementation of the process in accordance with the recommendations of the relevant laws, regulations, and best practice recommendations.

By asking questions, "What is the risk?", realistically placed three subquestions that should determine which adverse event can happen, how often it happens and if it happens what negative consequences can result. Assessment of risk is very subjective process and it is necessary to respect certain principles when assessing the risk in order to reduce subjectivity at the lowest possible level. Considering the criteria on the basis of which the risk assessment is performed, the methods used for risk assessment can be divided into: quantitative, qualitative, and semiquantitative (combined). Assessing risk of an unwanted event can be made on the basis of: quantitative or qualitative assessment of the probability of occurrence of this adverse event (Table 9.4) and a quantitative or qualitative assessment of negative effects (damage) that may cause the adverse event (Table 9.5) [27]. Also, there are semiquantitative or combined methods for the assessment of risk, all the methods that are applied in the field of safety and health at work.

Risk assessment is done by a multidisciplinary team. This team should have a good knowledge of the company organization, work process, material properties used in the process, hazards and other technical parameters, laws, regulations, standards, scientific methods, and techniques. Among other things, the team has the task of carrying out the following activities:

- Determine the risk management policy;
- Specify the criteria for assessing the level of risk presented (in numbers/in text);
- Determine measures for risk management;

Table 9.4 Probability assessment for unwanted events [27].

Probability of occurrence		
Qualitative	**Quantitative**	**Criterion**
Very rare	<0.1	The estimated probability of occurrence does not exceed 10%, i.e., the probability of occurrence exceeds a period of 10,000 years
Rarely	>0.1 or ˂0.4	The estimated probability of occurrence does not exceed 40%, but is greater than 10%, i.e., the probability of occurrence exceeds a period of 1000 years
Probably	>0.4 or ˂0.8	The estimated probability of occurrence exceeds 40%, but is higher than 80%, and the probability of occurrence exceeds a period of 10–1000 years
Very likely	>0.8	The estimated probability of occurrence exceeds 80%, i.e., the likelihood of occurrence is less than 10 years

Table 9.5 Assessment of the negative consequences that can cause an adverse event [27].

Consequences of occurrence		
Qualitative	**Quantitative**	**Criterion**
Meaningless	<0.1	Minimal or irrelevant consequences
Insignificant	From 0.1 to 0.4	They have little effect on the process being analyzed
Substantial	From 0.4 to 0.8	They greatly contribute to the increase in costs
critical	From 0.8 to 0.9	They seriously threaten the process, the staff
Disastrous	>0.9	Catastrophic consequences for business, environment, and people

- Calculate risk;
- Estimate rank/level of risk (very small, small, medium, large, and very large),
- Estimate the probability of an error occurring:
  - up to 25% of cases/low probability—0,
  - over 70% of cases/high probability—1, and
  - from 25%–70% of cases/mean probability—0.5;
- Estimate error/cause (small/medium, medium/large);
- Introduce all stakeholders to the risk;
- Monitor changes in risk intensity and take measures to prevent and stop the occurrence of risks;
- Establish a system for early identification of sources of danger.

Risk management is carried out through the activities of the subprocess assessment and risk control (Fig. 9.1).

The concept of the risk management process begins by establishing the appropriate context (environment) in which the project implementation began (identification of stakeholders, understanding of goals, understanding of project output, defining areas with limits of risk management activities, defining links and possible overlaps with other projects, organizational and strategic constraints, etc.).

Risk identification is the next step, which is necessary to cover the impact of risk on all project objectives (costs, time, quality, statutory and legal compliance, safety, responsibility, health protection, environmental protection, etc.).

The purpose of risk assessment is to conduct all necessary analyses and assessments of identified risks, with the aim of determining the necessary treatments to overcome them. This assessment can be carried out quantitatively or qualitatively, whereby preliminary qualitative analysis can be carried out in the initial stages of project design, while quantitative analysis is only applicable in the presence of certain databases. The risk assessment includes comparing it with certain criteria, as well as defining the initial priorities for treating them, with the aim of avoiding them completely, reducing the likelihood of their occurrence, reducing the consequences of their occurrence, transferring or sharing the risk or retaining it, and making the necessary recovery plans and repercussions [25]. A simplified view of the relation between risk analysis and other risk management activities (selection, implementation, and monitoring of adequate control measures, risk mitigation measures, etc.) is given in Fig. 9.2.

Risk analysis, performed using specific algorithms (Fig. 9.3), is essentially a structured process that identifies both the likelihood and extent of the undesirable consequences that arise in order to provide answers to three key questions:

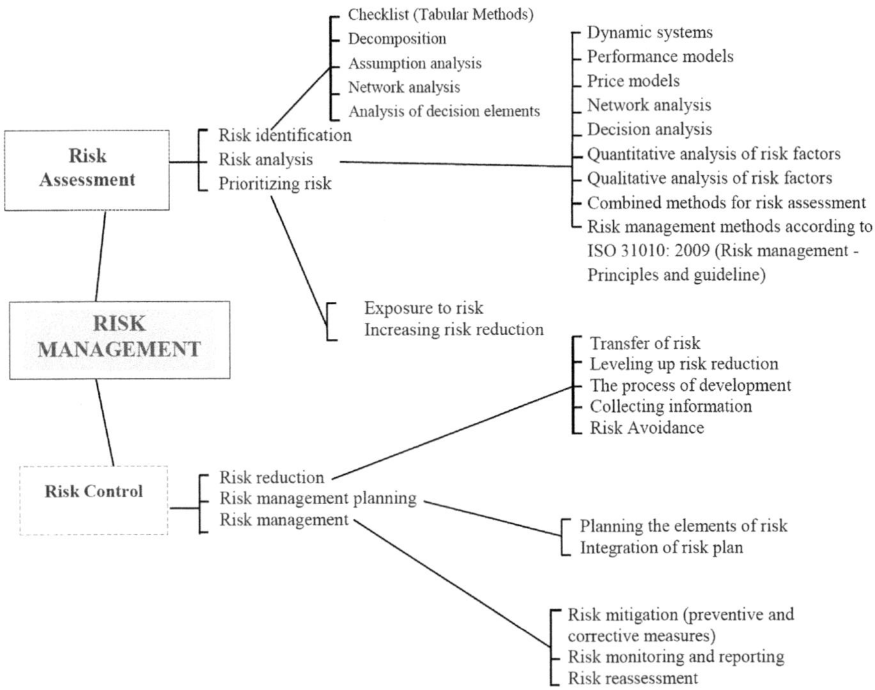

FIGURE 9.1

Demonstration of risk management process.

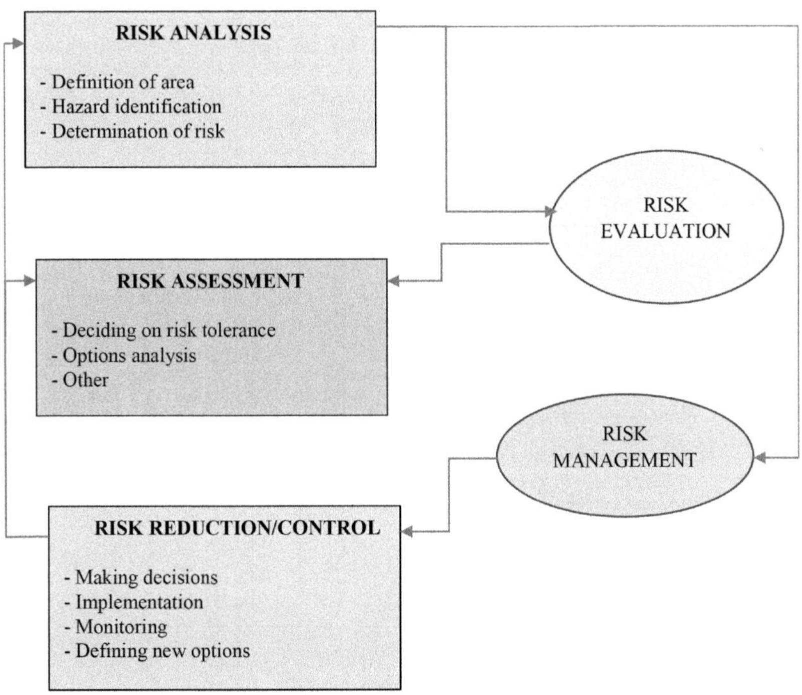

FIGURE 9.2

The simplified view of the relation of "risk analysis and other activities of risk management" [6].

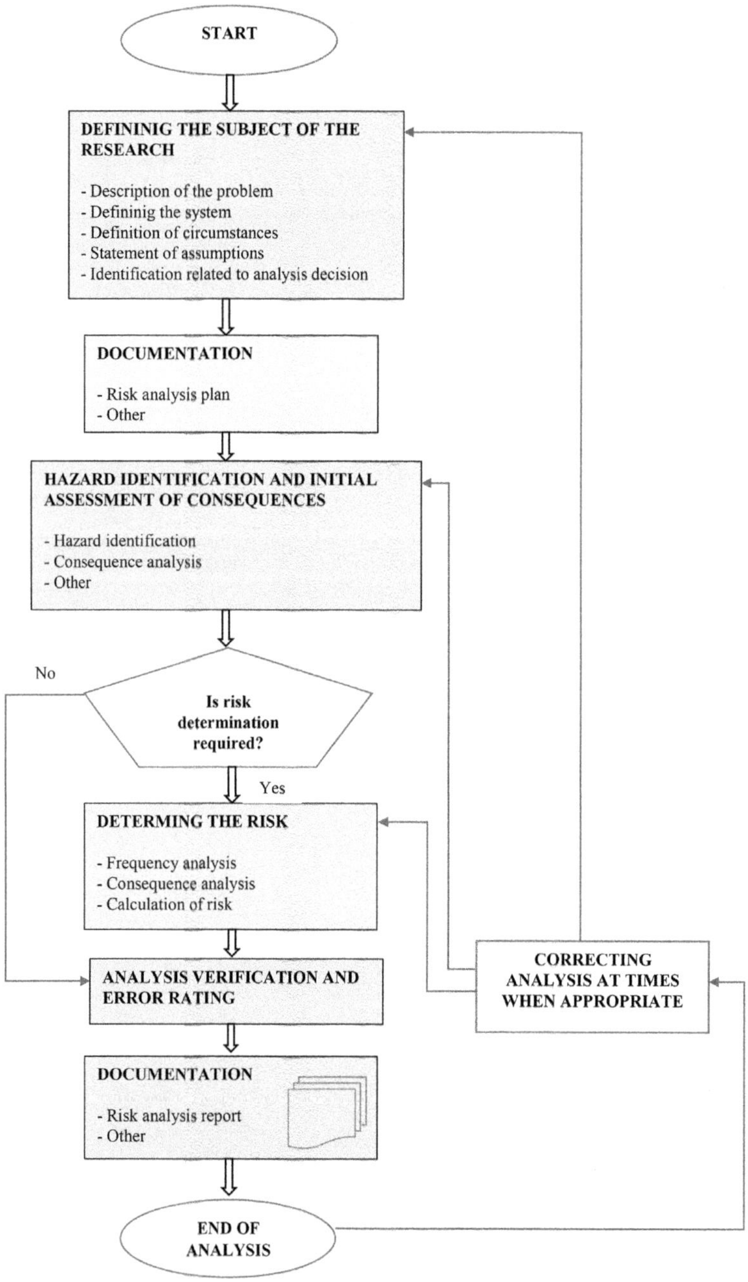

FIGURE 9.3

The algorithm of the process of risk analysis [6].

- What can trigger an unwanted (wrong) course? The answer should be sought by means of hazard identification.
- What is the probability of an adverse event occurring? The answer should be sought by conducting a frequency analysis.
- What are the consequences of an adverse event? The answer should be sought on the basis of consequence analysis.

Risk is analyzed by combining the likelihood and the impact of a cause. Many methods of reliability analysis, such as types and consequences of failure analysis, failure tree analysis, and importance analysis in terms of reliability of technical system units, can also be successfully applied to determine system security characteristics, such as: primary and secondary events, peak event, the likelihood of peak events, a minimum set of sections, the degree of criticality, and type of dismissal whole system [26].

The correct choice of risk assessment methods enables the implementation of adequate measures that will ensure a safer workplace and working environment, as well as a lower probability that may cause occupational diseases and injuries of employees. Risk assessment technique includes interviews with experts in the area of interest with the use of expert knowledge of multidisciplinary working groups with the use of simulation methods. Risk assessment at the workplace is the systematic recording and evaluation of all hazards in the work process which can give rise to occupational injuries, diseases or damage to health and identifying opportunities, and ways to prevent, eliminate, or reduce risks. Risk assessment is primarily an empirical process of making engineering decisions based on knowledge and experience to improve occupational safety and health.

The risk management approach involves identifying, assessing, and controlling it. There are three options that are not mutually exclusive: risk reduction, risk transfer, and risk acceptance. Risk reduction is a process in which appropriate countermeasures are sought, based on the risk analysis conducted, and security controls are implemented to protect the organization's resources. The process seeks to reduce the likelihood of a hazard and/or its impact on the process.

The rest of the risk includes activities to provide control and a logical framework for counteracting the consequences. If it proves more profitable, the risk can be transferred to a third party (e.g., insurance company). It is also possible that implementation of countermeasures or risk transfer may not be cost-effective. In this case, the company may decide to accept the risk or the resulting costs. The only approach that is not acceptable in risk management is to ignore risk. It should be noted that risk management is a continuous process and that the relation between resource values, vulnerabilities, and threats changes over time. Risk control represents a set of activities to minimize, mitigate, or eliminate risk, through the reduction, planning, and resolution of causes of risk.

The process of risk assessment of strategic plans and processes requires management to make the most optimal decisions based on risk assessment, key professional processes. Based on the risk management plan, categorization of nonconformities into small, large, and critical is performed. A small discrepancy (occurrence of up to 25% of cases) is isolated, and it is individual and concerns a requirement of the standard and threatens the system. Major discrepancies (occurrence of over 70% of cases) are the lack of some standard requirement completely or a large number of minor discrepancies in some part of the project, process, and system, which threatens them. Critical noncompliance (occurrences of 25%–70% of cases) is one that can result in material, security, and environmental threat to human beings, goods, and information.

Statistical, engineering, and managerial methods are used to improve quality, and eliminate or minimize the impactful error in the system, process, project, product, and service. Statistical methods are used to collect and process data. Engineering methods are used to improve the quality of the process, by analyzing the causes and consequences of the error that affects the quality and reliability of the process. Management methods are used to plan, organize, manage, and control the business system process. The analysis of the quality of the environmental and work process is done comprehensively on the basis of relevant indicators, using legal regulations and standards harmonized with international standards and EU directives. The calculation of synergistic effects is done on the basis of literary calculations and practical experience.

As a recommendation to select an appropriate method for analyzing the impact of a fault on quality, it is in a function of the area of work, as shown in Fig. 9.4.

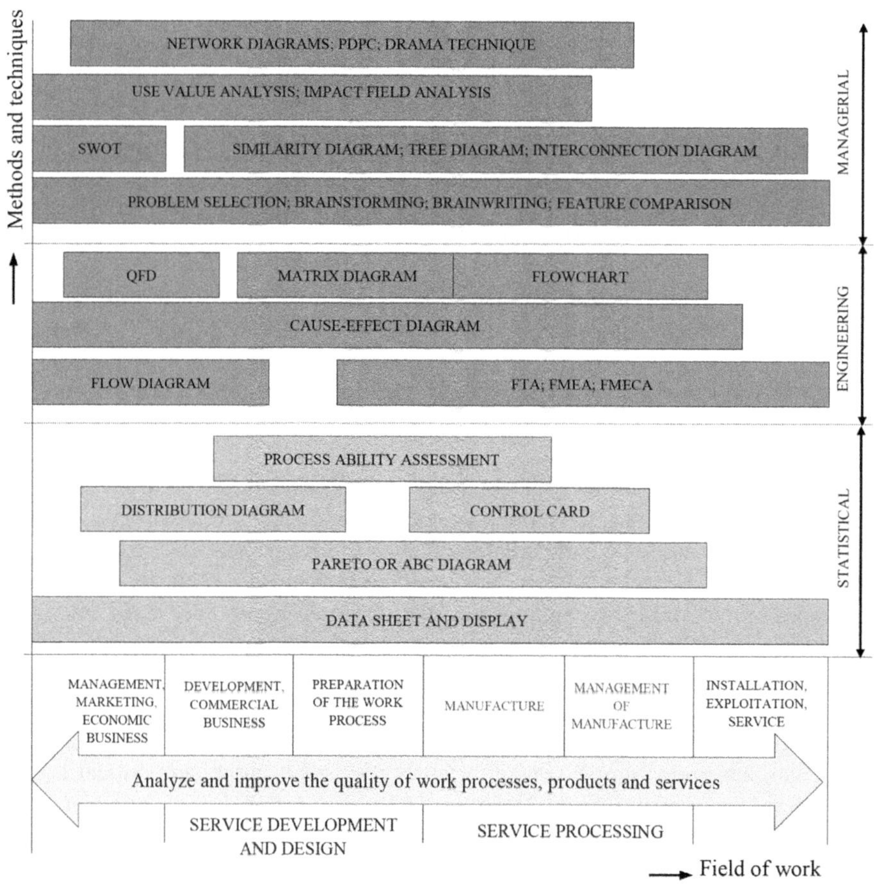

FIGURE 9.4

The presentation of recommendations for the selection of an adequate method for analyzing the impact of an error on quality is in a function of the area of work.

In power plant installations of particular importance is risk assessment of environmental pollution, which comprises:

- identification of danger (hazard), the identification of negative consequences (effects) that polluting substance can cause, and
- dose determination (concentration of pollutants received) and the estimation of the expected effects (intoxication response), including the ratio of the dose which is a function of exposure and the occurrence or severity adverse effects.

The environmental impacts that occur during exploitation are due to the existence and use of a TPP in a given space. They are mostly permanent in nature with a tendency for spatial and temporal increase of influence, so it is necessary to reveal their existence and nature in a timely manner. Conducting an environmental impact analysis aims to direct negative activities in the positive direction to additional activities, and to predispose potential negative impacts. The area of direct influence is the space of direct occupation of the TPP facility and associated facilities. Construction works are being carried out in this area, which occupy and alter the habitats of humans, plants, and animals. The area of direct impact of a TPP is an area generally within a narrower range. The area of indirect influence is the space where the object is not built, but it is felt as a result of the construction of the object in question.

The methodology of assessment of the environmental impact consists of the following steps:

1. Assessing the current situation based on the results of the measurements carried out earlier for the quality of air, water, and land (determining the existing situation from the point of impact of certain areas closest to the receptor based on previous measurements of individual parameters, creation of initial existing pollution status);
2. Realization of the environmental impact study of the TPP, which is a predicted measurement for determining the so-called "zero" environmental condition at the macro- and microlocation concerned (air, water, and soil quality, noise level, and other characteristics according to the decision of the line ministry) for at least 1 year (fixed station at microlocation and mobile station up to 30 km in diameter, with change in locations every 15 days);
3. Phase of TPP (assessment of the impact produced during the construction of the thermal machine, impact assessment of the resulting increase in traffic rate during the construction of the thermal, impact assessment for performing the necessary building-architectural work and installation work on site plant);
4. Phase of normal operation (exploitation) of TPP (impact assessment produced by TPPs during stationary and nonstationary mode, assessment of the impact resulting from the increase in traffic importance for the operation and maintenance of power plants, assessment of the impact produced by the auxiliary installations and equipment in TPP);
5. Phase of reconstruction, revitalization, and modernization of the TPP after the end of the basic working life and creation of conditions for the extended working life (assessment of contributions and environmental effects resulting from the reconstruction and modernization of the TPP);
6. The phase of removal or change of purpose of the object.

During construction, there are impacts that are a consequence of the construction of the building and are mostly temporary. The consequence is the presence and operation of humans and machines, as well as technology and construction organization. As a rule, negative impacts result from the excavation/deposition, transport and installation of large quantities of building materials, as well as the permanent or temporary occupation of space and all related activities.

Construction of a TPP can be divided into three phases: preparatory works (realization of the recording baseline of the environment for a period of at least 1 year, drafting of necessary project documentation, obtaining necessary approvals and permits necessary to start construction, implementation of the tender procedure for purchase of equipment and selection of contractors, supervisory authority), construction works (with training working personnel), and installation of equipment and final construction works on TPP (test mode). During the construction of the TPP and subsequent exploitation, characteristic influences on the following environmental parameters can occur: air quality, water quality, soil quality, noise level, vibration and radiation intensity, flora and fauna quality, population health, meteorological parameters and climatic characteristics, quality ecosystems, population, concentration and migration of the population, quality of purpose and use of the land (built and undeveloped areas, use of agricultural land), natural assets of special value, cultural assets, material assets including cultural and historical and archeological heritage, quality of landscape features of the area. Impacts during construction are most often local and of temporary nature and generally can be reduced by good organization of construction works. They will last nearly like the life of the building. Impacts during exploitation are more significant and last for the lifetime and lifetime of the TPP (realization of monitoring and waste management, according to the obtained environmental permit, optimization of the operating regime with the aim of reducing the negative impact on the environment, realization of activities on improving the quality of input raw materials, etc.).

An overview of the process risk aspects is given in Fig. 9.5.

The model includes looking at the requirements of the following families of ISO standards: ISO 9001—Process quality management system, ISO 14001—Environmental management system, and ISO 18001—Safety and health management system. ISO 31000 standard—risk management system,

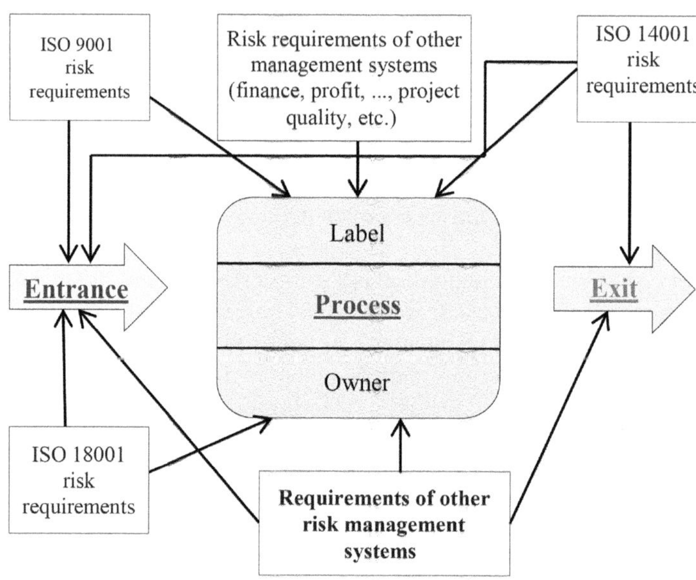

FIGURE 9.5

Outline of the risk perception process.

and it includes identification, analysis, and evaluation of risks. The methodology is based on the previously applied process approach and the fact that the following actions are taken with the identified business processes:

P (Plan)— planning and establishing goals and processes necessary to deliver results in accordance with customer requirements and organization policies;
D (Do)— application of these processes;
C (Check)— monitoring and measuring processes and products with respect to the set policy, goals, and requirements;
A (Act)— taking actions for further improvement of the process.

The interaction of PDCA methodology and process approach is the essence of the ISO 9001 quality management system.

In this model, which is also shown, among other things, in ISO 9000, the individual processes are not shown in detail, but together they represent the general process of "product realization." The model presented also takes into account other principles of the quality management system according to ISO 9001, for example, the principle of customer focus, because they are the ones who should, by their needs and expectations for a particular product or service, indicate the direction and the goal that they want to achieve by constantly improving the business processes of the organization. Ongoing and applying PDCA methodology to individual processes indirectly condition the application of the same methodology to the entire management system. In this case, the steps are as follows:

P—establishing of appropriate policies and objectives and document management processes and systems through the documentation of the management system.
D—the introduction and operational use of policy, controls, processes, and documents of the management system.
C—measuring of function of the system in practice, and a comparison with the set policies and goals through independent assessment and management reviews.
A—implementation of preventive actions, corrections, and corrective actions for individual business processes in order to establish constant improvement established system.

The risk management model is based on the ISO 31000 standard and is applicable to all companies regardless of activity, size, and ownership structure. The model is presented in Fig. 9.6.

6.1 Quantitative risk assessment methods

Quantitative risk assessment criteria use numerical values for description of probability and consequences of events (the value of resources is most often displayed in monetary units, usually on an annual basis). Values of certain resources are not always possible to express financially, and as a result it may appear that the figures do not represent the real situation.

A common example of such assessment within analysis of the TPP is an assessment in the field of security and health at work (risk assessment of noise and other physical or chemical hazards), as well as the environmental impact when a level of permitted exposure is clearly defined, as well as level of increased exposure and illicit exposure level for both humans (working staff) and the environment (emission of pollutants, impact on land, wastewater and their impact, etc.). In order to define the probability of risks and consequences that may arise in the form of numerous values, it is necessary to

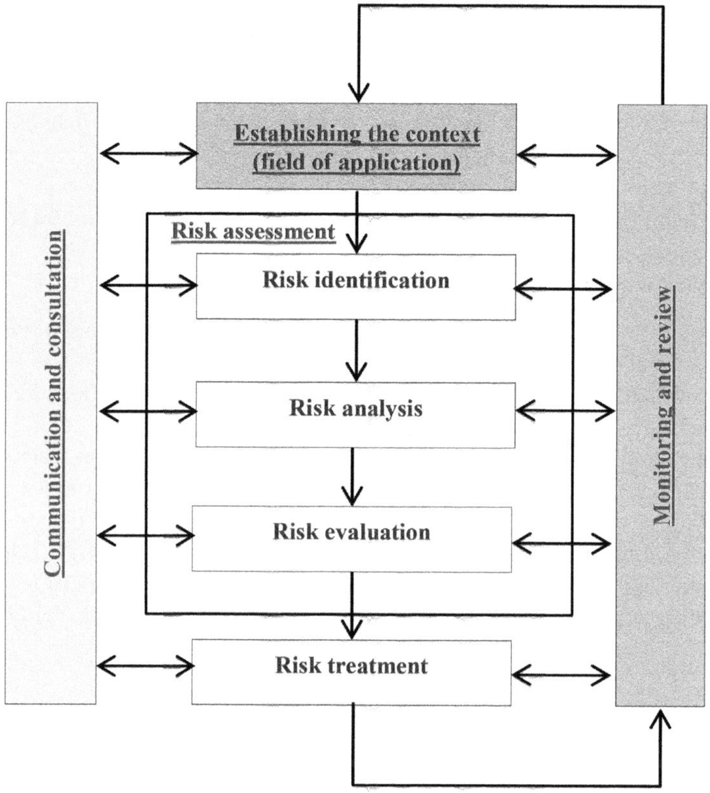

FIGURE 9.6

The risk management model is based on the ISO 31000 standard and is applicable to all companies, regardless of activity, size, and ownership structure.

conduct deeper analyses, to have appropriate accident statistics as well as monitoring plans during plant operation. Therefore, in the area of occupational safety and health, priority is given to qualitative risk assessment, while quantitative is applied primarily in high-risk cases. This approach is not adequate for all processes in a TPP (values taken from book value do not always represent the true valuc of thc rcsourcc).

Recent experiences of developed EU countries suggest that quantitative risk assessment should be introduced wherever possible and given much greater importance. There are already a number of new forms, recommendations, and tables indicating how a quantitative risk assessment in the field of occupational safety and health can be carried out relatively effectively and simply.

The full quantitative assessment is not always possible in case of lack of information about the system or the activities being analyzed, incomplete information about the dismissals, the impact of human factors, etc. Some elements of risk cannot be quantified by probability distribution; it is then necessary to carry out by assessment based on quantitative consideration of the nature of what is being protected (humanity, environment), the severity of the injury or damage (e.g., slight, serious, catastrophic), or the extent of the damage (one or more persons). It should also be noted that the magnitude of the damage suffered can be defined differently depending on the situation.

Consequence analysis is given as the likelihood of affecting people, the environment, or property, should an adverse event occur. In general, the consequences of different types of risks are expressed in safety (e.g., fatal, harmful), in health, in financial, and in environmental terms. Finally, the risk must be expressed in an appropriate form. Generally, the probability of an event is presented as its frequency per unit of time or activity, while the consequences are presented as a numerical loss (financial, lost business days, lost financial profit, etc.).

The methodology of risk assessment takes place the following order [27]:

- Addressing each adverse event (risk situation) based on individual risk determination forms.
- Determining the quantitative value of the probability of an adverse event P occurring appropriately and also on the basis of a realistic criterion (probability is expressed as a decimal number from 0 to 1, where 0 indicates an impossible event and 1 event to be realized with a probability of 100%).
- Determining quantitative values due to the occurrence of each adverse event P in an appropriate manner and based on a realistic criterion.
- Using the formula $R = V \cdot P$, the risk factor for each identified adverse event is determined.
- Based on the values obtained, the level of risk for each identified adverse event is determined.

6.2 Qualitative risk assessment methods

Qualitative risk analysis is a more subjective approach than quantitative risk assessment. In this approach, resources, risks, and countermeasures are viewed relatively with respect to the system. However, its implementation does not require a thorough knowledge of the material values of individual resources. For their evaluation it is necessary to know the importance for individual business processes to the technical system that is the subject of research. The result obtained is only a relative relation between the value of the damage caused by the action of a hazard and the introduction of countermeasures, bearing in mind that this assessment is subjective and therefore subject to error. To describe the likelihood of an adverse event, the qualitative risk assessment criteria use words such as: infrequent, unbelievable, possible, probably or almost surely likely possible, often, etc. To describe the negative consequence of an adverse event, qualitative risk assessment criteria use words such as: fatal, serious, small or negligible, catastrophic, etc. In practice, qualitative scales with three to seven qualitative descriptions are optimally used. Methods with less than three qualitative descriptions for risk factors are not precise. Methods with more than seven qualitative descriptions for risk factors lead to significant subjective difficulties, related to the inability of the participants in the risk assessment team to recognize quite accurately the qualitative description of risk factors. Qualitative risk assessment involves the use of nonnumerical and/or qualitatively described data. The risk matrix methods (risk ranking matrix) are part of a qualitative method for risk assessment. There are also a number of modified methods for qualitative risk assessment, resulting from the elimination of shortcomings of the previously used type methods, which may result in unreliable risk assessment results. Risk ranking is based on a matrix, which has consequences for its axes and ranks of probabilities. Participants in the risk assessment team often use in their work a risk matrix to establish a logical link between consequences and likelihood in assessing risks for previously identified hazards. They are also used as a

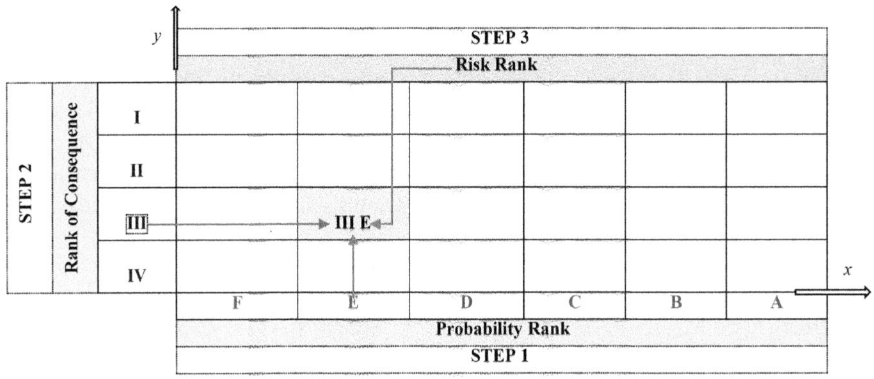

FIGURE 9.7

Formation of risk matrix [27].

uniformly defined way of determining the degree or level, of the individual risks being assessed. Risk matrix given in Fig. 9.7 is formed by applying probability ranks on the x-axis (step 1) and the consequence ranks on the y-axis (step 2) and then determining the risk rank (step 3) [27].

Typical qualitative methods for risk assessment are: 4 × 6 risk matrix (MIL-STD-882C), 5 × 5 risk matrix (AS/NZS 4360: 2004), and 3 × 3 risk matrix (Occupational Health and Safety Assessment Series—OHSAS standard). Thus, matrix 4 × 6 (MIL-STD-882C) contains four levels and qualitative job descriptions (I, II, III, and IV), which is related to occupational diseases/injuries, loss of equipment, and work time and environmental impact. A qualitative description and definition of the likelihood of an adverse event occurring is presented at six levels (A, B, C, D, E, and F). When using this matrix, three qualitative descriptions of the risk level (increased, medium, and low risk) are recognized. Risk is considered unacceptable if assessed as increased and acceptable if it belongs to medium or low risk. Recommendations from the European Agency for Safety and Health at Work for the implementation of the OHSAS standard recommend a 3 × 3 matrix, with three levels for qualitative description of probability (unbelievable, likely, and very likely), as well as three levels for describing the consequence (moderate, medium, and large). Risk also has three levels expressed in qualitative description: small, medium, and large. The advantage of all risk matrices is reflected in preventing the acceptance of unacceptable risk by allowing operational, engineering decisions to be taken to reduce risk to an acceptable level.

Limits in applying risk matrices are: possibility of application only for identified risks/hazards (there is not a tool for identification of danger/damage), a great degree of subjectivity in the risk assessment and the possibility of only a comparative analysis of risk levels. The positive features of the application of risk matrices are that it enables independent determination of the probabilities and consequences of risk and provides a qualitative definition of risk and its severity.

6.3 Semiquantitative (combined) methods for risk assessment

Semiquantitative (combined) risk assessment methods are widely used in practice, as it is often not possible to estimate the likelihood of an adverse event (especially for rare events) and the magnitude of the consequence (different consequences for different conditions). Assessing and ranking these sizes is based on the experience and knowledge of the participants in the risk assessment team. Qualitative scales with a number of qualitative descriptions for likelihood and consequence are the basis for evaluating a risk measure, which is most often determined as the product of the level of probability ranking and potential adverse effects. Each quantitative measure of risk is joined by a qualitative interpretation, that is, a qualitative description and an appropriate ranking. Common to all semi-quantitative risk assessment methods is to use the ranking approach to move qualitative to quantitative assessments of individual risk factors. To each degree of qualitative assessment, we add a rank, that is, some conditional numerical value. The ranking approach is distinguished by its simplicity, but also by its inability to detect slight differences or shortcomings.

The choice of method to be applied for risk assessment is not limited, but it is important to fulfill all the conditions for monitoring the implementation of the assessment. Semiquantitative (combined) methods are predominantly used for risk assessment. There are three approaches to risk assessment in semiquantitative methods, namely: the matrix method of risk assessment (based on a combination of matrix and table formation), the tabular method of risk assessment (based on the formation of tables (patterns) of all elements of risk assessment, as well as the risk itself), and a graphical method of risk assessment.

Matrix method of risk assessment by semiquantitative method is based on the forming of matrix based on a combination of tables and the submatrix, for the risk assessment as production of likelihood for formation of an unwanted event and its consequences. One of the recognized semiquantitative methods for the risk assessment is a matrix of 5 × 5 based on recognized and known methods of AUVA (Allegemeine Unnfall Versicherungs Anstalt—method of Austrian associations for pulp and paper production) and BG (Berufs Genossenschaften—method of German professional engineers). More steps need to be taken to establish a 5 × 5 matrix for risk assessment. Thus, the KINNEY and PILZ methods for danger/damage recognition use the danger/damage list from EN ISO 14121-1: 2007 and EN ISO 14121-2: 2007.

In addition to the matrix and tabular method, some of which were introduced by Kinney and Wiruth, a graphical method was developed to analyze the risks and cost justification of the corrective measures applied. This graphical method consists of two nomograms formed [27,28]: a risk analysis nomogram and a cost justification nomogram.

7. A new way of thinking about problems

Increasing the demands of users of technical systems require more and more efforts to ensure their useable quality, i.e., properties such as: exploitation reliability and safety, as well as stability of technological processes. Modern businesses that aim for the competitiveness of their products are recognized for their focus on reducing risk in their work processes. Such a concept is called the concept of risk analysis and management. The focus of enterprises to identify and analyze risk arises for many reasons, such as user requirements, philosophy of continuous quality improvement, Kaizen [29], competition, and others. Requirements, needs, and expectations in terms of the properties of the technical system are clearly stated in the terms which are shown in Fig. 9.8.

The tendency of the management structure and the specialists in enterprises to minimize the risk in certain projects, processes, and maintenance of technical systems encourages research in the field of

STEAM TURBINE SYSTEM USER REQUIREMENTS	
USER SEARCHES: ⟹	THAT IS IT:
Yes he can buy the system	Low life cycle costing system
That the system works	System available to be put into operation in no time
Make the system work when needed	System reliable
To keep the system in working order	System suitable for maintenance
That the system has a set of elements necessary for proper functioning	System with elements of logistical support

FIGURE 9.8

User requirements regarding the properties of technical systems.

security engineering [30], with the aim of more successful risk management [31]. Systematic analysis of quality, reliability, and safety allows reducing the risk of technical systems. Problems have been carefully studied to answer questions about the occurrence of a possible adverse event and the consequences it may cause. Of course, while studying the problem, it was assumed that someone was responsible (someone was blamed for the problem), and then action was taken. Today, this rule of conduct has changed. The focus rule is to prevent the problem from occurring. A comparison of two ways of thinking (two rules of behavior) is shown in Fig. 9.9.

Generally, the new rule, which requires early consideration of the process of using and maintaining technical systems, shapes the needs, requirements, expectations, and priorities of users in the early stages of the system life cycle, resulting in lower levels of risk in work processes.

Because of the multitude of data, because of their better visibility, easier inference, giving risk assessments and corrective measures, the Failure Modes, Effects and Criticality Analysis – FMECA method has been developed [32]. It is a means of rationalization through systematization according to the principle: "it is always better and more economical to identify the causes and prevent the failures than to detect and eliminate them later, or to bear the costs of the consequences of the failure."

THE OLD WAY ⟹	THE NEW WAY
Troubleshooting	Preventing problems
Loss control	Exclusion of losses
Quantifying reliability	Reducing unreliability
Sequential Engineering	Competitive Engineering
Three sigma methodology	Six sigma methodology

FIGURE 9.9

Old and new way of thinking about problems.

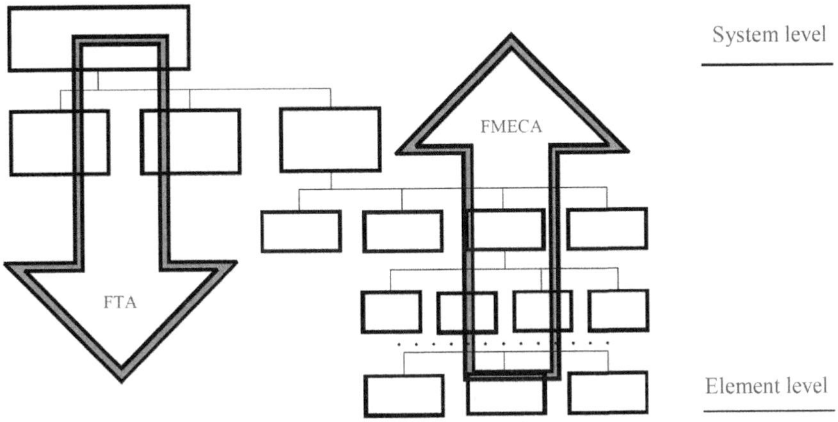

FIGURE 9.10

Top-down approach and bottom-up approach (path) in analysis: FTA and FMECA.

FMECA is associated with fault tree analysis (FTA) and the first to become popular with reliability engineers. Both methods are logically very similar, though they look quite different. Some reliability analysts debate whether it is better to conduct a top-down (FTA) or bottom-up (FMECA) analysis (Fig. 9.10). In practice, it is most appropriate to use both ways together, which means that FTA and FMECA are logically equivalent methods.

User requirements greatly encourage the motivation to implement FMECA. For example, all leading car companies have established their standards for certification of suppliers, even before certain international standards (such as Ford Q101; General Motors—Goals for Excellence; Chrysler—Five Star). In these standards, car companies require the FMECA implementation program from their suppliers (Chrysler—1986, Ford—1992, General Motors—1988). The same situation is in other sectors of the economy, such as the production of semiconductors, computers, medical devices, and aviation.

7.1 Identification of dangers in the work of thermal power plants

Danger can be defined as a set of conditions that can cause an unwanted event (injury or damage). The danger may be of different kind: natural (floods), technological (structure), sociological (war), and lifestyle (smoking). Danger identification and hazard scenarios are of particular importance for risk analysis, requiring a thorough examination and understanding of the system, with the development of technologies gaining in complexity. The objective of the analysis and the number of available system information should guide the selection of an adequate risk assessment method. Once the risk assessment method is selected, then the hazards should be identified and the possible sequences of adverse events and the factors that may lead to it should be described. The method determines the process of hazard identification. Some of the methods are qualitative, while others can provide quantitative estimation.

There are two approaches that can be used to analyze the causal links between element failures and technical system failures. These are inductive and deductive analyses. Inductive analysis begins with a set of states *in the failure* of the elements. The process is carried out by identifying (determining) the possible consequences, i.e., through the approach of "what will happen if." The failure tree analysis is an example of deductive analysis, i.e., the approach of "what can cause this." This analysis is used to identify (determine) the causal relation that leads to certain types of technical system failures.

The failure tree is a method by which a specific type of technical system failure can be expressed through the types of element failures and operator actions. The type of failure of the technical system under consideration is called a "peak event," and the failure tree that develops further displays the events that cause it. In this way, the events displayed in the tree are defined until the lowest events are analyzed. This process of tree development ends when types of element failures, referred to as basic events, occur. The failure tree analysis also includes the collection of data on the probabilities of occurrence of the underlying events.

A top event is an event whose possibility (or probability) of occurrence should be determined. Choosing a peak event is the first step in this process. The peak event can be selected based on a previous hazard analysis. It is important that the top event and the boundaries of the technical system are selected, taking care that the analysis is not too complicated or too scarce to provide the required results.

Each failure tree considers one of the many possible types of technical system failures, meaning that more than one failure tree can be constructed during the evaluation of any system. For example, when evaluating the security of protection of a technical system, the peak event most often refers to the failure of the protection system when a request to complete the requested task occurs. This peak event leads to the development of the failure tree to model the cause of this situation. It is possible to include different levels of redundancy and changes in the form of security protection of technical systems so that the probability of their failure at the required moment is very low.

The analysis of failure tree methodology uses a deductive approach to show the strengths and weaknesses of design. It is a "top-down" approach, as opposed to a "bottom-up" approach in the FMECA method. Therefore, it starts from the peak event and goes from the top down, determining the different paths in which failure types can logically cause a specific peak event to occur.

The usual procedure for analyzing a failure tree involves the following characteristic steps [33]:

1. Defining the technical system, its function of purpose, substrates, and rules for use in concrete analysis,
2. Development of simple block diagrams (hierarchical and functional block diagrams, block diagrams in terms of reliability) of the technical system, showing inputs, outputs, and connections,
3. Identification of problems and condition limits (description of the problem, i.e., unwanted peak event, and defining the position of the technical system limits),
4. Defining a specific influential peak event (type of failures at the technical system level, as the ultimate consequence of types of element failures),
5. Constructing a failure tree for the peak event, to the fullest detail possible, using the rules of prescribed logic,
6. Conducting qualitative analysis (determining the minimum set of cross-sections),

7. Collecting basic data, such as failure rates, mean running times, or probabilities of element failure types,
8. Conducting quantitative analysis (determining the probability of occurrence of a peak event),
9. Control of a fully engineered failure tree,
10. Providing recommendations for any corrective actions during the use and maintenance phase or for changes to the system design phase,
11. Documenting a concrete analysis of the failure tree and the results obtained.

Using FMECA in complex systems security engineering, although it may take too much time and cost, does not involve other potential problems besides equipment failures, such as personnel errors (human factor). In many types of technical systems, unit failures can cause system interruptions but not a security breach. In some situations, it is necessary to have an analysis method that focuses on the possibility of an event occurring, which shows the complex relationship of the cause of the event. For this reason, FTA was developed at Bell Telephone Laboratories at the request of the US Air Force. The US Air Force wanted to know the possibilities and probabilities of reckless or unauthorized launch of Minuteman missiles and reckless or unauthorized handling of nuclear installation [34]. Although failure tree analysis has been developed to quantify probabilities, it is usually much more used in qualitative form [35]. Since different factors can be presented in a systematic way, this can be investigated in any situation. Quantitative analysis and results are nevertheless desirable in many applications, but to perform quantitative analysis, qualitative analysis must first be performed [33].

Preliminary hazard analysis (PHA) is a rough inductive and qualitative method for identifying a potential hazard. Follow-up lists of potential hazardous elements and situations provide assistance with the implementation of PHA. Successful analysis requires the formation of a team made up of experts who are "familiar" with the system. Each identified adverse event is analyzed separately to describe the possible causes, consequences, and probabilities. The consequences can also be isolated example, the ones that have an impact on the environment, human health, and the economy, and therefore the opposing opinions. After that, consequences and probabilities are ranked by their severity. The analysis produces a preliminary qualitative document on possible adverse events with respect to identified sources of risk. PHA does not identify specific components that may cause large-scale damage, but may serve as a basis for future analysis with one of the methods (FMEA, FMECA, and HAZOP).

Errors, failures, or any other type of mismatch may occur at all stages of the technical system life cycle, resulting in increased system life cycle costs. Failure to resolve the discrepancies will result in poor quality, customer dissatisfaction, and high risk.

In the general case, the types of failures are divided according to the consequences into: types of failures that are detected in the stages of creating a technical system and types of failures that are detected by users of the technical system.

The aim of each manufacturer is to take measures to prevent the occurrence of failures at all stages of the technical system's life cycle.

The FMECA method was developed to evaluate the occurrence and impact of failures on a technical system or process, as well as to determine the possibility of detecting and eliminating potential failures before delivering the system to the user. However, the FMECA method is primarily designed to evaluate system and process designs, with the aim of preventing or mitigating the effects of errors and system failures on users. In addition, the FMECA method can also be applied in the following cases: when modifying a design of a technical system or process, when changing conditions of use of existing technical systems in relation to design conditions, when analyzing problems in the use and maintenance of a technical system or in the production process, and in analyzing units influences on the safety of users of the technical system or on the environment.

Therefore, the purpose of the FMECA method is to enable designers to: timely detect and prevent potential failures in the technical system, avoid or mitigate risks in the project, prevent the cost of possible early withdrawal from use due to the failure of the technical system, prevent loss of reputation in the market, as well as preventing loss of market share.

The following factors are considered as the basis for evaluating the solutions in the design, technological process, and maintenance of the technical system: possible failures of the technical system and the probability (frequency) of occurrence of potential failures, possible consequences of failures manifested to the user or the environment, as well as the possibility of detecting failures and preventing to quit the company and reach the customer.

Today, quite a large amount of practice has been accumulated in the application of various failure analysis procedures, collectively referred to as FMECA. Common to all FMECA methods are the following aspects: occurrence probabilities, diagnostic capabilities, and severity of performance in assessing the degree of failure criticality in technical systems.

The essence of the application of the FMECA method is the implementation of the following activities [36]:

- determination of all possible types of failures in the technical system that may result from a fault in the system or process design or during maintenance during the system utilization phase;
- identifying all possible consequences of each potential type of failure;
- identifying all possible causes of each potential failure;
- defining control and diagnostic measures;
- determining, for each pair, the "possible failure type—possible cause of the failure type," of the following bases for assessing the degree of criticality:
  - probability of failure type (PF),
  - severity of failure demerit value (FDV),
  - probability of failure remedy (PFR),

  where the evaluation of these substrates is usually performed using a scale of 1–10;

- Risk Priority Number (RPN) rating, for each pair, "possible failure type—possible cause of failure type," using the expression:

$$RPN = PF \cdot FDV \cdot PFR; \tag{9.3}$$

- evaluation of the estimated degree of criticality;
- proposal of potential corrective measures;
- appointment of a responsible function for the implementation of corrective measures;
- verification of corrective measures applied;
- repeat the process of determining the substrates PF, FDV, and PFR; and
- evaluation of the new value of the RPN criticality level.

PF, FDV, and PFR substrates are most commonly rated 1 to 10 (though other intervals may be used). The RPN criticality rating thus evaluated is compared with the predetermined allowable RPN_{dose} values. The solution is evaluated as satisfactory, if the RPN is $<RPN_{dose}$, and if this is not the case, then the appropriate corrective measures provide the target. In the absence of a predetermined value of RPN_{doses}, coarse criteria can be used to evaluate the solution as a function of PF, FDV, and PFR values.

According to the definition given in the International Electrotechnical Vocabulary [37], criticality of the type of failure means a set of characteristics, which determines the consequences of certain types of failure. The classification of types of failures according to the degree of criticality aims at isolating the types of failures (and eliminating the causes of their occurrence), which can lead to more severe consequences for human health and life, the environment, and technological equipment. These types of failures are catastrophic, are called failures, and are subject to preventive detection when designing and eliminating them according to priority.

The FMECA method is a documented method. The FMECA form is used to document the process. Entering data into the form starts practically from defining the subject of analysis and runs in parallel with other stages of the procedure. Data entry includes both proposed and applied corrective measures as well as the function within the company responsible for the effects. The FMECA procedure is not considered concluded unless corrective measures have achieved the effects in terms of reducing the PF, FDV, and PFR substrates, or reducing the RPN criticality. In this case, new corrective measures must be proposed and implemented until this reduction is achieved. The implemented corrective measures without planned effects significantly slow down the implementation of the FMECA procedure, so it is the obligation of the FMECA team to carefully study their final effects during the proposition of corrective measures and timely plan additional corrective measures.

Hazard and operability analysis (HAZOP) is a qualitative and inductive method for systematic analysis of the ways in which may arise deviations in the system, and for the analysis of potential risks and deviations. Based on the flow chart of the system and a set of leading words or scenarios, the analysis results in the identification of hazards or operational problems.

Probabilistic risk assessment (PRA) is one of the most important analytical methods for identifying and analyzing project risks and complex systems. The PRA process begins by identifying a set of initial events that "trigger" the system. For each such event, the analysis determines the next event that leads to the realization of the adverse (peak) event. The magnitudes of the consequences for the scenarios, as well as their occurrence probabilities, are then determined. Finally, they integrate and represent the risk profile for a given system. Fig. 9.11 shows the aforementioned methodology. Determining total risk based on a set of scenarios provides a starting point for identifying and ranking risk shares [24–27].

In the process industry, risk management has traditionally focused on considering the likelihood of specific events or occurrences. A structured approach is introduced to identify failure scenarios and conceptualize mathematical tools in numerical risk assessment—Probabilistic Safety Assessment (PSA). Systems in the process industry are usually well defined and allow sophisticated analysis tools to be applied. The methods used to identify critical events or event sequences are FMECA, HAZOP, and risk matrix.

Methods for determining the likelihood of events and the effects of potential activities include the analysis of a failure tree and an event tree. Also, some of the measures of component importance (Fussell-Vesely, Birnbaum) are very useful in trying to increase system reliability.

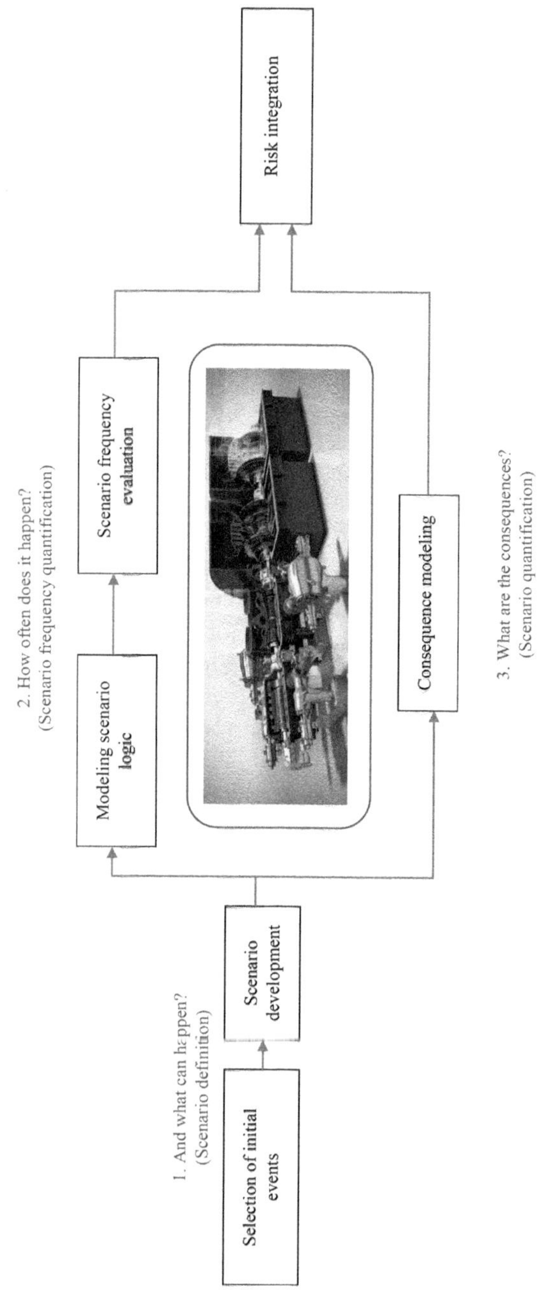

FIGURE 9.11

Implementation of risk concept in PRA.

7.2 Indeterminacy in the safety tasks

When developing complex technical and artificial intelligence systems, it is often necessary to make decisions under uncertainty. Due to the nature of indeterminacy, it is in principle impossible to absolutely predict the consequences of certain activities, technical solutions, failures, etc. The application of quantitative models is oriented to the use of the concept of probability to describe indeterminacy of different nature.

The so-called the Bayesian approach to reliability and safety assessment has proven to be very promising [25]. In the Bayesian approach, indeterminacy is seen as a probability that can be interpreted as relative frequency, as a level of conviction, or otherwise. In order to solve the uncertainty problem in this approach, it is necessary to provide a set of prior probabilities that describe the basic set. Prior probabilities can be determined using frequencies or statistical analyses. Such statistical analyses assume in advance that adequate data are available to describe the basic set. If such data are not available, then prior probabilities are given as subjective estimates by experts. The result of the analysis represents a set of posterior probabilities.

On the other hand, the fuzzy set theory is a suitable mathematical apparatus for treating indeterminacy. The fuzzy set represents such a set (interval) of values with the corresponding membership function given at this interval. For the term fuzzy, there is no standard translation for us, and in many works we find translates as: fuzzy [26], vague, indistinct, dim, fluffy, fibrous, and imprecise. However, most often this term is not translated but used in its original form, i.e., as a fuzzy set. In the case where the starting data for the FMECA analysis are not given in a dotted line, the problem of estimating the degree of criticality can be successfully solved by setting the interval of change with the corresponding membership functions. This approach can be implemented with expert systems designed for FMECA analysis.

The safety-relevant elements displayed by FMECA can have dangerous impacts that require special control procedures to ensure an acceptable low failure rate. In the literature, several definitions of security can be found. According to Niczyporuk [38], security is: "The ability of an object, when working and at failure, to cause no adverse effects on the life and health of humans and the environment." Safety is also defined as the absence of unacceptable risk (individual injury), and risk is a combination of the possibility of some undesired failures and their consequences in a given system. Thus, all the constituent elements of a technical system whose failures cause someone (something) to be injured (damaged) or dead (destroyed) must be considered as a significant aspect of safety.

Elements relevant to the performance of the objective function are elements that are uncertain, with the occurrence of a failure type having an effect on the performance of the objective function, which require control to achieve the business objective. The consequences of failure of this type of elements can have a significant impact on revenue, maintenance costs, working capacity, reputation, utility benefits, etc.

8. Modeling of the quantitative risk assessment (analysis) system

8.1 Indeterminacy and its measurement

Peak events, which are composed of a series of events, and each event are defined by parameters. Suppose a basic event is with probability of occurrence of p. This value is not known exactly, but is estimated on the basis of data or expertise. The uncertainty of p is quantified by the probability distribution: the mean is the best estimate for p, the dispersion of the distribution is a measure of uncertainty for p, i.e., greater or less uncertainty reflects, respectively, greater or less uncertainty of probability p. For all basic events, random probability samples are taken based on the uncertainty of the distributions. These probability values are then used to calculate the probability of a peak event. If these samples and the peak event are calculated multiple times, the distribution of the uncertainty of the peak event can be determined empirically. The mean value of distribution induces the probability of a peak event, and the dispersion quantifies the uncertainty of the probability. Monte Carlo method is used to describe the analysis using multiple repetition of random sampling. By using a random number generator, multiple quantification of the peak event is realized.

Uncertainty analysis requires knowledge of probability distributions (lognormal, exponential, normal, beta, etc.).

Quantification of uncertainty at the technical system level must be part of the final decision-making process. Knowledge of indeterminacy (uncertainty) in the overall risk is very important to decide on the basis of more information.

8.2 Risk scenario ranking

Scenario rankings give us insight into the importance of a failure group, but not the failure of individual components. An event (e.g., component X failure) that occurred in the structure of a low probability scenario may be neglected in defining a dominant risk scenario. If the proportion of low probability scenarios in the overall risk is comparable with a less dominant risk scenario, then the ranking of the scenario will not capture the importance of component X risk.

For this reason, we have the several quantitative measurements of significance of individual events and parameters in the PRA. The essence of these measurements is to define changes in the quantitative risk matrix caused by a change in the probability of occurrence of events in a risk model. Based on calculated significance measures, the events of one risk model are ranked based on their relative values of significance measures.

The information generated through this process is often used in risk-based decision-making and serves as a starting point for risk reduction activities and efforts, such as redesign of hardware components, additional redundancy, etc.

Significance measures are strictly formulated to determine the sensitivity of risk metrics to changes in the likelihood of occurrence of underlying events. In doing so, the risk metric takes the following form:

$$R = f(x_1, x_2, \ldots x_i, x_j \ldots x_n), \tag{9.4}$$

where x_i is the basic event and with the probability of occurrence of p_i.

8.3 Risk assessment methods according to ISO 31010: 2009

According to the norms of ISO 31010: 2009 risk represents any condition which deviates from the expected conditions. They can have different aspects such as financial, health, safety, and environmental aspects and may be at different levels such as the strategic level of organization, project, product, or process. According to the stated norm, the risk is treated as a reference to potential events and consequences. It can also be expressed as a combination of the consequences of a particular event and the associated probability of occurrence [39].

Risk Management Standard ISO 31010: 2009 (Risk management—Principles and guideline) divides risk assessments into three parts, namely, risk identification, analysis, and evaluation. Identification represents finding, identifying and describing of the risk while the risk analysis involves understanding the nature of the risk and determining the risk level. The risk evaluation is a process of comparing the results obtained from risk analysis in order to determine the impact of the risk in the observed process. The possibilities of applying certain risk assessment methods in different phases of risk assessment can be assessed as: very applicable, applicable and inapplicable, as indicated in the Table 9.6, where there are mentioned only risk assessment methods, which are commonly used, as well as those on the basis of which emerged many other methods that use the same or a similar approach in the assessment of risk.

8.4 Environmental risk assessment methods

Environmental risk assessment helps to reduce the impact of large-scale environmental disasters and is the basis for planning preventive actions, both by local governments and public services.

At the macro-organizational level, environmental risk assessments can significantly contribute to reducing further environmental degradation by applying an integrated approach to product life cycle planning and monitoring. Monitoring the environmental effects of the product life cycle is an imperative for environmentally responsible business. Also, environmental risk assessment is significant in the process of selecting materials in nanotechnology as one of the fastest growing manufacturing industries.

8.4.1 Possible impacts of the thermal power plant on the health of the population of the locality in question

The evaluation of the impact on human health of technological systems for electricity production is carried out with regard to three components: the spatial component, the temporal component, and the type of impact, i.e., effects on environmental pollution. Considering of spatial components mainly uses model that ranks the environmental impacts at local, regional, fluvial, continental, and global. From the point of view of the size of the space affected by the environmental impact of the TPP, local effects may be the emission of combustion products to a smaller or larger area in the TPP environment, while the global effects include the propagation of combustion products to very large supraregional surfaces, with occurrence of acid rain and greenhouse effect. Considering the time component, environmental and human health impacts are most often considered short-term, medium-term, and long-term. Sometimes the time component is also considered from the aspect of the TPP life cycle stages, so the period of preparation and design, the period of construction (with temporary impacts of mechanization and technological procedures of construction), the period of trial commissioning (with a high risk of occurrence of accidents), the period of regular exploitation (with permanent environmental and human health impacts), the extended working life (the period with a slightly more favorable impact to incorporate new or improved technologies) and the period of shutdown and removal of TPP (returning to the period before construction of the plant) are observed. When analyzing the effects on human health, it is important to keep in mind the definition of health established by the World Health Organization (WHO) that reads: "Health is a state of complete physical, mental and social well-being, not just the absence of illness, weakness and infirmity." Increased awareness of the impact of the polluted environment on health has recently become very large, which has been contributed by both the rapid

Table 9.6 Possibilities of applying individual risk assessment methods at different stages of risk assessment ISO 31010: 2009 (Risk management—Principles and guideline).

Tools and techniques	Risk assessment process				
		Risk analysis			
	Risk identification	Consequences	Probability	Risk level	Risk evaluation
Brainstorming	SA[a]	NA[b]	NA	NA	NA
Structured or semistructured interviews	SA	NA	NA	NA	NA
Delphi	SA	NA	NA	NA	NA
Checklists	SA	NA	NA	NA	NA
Primary hazard analysis	SA	NA	NA	NA	NA
Hazard and operability studies (HAZOP)	SA	SA	A[c]	A	A
Hazard analysis and critical control points (HACCP)	SA	SA	NA	NA	A
Environmental risk assessment	SA	SA	SA	SA	SA
Structure "What if?" Technique (SWIFT)	SA	SA	SA	SA	SA
Scenario analysis	SA	SA	A	A	A
Business impact analysis	A	SA	A	A	A
Root cause analysis	NA	SA	SA	SA	SA
Failure mode and effect analysis	SA	SA	SA	SA	SA
Fault tree analysis	A	NA	SA	A	A
Event tree analysis	A	SA	A	A	NA
Cause and consequence analysis	A	SA	SA	A	A
Cause-and-effect analysis	SA	SA	NA	NA	NA
Layer of protection analysis (LOPA)	A	SA	A	A	NA
Decision tree	NA	SA	SA	A	A
Human reliability analysis	SA	SA	SA	SA	A
Bow tie analysis	NA	A	SA	SA	A
Reliability-centered maintenance	SA	SA	SA	SA	SA
Sneak circuit analysis	A	NA	NA	NA	NA
Markov analysis	A	SA	NA	NA	NA
Monte Carlo simulation	NA	NA	NA	NA	SA
Bayesian statistics and Bayes nets	NA	SA	NA	NA	SA
FN curves	A	SA	SA	A	SA
Risk indices	A	SA	SA	A	SA
Consequence/probability matrix	SA	SA	SA	SA	A
Cost–benefit analysis	A	SA	A	A	A
Multicriteria decision analysis (MCDA)	A	SA	A	SA	A

[a]*SA—strongly applicable.*
[b]*NA—not applicable.*
[c]*A—applicable.*

industrial development in the previous century (and with it came the rapid increase in environmental pollution), as well as the growing public interest in the results of research and scientific research evidence on the impact of environmental pollution on human health. The main health consequences arising from the air emissions from TPP, according to the intensity of the consequences, are most often classified into four groups: irritations, disorders and nausea, reduction of organ function, then clinical diseases, as well as the most undesirable consequence—death. In doing so, humans are exposed to environmental pollutants through the air, water, drink, and food they bring into the body, as well as through materials that come in contact with the skin. This exposure is high, encompassing the exterior and interior of a residence in a residential or industrial facility, the extent of the health consequences depending on the intensity of exposure (concentration of pollutants), and the time of exposure. When considering the intensity of exposure, one should take into account that a person is most often exposed to the same pollutant from multiple media (food, water, and air). Occasional exposure to pollutants generally has negligible health consequences, whereas prolonged exposure to the same intensities can have very significant consequences. On the other hand, in addition to exposure, the health consequences depend on the body's ability to absorb the pollutant, its ability to disperse or concentrate it within the body, its ability to metabolize it, and to eject it from the body, introducing a new term known as the dose received pollutants, which determine the likelihood of a health consequence, as well as the type and intensity of impairment of human health. For these projects, such as TPP or other power plants with higher power, and that can have a significant impact on the environment considering their nature, size, or location, assessment of the environmental impact must be carried out in order to obtain a decision on approving studies of environmental impact in accordance with the procedure defined by the law on environmental protection.

8.4.2 Measures to be taken in the event of a major accident (in the event of an accident)

According to the adopted EU Directive 2012/18 ("Seveso III Directive"), accident is the occurrence of high emissions, fire, or explosion occurred suddenly in industrial activities, which endanger people or the environment, immediately or after a certain period of time within or outside the borders of the enterprise, and that includes one or more hazardous chemicals. The objective of this Directive, which is in effect from June 1, 2015, is the prevention of large-scale accidents that may occur due to the presence of hazardous materials in a facility, limiting the effects on people and the environment. It defines the obligations of these facilities with respect to the establishment of the systems of accident risk management that may occur on them. Entities should take measures to prevent large accidents and limit their consequences for the humans and the environment.

For the analysis of the impact in the case of accidents, it is important to analyze and define the reference accidents on TPP, the likelihood of an accident occurring, then the extent of the adverse impact of the incident, as well as the risk of the accident. Based on the results of the aforementioned analyzes, it is necessary to predict accident prevention activities, then accident response activities as well as accident elimination activities. Generally, at coal TPP, potential accident sites are: chemical and liquid fuel storage facilities, existing wastewater treatment plants, flue gas treatment plants, ash dumpsites, and technical gas storage facilities.

The design of individual TPP systems predicts measures to reduce the likelihood of an accident, as well as measures to reduce the consequences after its occurrence. Some of these measures are listed under the section relating to technical protection measures, management security measures, and information system security measures. The intensity of the possible consequences of an accident depends on the amount of dangerous substances present. Given that these quantities are not defined at the level of this document, a more detailed analysis of incident situations will be carried out in the next stages of the project. Taking into account the existing experience in similar projects, it can be said that the TPP capacity provided here does not pose a serious risk in the event of an accident, concerning, first and foremost, the nature and quantity of the hazardous substances present. The following are brief considerations of the impact of TPP on particular segments of the environment.

For new plants, operators are required to submit a notification to the competent authority within a reasonable time prior to the start of construction or operation. Operators should notify the competent authority immediately if there is a significant change in the nature of the hazardous materials, if there is any change in the process of disposing it, or if the drive is closed permanently. TPP, as possible source of accidents of larger scale, can lead to: fire qualifying failures on electrical installations, accidental situations on the system of wastewater, accidental situations in the operation of the steam circulating fluid bed (CFB) boiler, and accidents on the limestone system.

In accordance with the ordinance, which treats the contents of the report on the state of security, content information about the security measures and internal content of external emergency plans in order to protect the environment and the local population, as well as the working environment, and having in mind the location and use of the building, physical and chemical properties of materials with which it will be manipulated during the execution of the works during the construction and operation of the facility, and possibility of accidents, it is necessary to make "large-scale accident prevention plan" for the construction phase and the phase of utilization of the power plant. In doing so, the responsible person of the plant, where there are dangerous substances, must take all precautions necessary to prevent accidents of larger scale and limit their impact on people and the environment. Responsible person at any time at the request of the Ministry must show, especially during inspection, that he has taken the necessary measures prescribed by this law. For the plants in which dangerous substances are present in the prescribed quantity, the responsible person must make a "Report on the State of Security," which must contain all the necessary data. The responsible person must carry out the audit report on the security state of every 5 years, and when necessary report changes before or on the initiative of the responsible person, or at the request of the competent authority because the factual situation has changed or new technologies have appeared related to the security issue. The operator is obliged to develop and prepare "internal intervention plan," which contains measures that should be taken to the plant in the event of an accident of larger scale and submit it to the competent authority for emergency management, to create "external intervention plan" for measures that should be taken out of the plant.

The purpose of the "Plan for the prevention of accidents of bigger scales" is to control the accident in such a way that their effects are reduced to the lowest possible measure and to limit bad impact on people, then to pass on the necessary information to the public and the relevant services and bodies located in the area, as well as to enable the remediation and cleaning of the environment following large-scale disasters. In doing so, the responsible person shall ensure that the internal emergency plans within the facility are drawn up in collaboration with the employees of the installation, including relevant long-term contractors, and that they consult with the public and civil protection when drafting, updating, or modifying external emergency plans. Internal and external intervention plans must be implemented without delay in the event of a major accident or in the event of an uncontrolled accident that could lead to a major accident. The responsible person shall check and, if necessary, amend the internal or external plan of intervention, taking into account changes that have occurred in the operation of the plant, the plans of intervention, or in new technological knowledge, within a period of 3 years from the development of the plans and authority responsible for issuing environmental permits and competent inspection. During construction, the contractor is required to develop a "Fuel and Lubricant Leakage Intervention Plan" which includes the development of an emergency cleanup program in the event of unforeseen leaks or leaks of fuel, oil, chemicals, or other toxic substances. The intervention plan should include the following:

- Response teams with clearly defined duties and responsibilities;
- Training members of the leak response team, leak prevention and cleaning measures, and handling of toxic substances;
- Establishment of a leak reporting process, which includes providing information to the competent authorities, or informing the public in the event of a disaster;
- Store and maintain equipment (absorbent materials, absorbent pads, pumps, buckets, and collection tanks) to respond to emergencies;
- The possibility of emergency assessment areas of accident and plan of operations for quick action with documenting the characteristics and quantities of oils, fuels, and chemicals used and stored, frequency of delivery, methods of operation, the proximity of surface water flows;
- A protocol for informing the public in the event of an accident and what procedures must be followed to avoid health and safety risks;
- The implementation of procedures to ensure that contractors with which it subcontracted accept the "contingency plans highlighting and dealing with emergencies," and that the transport of toxic substances must be registered.
- Plan for the unexpected highlight and action in emergencies and the transport of toxic substances must be registered.

Fire measures include the following:

- Urgently call the fire unit of the replacement unit of the TPP;
- Make a call for supplies and other fire units in the immediate area;
- Call an ambulance service for injuries;
- Perform publicity;
- If necessary, suspend the operation of the replacement unit of the TPP.

Measures in the event of an accident on a wastewater system include the following:

- Dispose of all wastewater into a large retention basin of the required capacity;
- Perform extraordinary quality control of the water at a given site, in particular, parameters including fluorides, chlorides, and sulfates;
- Perform extraordinary control of groundwater regime and quality;
- Perform publicity.

Accident measures at the boiler plant include the following:

- Stop the boiler;
- Wait for its cooling;
- Eliminate the resulting malfunction.

Accident measures on the limestone system include the following:

- Suspend the operation of the replacement unit of the TPP;
- Measure the concentration of particles in the atmosphere;
- Call an ambulance service for injuries;
- Contact the hydrometeorological institute for information on wind direction and speed;
- Sampling the earth for analysis;
- Prepare an emergency plan for remediation.

Accident measures at the desulphurization system, the deNOx plant, and the particulate and noncombustion separation plant include the following:

- Suspend the operation of the replacement unit of the TPP;
- Measure the concentration of SO_2, NOx, and particulate matter in the atmosphere;
- Call an ambulance service for injuries;
- Contact the hydrometeorological institute for information on wind direction and speed,
- Make the public aware.

9. Examples of risk modeling in thermal power plants

The estimation of the frequency of failures is made on the basis of data on previous failures and cannot be considered reliable. Also, information from equipment manufacturers is often unavailable or difficult to use in a real work environment. From a security standpoint, it is necessary to use a different and more complex analysis of data than this recently presented. Particular attention should be paid to distinguishing between the individual components of the system in terms of risk exposure, since in fact a small number of components significantly affect the amount of overall risk.

The goal of the overall engineering work in the field of maintenance has been and remains the reduction of security risks to the minimum level. Name of maintenance on the basis of the assessment of risks should not be brought into connection with the increasing risk and increasing the potential for adverse effects on people, equipment, and the like. But in fact the opposite, the word is reducing the overall risk to a minimum. Decisions that may have an effect on reducing employee safety, environmental pollution, and generally leading to adverse environmental consequences should not be made solely on the basis of a risk assessment as a maintenance expense. Only when we are sure that it made the decision not to increase the risk of plants per environment can make decisions on the basis of risk just as cost.

9.1 The role of statistical analysis of safety in the exploitation phase of technical systems

The role of statistical analysis of safety in the exploitation phase of technical systems is different than in the design phase. This is explained by the fact that by that time the technical system was already in place, and therefore the security analysis aimed to clarify whether the technical system under operation meets the safety requirements. If the technical system does not meet the established security requirements, it is very important in that case to discover the fact quickly enough until the accident occurred. In this case, urgent development of corrective action aimed at enhancing security is required.

In conditions of limited resources, corrective measures should be clearly (precisely) reasoned so as not to allow waste of resources. Subjectivism in the formulation and decision-making of management decisions and corrective measures relating to the safety of technical systems creates major economic losses, which is why it is very important to develop scientific approaches to operational security management using statistical analysis in the exploitation phase [40].

Finally, it must be ensured that the corrective measures have proved effective. This whole set of issues is considered within the framework of the statistical analysis of safety in the exploitation phase and is an important element of operational safety management in the exploitation of technical systems. The problems of analyzing the safety of technical systems in the exploitation phase, integrated by common directivity, can be realized in one procedure shown in Fig. 9.12.

Safety-related events are divided into: disturbances, breakdowns and accidents. In practice, in the exploitation of technical systems, disturbances occur most often, less frequently – breakdowns and very rarely – accidents. This setting is illustrated by data relating to rotor dredges used on coal surface mines. Based on testing [41], in the period from January 1, 2006, to December 31, 2006, for the reliability assessment of the rotor excavator SRs 1200 × 24/4 × 0 (400 kW) + VR on the surface mine Land D, Kolubara, Serbia, the results were obtained regarding the frequency of occurrence of failure types (disturbances and accidents) in rotor dredge construction groups, as shown in Fig. 9.13. The result of these tests shows that in the observed period of one calendar year, 373 disturbances and accidents occurred in all units (groups of construction) of the rotary excavator and not a single accident.

The analysis of this result confirms the conclusion that the number of disturbances in the operation of the rotor dredger significantly exceeds the number of accidents. The same image is characteristic of other potentially dangerous technical systems. Disorders carry a great deal of information about the safety of technical systems, which is why their analysis, including statistical ones, makes it possible to effectively detect failure factors. As highlighted in the International Civil Aviation Organization (ICAO) Manual for the prevention of airplane accidents, the most important features of a disturbance information are: the similarity of the disturbance and the accident, which allows the detection of the same accident factors, as well as in accidents, excluding serious consequences as well as disturbances they occur much more frequently than accidents (by various estimates, by 10–100 times) and are therefore a comprehensive source of information on accident factors [42].

Therefore, in the exploitation phase, the most comprehensive channel of information is the information on disturbance of exploitation conditions, whose number exceeds the number of serious accidents and accidents by several times. Operational management of operational safety is appropriate to guide the results of the disturbance analysis. In doing so, the analysis should take into account that in conditions of limited time and financial resources, the need and urgency of developing corrective measures depends on the degree of risk of the disorder. This is conditional on the justification of identifying the indicators characterizing the severity of the disturbance and the appropriate methods of evaluating the indicators introduced against the exploitation data.

It should be noted that the ordinary reliability indicators used to analyze the safety during the exploitation phase of technical systems are not sufficient to fully describe the safety, since they do not define the severity of disturbances in the exploitation of the system. However, the degree of possible proximity (similarity) to a disruption is an important feature. Therefore, when evaluating the safety of technical systems in operation, for example, based on the mean specific number of disturbances n, from the fact that the number of disturbances n_A for technical system A is less than the number of disturbances n_B for technical system B ($n_A < n_B$), it does not follow that technical system A is safer than technical system B, since it can be shown that the degree of "proximity" (disruption) of disruption

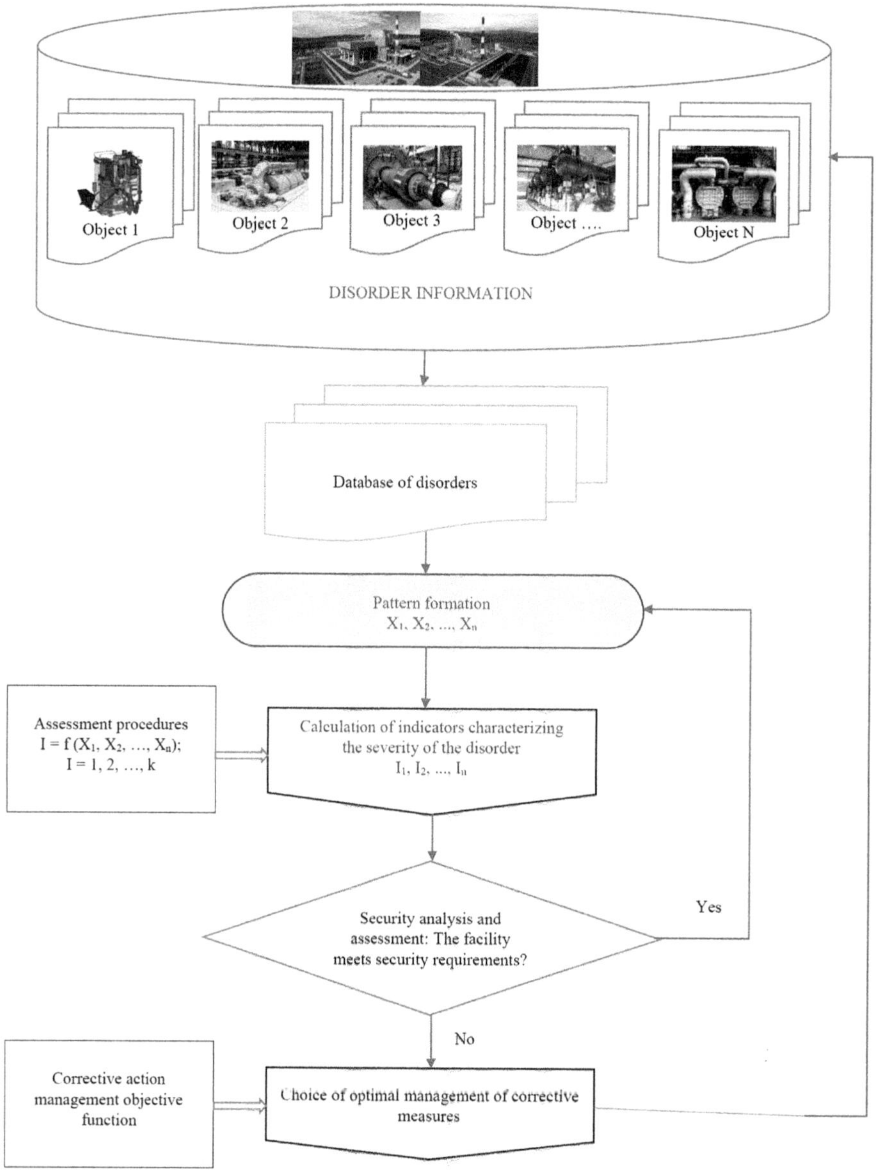

FIGURE 9.12

Algorithm for analysis of safety of technical systems during exploitation [40].

to technical system A is much higher than that of technical system B. Therefore, traditionally used reliability indicators are clearly insufficient for a complete and correct description of the safety of the technical systems in operation. This encourages the introduction of new indicators for assessing the safety of technical systems in the exploitation phase.

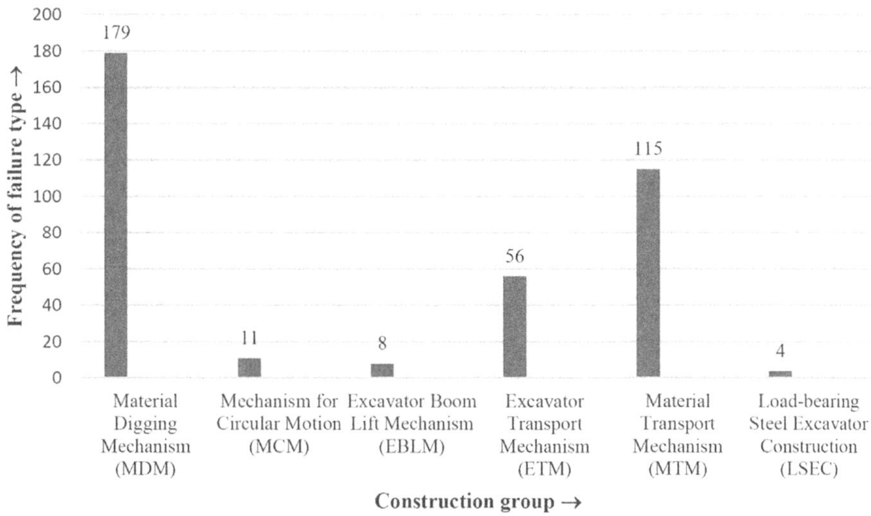

FIGURE 9.13

Frequency of failure types across units (construction groups) for SRs 1200 × 24/4 × 0 (400 kW) + VR.

9.1.1 Accident risk assessment based on the event tree method

In order to correctly introduce complex risk category indicators that describe the safety of exploitation, it is advisable to refer to the disorder development model, which is represented by a right dichotomous (divided into two opposites, contrary, mutually conditioned and assumed parts) event tree [43].

An example of the calculation of the accident risk is considered and it is associated with a wagon crash, a railway composition for transporting oil from a surface mine to a TPP, from rails. Let the disturbance of exploitation be associated with the existence of an occasional defect in the movement of the railway composition. The actual event scenario in the form of the event tree is shown in Fig. 9.14. Of course, the severity of this disorder (the safety of movement of the railway composition) should be evaluated in the form of the likelihood of its eventual transition into an accident. The higher is this probability the greater is the severity of the disorder under consideration and the lower is the safety of movement of the railway composition. Potential accident scenarios, which could be realized in the process of developing this disorder, are analyzed. To this end, any entity that has worked without failure in the process of developing this disorder should be viewed as a potential source of the initial accident event.

One of the possible accident scenarios leading to an accident associated with the fall of the railway composition wagon from the rails is shown in Fig. 9.15 (dashed by a dashed line). This scenario assumes the resonance during the movement of the railway composition.

In this case, the conditional probability $Q_i\ (I_0)$ of the realization of this accident scenario (E_1) will take the form:

$$Q_1(I_0) = P(E_1 / I_0), \tag{9.5}$$

where $P(E_1/I_0)$ is the probability of realization of this accident scenario (in the observed case, the probability of resonance occurrence).

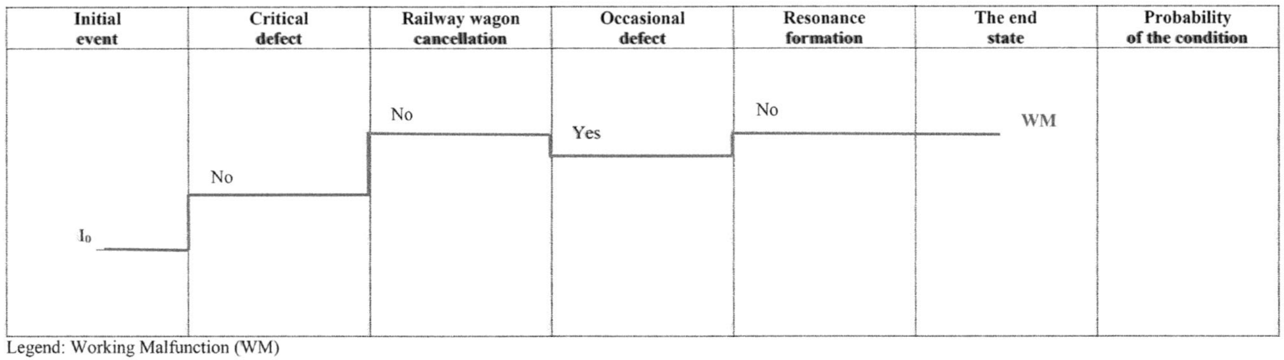

FIGURE 9.14

An event tree describing an exploitation disorder associated with the existence of an occasional defect.

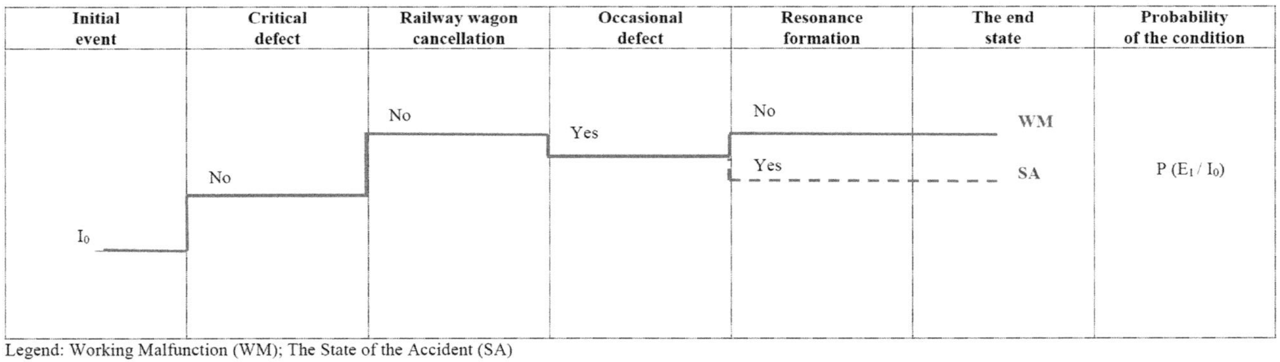

FIGURE 9.15

Event tree described by the crash scenario—resonance occurrence during the movement of the railway composition.

Another accident scenario for the initial disruption of the operation is caused by the occurrence of railway wagon failure. For this case, the event tree, which describes the crash, is shown in Fig. 9.16.

In this case, the conditional probability of the realization of this accident scenario (E_2) takes the form:

$$Q_2(I_0) = P(E_2 / I_0), \tag{9.6}$$

where $P(E_2/I_0)$ is the probability of realization of this accident scenario (in the observed case, the probability of railway wagon failure).

The third accident scenario for the initial disturbance of exploitation is caused by a critical defect. For this case, the event tree, which describes the crash, is shown in Fig. 9.17.

In this case, the conditional probability of this scenario (E_3) being realized is:

$$Q_3(I_0) = P(E_3 / I_0), \tag{9.7}$$

where $P(E_3/I_0)$ is the probability of realization of this accident scenario (in the observed case, the probability of a critical defect).

Therefore, for the disorder under consideration, three accident scenarios are possible, shown in Fig. 9.18. In this case, the severity of the particular disorder is estimated by the probability of the disorder's occurrence in the accident. Since the crash scenarios are not compatible (they are not concurrent), in this case the conditional probability of a failure occurrence $Q(I_0)$ for the initial event I_0 is determined as the sum of the probability of realization of the crash scenario:

$$Q(I_0) = Q_1(I_0) + Q_2(I_0) + Q_3(I_0) = P(E_1 / I_0) + P(E_2 / I_0) + P(E_3 / I_0). \tag{9.8}$$

If L is possible for an arbitrary disorder, then in each case, by the formulas analogous to Eqs. (9.5)–(9.7), the corresponding conditional probability Q_i ($i = 1, 2, \ldots, L$) can be calculated transition of the disorder to the accident. In other words, each disorder is assigned a corresponding L probability of a breakdown.

In this case, it is logical to form several indicators that describe the severity of the disorder and reflect the likelihood of its transition to a disaster:

- overall rating of the disorder:

$$R_\Sigma = \sum_{l=1}^{L} Q_l, \tag{9.9}$$

- maximum disorder rating:

$$R_{\max} = \max_{1 \le l \le L} Q_l \tag{9.10}$$

- mean disorder rating:

$$R_{av} = \sum_{l=1}^{L} Q_l / L = R_\Sigma / L \tag{9.11}$$

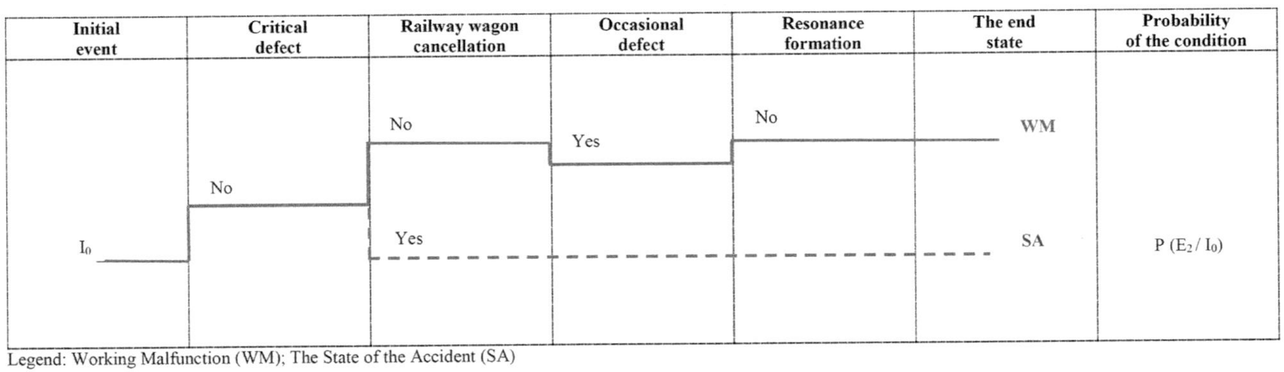

FIGURE 9.16

Event tree describing the accident scenario—railway wagon failure.

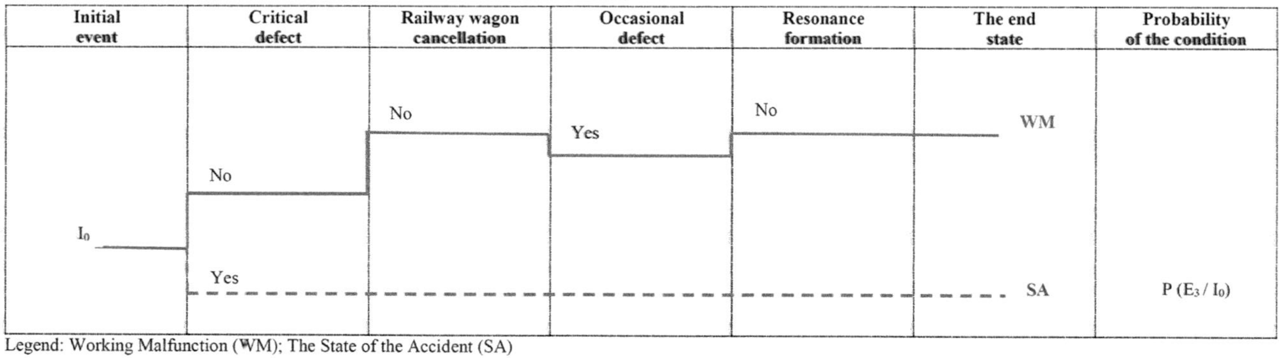

FIGURE 9.17

Event tree described by crash scenario—critical defect.

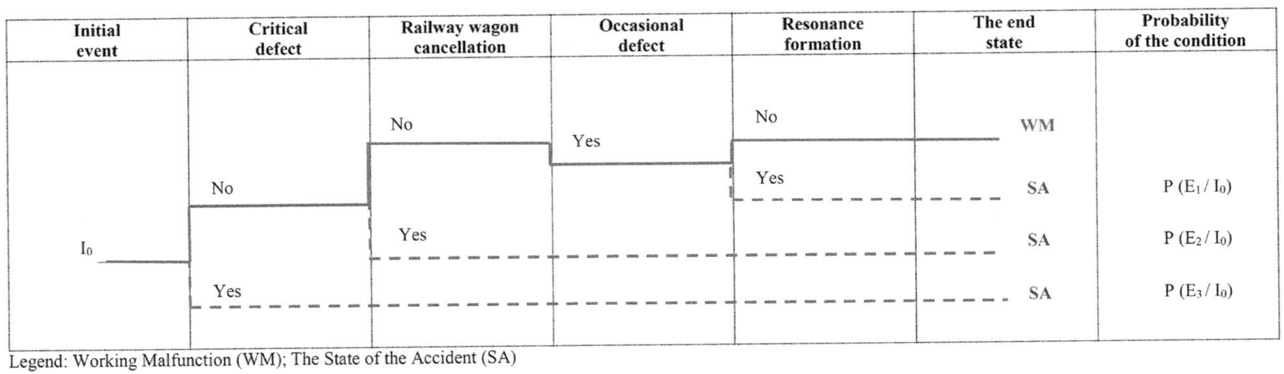

FIGURE 9.18

Event tree with three possible crash scenarios.

The magnitude of $Q(I_0)$, calculated by Eq. (9.8), represents the overall rating of the disorder.

The set of indicators (9.9) and (9.10) describes the severity of a particular disturbance in the form of the probability of a possible disruption to the accident, thus reflecting the level of safety of the technical system during operation, at a given time t of the occurrence of the disturbance.

On the other hand, the value of the probability Qi and the realization of the l-th hypothetical accident scenario can be calculated by the formula:

$$Q_l = \prod_{j=1}^{m_l} q_{jl} = \prod_{j=1}^{m_l} (1 - p_{jl}) \tag{9.12}$$

where q_{jl} (p_{jl}) is the probability of failure (failure-free operation) of the whole $j = 1, 2, \ldots, m$, that could not perform the function of the target (which could cancel) in the process of transition of the disorder to the observed accident l scenario.

It should be noted that the introduced indicators $R\Sigma$, $R_{\max}$, and R_{av} are not identical and reflect different aspects of the severity of the disorder. The indicator $R\Sigma$ shows the likelihood of a disruption occurring in an accident in (by) all potential accident scenarios and has integral significance in this respect, including unlikely accident scenarios. The $R_{\max}$ indicator shows the likelihood of a disruption occurring in an accident in the (most) probable accident scenario. The R_{av} indicator shows the average probability of a break-in, which corresponds to a single accident scenario.

In accordance with the law of large numbers, accident scenarios with a low probability of realization of Q_l in a single disorder, in a series of disorders, can occur with a rather high probability. Therefore, for technical systems that have considerable time in operation, it is useful to use the indicator $R\Sigma$, which takes into account all, including unlikely, potential accident scenarios.

If only one failure scenario is possible for a particular disorder, i.e., $L = 1$, then $R\Sigma = R_{\max} = R_{av}$. For arbitrary disorder, it follows from Eqs. (9.9)–(9.11) that $R\Sigma \geq R_{\max} \geq R_{av}$. The two disturbances A_1 and A_2 of the normal exploitation of the technical system are completely equivalent in terms of safety, provided that the following conditions are simultaneously fulfilled:

$$R_\Sigma(A_1) = R_\Sigma(A_2),\ R_{\max}(A_1) = R_{\max}(A_2),\ R_{av}(A_1) = R_{av}(A_2).$$

Thus, the previously introduced indicators (9.9)–(9.11) allow the comparison of disturbances of the same technical system or disturbances of technical systems of the same type according to the degree of their severity and the evaluation of the effectiveness of corrective measures for increasing the safety of the technical system by comparing indicators showing the severity of the disturbances before and after the introduction of measures.

A recapitulation of the consideration of indicators (9.9)–(9.11) is shown in Table 9.7.

If K ($K \geq 1$) of disturbance was registered during the exploitation period Δt, in that case on the basis of indicator (9.12) an indicator can be formed describing the level of safety of the technical system in the observed exploitation period:

$$R_\Sigma[\Delta t] = \sum_{k=1}^{K} R_{\sum k}, \tag{9.13}$$

where $R\Sigma_k$ ($k = 1, 2, \ldots, K$)— values of the total rating of the k-th disorder.

Table 9.7 Description of safety indicators.

Indicator describing the severity of the disorder (disorder rating)	Label	Basic formula for the budget	Description and scope
Overall	$R\Sigma$	(9.9)	Describes the likelihood of a disruption occurring in a disaster in (by) all disaster scenarios. It is recommended for assessing the severity of long-term disruption of technical systems
Maximum	R_{max}	(9.10)	Describes the likelihood of a disruption occurring in an accident in the (most) probable accident scenario. It is recommended for assessing the severity of disturbances of particularly dangerous technical systems
Medium (average)	R_{av}	(9.11)	Describes the likelihood of a disruption occurring in an accident, which corresponds to a single accident scenario

Indicator of form (9.13), unlike indicator (9.9)–(9.11), does not describe the severity of the disturbance of the technical system, but its safety during the exploitation period Δt. Obviously, at $K = 1$, $R\Sigma\ [\Delta t] = R\Sigma$. The previously introduced indicators of the safety of the technical system during operation are complex, as they take into account possible initial events, scenarios of disruption to the accident, and possible end states. Rating indicators allow for the implementation of risk diversification procedures, which in this case means listing all initial events that initiate disruption of the exploitation of technical systems and the corresponding values of the rating of the disorder. These data can be used for the purposes of insurance and reinsurance of the risks of potential hazardous technical systems and for evaluating the investment attractiveness of hazardous production.

9.2 Failure tree and block diagram in terms of reliability

The formation of a failure tree for a complex technical system involves the precise knowledge of all the functional connections of the elements, the causes of their types of failures, and also the consequences of those types of failures. Functional connections are obtained through the formation of hierarchical-functional schematics of the technical system and block diagrams in terms of reliability, which are the block diagrams shown in Fig. 9.19. A more structured approach takes into account primary and secondary types of failure, i.e., basic types of failure, etc.

After the failure tree is formed, its qualitative and quantitative evaluation is performed and the probability of the resulting complete failure of the technical system is calculated based on the known information about the reliability of the elements, i.e., based on the information on the probabilities and intensities of their failure types, certainty, etc.

Two graphical reliability calculation models can be proposed for the technical system, one of which is a block diagram in terms of reliability and the other a typical tree.

In practice, it is possible to select a model of the technical system structure using a failure tree or using block diagrams in terms of reliability. When the failure tree is limited to OR gates and AND gates, both models give the same result, so it is possible to convert a particular failure tree into a

(A)

FIGURE 9.19

Three types of system block diagrams.

concrete block diagram in terms of reliability and vice versa. This is illustrated by the links shown in Fig. 9.20, which provide examples showing how block diagrams translate into failure tree type diagrams in terms of reliability.

If the logical variable corresponding to the state of working capacity of the i-th element is indicated with C_i with S the working state of the technical system, then the trees shown in Fig. 9.20 correspond to the system failure trees. The line above the logical variable indicates its negation, that is, the opposite event.

In a block diagram in terms of reliability, a "link" by a block (row) means that the element shown as a block is working. This indicates that some type of failure or a specific set of failure types of a particular element does not occur. In a particular failure tree, one type of failure or a specific set of failure types of a particular element can occur as a basic event. A specific peak event occurs if element 1 or element 2 or element 3 or element n fails.

In the same way, a concrete parallel connection in terms of reliability can be depicted as a failure tree where all basic events are bound by a single gate. A specific peak event occurs, i.e., a parallel connection in terms of reliability fails if element 1 and element 2 and element 3 and element n fail.

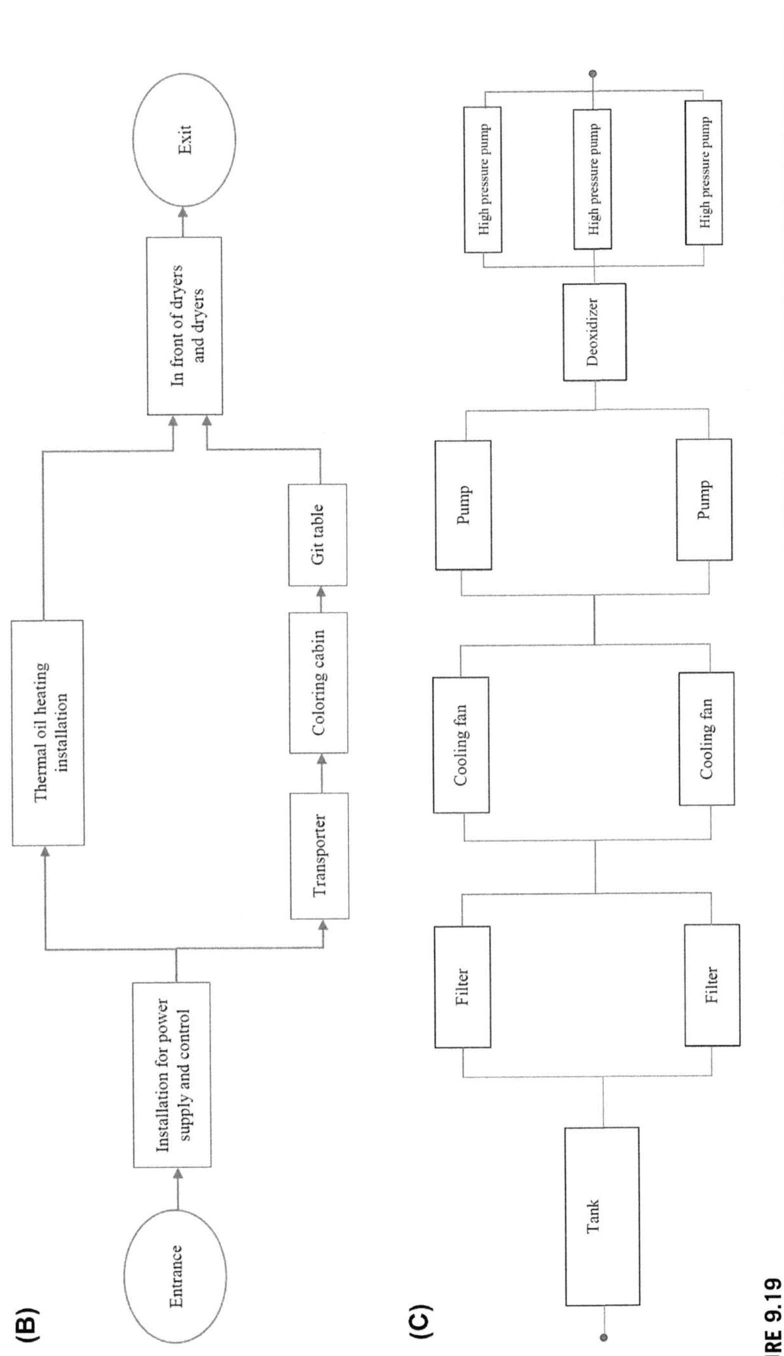

FIGURE 9.19
Cont'd.

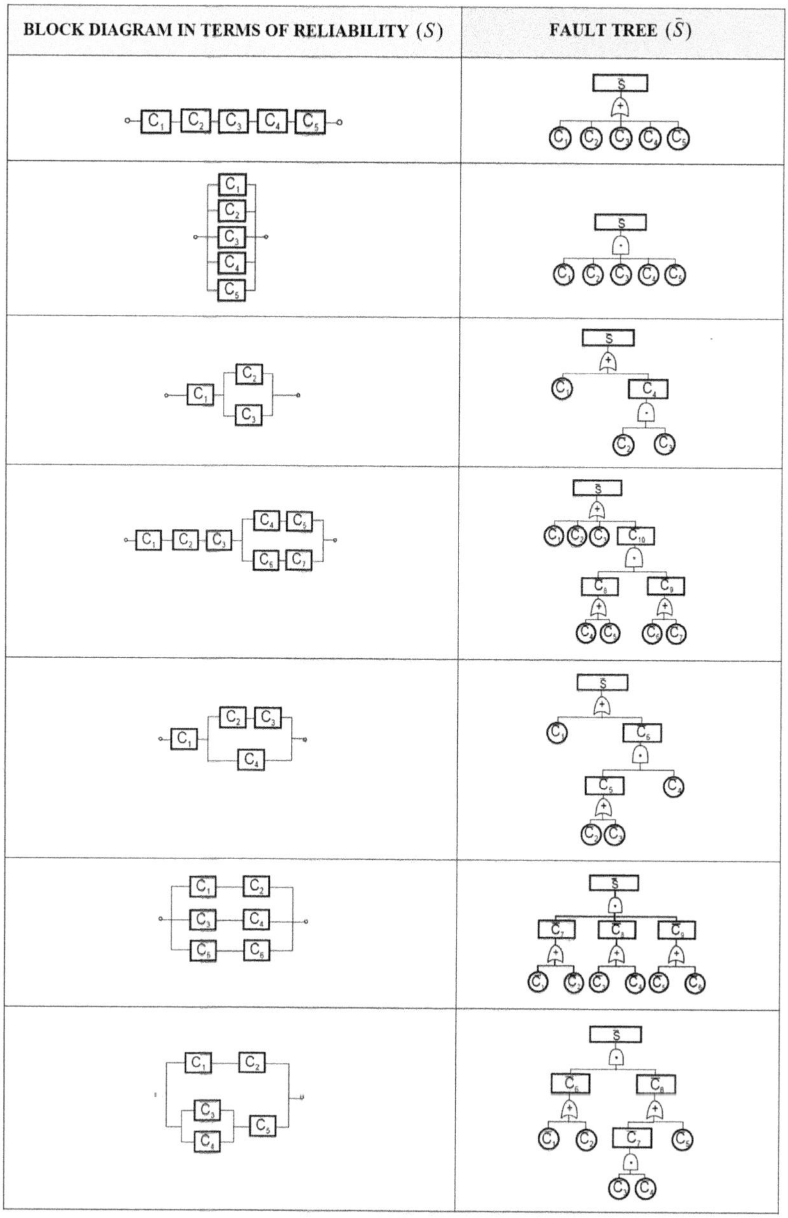

FIGURE 9.20

The link between the block diagram in terms of reliability and the failure tree [33].

10. Result discussion

Direct research into risk and safety in the operation of TPPs is, as a rule, difficult and sometimes impracticable, especially at the stage of their planning, design, and installation and commissioning. To facilitate the study of such systems, a modeling method is often used, with the real plant being replaced by its model and then the results obtained from the model transferred to the real system. The level of simplification of the real system affects the validity of the model, while the validity of the model shows how well the real system is represented across the model. During the exploitation phase of TPPs, information containing disturbance conditions, whose number exceeds the number of serious accidents and accidents by several times, is important. Operational management of the operational safety of the TPP should be directed toward the results of the disturbance analysis, and the conducted analyses should take into account that, in conditions of limited time and financial resources, the need and urgency of the development of corrective measures depends on the degree of danger of the disturbance. This is conditional on the justification of identifying the indicators characterizing the severity of the disturbance and the appropriate methods of evaluating the indicators introduced against the exploitation data.

The goal of overall engineering work in the field of maintenance was and remains to minimize safety risk. For the analysis of the impact in the case of accidents, it is important to analyze and define the reference accidents on the TPP, the likelihood of an accident occurring, then the extent of the adverse impact of the incident, as well as the risk of the accident. Based on the results of the aforementioned analyses, it is necessary to predict accident prevention measures, then accident response measures, as well as accident prevention measures. Generally, at TPP on coal the potential accident sites are: chemical and liquid fuel storage facilities, existing wastewater treatment plants, flue gas treatment plants, ash dumpsites, and technical gas storage facilities. The design of individual TPP systems predicts measures to reduce the likelihood of an accident, as well as measures to reduce the consequences after its occurrence facilities, existing wastewater treatment plants, flue gas treatment plants, ash dumpsites, and technical gas storage facilities. The design of individual TPP systems predicts measures to reduce the likelihood of an accident, as well as measures to reduce the consequences after its occurrence.

In order to perform quantitative risk analysis, a model of a real system must be made. The model is a simplified representation of the real system and contains descriptions of the relations between the observed quantities. There are several different methods for modeling a system in order to calculate its reliability. Two of these methods are functional and hardware modeling. A functional model of a real system is a model that first describes and then logically links the functions necessary for the operational work of the system. Hardware modeling uses the equipment required to operate the system to show the logical connections between the individual components and describe the functions necessary for the system to function. Often the two modeling methods are combined.

For maintenance, it is important to determine the consequences of failures by system and environment, not just the types of failures of elements and the frequency of their occurrence. By analyzing the consequences of failure of all constituent units in a technical system, it can be concluded that there are two types of consequences: those that affect safety and those that affect the performance of the objective function [27].

Failure engineering analyses provide insight into the types of failures that may occur in the whole (technical system). Thus, each unit from the list of constituent units in the technical system must be analyzed from the aspect of failure monitoring, and the consequences of failure, in particular, must be registered. The most commonly used engineering procedure to perform this task is the FMECA method. It is an extremely detailed method that is mainly focused on design and partly on maintenance options. The results of these analyses of all the constituent parts of the technical system are divided into two groups according to the significance of the consequences of the failure. Thus, there are elements important for security and elements important for performing the function of the goal.

11. Conclusion

In quantitative risk analysis, we are interested in the likelihood of an adverse event occurring. To this end, models were developed, for example, a failure tree, with probabilities of basic events as parameters. These probabilities are subjective and express the uncertainty of quantity. By developing and using probabilistic models, it is possible to associate uncertainty with each random variable in the model. The indeterminacy analysis allows one to calculate the indeterminacy in the probability (frequency) of occurrence of a peak event as a result of the indeterminacy in the probabilities (frequencies) of occurrence of the underlying events. Knowledge of reliability and event distribution is required to perform the analysis. Ranking risk scenarios based on their occurrence frequencies gives us limited insight into the overall risk allocation given the importance of the proportion of individual events as well as the failure of components in the overall risk.

References

[1] Encyclopedia Britannica Online, https://www.britannica.com/topic/Britannica-Online.

[2] P. Galbraith, N. Clatworthy, Beyond standard models-meeting the challenge of modeling, Educ. Stud. Math. 21 (2) (1990) 137–163.

[3] K.A. Cheng, Teaching mathematical modelling in Singapore schools, Math. Educat. 6 (1) (2001) 62–74.

[4] M.J. Willis, M.T. Tham, Advanced Process Control, School of Chemical Engineering and Advanced Materials, Newcastle University, 2009. http://lorien.ncl.ac.uk/ming/advcontrl/apc.htm.

[5] I. Mihajlović, et al., Kinetics and mechanism of As_2S_2 oxidation, J. Serb. Chem. Soc. 70 (6) (2005) 869–877.

[6] L. Papić, Z. Milovanović, Maintenance and Reliability of Technical Systems, DQM Monograph Library Quality and Reliability in Practice, Book 3, Prijevor, 2007 (In Serbian).

[7] U. Schindel, et al., Modeling and simulation, Microprocess. Microprogram. 10 (2–3) (1982) 139–151.

[8] K. Gaska, A.J. Wandrasz, Mathematical modeling of biomass fuels formation process, Waste Manag. 28 (6) (2008) 973–985.

[9] K. Großmann, et al., Methods for modelling and analyzing process chains for supporting the development of new technologies, Procedia Mater. Sci. 2 (2013) 34–42, https://doi.org/10.1016/j.mspro.2013.02.005.

[10] M. Avalle, et al., Experimental and numerical characterization of a mechanical expansion process for thin-walled tubes, J. Mater. Process. Technol. 214 (5) (2014) 1143–1152.

[11] R. Jovanović, et al., Sensitivity analysis of different kinetic factors for numerical modeling of Serbian lignite DE volatilization process, Int. J. Heat Mass Tran. 72 (2014) 489–500, https://doi.org/10.1016/j.ijheatmasstransfer.2014.01.036.

[12] J.B. Johnson, K. Omland, Model selection in ecology and evolution, Trends Ecol. Evol. 19 (2) (2004) 101−108.

[13] A. Demodoran, Equity Risk Premiums (ERP): Determinants, Estimation and Implications, New York University, 2008, https://doi.org/10.2139/ssrn.1274967.

[14] W.R. Knechel, Auditing: Assurance and Risk, second ed., South-Western College Publishing, Cincinnati, OH, 2000. ISBN-10: 0324022131, ISBN-13: 978-0324022131.

[15] I.A. Moosa, A critique of the advanced measurement approach to regulatory capital against operational risk, J. Bank. Regul. 9 (3) (2008) 151−164, https://doi.org/10.1057/jbr.2008.7.

[16] A. Champan, Regulating chemicals - from risks to riskiness, Risk Anal. 26 (3) (2006) 603−616.

[17] R.F. Knight, D.J. Petty, Philosophies of risk, shareholder value and the CEO, in: R.J. Chapman (Ed.), Simple Tools and Techniques for Enterprises Risk Management, John Wiley, London, 2001.

[18] V. Gajović, G. Radivojević, Risk management in the insurance by AHP method, Tech. Manag. 64 (4) (2014) 687−693 (In Serbian).

[19] S. Kaplan, J. Garrick, On the quantitative definition of risk, Risk Anal. 1 (1) (1981) 11−27, https://doi.org/10.1111/j.1539-6924.1981.tb01350.x.

[20] B. Bemeleit, et al., Methods of Risk Assessment and Their Suitability in a Logistic Environment, Symposium on Risk Management and Cyber Informatics, SCI 05, 2005, pp. 425−431. https://www.researchgate.net/publication/237330932_Methods_of_Risk_Assessment_and_their_suitability_in_a_logistic_environment.

[21] B. Bemeleit, J. Schumacher, Risk management for transportation of sensitive goods, in: Proceedings of 10th International Symposium on Logistic, 2005, pp. 492−498. www.researchgate.net/publication/257416714_Risk_Management_for_Transportation_of_Sensitive_Goods.

[22] V.M. Gajević, Risk Modeling in Logistic Processes and Application in Transportation Insurance, Ph.D. Thesis, University of Belgrade, Faculty of Transport and Traffic Engineering, Beograd, 2015 (In Serbian).

[23] Z. Milovanović, Monographs: Energy and Process Plants, Tom 1: Thermal Power Plants - Theoretical Foundations, University of Banja Luka, Faculty of Mechanical Engineering Banja Luka, Banja Luka, 2011 (In Serbian).

[24] Z. Milovanović, Monographs: Energy and Process Plants, Tom 2: Thermal Power Plants - Technological Systems, Design and Construction, Exploitation and Maintenance, University of Banja Luka, Faculty of Mechanical Engineering Banja Luka, Banja Luka, 2011 (In Serbian).

[25] Z. Milovanović, Modified Method for Reliability Evaluation of Condensation Thermal Electric Power Plant, Ph.D. Thesis, University of Banja Luka, Faculty of Mechanical Engineering Banja Luka, Banja Luka, 2000 (In Serbian).

[26] Z. Milovanović, Optimization of Power Plant Reliability, University of Banja Luka, Faculty of Mechanical Engineering Banja Luka, Banja Luka, 2003 (In Serbian).

[27] B. Ilić, Applied Risk Modeling Methods/Applied Experimental Modeling Methods, Script, College of Professional Studies in Novi Sad, Novi Sad, 2018 (In Serbian).

[28] G.F. Kinney, A.D. Wiruth, Practical Risk Analysis for Safety Management, Naval Weapons Center, China Lake, California, USA, 1976, pp. 9−12.

[29] M. Imai, Gemba Kaizen. A Commonsense Approach to A Continuous Improvement Strategy, second ed., Mc Graw Hill, New York, 2012. ISBN13: 9780071790352.

[30] B.S. Dhillon, Engineering Safety, Fundamentals, Techniques and Applications, World Scientific Publishing, New Yersey, 2003.

[31] J. Reason, Managing the Risks of Organizational Accidents, University of Manchester, Manchester, 1997.

[32] D.H. Stamatis, Failure Mode Effect Analysis: FMEA from Theory to Execution, second ed., American Society for Quality, Milwaukee, 2003, 487 p.

[33] L. Papić (Ed.), Fault Tree Analysis (FTA), Theoretical and Applied Aspects, DQM Research Center, Čačak, 2004 (In Serbian).

[34] W. Hammer, Product Safety Management and Engineering, second ed., Willie Hammer, 1993.
[35] H. Kumamoto, E.J. Henley, Probabilistic Risk Assessment and Management for Engineers and Scientists, second ed., Institute of Electrical and Electronics Engineers, Inc., New York, 1996.
[36] L. Papić, M. Khan, Quality assuranace on the basis of failure modes, effects and criticality analysis: review, communications in dependability and quality management, An Int. J. 1 (1) (1998) 101–113.
[37] International Standard IEC 50 (191), International Electrotechnical Vocabulary (IEV), Chapter 191, Dependability and Quality and Service, International Electrotechnical Commission, Geneva, 1990.
[38] Z.T. Niczyporuk, Risk assessment in mining technologies, Arch. Min. Sci. 42 (1) (1997) 163–175.
[39] International Standard ISO 31000, Risk Management - Principles and Guidelines, 2009.
[40] L.N. Aleksandrovskaya, et al., Statistical Methods for Safety Analysis of Complex Technical Systems, Logos, Moskva, 2001 (In Russian).
[41] M. Pantelić, et al., Maintainability and Safety Engineering of Excavator Units, DQM Research Center, Prijevor, Serbia, 2011 (In Serbian).
[42] Human Factor, Issue 12, Human Factors Role in Aviation Maintenance and Inspection, Circular ICAO, International Civil Aviation Organization, Montreal, 1995.
[43] L.N. Aleksandrovskaya, et al., Mathematical Bases of Technical Systems Risk Management, vol. 2, Probability-Statistical Methods of Risk Estimate AIR, Moscow, 2019 (In Russian).

CHAPTER

10 Analysis of the technical system reliability assessment with the application of technical diagnostics

Dejan Lj. Branković[1], Zdravko N. Milovanović[1], Ljubiša R. Papić[2]

[1]*Department of Hydro and Thermal Engineering, University of Banja Luka, Faculty of Mechanical Engineering, Banja Luka, Republic of Srpska, Bosnia and Herzegovina;* [2]*DQM Research Center, Čačak, Serbia*

Notations

t_r	working hours
t_f	failure hours
t_i	working hours till the failure
t_{isr}	average running time
$R(a)$	reliability function
$F(a)$	unreliability function
σ	dispersion
$F_o(t_i)$	cumulative functions of the observed data of the state of the real system
$F_E(t_i)$	cumulative function indicated in the assumed distribution
D_i	difference between the cumulative functions of the observed data of the state of the real system and the values of the cumulative function indicated in the assumed distribution
D_{CR}	critical difference value of the cumulative functions
D_{max}	maximum difference value of the cumulative functions
λ	intensity of failure
$f(t)$	failure probability density function
β	shape parameter of the meter distribution
η	scale parameter of the meter distribution
γ	position parameter of the meter distribution
$F_e(t)$	cumulative probability
t_{cum}	cumulative operation conditions number
t_{mid}	midinterval hours
n_i	month

1. Introduction

The basis for the development and survival of every serious state and society as a whole is the creation and maintenance of production business systems. Manufacturing industrial systems are complex structures whose function is to create new products of a certain useable value and thus to provide a social breakthrough as a whole. If we talk about manufacturing industrial systems, we can say that they

https://doi.org/10.1016/B978-0-12-819582-6.00010-1

represent a set of interconnected subsystems that have a common function of purpose—production and creation of new useable value. For the industrial system to be functional, it is necessary to ensure the interconnectedness of all parts of the system in terms of the realization of the set tasks. The fulfillment of the goal function in the production system is possible if we have a functional technical system of production, as well as other systems, such as the system of integral support, in which beside logistics, control, and procurement there is also a maintenance service.

The specificity of an industrial production system is that it is a system that is dynamic and oriented to the creation of a new, useable value, or a specific semiproduct or product and whose performance (efficiency and effectiveness) can be uniquely defined and measured through production indicators. Production systems are cyber systems because they are simultaneously:

- dynamic (there is a continuous flow of information and resources that result in a specific product);
- manageable (there is a clear organization and rules by which the system should be managed);
- open (it would be impossible to imagine the isolation of such a system that constantly seeks to communicate both internal units and systems with the environment—input materials and information, suppliers, customers, society, politics, competition, etc.);
- self-managing (in addition to being managed by the highest structures, the system also acts as a separate, manageable, "living," and evolving entity).

The main characteristic of production industrial systems is their working conditions in which the systems fulfill the intended function of the goal, that is, fulfill their projected role. Production systems can be found in two states over time: operating conditions and failure conditions. The tendency is to minimize the real-time conditions of failure in the failure, and in this endeavor the maintenance system of production systems plays a key role. The very creation of a production system is conditioned by the need to define certain procedures that ensure the elimination of possible causes of system failure.

2. Theoretical settings of industrial system maintenance

In theory, there are numerous definitions of maintenance as an extremely widely viewed term. In essence, maintenance is a description of the activities that cause the set goals to be met by this term. One of the first definitions was presented in 1963 at the Organization for Economic Cooperation and Development (OECD) Congress:

"... maintenance is the function of an enterprise entrusted with the ongoing control of facilities and the carrying out of certain repairs and audits, thereby enabling the continued functional capability and preservation of manufacturing facilities, ancillary facilities and other equipment ..."

According to Ref. [1], the term "maintenance" means the implementation of all measures necessary for a machine, plant, or the whole plant to function properly, developing performance within the prescribed limits, i.e., with the required performance and quality, without failure, and with the prescribed environmental protection, assuming good security for all conditions, that is, with the necessary logistical support. Modern understanding of maintenance implies an orderly system of activities aimed at ensuring stable conditions of functioning of the production system. Maintenance involves taking measures and actions that are oriented to prevent any foreseeable anomalies that may occur in the process and lead to a violation of the quality requirements of the final product or the occurrence of a failure condition on parts of the technical system. In other words, maintenance is a set of segments or elements that jointly provide maintenance of a technical system in accordance with the set requirements and criteria.

Given the importance of maintaining or restoring the condition or performance, there is a basic division of type or strategy of maintenance into preventive and corrective maintenance (Fig. 10.1).

Corrective maintenance is defined as a series of measures to bring a system to a functional state after a system failure occurs or a failure is uniquely diagnosed. European Standards (EN standards) 13306 (2001) [2] defines corrective maintenance terminology as:

"... maintenance performed after the failure is detected and intended to set the system in a state in which the required functions can be provided ..."

According to European Standards (EN standards) 13306 (2001), preventive maintenance is defined as:

"... maintenance at previously planned intervals, in accordance with the criteria described above in order to reduce the likelihood of failure or degradation of system parts ..."

Preventive maintenance can be understood as a series of activities aimed at preventing failures or failures. Corrective maintenance refers to activities undertaken after failure or apparent failure. Maintenance development has been refined and modified with the development of new technical systems. The concept of elimination of redundancies after they occur is increasingly abandoned as they have proven to be an extremely large expense for a continuously operating production and business system that, with increasing complexity of system elements, has evolved into a very expensive system of functionally tightly integrated parts. It is not uncommon for the failure of just one small part or subassembly to fail a complete system; this further represents a significant loss of production and therefore of business. Modern business systems are making great efforts and significant investment in both equipment and maintenance of the production system, as these are two closely related elements.

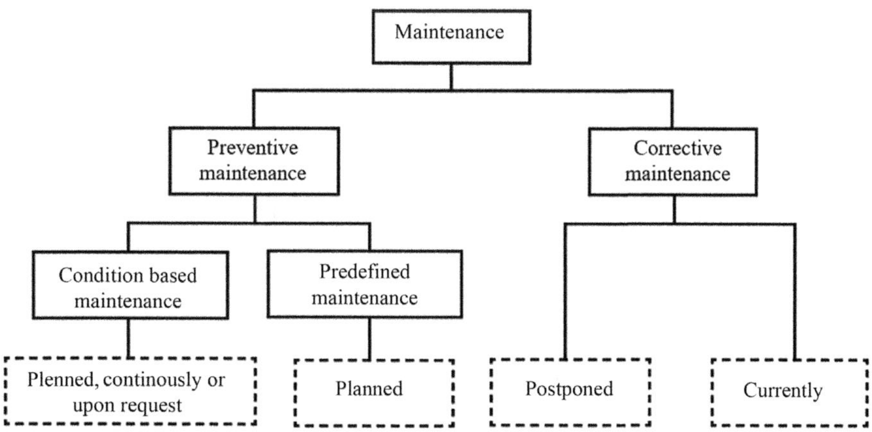

FIGURE 10.1

Maintenance division according to EN 13306 (2001) [2,3].

Maintenance is seen as:

- complex process of monitoring the behavior and knowledge of the complete technical and technological production cycle—if the nature of the technological process is not sufficiently known, it is not possible to design an appropriate form of maintenance;
- predicting the behavior of system elements in the current and future times—the basis of the prediction is experience, monitoring history, and the potential use of technical diagnostics to monitor and record operating parameters;
- developing a plan for preventive inspections, repairs, replacement of worn-out parts, or parts for which a (pre)cancellation condition is foreseen—the basis for the prevention of cancellation conditions;
- planning of available and future required resources (skilled workers, materials, parts, etc.)—material and human resources are the basis for quality maintenance;
- planning of a network of service and logistical support—services of specialized companies for the service of specific and purposefully produced technical systems;
- construction and planning of interactions with other parts of the production and business system (emphasis on the manufacturing sector)—technology link—maintenance, etc.

The notion of maintenance as an "accidental cancellation activity" has been overcome. This method of corrective action at the moment of the already stalled production proved to be financially unfavorable because it disrupts the continuity of production with an uncertain outcome at the time of solving the already encountered problem.

2.1 Maintenance concepts of industrial systems

The function of maintaining a modern industrial system is to provide a high degree of efficiency. In terms of maintaining a production system, effectiveness is closely linked to the notion of efficiency. Efficiency means doing things the right way, fast, and quality, and being efficient means doing the right things. The proper choice of the proper maintenance concept is of great influence on the efficiency of the industrial system. Industrial production systems operate in a continuous mode so that stable working conditions are extremely demanding work. Practice shows that the right choice of maintenance concept can significantly improve the conditions that ensure the continuity of the plant.

There are several specific maintenance concepts that can be found in the literature:

- corrective maintenance;
- preventive maintenance;
- condition-based maintenance (CBM);
- terotechnology maintenance;
- logistics maintenance;
- scheduled maintenance with overhauls (so-called legal maintenance);
- maintenance as required;
- total productive maintenance (TPM);
- reliability-based maintenance (RBM);
- expert systems;
- self-preservation.

Historically, the development of maintenance concepts has been accompanied by the development of technical systems and all associated infrastructure, computer support, and positive experiences in the field of maintenance and maintenance monitoring. Unlike corrective maintenance in the past and action to eliminate failures when they have already occurred, today we have the development of expert systems and modern self-determination systems in situations of disturbance of optimal conditions of exploitation (Fig. 10.2).

In the case of industrial production systems, the choice of an appropriate maintenance concept depends on a number of factors:

- nature of technological process;
- complexity degree of the technical system;
- organizations of production and business system;
- economic justification, that is, assessing the cost-effectiveness of investing in more modern system than expected;
- financial capabilities of the business system;
- available human resources in maintenance (degree of expertise), etc.

If we are talking about industrial systems for the production of hygiene paper, they are characterized by continuous production processes with high reliability requirements of the technical systems of which they are composed. In this case, the basis of the production process is the technical paper machine system (Fig. 10.3).

Paper machine is a complex technical and technological system whose elements are interconnected and whose mutual function depends directly on the conditions of exploitation, environmental conditions, method of manipulation, etc. Often, failure of only a small number of components causes the complete production process to be stopped. One of the most important features of a paper machine

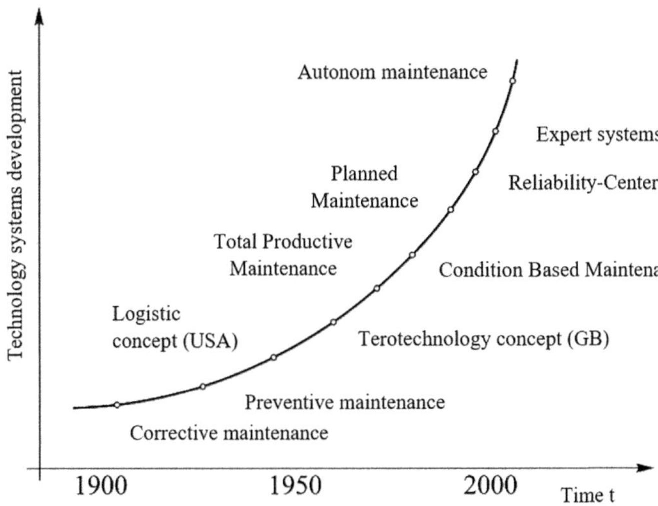

FIGURE 10.2

An overview of the development of maintenance procedures and concepts [1].

FIGURE 10.3

Technical paper production system—paper machine.

system is the requirement for a continuous system of work. The basis of the machine is a drying roller that is heated to operating temperature and any production downtime disrupts both technological conditions and high costs in providing energy—the consumption of steam or gas for heating. In addition to the high consumption for heating, there is also significant consumption of electricity (a large number of high-voltage and low-voltage motors) as well as air (operation of pneumatic control systems, a series of actuators, and the like). All this sets very high requirements for the readiness of the plant and for ensuring the highest possible reliability of the technical system. Under these conditions, defining and establishing maintenance system is an extremely complex task.

2.1.1 Maintenance concept according to the state of technical system of the paper machine

If we look at industrial systems for the production of hygiene paper, we can conclude that these are specific technical systems that have been developed by specialized equipment manufacturers. Since most of the equipment, machines, subassemblies, and assemblies of certain technological units are involved, the purchase price and the maintenance of such equipment are very expensive. The legitimate question arises: Which concept of maintenance gives the optimum between meeting the requirements for high reliability on the one hand and the most economical system of maintenance organization?

For modern industrial systems operating in a continuous mode, such as technical systems for the production of hygienic paper, the application of the concept of maintenance according to the state gives satisfactory results. This modern concept involves the use of modern devices for monitoring and analysis of currently measured state quantities that characterize the state of the equipment within the technological or work process.

Terminologically, maintenance by condition is defined according to European Standards (EN standards) 13306 (2001) as:

"… preventive maintenance which include a combination of condition monitoring and/or inspection and/or testing, analysis and the ensuing maintenance actions …" [2].

There are numerous papers in the field of development of maintenance concepts by condition. Jong-Ho and Hong-Bae in their research [4] analyze several aspects of the maintenance concept according to the condition: definitions, international standards, procedures, and techniques of maintenance with display of specific case studies by introducing case studies. In cases of high-value industrial property, the activities of operating and maintaining production systems require significantly higher costs and much more effort than in the activities/stages of equipment installation. The use period of industrial equipment is much longer than the installation period and during the use of equipment there are situations of potential risk/system failure that can have catastrophic effects on the industrial system. More recently, with the advent of information and communication technologies, one can obtain an overview of information on the status of assets during the period of use. New sophisticated technologies provide new solutions to improve the efficiency of a property's operations. One of the basic problems of implementing a maintenance-by-condition concept is defining a condition diagnosis of assets based on systematic monitoring, predicting the timing and conditions of system failure, and planning appropriate maintenance measures to avoid system failure.

An overview of the concept of maintenance with definitions of different terms and of the history and current development, applications, and challenges of exploring the concept of preventative maintenance is presented by Prajapati, Bechtel, and Ganesan in Ref. [5]. Different maintenance strategies are presented with their advantages and disadvantages over each other. The analysis involves the application of different methodologies and techniques that are applied in practice. Maintenance concepts by condition have found practical application in various fields such as automotive, manufacturing, aviation, medicine, and the like. Conducting the concept involves different disciplines such as data collection and retrieval, artificial intelligence, statistical methods of data processing, and others. These disciplines make it possible to predict future conditions/events based on a database of previous time interval and current behavior of the observed system. These predictions of the future behavior of the system are made possible by the application of certain software solutions for the processing and implementation of forecasting techniques. The social aspect of implementing an advanced maintenance system is reflected in the reduction of operating costs of the complete business system.

Rastegari and Marcus give an account of the maintenance concept implementation according to the state in the Volvo industrial system in Sweden [6]. Survey data were collected during the implementation of a pilot project to introduce a maintenance system according to the condition. The aim of the pilot project is to introduce a state monitoring system for some of the key critical components within the heat treatment process. Two on-line monitoring techniques were applied to the position of the electric motor bearing: vibration analysis and the shock pulse method (SPM). The chapter describes the elements of the process, implementation of the new system, monitoring, and how the results of monitoring are analyzed. The measurement data at the positions of the heaters of the heat treatment furnace fan were analyzed as well as the failure data of the observed system. By obtaining a database of system behavior, it is possible to evaluate in what direction to develop the process of maintaining these positions, control costs and significantly affect the overall effectiveness of the system. A very important element is the influence on the protection of human environment, which should also be given considerable attention.

One example of the transition from the usual corrective maintenance method to a maintenance system by state, by introducing a diagnostic vibration monitoring system, is presented in Ref. [7]. Radionov, Evdokimov, Petukhova, Shokhina, and Yabbarova provide typical examples of rotating elements' failure installed in an existing production process. The analyses include failures caused by mechanical damage to the pumps, damage to the bearings of the electric motors, couplings, and analysis of the direct vibration spectrum at the aforementioned positions. The second type of analysis is related to rotating elements with roller bearings and damage to the shaft sleeves. These analyses include spectral analysis of vibrations. The work also includes an overview of the types of failures in the case of induction motors. In addition to stationary systems for vibration diagnostic measurements, the operation and benefits that can be achieved with portable vibration diagnostic analyzers are described.

2.1.2 Technical diagnostics as a precondition of maintenance according to state

The basic purpose of introducing a maintenance concept by condition is to avoid equipment failure conditions for the application of corrective actions by the maintenance service. The basic precondition for applying the maintenance according to state is the introduction of a system of technical diagnostics. Condition maintenance has the function of converting measurable energy input quantities (temperature, vibration, noise, etc.) into the desired shape or effect (element state, comparison with allowable values for safe operation, forecasts for failure occurrence, etc.) in space and time. Maintenance system according to state uses the characteristics and parameters of the monitoring technique (vibration, sound, heat, etc.) in terms of detecting performance disruptions or changes in performance of the observed system.

In recent decades, the term technical diagnostics has come into all branches of technology, especially in electrical engineering and mechanical engineering. This term means all activities that are used to evaluate the current state or provide a forecast of the behavior of a machine, device, equipment, etc., without disassembly/dismantling or destruction over a period of time.

The technical diagnostic process includes the following activities:

- research into technical diagnostics;
- analysis of measurement, test, and control results;
- making conclusions about the condition and features of the technical system;
- forecasts of the future state of the system, defining recommendations for the sustainability of that state.

The basic task accomplished by technical diagnostics is timely forecasting of malfunction of parts of the technical system. The implementation of technical diagnostic measures increases the overall degree of reliability, availability, and effectiveness of the complete technical system.

Jardine, Joseph, and Banjević demonstrate the development of an optimal maintenance program based on vibration monitoring of key bearing positions of the food production system [8]. Statistical analysis of the measured diagnostic results is performed using the EXAKT software package. The system identifies the main vibration signals/values that are necessary for risk assessment. After determining the risk curve using a proportional hazard model and obtaining cost data, the optimal maintenance model is identified. During vibration monitoring, it is possible to identify and perform key measurements/tests that benefit the maintenance management system and significantly reduce the cost of additional analysis and testing of equipment used and observed in the manufacturing process.

The application of the diagnostic model of vibration analysis for the case of centrifugal pumps with power from 15 to 300 kW is shown in Ref. [9]. Ugechi, Ogbonnaya, Lilly, Ogaji, and Probert provide the results of an analysis that, among other things, showed a corresponding value of vibration levels from 0.9 to 2.7 mm/s for safe operation of the plant, while for vibration values from 2.8 to 7.0 mm/s continuous measures of the system state are necessary as it goes into the state of increased risk. The diagnostic model that is described uses computer support within the artificial intelligence and neural network models and is applicable to other rotational elements within the specified power ranges. In this case too, the correlation between achieving high productivity of the industrial system and achieving lower costs of production with the application of maintenance policy according to the state and application of modern achievements in the field of technical diagnostics is shown. Ensuring the degree of reliability and availability of production system equipment is an enduring goal in the development of maintenance concepts.

Studying vibration problems with rotating machines, Criqui [10] argues that a large number of vibration problems are caused by a number of causes, the most common being imbalance, misalignment, and resonance. However, monitoring the operation of rotary machines often leads to the conclusion that it is a vibration, and in fact it is some completely different external disturbances. It should always be borne in mind that vibrations are not a problem in themselves, but are a "symptom," i.e., an indicator that there is a problem, and therefore a solution. In addition, it should be taken into account that vibrations are almost always measured indirectly, so that the measurement results are more susceptible to external influences and are less accurate [10].

Lifson et al. [11] conducted vibration studies on rotary machines. They state that vibrations can be the result of various phenomena and can affect different parts of the machine, and that most problems caused by vibration can be classified into: structural fractures caused by material fatigue or dynamic overload; wear, friction, or fatigue of bearing surfaces, gears, couplings, etc.; and performance degradation due to internal machine clearances [11].

Studying vibration problems of complex technical systems Yevgenevich [12] states that the highest diagnostic accuracy of vibration can be achieved by using several methods. However, the possibility of practical application of different methods for solving complex systems of vibration diagnostics is in the operational phase. In this case, the diagnostic results obtained by the various methods are complementary to others, as are the conclusions of further use, repair, or replacement after a comprehensive analysis of the information available. Thus, it points to the need to introduce and apply automated diagnostics on complex machines, as this more effectively identifies the main failure stages of the bearing diagnostic system, as well as the choice of optimal maintenance and strategy in estimating the remaining life of the system [12].

The problem of applying the technical diagnostic system within the hygienic paper processing industry is mostly covered by the research of companies specialized in supporting the technical systems designed for this purpose production. An overview of the implementation of Bruel and Kjaer Bearing Damage Detection Systems as part of a paper machine system is described in Ref. [13] by Brown and Jensen. The paper presents a number of advantages of installing an early bearing damage detection system. Detecting the possibility of damage well before the occurrence of a critical level allows for a significant reduction in downtime and save on production costs. The analysis of the results of the application of diagnostic systems is presented through seven case studies of concrete events in the production process of the Alma paper machine in Quebec, Canada. Special cases of fault detection on the drying rollers, felt rollers, and feed pump, as well as interventions at the mentioned positions, prevented the occurrence of far more serious damage to the equipment. The analysis describes in detail the equipment used and how it is applied in specific events (e.g., narrow spectrum analysis using narrow spectrum, 3D amplitude display, vibration trend analysis, etc.).

One of the applicable systems within the paper processing plant is the robot support system. In Ref. [14] Wagash, Semyon, Balazh, and Varga point to the need to use proactive and intuitive maintenance as a key instrument for maximum operational reliability and usability of industrial robots using the minimum cost and human resource requirement. The use of diagnostic methods ensures the robot's operability, life extension, and energy savings. In this case, vibration measurement is realized with three types of acceleration sensors. The evaluation of the results was provided through the Scope Win computer software. Fast Fourier Transformation (FFT) was applied to analyze the vibration spectrum. The aim of the paper is to propose a comprehensive methodology for measuring and estimating vibrations for industrial robots and serve as a way to obtain additional recommendations and draft procedures for the maintenance and diagnosis of industrial robot failures.

The implementation projection of the results of the introduction of the maintenance concept according to the condition on a paper machine at SCA in Collodi, Italy, is presented in an article [15]. The activity of reconstructing the maintenance concept came about after numerous problems on machines and system components that caused frequent and prolonged plant downtime. The concept of corrective maintenance was abandoned and the implementation of the maintenance concept according to the condition that entailed the installation of technical diagnostic measures at the predefined critical points was started. External Italian company ISE Industrial Service Engineering has been selected as a strategic partner in the field of introduction of diagnostic systems and processing of measurement results. It was emphasized that the proper maintenance procedure required the introduction of appropriate technology, and the correct choice of such technologies resulted in the success of the maintenance policy as a whole. This fact is of fundamental importance. The policy of reconstruction of the maintenance system has been carried out through several stages: analysis of critical components of the system, inspection by experts in terms of complex measurements, formation of reports—lists of collected data, their analysis, and recommendations for their elimination. Certainly, an indispensable part and cost estimation is to make a profit by realizing a certain activity. Implementing an effective maintenance plan based on predefined conditions determines the immediate benefits of reducing manufacturing losses and repair costs, optimizing machine availability.

A large number of research papers can be found in the literature by specialized firms that support paper machine maintenance systems. The difference between the application of innovative programs for maintaining the performance of paper rolls of paper machines supported by continuous monitoring systems in relation to the traditional monitoring and corrective action is presented in an article [16]. The key parameters for monitoring the functionality of the paper machine rollers are the temperature of the bearings, the state of vibration, and the quality of the surface of the roller liner. The result of research and monitoring of economic parameters within the maintenance evaluation gives a positive result in terms of a longer working period without the risk of failure. This approach has significant advantages not only in comparison to corrective actions but also in the case of the usual preventative maintenance approach when parts of the paper machine are serviced at predetermined intervals regardless of remaining usability.

The importance and main reasons for implementing preventive maintenance systems in hygienic paper production systems with particular emphasis on paper machine systems were demonstrated in a case study [17]. The paper shows the development of paper mills since the 1990s in terms of available system maintenance solutions, the emergence of crisis economic periods, along with efforts to increase the profitability, and non-need of investment and preventative maintenance solutions in terms of reducing cancellations and reducing costs. The new maintenance organization entails new concepts in a wide range of activities covering the maintenance of industrial systems and relates to:

- defining a strategy within the set requirements of allowable costs, availability, performance, and objectives;
- computerized maintenance management system (CMMS);
- safety, health, and environmental (SHE);
- assessment of plant risk and criticality;
- consistent planning of activities based on the state of the plant assessment;
- implementation of the TPM concept and involving all users in the processes of maintaining the production system;
- establishing a central, process-oriented maintenance plan—plan of maintenance (PM) tasks, which is the basis for further planning process.

The end result of this way of understanding the new concept of maintenance within the manufacturing process for the production of hygienic paper is the greater reliability of the plant. Production downtime is no longer a threat to production, and production line speeds and production speeds are steadily increasing. The application of PM is carried out without increasing the manpower resources which in fact becomes a comprehensive maintenance trend and a regular work procedure/routine.

A very interesting reflection on the role of monitoring in terms of maintaining the operating costs of a manufacturing industrial system is presented in Ref. [18]. This paper looks at the monitoring of the situation as an activity of finding a potential problem that undoubtedly causes a great involvement of the maintenance service, higher costs, and a burden on maintenance personnel. In this analysis, primacy is given to the economic aspect that accompanies the process of implementation of preventive and/or maintenance by condition. The main importance of maintenance monitoring by condition is its role in eliminating all potential anomalies, which is its main benefit. The best result can be obtained if a limit can be determined when the strategy set is a positive balance between total maintenance costs, i.e., system performance relative to achieving the desired level of reliability at minimal cost. Any maintenance strategy replacement activity must accompany the final cost projection, in which case it becomes a meaningful and purposeful activity.

Monitoring systems and technological process control capabilities using methods and devices of technical diagnostics within the paper production system are of great importance in the process of controlling the chemical processes and components used. Research by the group of authors Andrea, Sauro, and Nicola, which describes process monitoring via signal analysis of fault detection and isolation (FDI) procedures to control the use of chemical components, is presented in Ref. [19]. The controls of the production process with advanced technologies make it possible to comply with the safety and environmental standards and to a large extent depend on the quality of the data provided by the sensors. Chemical processes within the manufacturing process are known to be well equipped with sensors for measurement, such as temperature, flow, pressure, and electrical size sensors. The availability of many sensors provides the ability to detect and identify errors, since sensor measurements under normal conditions are correlated. These correlations are mainly due to the physical and chemical principles governing process operations, such as mass and energy balances. The paper industry is a manufacturing sector with a high degree of automation and, since the modern paper mill has hundreds of sensors and actuators connected to its automation systems, it is evident that some systematic methods are required for data processing.

The application of modern diagnostic systems and the results of monitoring the performance of the system and the behavior of the bearing positions of the position of the press paper machine are shown in Ref. [20]. Within a few, specifically 15 typical practice cases, the impact of using diagnostic devices in terms of obtaining results that help in making key decisions about the operation and use of production process equipment is presented. The application of diagnostic devices refers to the procedures and control of mounting large bearings of press paper machines, defining radial and axial clearances and software support that provides the required level of accuracy and reliability. One part of the research and education also relates to the control of the condition of the bearing in the processes of the audit grinding of the press lining, which represent regular periodic service activities. The purpose of diagnostics is to give instructions and opinions on the possibility of increasing the working speed of the paper machine and the possible consequences for the production process and the condition of the equipment used. The basis of the inference process is the control of the bearing behavior of the bearings and the conditions that define the bearing seats. As these are very demanding positions, one part of the research is also related to the results of diagnostic measurements and load calculations of specific press rolls of paper machine with the possibility of visualizing roll deformations under the influence of workloads.

The use of modern diagnostic devices in the process of continuous operation of the paper machine system can significantly affect the prevention of failure conditions and the occurrence of large maintenance and production costs. Costain, in a case study [21], cites examples of analysis results from measurements of diagnostic systems installed at different paper machine working positions, such as problems with the roller drive of a roller coaster of a paper machine producing graphic paper, the wrinkling of a roller lining in a V-nip construction resonance of the support structure of the drying roller in the paper graphic machine, and the vibration of the guide roller for winding the natron. All the cases described represent a wide range of different problems that can occur on paper machines during the production process. Early diagnosis and detection of potential and causes of already occurring problems is possible only if adequate sophisticated diagnostic equipment is in place. By analyzing the results of measuring physical performance that each individual disorder causes, it is possible to draw conclusions and make suggestions for eliminating the problem or, ultimately, avoiding major adverse consequences.

Timely information about the potential cancellation situation opens the possibility of realization of planned repairs and realization of significant savings of both production and complete business system. The cost of investing in modern diagnostic equipment is far less than the positive financial result of the business system due to the increased efficiency of the production system.

2.2 Reliability as a measure of the effectiveness of a maintenance system

If we want to evaluate a measure of the success of an organization and the operation of a maintenance service on the operation of a production industrial system, we can monitor the level of reliability of the production technical system observed over a period of time. The reliability of the system is the ability to keep the system in working condition during the time of exploitation. This magnitude is defined by reliability theory, which is the scientific discipline of recent times. The basis of the theory of reliability is the theory of probability that predicts changes in the system operation for a certain period (Fig. 10.4).

Because reliability is:

$$R = \int_{t=0}^{t_{gr}} (\text{working ability})dt \tag{10.1}$$

it can be concluded that reliability can only be quantified after the failure condition occurs.

System reliability is the likelihood that the system, after repeated use, will successfully perform the function of the criterion in the projected duration and given environmental conditions, or the likelihood that the single-use system will successfully perform the project task in a given time and given environmental conditions.

The complementary function $F(a)$ is an unreliability function:

$$R(a) = 1 - F(a) = \int_{0}^{+\infty} f(t)dt \tag{10.2}$$

If we look at the technical system at the time of exploitation, the relation between the reliability and unreliability functions can be represented graphically in Fig. 10.5.

It can be seen that the reliability of the technical system decreases during exploitation time as the level of unreliability increases.

There are numerous papers that describe the importance and role of maintenance in relation to the reliability of technical systems. Thus, Papic and Milovanovic provide a complex analysis of the basics of maintenance of technical systems, methodology and application of methods of improving the process of maintenance and reliability of technical systems, and the relationship of the effects of investment in the components of the production cycle according to the results of increasing efficiency

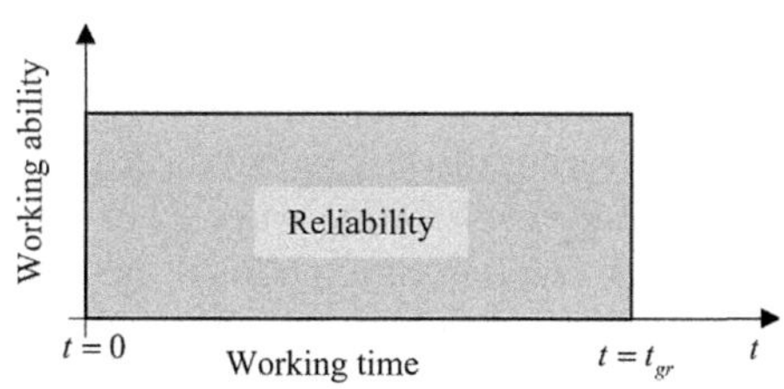

FIGURE 10.4

Reliability graph [22].

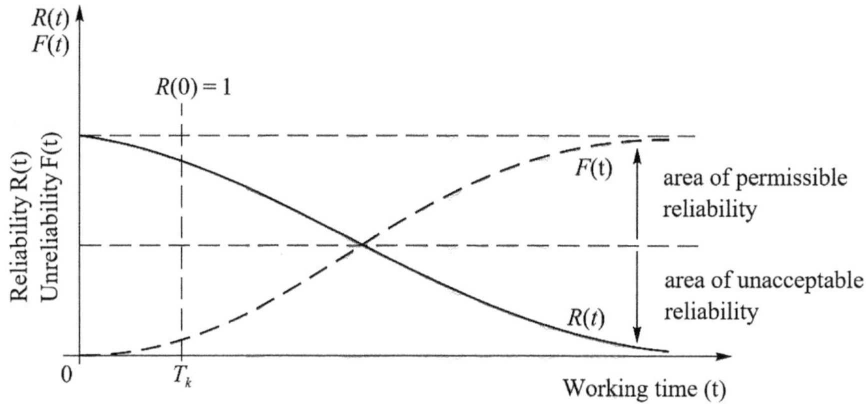

FIGURE 10.5

Graphic representation of the reliability and unreliability function [22].

(reliability, availability, and maintenance convenience) ([1]). A well-chosen concept of maintenance, with proper organization, programming, and implementation of certain maintenance activities during operation, with good staff training and quality management in maintenance, also contributes to improving the economic performance of the business system. Research aimed at increasing the level of reliability and managing reliability over the life cycle of a system aims to define a set of protection measures and optimize them from the aspect of simultaneous provision of economic exploitation and implementation of complex regulations related to the protection of human environment and safety of both the business system and the environment.

Bulatović in his book [23] presents a summarization of research and analysis of the reliability of technical systems on a concrete example of a change in operational readiness and reliability at the company Alumina Production Factory in Podgorica. Contribution to scientific research is the method of incorporating state parameters and probability of failure into the expert system. Complex maintenance problems from the point of view of determining and monitoring the status of systems in the function of predicting and preventing failures are effectively addressed by the use of expert systems.

Defining and forecasting reliability indicators are of utmost importance in preventive maintenance of complex systems. Only on the basis of timely assessments can corrective action be taken in order to further development of failures and to prevent major damage. Milovanovic proposed a modified method for estimating the optimal reliability of a condensing thermal power plant (2000), providing a good basis for further work on refining and improving the accuracy of the estimated values, with the introduction of technical diagnostics and a modern information management system [24]. Using benefits of a modern system reengineering directions and structural or reliability-centered maintenance (RCM)—maintenance according to reliability, ways of using the best methods in reliability analysis of complex systems—a block diagram of a modified method for a 300 MW basic reference block has been formed. This method also involves calculating the level of risk and, consequently, indirect ways to maintain these systems or their components. The choice of the most appropriate maintenance task is achieved through the use of a flow chart of decision-making activities, which takes into account the technological capabilities of the proposed tasks and the verification of their realization. The activity flow diagram of this method incorporates modern knowledge regarding the application of the TPM methodology, i.e., implementation of the concept of TPM, similar to the maintenance methodology according to the condition, with the fact that maintenance requires a special user attitude toward the system and a "total" responsibility for the quality of maintenance procedures of all employees within the thermal power plant. In doing so, the actual maintenance costs are minimally possible for each specific situation. The dependence of the cost of generating electricity on a thermal power plant on the level of reliability needs to be considered from two aspects: the thermal power plant and the user. In both cases, the point of minimum cost determines the optimal reliability of both the thermal power plant and the user.

The main task of the maintenance department is to provide the highest possible level of reliability so that the technical system ensures the normal function of the entire production system. For this reason, there is a constant effort to optimize and improve the maintenance system, minimize failure conditions, and provide conditions for continuous operation. Case studies show that even a small increase in the level of reliability can result in significant financial results.

3. Application of technical diagnostic measures at critical positions

3.1 Defining critical paper machine positions

This study analyzes the impact of the operation of the technical diagnostic system within the paper machine system number 2 of the production industrial system for the production of hygienic paper SHP Celex, Banja Luka, Bosnia and Herzegovina, on the assessment of the reliability value of the technical system. The production technical system is a paper machine with a series of connected elements within the technological process of production of hygienic paper from 15 to 40 gr/m^2.

The basis of the introduction of a diagnostic system for monitoring the elements of the production process is to define the critical components of the system whose failure can lead to a complete system shutdown. The criticality of a job position can be analyzed from different starting points. Considering the nature of the function of the maintenance service, the dominant issue is the length of the cancellation condition, the availability of a spare part within the critical position, the amount of costs related to the spare parts within the critical position, the quality of the final product, and the like.

By monitoring and analyzing the most significant paper machine failures in the period 2004–15, it was concluded that there are critical press positions in terms of increased costs for repairing damage and replacing large bearings (Fig. 10.6).

The justification for investing in the controls of the technical parts of the production system is to avoid failure conditions characterized by an uncertain outcome of repair, a permanent shortage of time in terms of filling production volumes, and the use of high-value replacement parts. On the one hand, we have diagnostic equipment with significant investment costs, and on the other hand, we may face cancellations and high maintenance costs. The goals of production and the continuation of the continuous working regime certainly give priority to monitoring and ensuring the stability of the production process.

Considering its importance in the production process, critical points had to be provided with a continuous system for monitoring the performance of bearing places. In December 2015, a modern online monitoring system was installed to monitor the vibration of the suction and crushed-press bearings. The Octavis diagnostic system manufactured by IFM Electronic is installed (Figs. 10.7 and 10.8).

FIGURE 10.6

Critical positions of paper machine - press rolls.

(A) (B)

FIGURE 10.7

Octavis system control cabinet - (A) signal lamps and (B) interior.

(A) (B)

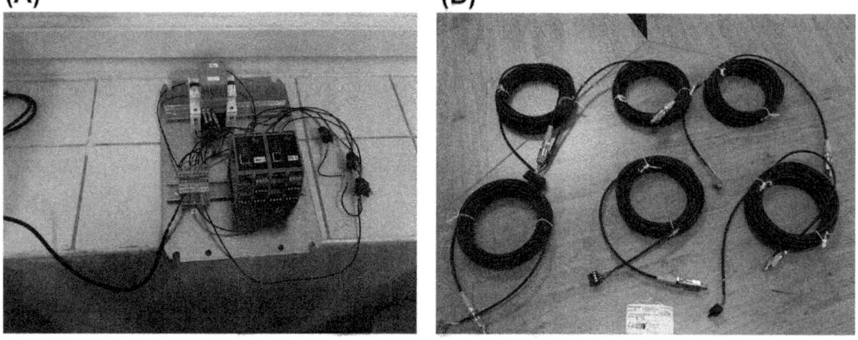

FIGURE 10.8

(A) VSE100 monitoring units and (B) vibration sensors.

The aforementioned diagnostic system is a modular system for monitoring the state of vibration and condition of bearings by applying high-frequency analysis of the collected signals. On-line local monitoring units provide automatic, periodic monitoring of the condition by applying permanently fixed sensors to the elements of the technical system being monitored. They provide much more frequent parameter monitoring and therefore earlier problem detection. The basis of the system is the connection to the database of vibration analysis software. The four-channel monitoring programmable logic controller (PLC) unit VSE100 continuously analyzes the signal obtained from four VSE001 acceleration sensors and compares the measured results with the values of the preset measurement structure. All measured values are automatically transmitted via Ethernet to the control system with software VES003 installed where they are archived as long-term trends. Since the sensors are permanently attached to the equipment, measurements on-line to local units are performed much more frequently than in the case of transmission systems. This provides early detection of problems on parts of technical systems and additional trending of data for repair planning. As a result of measurement, a frequency analysis of real-time oscillations is obtained.

The oscillation values are collected using detection sensors that are mounted at critical press positions (blind drilled and suction press roll) in radial direction on the drive and guide side and axial direction (sensor located on the drive side) (Figs. 10.9 and 10.10).

An overview of the arrangement of sensor positions is shown in Table 10.1.

The Octavis system continually collects information on vibration values at the bearing positions of critical press positions in real time. Generally, with permanently attached dynamic and static sensors, programmable measurements with alarm points, diagnostic system:

- "senses" changes in critical positions very early and automatically alerts work and maintenance personnel to device problems;
- automatically "captures" data at each transient state of starting or stopping the device (e.g., changing operating speed, etc.);
- provides real-time display in control rooms or measurement sites with visual monitoring of steady-state and transient conditions;
- provides a very powerful view of the analysis of short- and long-term situations and data collected during transitional periods;
- compares the data collected against the programmable criterion and automatically displays the exceedance of the criteria.

(A) (B)

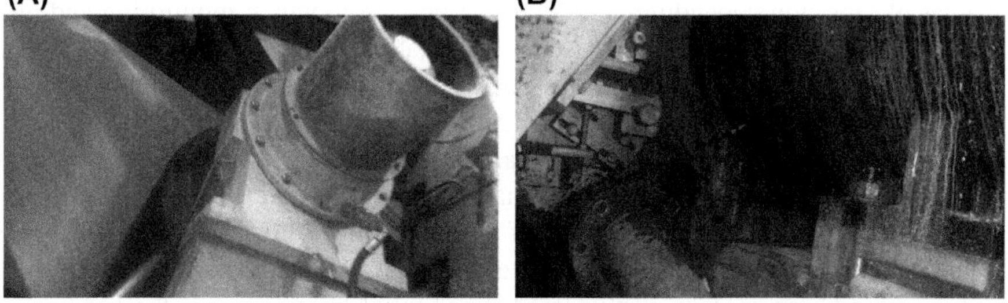

FIGURE 10.9

Vibration sensors on the service side - (A) blind drilled press roll and (B) suction press roll.

FIGURE 10.10

Vibration sensors on the drive side in radial and axial direction - (A) suction press roll and (B) blind drilled press roll.

Table 10.1 Arrangement of sensors for measuring vibration at critical positions.

Monitoring unit	Sensor number	Sensor position
Suction press roll		
PLC-1	Sensor 1	Operated side—radial direction
PLC-1	Sensor 2	Drive side—radial direction
PLC-1	Sensor 3	Drive side—axial direction
Blind drilled press roll		
PLC-2	Sensor 1	Operated side—radial direction
PLC-2	Sensor 2	Drive side—radial direction
PLC-2	Sensor 3	Drive side—axial direction

All measurement values are displayed on the control monitor as a visual display. The graph shows the limit values of vibration depending on the type of bearing to which monitoring and measurement refer (Fig. 10.11). If there is a sudden increase in the vibration value, it is detected on the alarm signal lamp on the control cabinet. This is a signal to operators and maintenance staff to check the status of the operating parameters and to identify the source of the potential anomaly.

With the instantaneous display of vibration values, the Octavis system stores the results obtained over time. All values represent the trend of behavior of the system, i.e., critical positions in time (Fig. 10.12). Based on the analysis of the trend of vibrations, it is possible to make an assessment and predictions of the situation in the short-term future time. An assessment can be made if, in addition to the trend of the monitoring system, it possesses experiential knowledge of the behavior of specific elements of the technical system.

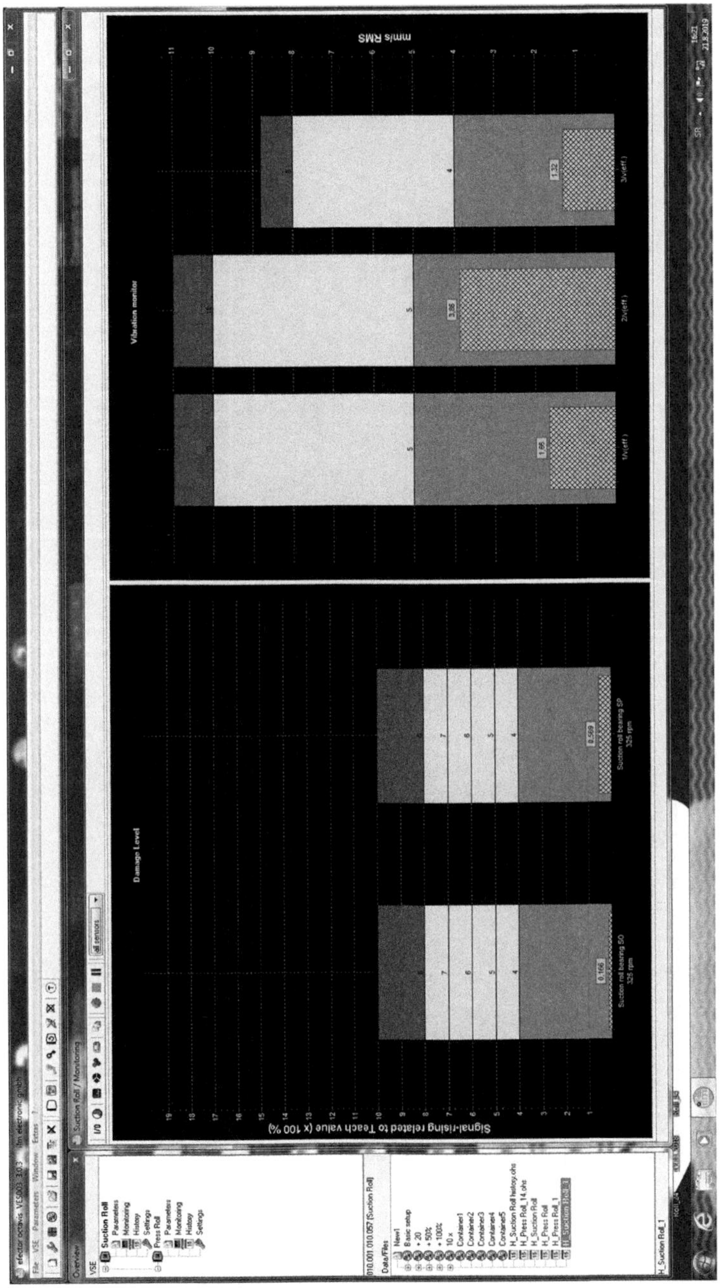

FIGURE 10.11

Visualization of Octavis system vibration measurement values.

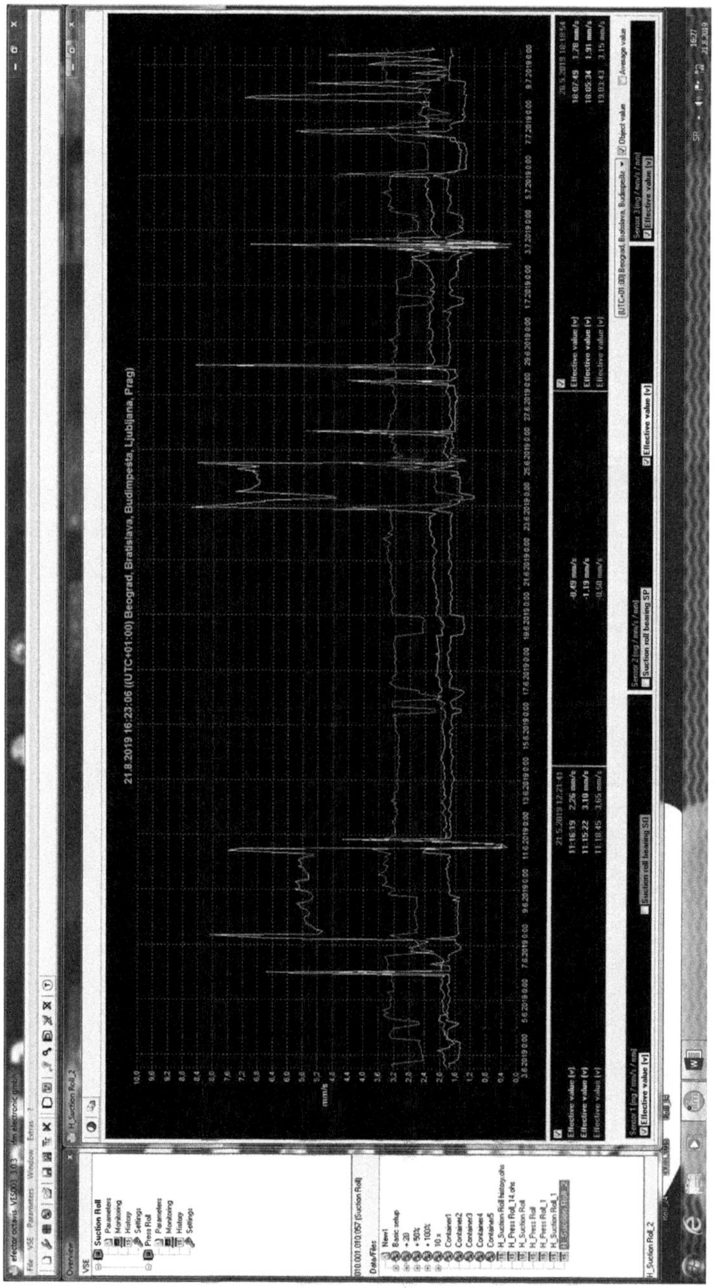

FIGURE 10.12

Display of the trend of vibration values of critical positions of the Octavis system.

General reports show locations or critical positions whose vibration and process levels exceed set alarm limits that are aligned with optimal, theoretical values. Current data are compared to the baseline data for analysis and are trending to show real-time changes in critical positions.

Trend prints provide early warning of problems on machines and are used to plan the most appropriate repair time.

A well-organized critical position monitoring program often uses a combination of time domain and spectral analysis to identify the state of change as accurately as possible. This combination is due to the fact that there are certain types of data that are not well transformed. Large hardware and software memory is available today, so it is advisable to use both time and frequency domains for all measurements.

Measurements transferred to vibration analysis software are quickly controlled to check for deviations from normal conditions. In the actual operating conditions of technical systems, there are many sources of vibration that, when summed up or acted upon at the same time, can cause very complicated causes in the time domain. For this reason, the software part of the diagnostic system uses the FFT program, which has the ability to "break" and restore complex inputs to individual components that are easier to recognize and identify the dominant cause of problems in critical positions. FFT takes data from one domain and transfers it to another domain without any loss of information. In the case of vibrations, there are two domains: the time domain (amplitude vs. time) and the frequency domain (frequency vs. time). FFT is a tool for breaking complex weather data into their frequency and amplitude components. Overall, FFT and time domain spectrum serve to aid in the identification, localization, and analysis of vibration changes.

As specific device problems are known to generate vibrations at specific frequencies, this information can be used to diagnose and detect causes of excessive vibration. There are two steps involved in analyzing the FFT spectrum:

- gathering useful information on the behavior of critical positions;
- spectrum analysis.

Before analyzing the spectrum, the components that can generate the vibration must be known. In the particular case of the Octavis diagnostic system, the known data are: types of critical position bearings for pressed paper machines, roller rotation numbers, geometry and bearing method, crimping system, and load distribution. This input database and advanced software enable the analysis of the oscillation frequency spectrum and the processing methods of the obtained signals, such as the accelerator envelop method, which provides information on the current state of the monitored quantities (vibration values and bearing condition) (Fig. 10.13).

Acceleration enveloping or "demodulation" is a signal processing technique that greatly improves the ability of experts to determine the state of rotary equipment that is the subject of a particular Octavis monitoring case. This signal processing method allows the detection of repeated low-energy pulse signals, such as bearing damage, much earlier than traditional analysis techniques. The acceleration enveloping process first filters the low-frequency vibration of the device and looks for harmonics of repetitive signals that occur in the higher frequency ranges previously specified by the user. These fault signal harmonics are present in the early stages of damage and usually occur at high frequencies. Once the harmonics are detected, the program together sums and sums them amplified or sums the signal in the frequency range of the fundamental error frequency. The result is a frequency spectrum with a much more conspicuous and sharpened representation of the fundamental damage

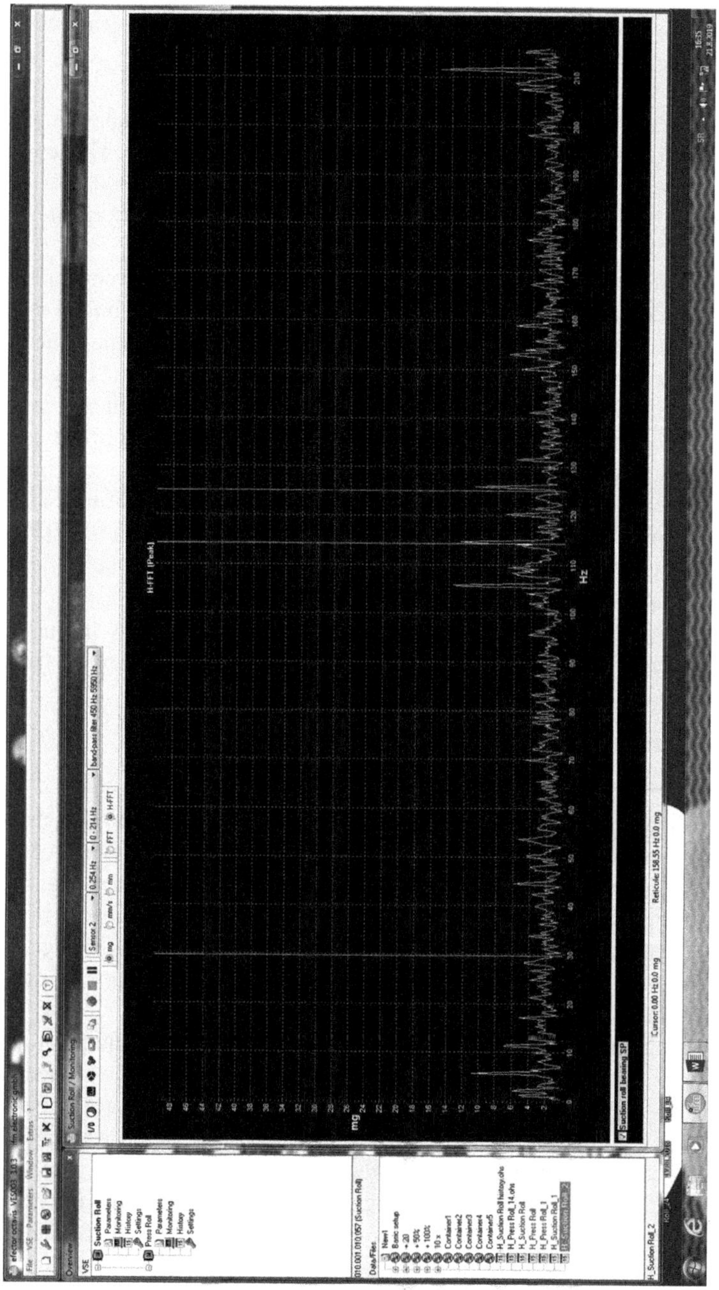

FIGURE 10.13

Real-time critical positions of Octavis critical positions—spectral analysis.

frequencies. Enveloped acceleration measurements give the best results when the data acquisition sensor is positioned as close as possible to the load zone (such as the vibration measurement position of critical presses of the machine press in the radial and axial direction) because impulse damage signals are more strongly generated in the load zone.

On early identification of problems with the presses of critical press positions, the maintenance team has the opportunity to apply proactive corrective measures that can extend the life of very expensive bearings of critical press positions and save on the costs and operations of the complete industrial system.

4. Analysis of the reliability assessment of the technical system

The subject of the analysis and assessment of the reliability of a real technical industrial system is to determine the parameters of the state of the production system while maintaining performance under the influence of many factors that affect changes in the state of the system (and related to the engagement of the maintenance service). The goal of obtaining reliability information is to predict the moment of failure. Reliability was analyzed by summarizing the data for two time intervals of monitoring—before and after the installation of the technical diagnostic system. The monitoring interval was taken to be equal to 16 months prior to the installation of the technical diagnostic system and equal time to 16 months after the installation of the diagnostic system. The aim of the study was to show the degree of influence of the application of the maintenance concept according to the state on the reliability level of the technical system. Reliability assessment involves defining a database of relevant data on the behavior of a real industrial system, defining the parameters of the distribution law, checking the distribution law, and displaying a diagram of the density of the run-time function until a failure occurs. Conducting the Kolmogorov–Smirnov test is a test of whether the data related to the state of the parameters of the industrial system belong to the distribution in the previously assumed hypothesis. In the specific case, the Kolmogorov–Smirnov test was applied for reasons of continuous distribution and relies on the estimation of the maximum difference between the value of the cumulative function of the observed data of the state of the real system and the value of the cumulative function in the assumed distribution.

4.1 Reliability analysis of the production system up to the time of technical diagnostic installation

The input data for the analysis of the reliability elements were used: hours of paper machine system operation, downtime caused by maintenance, and hours of operation until failure condition occurred (Table 10.2).

The analysis of the set hypothesis on the law of distribution of reliability is represented by checking the law of distribution, checking the correctness of the set hypothesis, defining the parameters of the law of distribution, and displaying a diagram of the function of the time density function until the state of failure occurs.

4.1.1 Assumption number 1—check for normal distribution

Suppose that the operating time of the equipment until the condition of failure occurs is in accordance with the laws of normal distribution.

Table 10.2 Reliability research of the real system up to the date of installation of technical diagnostics.

Months		Total availability (hours)	Planned maintenance downtimes (hours)	t_r Degree of use—total working hours (hours)	t_f Downtimes caused by maintenance (hours)	t_i Hours of operation to the occurrence of failure
2014	August	744	18	724.25	11.5	62.9
	September	720	19	643.25	43.25	14.8
	October	744	14	698	18.5	37.7
	November	720	14	621.75	62.25	9.9
	December	744	21	700.25	20	35
2015	January	744	42	704.25	10.5	67
	February	672	18	534	76.25	7
	March	744	18	729.5	6.5	112.2
	April	720	18	664	23.5	28.2
	May	744	19	717.75	2.75	261
	June	720	18	696.5	1.25	557.2
	July	744	157	718.25	4	179.5
	August	744	18	719	6.5	110.6
	September	720	19	664	25.5	26
	October	744	19	694	21.75	31.9
	November	720	18	689	11.25	61.2
Total:		11688	450	10917.75	345.25	1602.1

Dispersion for the observed number of samples:

$$\sigma_1^2 = \frac{1}{16}\left[(\bar{x}_1 - 11.5)^2 + (\bar{x}_1 - 43.25)^2 + (\bar{x}_1 - 18.5)^2 + (\bar{x}_1 - 62.25)^2 + (\bar{x}_1 - 20)^2 + \cdots\right] \quad (10.3)$$

$$\sigma_1^2 = 538.31.$$

Dispersions:

$$\sigma_1 = 23.2.$$

4.1.1.1 Distribution testing: Kolmogorov–Smirnov test for the period until the installation of technical diagnostics

If the difference between the cumulative functions of the observed data of the state of the real system $F_o(t_i)$ and the values of the cumulative function indicated in the assumed distribution $F_E(t_i)$:

$$D_i = |F_o(t_i) - F_E(t_i)|, \quad i = 1, 2, 3, \ldots n, \quad (10.4)$$

then the maximum difference value of the cumulative functions $D_{\max}$ should be less than the critical value D_{CR} [25]:

$$D_{\max} < D_{CR} \tag{10.5}$$

and in that case the hypothesis set can be accepted. The D_{CR} date values are tabulated.

The analysis of hours of operation of the real system until the occurrence of the failure state (t_i) and the calculation of $F_o(t_i)$ and $F_E(t_i)$ are presented in Table 10.3.

The average running time t_{isr}:

$$t_{isr1} = \frac{\sum t_r}{\sum t_f} = \frac{10917.75}{345.25} = 31.62 \tag{10.6}$$

According to the table values for the Kolmogorov–Smirnov test (K–S test) for 16 measurements [25]:

$$D_{CR} = 0.328 \tag{10.7}$$

As is it:

$$D_{\max} > D_{CR}$$

the initial hypothesis should be rejected, i.e., the analyzed parameters of the real system do not correspond to the normal distribution.

Table 10.3 Numerical values of normal distribution reliability indicators until technical diagnostics is installed.

Month n_i	t_i—working hours till the failure	$F_o(t_i) = n_i/n$	$F_E(t_i) = (t_i - t_{isr1})/\sigma_1$	$D_i = \lvert F_o - F_E \rvert$
1	62.9	0.0625	1.34	1.28
2	14.8	0.125	0.75	0.85
3	37.7	0.1875	0.26	0.07
4	9.9	0.25	0.93	1.18
5	35	0.3125	0.14	0.16
6	67	0.375	1.52	1.15
7	7	0.4375	1.06	1.49
8	112.2	0.5	3.47	2.97
9	28.2	0.5625	0.14	0.70
10	261	0.625	9.88	9.26
11	557.2	0.6875	22.65	21.96
12	179.5	0.75	6.37	5.62
13	110.6	0.8125	3.4	2.59
14	26	0.875	0.24	1.11
15	31.9	0.9375	0.01	0.92
16	61.2	1	1.27	0.27

4.1.2 Assumption number 2—check the exponential distribution

Suppose that the operating time of the equipment until the condition of failure occurs is in accordance with the laws of exponential distribution. Intensity of failure (λ):

$$\lambda = \frac{1}{t_{sr}} = \frac{1}{31.6} = 0.03164 \quad (10.8)$$

Failure probability density function—$f(t)$:

$$F_E(t) = \lambda \cdot e^{-(\lambda) \cdot t_i} = 0,03164 \cdot e^{-(0.03164) \cdot t_i} \quad (10.9)$$

Cumulative function of density in occurrence of the state of work—reliability function:

$$R(t) = e^{-(\lambda) \cdot t_i} = e^{-(0,03164) \cdot t_i} \quad (10.10)$$

An analysis of the operating hours of the real system until the failure condition (t_i) and the calculation of $F_o(t_i)$ and $F_E(t_i)$ are presented in Table 10.4.

As it is:

$$D_{\max} > D_{CR}$$

the initial assumption should be rejected, that is, the analyzed parameters of the real system do not obey the exponential distribution.

Table 10.4 Numerical values of exponential distribution reliability indicators until technical diagnostics is installed.

Months n_i	t_i—working hours till the failure	$F_o(t_i) = n_i/n$	$F_E(t_i) = \lambda \cdot e^{-(\lambda) \cdot ti}$	$D_i = \lvert F_o - F_E \rvert$
1	62.9	0.0625	0.00432	0.05818
2	14.8	0.125	0.01989	0.10511
3	37.7	0.1875	0.00959	0.17791
4	9.9	0.25	0.02311	0.22689
5	35	0.3125	0.01045	0.30205
6	67	0.375	0.00319	0.37181
7	7	0.4375	0.02531	0.41219
8	112.2	0.5	0.00091	0.49909
9	28.2	0.5625	0.01296	0.54954
10	261	0.625	0.00001	0.62499
11	557.2	0.6875	0.00001	0.68749
12	179.5	0.75	0.00011	0.74989
13	110.6	0.8125	0.00095	0.81155
14	26	0.875	0.01389	0.86111
15	31.9	0.9375	0.01153	0.92597
16	61.2	1	0.00456	0.99544

4.1.3 Assumption number 3—checking the Weibull distribution

Suppose that the operating time of the equipment until the condition of the failure occurs is in accordance with the laws of the Weibull distribution. The Weibull distribution density function is:

$$f(t) = \frac{\beta}{\eta} \cdot \left(\frac{t-\gamma}{\mu}\right)^{\beta-1} \cdot e^{-\left(\frac{t-\gamma}{\eta}\right)^{\beta}} \tag{10.11}$$

where:

$f(t) \geq 0, t \geq \gamma$—is distribution density function.
$\beta > 0$—is shape parameter.
$\eta > 0$—is scale parameter.
$-\infty < \gamma < +\infty$—is position parameter of the meter distribution.
As it is:

$$f(t) = \lambda(t) \cdot R(t), \tag{10.12}$$

we notice the following sizes:

$$\lambda(t) = \frac{\beta}{\eta} \cdot \left(\frac{t-\gamma}{\eta}\right)^{\beta-1} \quad \text{– failure intensity,} \tag{10.13}$$

$$R(t) = e^{-\left(\frac{t-\gamma}{\eta}\right)^{\beta}} \quad \text{– reliability,} \tag{10.14}$$

The Weibull distribution for the case of two parameter distributions is:

$$F(t) = 1 - R(t) = 1 - e^{-\left(\frac{t}{\eta}\right)^{\beta}} \tag{10.15}$$

The calculated values of the cumulative probability $F_e(t_i)$ for the period up to the installation of the technical diagnostic system are shown in Table 10.5. Determining the values and testing the parameters of Weibull model for the period until technical diagnostic installation are shown in Tables 10.6 and 10.7.

Cumulative probability:

$$F_e(t) = \frac{t_i}{t_{cum}} \tag{10.16}$$

The parameters of the Weibull model are:

$$\beta = \frac{\sum X_iY_i - \frac{1}{n}\sum X_iY_i}{\sum X_i^2 - \frac{1}{n}\sum X_iX_i} = \frac{-120.91 - \frac{1}{15}(16,62 \times 124,9)}{1052.25 - \frac{1}{15}(124,9 \times 124,9)} = 1.42 \tag{10.17}$$

$$\eta = \exp\left[-\frac{\frac{1}{n}\left(\sum Y_i - \beta \sum X_i\right)}{\beta}\right] = \exp\left[-\frac{\frac{1}{15}(-16,62 - 1,42 \times 124,9)}{1.42}\right] = 8955.29 \tag{10.18}$$

Table 10.5 Weibull distribution parameters for the interval until technical diagnostics installation.

Month n_i	Interval (hours)	t_{mid} Midinterval (hours)	t_i Working hours till the failure	t_{cum} Cumulative operation conditions number	Cumulative probability $F_e(t)$
1	0/744	372	62.9	62.9	0.039
2	744/1464	1104	14.8	77.7	0.048
3	1464/2208	1836	37.7	115.4	0.071
4	2208/2928	2568	9.9	125.3	0.078
5	2928/3672	3300	35	160.3	0.1
6	3672/4416	4044	67	227.3	0.141
7	4416/5088	4752	7	234.3	0.146
8	5088/5832	5460	112.2	346.5	0.216
9	5832/6552	6192	28.2	374.7	0.233
10	6552/7296	6924	261	635.7	0.396
11	7296/8016	7656	557.2	1192.9	0.744
12	8016/8760	8388	179.5	1372.4	0.856
13	8760/9504	9132	110.6	1483	0.925
14	9504/10224	9864	26	1509	0.941
15	10224/10968	10596	31.9	1540.9	0.961
16	10968/11688	11328	61.2	1602.1	1

Table 10.6 The analytical determination of Weibull parameters for the interval until technical diagnostics installation.

Month n_i	$X_i = \ln(t_{mid})$	$Y_i = \ln\ln 1/(1 - F_e(t))$	X_i^2	$Xi \cdot Yi$
1	5.92	−3.22	35.04	−19.08
2	7.01	−3.01	49.14	−21.10
3	7.52	−2.60	56.55	−19.60
4	7.85	−2.51	61.62	−19.71
5	8.10	−2.25	65.61	−18.23
6	8.30	−1.88	68.89	−15.64
7	8.47	−1.84	71.74	−15.63
8	8.61	−1.41	74.13	−12.16
9	8.73	−1.32	76.21	−11.58
10	8.84	−0.68	78.14	−6.05
11	8.94	0.30	79.92	2.76
12	9.03	0.66	81.54	5.97
13	9.12	0.95	83.17	8.67
14	9.20	1.04	84.64	9.56
15	9.27	1.17	85.93	10.90
16	9.34			

Table 10.7 Test Kolmogorov–Smirnov until period of the installation of technical diagnostics.

Month n_i	$F_e(t)$	$F_t(t) = 1 - e^{-(t_{mid}/8955,29)^{1.42}}$	$\lvert F_e(t) - F_t(t) \rvert$
1	0.039	0.01086	0.02814
2	0.048	0.049888	0.00189
3	0.071	0.100015	0.02901
4	0.078	0.156079	0.07808
5	0.1	0.215172	0.11517
6	0.141	0.276304	0.1353
7	0.146	0.334117	0.18812
8	0.216	0.390606	0.17461
9	0.233	0.446874	0.21387
10	0.396	0.500422	0.10442
11	0.744	0.550869	0.193131
12	0.856	0.597982	0.258018
13	0.925	0.642327	0.282673
14	0.941	0.682444	0.258556
15	0.961	0.719123	0.241877

As it is:

$$D_{\max} < D_{CR} \tag{10.19}$$

the analyzed parameters of the real system correspond to the Weibull distribution and the initial hypothesis should be accepted.

4.1.3.1 Graphical interpretation of the Weibull distribution

The probability diagram of the Weibull distribution for the period before and after the installation of technical diagnostics will be verified by the graphical method.

The Weibull diagram is formed as follows:

- the uptime data until the occurrence of a failure is sorted in ascending order;
- from tables [25] (or by calculation) the central rank for each element is determined;
- center points, y-coordinates, and x-coordinates are displayed on the Weibull paper, which shows the failure data with the time parameter.

The center rank for a sample of 16 elements is:
For Rank 1:

$$1 = \frac{i - 0.3}{N + 0.4} = \frac{1 - 0.3}{16 + 0.4} = 0.4268 \tag{10.20}$$

The center rank value has been taken from the statistical tables [25].

Based on the previous procedure, data related to the Weibull diagram for the period up to the installation of technical diagnostics are shown in Table 10.8.

Table 10.8 Weibull data for the period until the technical diagnostics is installed.

Rank	$x = t_i$ Hours to the occurrence of failure	Median rank	Median rank %	1/(1-Median rank)	ln(ln(1/(1-Median rank)))	$y = \ln(t_i)$
1	7	0.0427	4.27	1.044605	−3.13182	1.94591
2	9.9	0.1037	10.37	1.115698	−2.21201	2.292535
3	14.8	0.1646	16.46	1.197031	−1.71566	2.694627
4	26	0.2256	22.56	1.291322	−1.36388	3.258097
5	28.2	0.2866	28.66	1.401738	−1.08556	3.339322
6	31.9	0.3476	34.76	1.532802	−0.85074	3.462606
7	35	0.4085	40.85	1.690617	−0.64418	3.555348
8	37.7	0.4695	46.95	1.885014	−0.45581	3.62966
9	61.2	0.5305	53.05	2.129925	−0.2796	4.114147
10	62.9	0.5915	59.15	2.44798	−0.11064	4.141546
11	67	0.6524	65.24	2.87687	0.055154	4.204693
12	110.6	0.7134	71.34	3.489184	0.222878	4.70592
13	112.2	0.7744	77.44	4.432624	0.398099	4.720283
14	179.5	0.8354	83.54	6.075334	0.590138	5.190175
15	261	0.8963	89.63	9.643202	0.818128	5.56452
16	557.2	0.9573	95.73	23.4192	1.148531	6.322924

Based on the obtained data, the coordinates of the points that approximate the direction in the probability diagram of the Weibull distribution are defined (Fig. 10.14).

Graphical direction estimation coefficient, i.e., the shape parameter β, compared to the analytically calculated value, is a satisfactory overlap. The graphical estimation of the scale parameter is determined as the value of $t = \eta$ for the value of the cumulative function $F(t)$. The resulting parameter value indicates that 50% of the elements will fail after 50 h of operation, which is also the mean time to failure.

An analysis and summarization of the previous data is presented in Tables 10.9 and 10.10 and in Figs. 10.15 and 10.16.

4.2 Analysis of the production system reliability after the installation of technical diagnostics

The methodology of analysis of reliability elements after installation of technical diagnostics is identical to the analysis of the previous period with the use of input data for that interval: hours of operation of the paper machine system, downtime caused by maintenance, and hours of operation until the condition in failure occurs (Table 10.11).

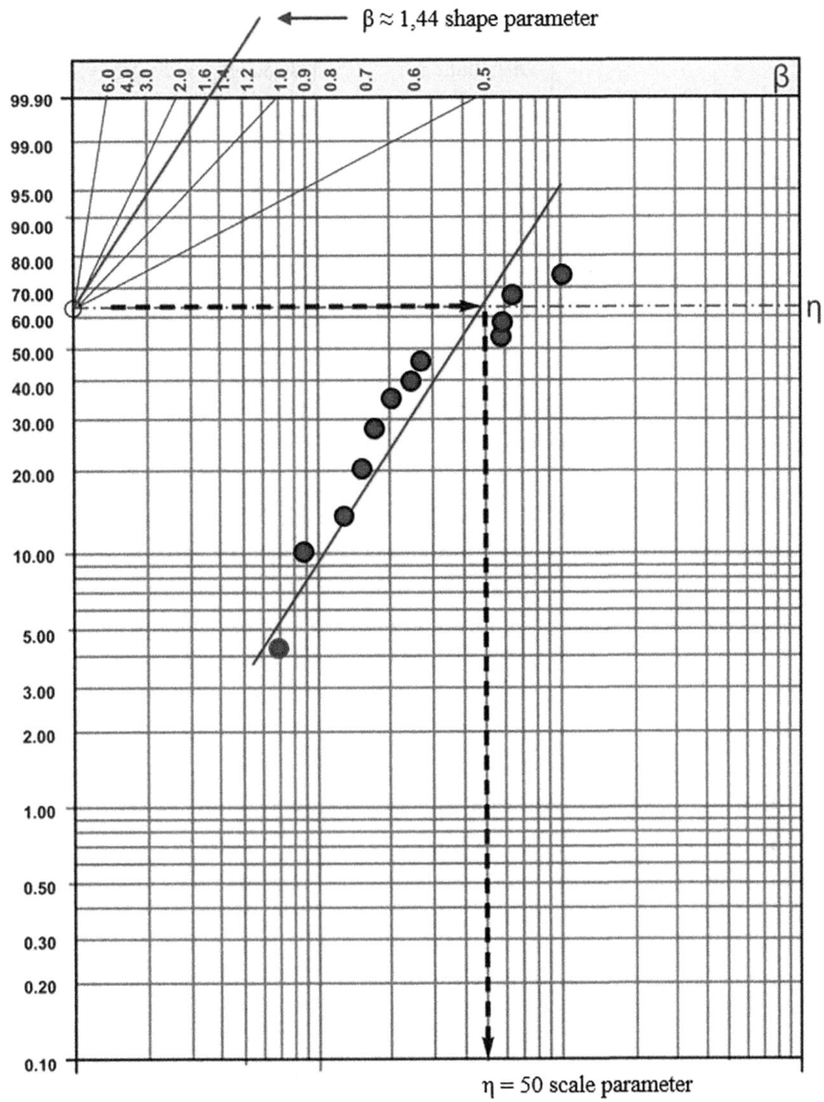

FIGURE 10.14

Probability diagram of Weibull distribution for the period until the date of installation of technical diagnostics.

As in the case of the previously analyzed period, the analysis is represented by hypothesis setting for the distribution law, verification of the correctness of the hypothesis that is set, determination of the parameters of the law of distribution, and diagram presentation of the function of the density of running time until the condition in failure occurs.

Table 10.9 Reliability statistics for the period untill the installation of technical diagnostics.

	Distribution	Distribution parameters	Komogorov–Smirnov test D_{max}	Note
The period until technical diagnostic installation	Normal	$t_{isr} = 31.62$ $\sigma = 23.2$	21.96	It is not accepted
	Exponential	$t_{isr} = 31.62$ $\lambda = 0.03164$	0.981	It is not accepted
	Weibull	$\beta = 1.42$ $\eta = 8955.293$	0.282673	It can be accepted

Table 10.10 Graphical interpretation of the Weibull distribution until the installation of technical diagnostics.

Month n_i	t_{mid} Midinterval hours (h)	Reliability $R(t_{mid})$	Unreliability $F(t_{mid})$	Failure rate $\lambda(t_{mid})$
1	372	0.768836	0.231164	0.0000416833
2	1104	0.660261	0.339739	0.0000658235
3	1836	0.598106	0.401894	0.0000815005
4	2568	0.553344	0.446656	0.0000938348
5	3300	0.51814	0.48186	0.000104258
6	4044	0.488643	0.511357	0.000113552
7	4752	0.464716	0.535284	0.000121513
8	5460	0.44381	0.55619	0.000128812
9	6192	0.424672	0.575328	0.000135801
10	6924	0.407552	0.592448	0.000142326
11	7656	0.392082	0.607918	0.000148462
12	8388	0.37799	0.62201	0.000154266
13	9132	0.36486	0.63514	0.000159872
14	9864	0.352951	0.647049	0.000165134
15	10596	0.341908	0.658092	0.000170174
16	11328	0.331625	0.668375	0.000175016

4.2.1 Assumption number 1—check for normal distribution

Suppose that the operating time of the equipment until the condition of failure occurs is in accordance with the laws of normal distribution. Dispersion for the observed number of samples:

$$\sigma_2^2 = \frac{1}{16}\left[(\bar{x}_1 - 20.75)^2 + (\bar{x}_1 - 10.25)^2 + (\bar{x}_1 - 15.5)^2 + (\bar{x}_1 - 9.75)^2 + (\bar{x}_1 - 48)^2 + \cdots\right] \quad (10.21)$$

$$\sigma_2^2 = 11908.93.$$

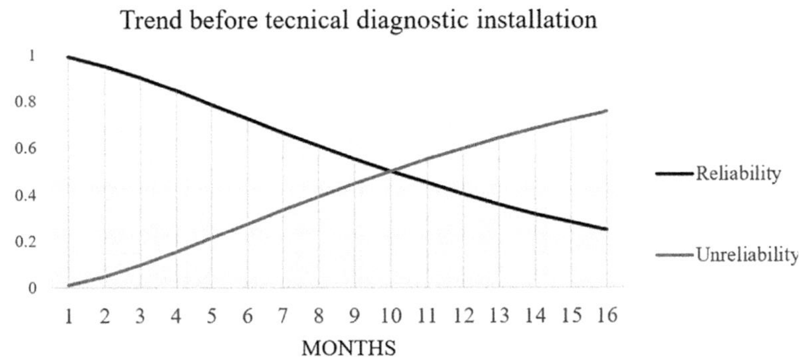

FIGURE 10.15

Graphic representation of the reliability and unreliability function to the installation of technical diagnostics.

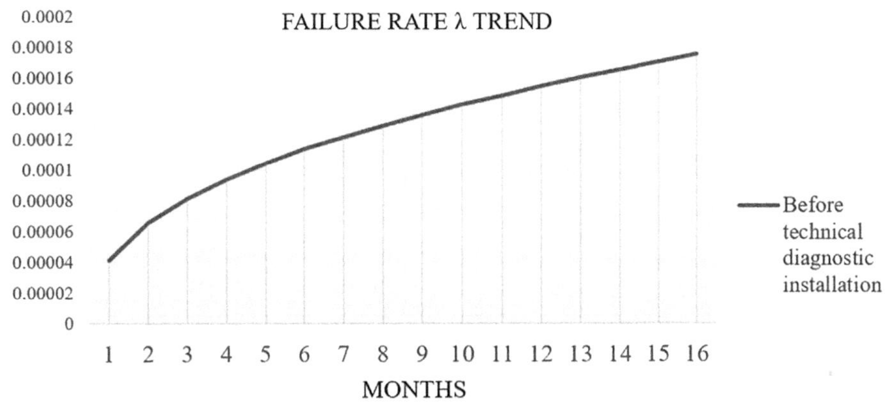

FIGURE 10.16

Graphic representation of the failure rate function until technical diagnostics is installed.

Dispersions:

$$\sigma_2 = 109.12.$$

4.2.1.1 Distribution testing: Kolmogorov–Smirnov test for the period after installation of technical diagnostics

By applying the same methodology, as in Section 4.1.1.1, we get an analysis of the hours of operation of the real system until the failure condition t_i. The calculated values of $F_o(t_i)$ and $F_E(t_i)$ are presented in Table 10.12.

Table 10.11 Reliability research after the date of installation of technical diagnostics.

Month		Total availability (hours)	Planned maintenance downtimes (hours)	t_r Degree of use—total working hours (hours)	t_f Downtimes caused by maintenance (hours)	t_i Hours of operation to the occurrence of failure
2015	December	744	19	590.5	20.75	28.46
2016	January	744	18	622.25	10.25	60.71
	February	696	17	679	15.5	43.81
	March	744	21	710	9.75	72.82
	April	720	16	640.5	48	13.34
	May	744	19	699.75	4.75	147.32
	June	720	15	691.75	14.25	48.54
	July	744	16	705.75	14.25	49.53
	August	744	18	724.75	5	144.95
	September	720	19	630.25	59.75	10.55
	October	744	233	609	8.5	71.65
	November	720	18	672.25	25.75	26.11
	December	744	19	732.5	2.5	293.00
2017	January	744	18	693.75	22.75	30.49
	February	672	18	631	8.75	72.11
	March	744	19	740	2.5	296.00
Total:		11688	503	10773	273	1409.3

The average running time t_{isr}:

$$t_{isr2} = \frac{\sum t_r}{\sum t_f} = \frac{10773}{273} = 39.46 \tag{10.22}$$

According to Table 10.12:

$$D_{\max} > D_{CR}$$

the initial hypothesis should be rejected, i.e., the analyzed parameters of the real system do not correspond to the normal distribution.

4.2.2 Assumption number 2—check the exponential distribution

Suppose that the operating time of the equipment until the condition of failure occurs is in accordance with the laws of exponential distribution. Failure intensity (λ):

$$\lambda = \frac{1}{t_{sr}} = \frac{1}{39.45} = 0.02534 \tag{10.23}$$

Table 10.12 Numerical values of normal distribution reliability indicators after installation of technical diagnostic.

Month n_i	t_i—working hours till the failure	$F_o(t_i) = n_i/n$	$F_E\ (t_i) = (t_i - t_{isr2})/\sigma_2$	$D_i = \lvert F_o - F_E \rvert$
1	28.46	0.0625	−0.10	0.163
2	60.71	0.125	0.19	0.070
3	43.81	0.1875	0.04	0.148
4	72.82	0.25	0.31	0.056
5	13.34	0.3125	−0.24	0.552
6	147.32	0.375	0.99	0.613
7	48.54	0.4375	0.08	0.354
8	49.53	0.5	0.09	0.408
9	144.95	0.5625	0.97	0.404
10	10.55	0.625	−0.26	0.890
11	71.65	0.6875	0.29	0.393
12	26.11	0.75	−0.12	0.872
13	293.00	0.8125	2.32	1.511
14	30.49	0.875	−0.08	0.957
15	72.11	0.9375	0.30	0.638
16	296.00	1	2.35	1.351

Failure probability density function—$f(t)$:

$$F_E(t) = \lambda \cdot e^{-(\lambda)\cdot t_i} = 0,02534 \cdot e^{-(0,02534)\cdot t_i} \tag{10.24}$$

Cumulative function of density in occurrence of the state of work—reliability function:

$$R(t) = e^{-(\lambda)\cdot t_i} = e^{-(0.02534)\cdot t_i} \tag{10.25}$$

An analysis of the operating hours of the real system until the failure condition t_i and the calculation of $F_o(t_i)$ and $F_E(t_i)$ are presented in Table 10.13.

According to Table 10.13:

$$D_{\max} > D_{CR}$$

the initial assumption should be rejected, that is, the analyzed parameters of the real system do not obey the exponential distribution.

4.2.3 Assumption number 3—checking the Weibull distribution

Suppose that the operating time of the equipment after the failure condition is in accordance with the laws of the Weibull distribution. We calculate the Weibull distribution density function according to the methodology given in Section 4.1.3.

The calculated values of the cumulative probability $F(t)$ for the period up to the installation of the technical diagnostic system are shown in Table 10.14. Determining the values and testing the

Table 10.13 Numerical values of the exponential distribution reliability indicator after technical diagnostic installations.

Month n_i	t_i—machine stops	$F_o(t_i) = n_i/n$	$F_E(t_i) = \lambda e^{-(\lambda)ti}$	$D_i = \lvert F_o - F_E \rvert$
1	28.46	0.0625	0.0123205	0.05018
2	60.71	0.125	0.0240878	0.100912
3	43.81	0.1875	0.0234851	0.164015
4	72.82	0.25	0.0228974	0.227103
5	13.34	0.3125	0.0223245	0.290176
6	147.32	0.375	0.0217659	0.353234
7	48.54	0.4375	0.0212213	0.416279
8	49.53	0.5	0.0206903	0.47931
9	144.95	0.5625	0.0201726	0.542327
10	10.55	0.625	0.0196678	0.605332
11	71.65	0.6875	0.0191757	0.668324
12	26.11	0.75	0.0186959	0.731304
13	293.00	0.8125	0.0182281	0.794272
14	30.49	0.875	0.017772	0.857228
15	72.11	0.9375	0.0173273	0.920173
16	296.00	1	0.0168938	0.983106

Table 10.14 Weibull distribution parameters for the interval after technical diagnostic installation.

Month n_i	Interval (hours)	t_{mid} Midinterval (hours)	t_i Operation conditions number	t_{cum} Cumulative operation conditions number	Cumulative probability $F_e(t)$
1	0/744	372	28.46	28.46	0.0201
2	744/1488	1116	60.71	89.17	0.0632
3	1488/2184	1836	43.81	132.98	0.0943
4	2184/2928	2556	72.82	205.8	0.146
5	2928/3648	3288	13.34	219.14	0.1554
6	3648/4392	4020	147.32	366.46	0.26
7	4392/5112	4752	48.54	415	0.2944
8	5112/5856	5484	49.53	464.53	0.3295
9	5856/6600	6228	144.95	609.48	0.4324
10	6600/7320	6960	10.55	620.03	0.4399
11	7320/8064	7692	71.65	691.68	0.4907
12	8064/8784	8424	26.11	717.79	0.5092
13	8784/9528	9156	293.00	1010.79	0.7171
14	9528/10272	9900	30.49	1041.28	0.7388
15	10272/10944	10608	72.11	1113.39	0.7899
16	10944/11688	11316	296	1409.39	1

parameters of Weibull model for the period after technical diagnostic installation are shown in Tables 10.15 and 10.16.

The parameters of Weibull model are:

$$\beta = \frac{\sum X_iY_i - \frac{1}{n}\sum X_iY_i}{\sum X_i^2 - \frac{1}{n}\sum X_iX_i} = \frac{-122.683 - \frac{1}{15}(-16,609 \times 124,934)}{1052.274 - \frac{1}{15}(124,934 \times 124,934)} = 1.28 \quad (10.26)$$

$$\eta = \exp\left[-\frac{\frac{1}{n}\left(\sum y_i - \beta\sum X_i\right)}{\beta}\right] = \exp\left[-\frac{\frac{1}{15}\left(-16,609 - 1,28 \times 124,934\right)}{1.28}\right] = 9798.65293 \quad (10.27)$$

According to Table 10.16:

$$D_{\max} < D_{CR}$$

the initial hypothesis should be accepted, i.e., the analyzed parameters of the real system correspond to the Weibull distribution.

4.2.3.1 Graphical interpretation of Weibull distribution after installation of technical diagnostics

According to the methodology described in Section 4.1.3.1, data related to the Weibull diagram for the period after installation of the technical diagnostics are shown in Table 10.17.

Table 10.15 Tabular concept of analytic determination of parameters for interval after technical diagnostic installation.

Month n_i	$X_i = \ln(t_{mid})$	$Y_i = \ln\ln 1/(1 - F_e(t))$	X_i^2	Xi Yi
1	5.918894	−3.8969	35.0333	−23.065339
2	7.017506	−2.72899	49.24539	−19.1506748
3	7.515345	−2.31216	56.4804	−17.3766734
4	7.846199	−1.84627	61.56284	−14.4862348
5	8.098035	−1.7785	65.57817	−14.4023144
6	8.299037	−1.2003	68.87402	−9.96130055
7	8.466321	−1.05352	71.67859	−8.91947136
8	8.60959	−0.91696	74.12504	−7.89466704
9	8.736811	−0.56856	76.33186	−4.96743259
10	8.847935	−0.54535	78.28595	−4.82520495
11	8.947936	−0.39346	80.06556	−3.5206584
12	9.03884	−0.34007	81.70063	−3.0738629
13	9.122165	0.233222	83.21389	2.127489791
14	9.20029	0.29451	84.64534	2.709580801
15	9.269364	0.444796	85.9211	4.122974612
16				

Table 10.16 Test Kolmogorov–Smirnov after period of the installation of technical diagnostics.

Month n_i	$F_e(t)$	$F_t(t) = 1 - e^{-(t_{mid}/9798.65)^{1.28}}$	$\lvert F_e(t) - F_t(t) \rvert$
1	0.0201	0.015077	0.005023
2	0.0632	0.060107	0.003093
3	0.0943	0.110625	0.01633
4	0.146	0.16394	0.01794
5	0.1554	0.218985	0.06358
6	0.26	0.273619	0.01362
7	0.2944	0.327002	0.0326
8	0.3295	0.378563	0.04906
9	0.4324	0.428706	0.003694
10	0.4399	0.475561	0.03566
11	0.4907	0.519805	0.0291
12	0.5092	0.561363	0.05216
13	0.7171	0.600217	0.116883
14	0.7388	0.636966	0.101834
15	0.7899	0.669426	0.120474

Table 10.17 Weibull data for the period after the installation of technical diagnostics.

Rank	$x = t_i$ Hours to the occurrence of failure	Median rank	Median rank %	1/(1-Median rank)	ln(ln(1/(1-Median rank)))	$y = \ln(t_i)$
1	10.55	0.0427	4.27	1.044605	−3.13182	2.356126
2	13.34	0.1037	10.37	1.115698	−2.21201	2.590767
3	26.11	0.1646	16.46	1.197031	−1.71566	3.262318
4	28.46	0.2256	22.56	1.291322	−1.36388	3.3485
5	30.49	0.2866	28.66	1.401738	−1.08556	3.417399
6	43.81	0.3476	34.76	1.532802	−0.85074	3.779862
7	48.54	0.4085	40.85	1.690617	−0.64418	3,.82388
8	49.53	0.4695	46.95	1.885014	−0.45581	3.902579
9	60.71	0.5305	53.05	2.129925	−0.2796	4.106108
10	71.65	0.5915	59.15	2.44798	−0.11064	4.271793
11	72.11	0.6524	65.24	2.87687	0.055154	4.278193
12	72.82	0.7134	71.34	3.489184	0.222878	4.287991
13	144.95	0.7744	77.44	4.432624	0.398099	4.976389
14	147.32	0.8354	83.54	6.075334	0.590138	4.992607
15	293	0.8963	89.63	9.643202	0.818128	5.680173
16	296	0.9573	95.73	23.4192	1.148531	5.690359

Based on the obtained data, the coordinates of the points that approximate the direction in the probability diagram of the Weibull distribution for the period after the installation of technical diagnostics are defined (Fig. 10.17). Graphical estimation of the direction coefficient, i.e., the parameter of the form β compared to the analytically calculated value, is a satisfactory overlap.

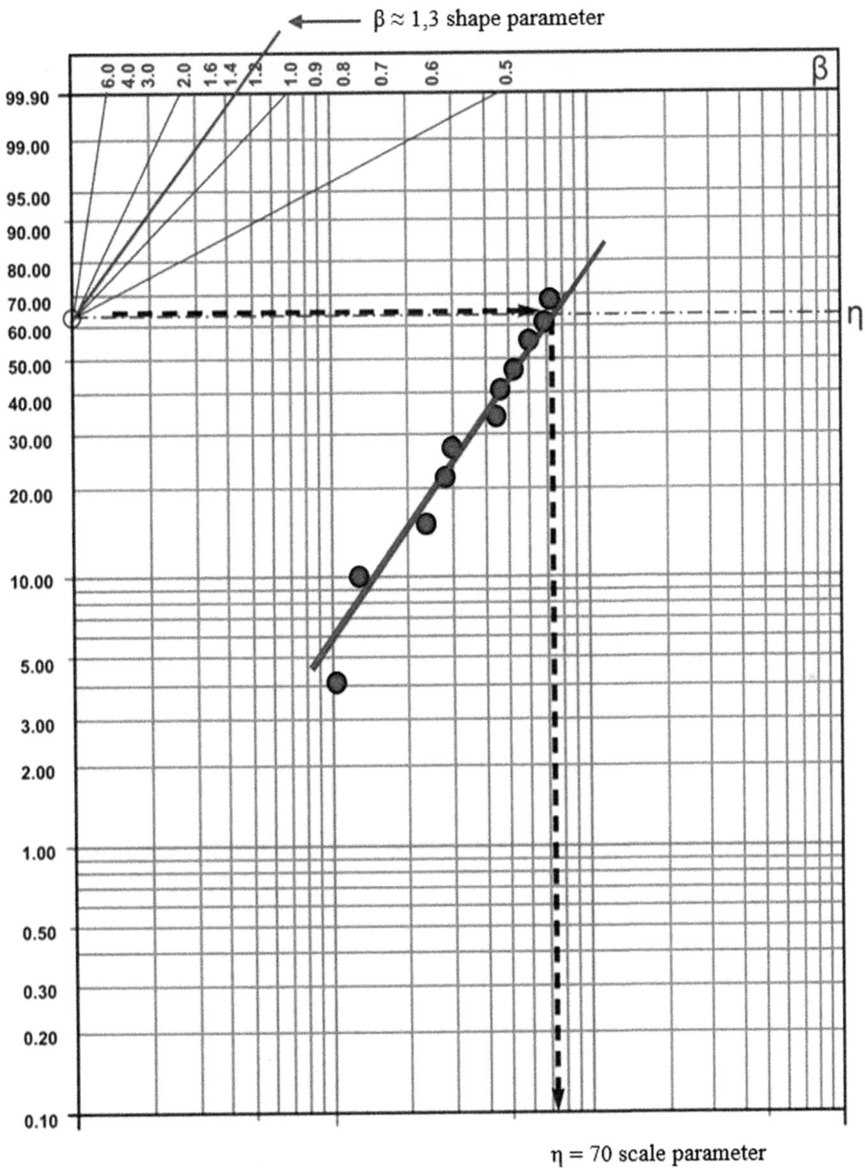

FIGURE 10.17

The diagram is likely to be Weibull distributions after the date of installation of the technical diagnostics.

The graphical estimation of the scale parameter is determined as the value of $t = \eta$ for the value of the cumulative function *F(t)*. The obtained parameter value indicates that 50% of the elements will fail after 70 h of operation, which is also the mean time to failure. If we compare the data for the period after the installation of technical diagnostics with the data obtained for the period up to the installation of technical diagnostics, we can see an increase in the hours of operation until the occurrence of a failure after the installation of technical diagnostics by an average of 20 h of operation.

An analysis of the previous data is presented in Tables 10.18 and 10.19 and Figs. 10.18 and 10.19.

Table 10.18 Reliability statistics for the period after technical diagnostic installation.

	Distribution	Distribution parameters	Komogorov–Smirnov test D_{max}	Note
The period after technical diagnostic installation	Normal	$t_{isr} = 39.46$ $\sigma = 109.12$	1.511	It is not accepted
	Exponential	$t_{isr} = 31.6$ $\lambda = 0.02534$	0.992	It is not accepted
	Weibull	$\beta = 1.28$ $\eta = 9798.65$	0.1204	It can be accepted

Table 10.19 Graphical interpretation of Weibull distribution after installation of technical diagnostics.

Month n_i	t_{mid} Midinterval hours (h)	Reliability $R(t_{mid})$	Unreliability $F(t_{mid})$	Failure rate $\lambda(t_{mid})$
1	372	0.984923	0.015077	0.0000522719
2	1116	0.939893	0.060107	0.0000710988
3	1836	0.889374	0.110626	0.0000817336
4	2556	0.83606	0.16394	0.0000896672
5	3288	0.781015	0.218985	0.0000962182
6	4020	0.726381	0.273619	0.000101789
7	4752	0.672998	0.327002	0.00010667
8	5484	0.621437	0.378563	0.000111036
9	6228	0.571293	0.428707	0.000115063
10	6960	0.524439	0.475561	0.000118699
11	7692	0.480195	0.519805	0.00012207
12	8424	0.438637	0.561363	0.000125216
13	9156	0.399783	0.600217	0.000128172
14	9900	0.363034	0.636966	0.000131007
15	10608	0.330574	0.669426	0.000133565
16	11316	0.300485	0.699515	0.000136004

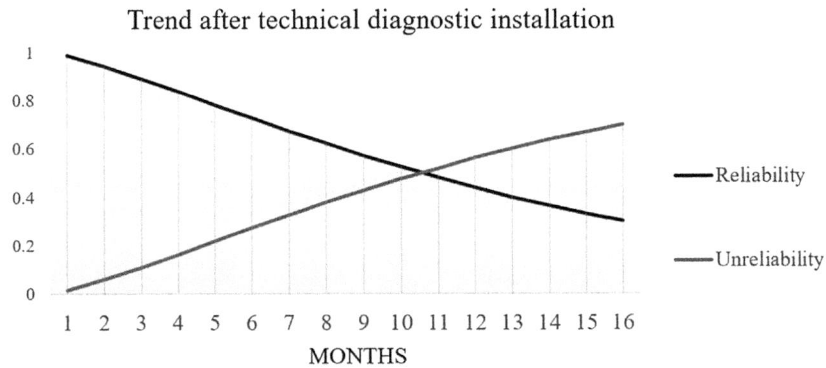

FIGURE 10.18

Intersection of reliability and unreliability functions for the period after installation of the technical diagnostics.

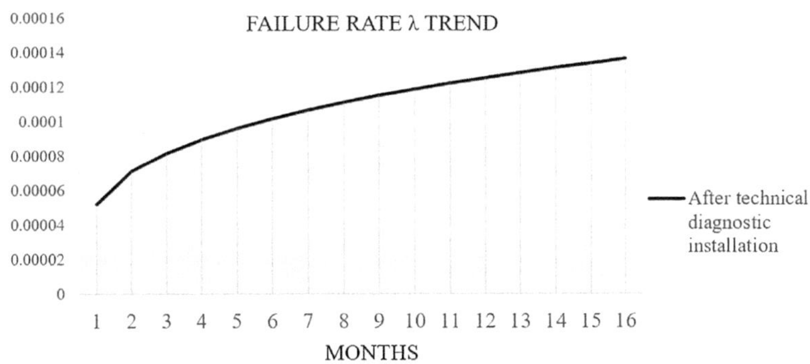

FIGURE 10.19

Graphic representation of the failure rate function after installing the technical diagnostics.

5. Results discussion

The results of the reliability analysis after the implementation of the technical diagnostic methods show a positive trend in increasing the level of reliability and reducing the failure rate of the production system. If we compare the reliability curves, we can see a milder trend of declining confidence levels after implementation of technical diagnostic measures compared to the period before implementation (Fig. 10.20). Like the trend in reliability, the trend of failure rates is slightly lower after implementation of technical diagnostic measures (Fig. 10.21).

This indicates that by preventing and reducing the total number of failure hours, the readiness of the technical plant increases and the technical system stays longer in reliable operating mode.

If we analyze the obtained values of the parameter η (Figs. 10.20 and 10.21), we can conclude that for the period prior to the installation of technical diagnostics the result of the parameter $\eta \approx 50$ indicates that 50% of the elements will fail after 50 h of operation, while for the period after the

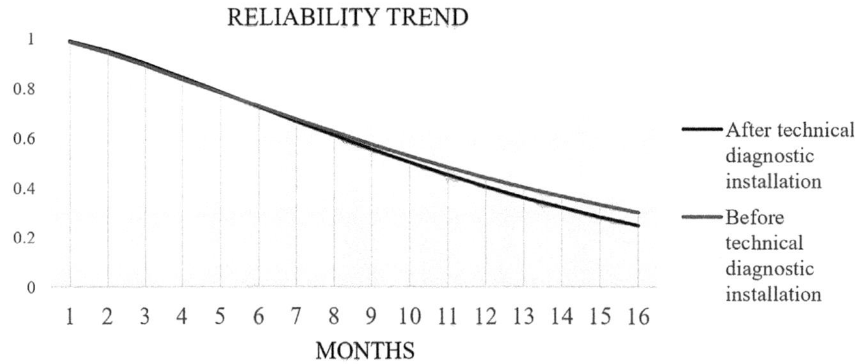

FIGURE 10.20

Comparison of the reliability trend using technical diagnostics.

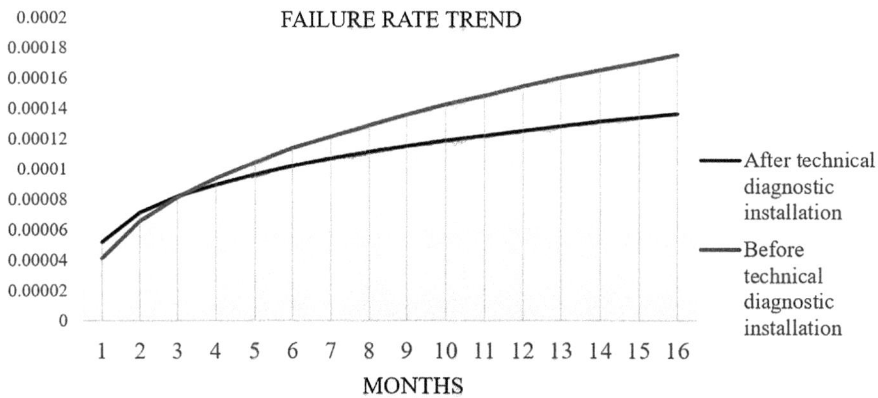

FIGURE 10.21

Comparison of the trend of failure rates using technical diagnostics.

installation of technical diagnostics we obtain the result $\eta \approx 70$, which is also the mean time to failure. An increase in the mean time to failure may result in an increase in hours of failure to failure after installation of technical diagnostics by an average of 20 h, which is also a very good result.

Comparing the cross-section graphs of the reliability and unreliability functions for the observed time intervals (Figs. 10.22 and 10.23), the extension point or the probability of failure can be observed by a little less than 1 month, which is very positive for continuous production systems.

Reliability analysis, as a representative indicator of the effectiveness of the industrial system, indicates an evident increase in the level of reliability, which can be related to the implementation of major investment activities—installation of technical diagnostic systems and introduction of monitoring of critical positions as a permanent activity of the maintenance service.

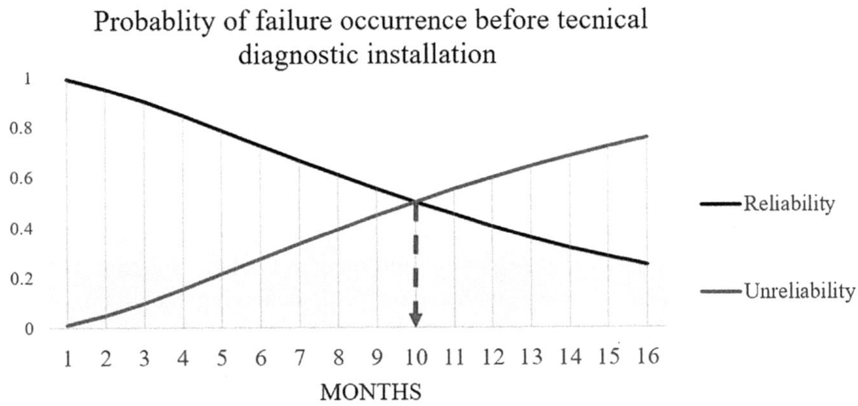

FIGURE 10.22

Probability of failure by the time of technical diagnostic installation.

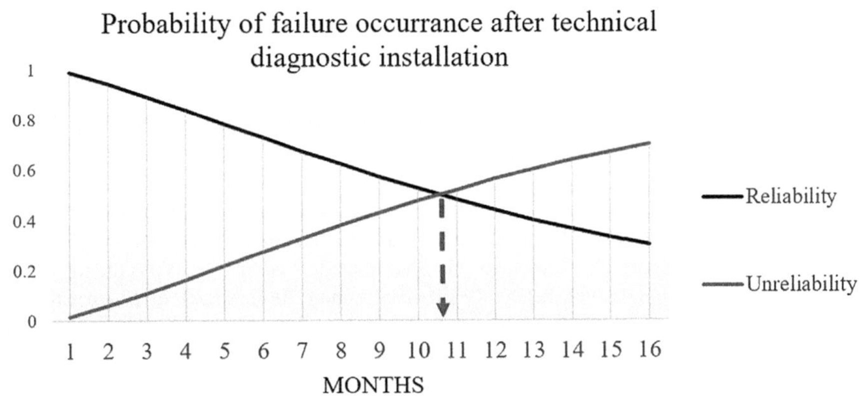

FIGURE 10.23

Probability of failure after installation of technical diagnostic installation.

6. Conclusion

The performance of a manufacturing industrial system can be significantly improved by applying the maintenance concept to the condition. The application of conventional failure remediation procedures at the time when they have already occurred gives way to repair planning methods and preventative maintenance activities, with the aim of eliminating the conditions of complete downtime of the production facility, and thus causing large losses in the operation of the complete system. Technical diagnostic methods, applied in appropriate operating conditions of system elements—machines—enable monitoring of operating parameters and early detection of any deviations from the optimal operating values of the process. If the right timely quality and useable information about the state of the system parameters is obtained and if the optimal values are known in advance, corrective action can be taken in a timely manner to prevent the conditions for failure and causing high costs of the production process and maintenance. Defining the real current state of the equipment using technical diagnostic methods creates the conditions for quality repair planning. A well-planned shutdown of the plant in order to prevent cancellation does not mean only saving time, increasing hourly efficiency, and higher production, but also lower costs of spare parts, optimization of warehouse and working capital, good coordination with external companies, optimal use of work equipment, rational use of human resources, and increasing the level of safety at work. The planning of maintenance activities ensures a higher quality fulfillment of the basic function of maintenance of production equipment, which is aimed at the main goal: achieving maximum availability and reliability of installed equipment at minimal cost, all with the aim of maximizing the profit of the business system.

References

[1] L. Papić, Z. Milovanović, Maintenance and reliability of technical systems, in: DQM Monograph Library Quality and Reliability in Practice, Book 3, Prijevor, 2007 (in Serbian).

[2] EN 13306:2001 Terminologie Maintenance, 2001.

[3] M. Bengtsson, Condition Based Maintenance Systems – an Investigation of Technical Constituents and Organizational Aspects, Malardalen University Licentiate Thesis, 2004. No.36.

[4] S. Jong-Ho, J. Hong-Bae, On condition based maintenance policy, J. Comput. Des. Eng. 2 (April 2015).

[5] A. Prajapati, J. Bechtel, S. Ganesan, Condition based maintenance: a survey, J. Qual. Mainten. Eng. 18 (2012).

[6] A. Rastegari, B. Marcus, Implementation of condition based maintenance in manufacturing industry, in: A Pilot Case Study, Conference on Prognostics and Health Management, PHM 2014, at Cheney, USA, 2014.

[7] A. Radionov, A. Evdokimov, O. Petukhova, G. Shokhina, L. Yabbarova, Vibrodiagnostic surveying of industrial electrical equipment, in: IEEE NW Russia Young Researchers in Electrical and Electronic Engineering Conference, 2016, 2016.

[8] A. Jardine, T. Joseph, D. Banjević, Optimizing condition-based maintenance decisions for equipment subject to vibration monitoring, J. Qual. Mainten. Eng. 5 (2006).

[9] C. Ugechi, E. Ogbonnaya, M. Lilly, S. Ogaji, S. Probert, Condition-Based Diagnostic Approach for Predicting the Maintenance Requirements of Machinery, Engineering, Scientific Research, 2009, 2009.

[10] A.F. Criqui, False and Misleading Sources of Vibration, Proceedings of the Twenty-Third Turbomachinery Symposium, Turbomachinery Laboratory, Texas A&M University, College Station, Texas, 1994, pp. 137–150.

[11] A. Lifson, H. Simmons, A. Smalley, Vibration Limits for Rotating Machinery, Mechanical Engineering, 1987, pp. 60–65.

[12] C.A. ЕвҐеньевич, Разработка сЏециальноҐо математическоҐо и ЏроҐраммноҐо обесЏечения для автоматизированной диаҐностики сложныч систем, Диссертация, Москва, 2007 (in Russian).

[13] D. Brown, T. Jensen, Machine-Condition Monitoring Using Vibration Analysis the Use of Spectrum Comparison for Bearing Fault Detection - A Case Study from Alma Paper Mill, Bruel and Kjaer application notes, Quebec, Canada, 2011.

[14] J. Semjon, V. Balaž, J. Varga, Methodology for the vibration measurement and evaluation on the industrial robot Kuka, in: The 23rd International Conference on Robotics in Alpe-Adria-Danube Region (RAAD), 2014.

[15] MaintWorld, Condition Based Maintenance in the Paper Industry, April 2014. http://www.lifetime-reliability.com.
[16] http://voith.com/en/twogether-article-31-en-63-roll_maintenance.pdf>.
[17] http://www.shiresystems.co.uk/downloads/whitepapers/papermill.pdf.>.
[18] http://www.lifetime-reliability.com/free-articles/maintenance-management/Dont_Waste_Your_Time_and_Money_With_Condition_Monitoring.pdf>.
[19] F. Francesco, G. Andrea, L. Sauro, B. Nicola B, Multi-scale PCA based fault diagnosis on a paper mill plant, IEEE Conference Publication, ʿhttp://ieeexplore.ieee.org/document/6059069/media>.
[20] SKF Global Pulp & Paper Segment, A Compilation of Issues 1-15 of the Technical Newsletter for the Pulp and Palper Industry, vol. 1, January 2011. No.1.
[21] http://www.bretech.com/reference/Case%20Studies%20on%20Paper%20Machine%20Vibration%20Problems.pdf>, A. K. Costain, Case Studies on Paper Machine Vibration Problems.
[22] D. Soldat, Maintenance Efficiency, Belgrade, 1993 (in Serbian).
[23] M. Bulatović, Maintenance and Effectiveness of Technical Systems, University in Montenegro – Faculty of Mechanical Engineering in Podgorica, 2008 (in Serbian).
[24] Z. Milovanović, Optimization of Power Plant Reliability, University of Banja Luka, Faculty of Mechanical Engineering Banja Luka, Banja Luka, 2003 (in Serbian).
[25] V. Zeljković, L. Papić, Reliability Testing, Lola Institute Belgrade, 2001 (in Serbian).

CHAPTER

Reliability assessment of replaceable shuffle-exchange network by using interval-valued universal generating function

11

Amisha Khati, S.B. Singh

Department of Mathematics, Statistics and Computer Science, G.B. Pant University of Agriculture and Technology, Pantnagar, Uttarakhand, India

1. Introduction

In today's high-tech world, everyone relies upon the proper functioning of machines and appliances for our day-to-day requirements as well as everyday safety, motility, and monetary welfare. When any of our communication system, electrical appliances, computer networks, nuclear power plants, transportation system, aerospace applications, etc., does not perform as expected, the results can be terrible, injury, or even loss of life. Thus it becomes very vital to assure their functioning, by carrying out the study on reliability, keeping in mind the existing possibilities. With the increasing cost and network complexity, the need to study reliability theory is also growing. Bisht S. and Singh S. B. [2] gave a method for finding the reliability, mean time to failure (MTTF), and signature reliability of complex networks by universal generating function (UGF) algorithm and concluded that if we made a little change in the considered complex network then its reliability got affected significantly. Interconnection networks play a very vital role in the performance of multiprocessors and parallel processing.

Multistage interconnection networks (MINs) are the natural consequences of the advances in the computer technology and are building a wholly new atmosphere by interconnecting a huge number of processors and memory modules through a number of stages to construct multiprocessors. Nowadays, MINs are used in several real-life applications like supercomputers, telephone switches, networks in industrial applications, wide area computer networks, and many more. Thus the idea, design, and performance of MINs are decisive factors at this point of time. Sharma et al. [11] analyzed the reliability and path length of some irregular MINs, in which each stage contains different number of switching elements (SEs). The reliability of the proposed networks in terms of their MTTF has been computed by them and then their reliabilities are compared with each other. Rajkumar and Goyal [10] attempted to compare and examine several network topologies of MINs on the basis of their reliability, cost-effectiveness, and fault-tolerance.

Shuffle-exchange network (SEN) is one of the most widely used practical MIN because of the small size of its SE and low network complexity. It consists of a unique path between any source and

The Handbook of Reliability, Maintenance, and System Safety through Mathematical Modeling.
https://doi.org/10.1016/B978-0-12-819582-6.00011-3

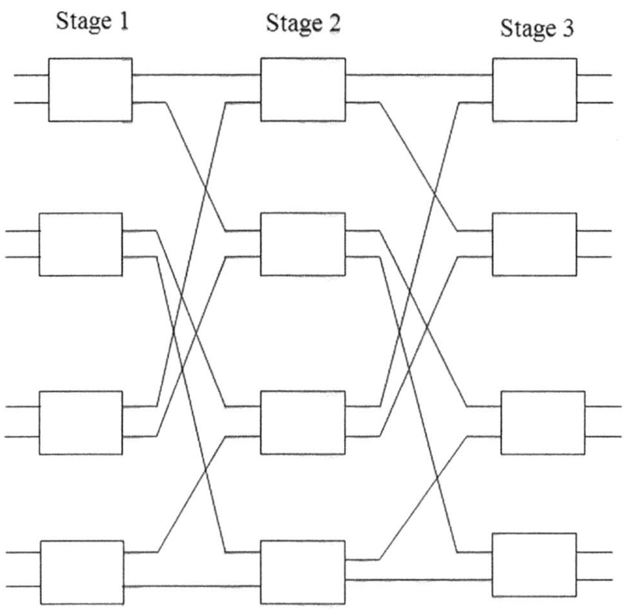

FIGURE 11.1

An 8 × 8 SEN.

destination. In an $N \times N$ SEN, there are $\log_2 N$ stages and each stage consists of $N/2$ SEs. In total there are $(N/2)$ $(\log_2 N)$ SEs in the SEN. An 8 × 8 SEN having three stages and four SEs per stage is shown in Fig. 11.1.

Due to the continuously increasing demand of SEN in the present era, various methods have been introduced by various researchers for their reliability analysis. Fard and Gunawan [5] introduced a modified SEN consisting of 1 × 2 SEs at the source, 2 × 2 at the intermediate stages, and 2 × 1 at the terminal, and evaluated the terminal reliability of SEN and modified SEN. The terminal reliability of modified SEN was found higher for network sizes greater than 4 × 4. Yunus and Othman [12] investigated six different types of SENs having extra stages. The reviewed SENs are SEN+, Irregular Augmented Shuffle-Exchange Network (IASEN), Irregular Augmented Shuffle Network (IASN), Generalized Shuffle-Exchange Network (GSEN), Improved Irregular Augmented Shuffle Multistage Interconnection Network (IIASN), and Irregular Modified Alpha Network (ALN). It was observed that a more redundant path is attained by enhancing the stages, which also reduce the latency and increase fault-tolerance by providing auxiliary links. However, increasing the network size increases the network complexity and can also increase the cost. Bistouni and Jahanshahi [3] gave a method to enhance the fault-tolerance and reliability of SEN by increasing the number of switching stages and concluded that the reliability of SEN with one extra stage (SEN+) is better than that of SEN or SEN having two extra stages (SEN+2), although the reliability of SEN+ has always been found more than the reliability of SEN+2. Bistouni and Jahanshahi [3] determined the reliability of SEN by using the method of reliability block diagrams (RBDs). Yunus and Othman [13] proposed a new network SEN with minus one stage (SEN−) and then compared it with SEN, SEN+, and SEN+2. The comparison was made on the basis of three parameters, namely, terminal, broadcast, and network reliability. The three reliability parameters, namely, terminal, broadcast, and network reliability of SEN− were found higher than the rest three considered networks. Bisht [1] figured out the terminal, broadcast, and network reliability of SENs by applying the method of UGF. Bistouni and Jahanshahi [4] analyzed the reliability importance of the SEs in SEN, SEN+, and SEN+2. They concluded that a high-reliable network can be reached by replacing the sensitive SEs and by using SEs having high reliability. From the above discussion one can easily observe that none of the researcher analyzes the networks incorporating the uncertainties in it while in the realistic situation it is quite possible to deal with such networks. Some of the reasons behind the uncertainties arising in the complex systems are:

(a) Temperature, humidity, etc., are some of the environmental factors which make the system and its components uncertain.
(b) In analyzing a complex system, sometimes it becomes very tedious and expensive to obtain a reliability data which is precise and accurate.
(c) When we use any system continuously, the performance of the system and its components degrades with time and thus the probability of the system and its components varies with time.

When the network has performance uncertainty then instead of obtaining the exact values of the probabilities of components, interval-valued state probabilities of the components are obtained. The UGF representation for interval-valued probability is known as interval-valued universal generating function (IUGF).

The IUGF of a component Gj with Mj states is defined as:

$$U_j(z) = \sum_{i=1}^{M_j} \left[p_i^j\right] \cdot z^{g_i^j}$$

where $\left[p_i^j\right] (j = 1, 2, \ldots, n; i = 1, 2, \ldots, M_j)$ is the probability interval and g_i^j is the performance of the network with respect to the state M_j.

Thus, in order to analyze the reliability of the network more precisely it becomes very essential to examine the network involving uncertainties. This will greatly enhance the plausibility of the reliability analysis of the network. IUGF is an approach used to analyze the reliability of a network possessing uncertainties. Li et al. [7] proposed a method to evaluate the reliability of multistate system (MSS) when the component's available data are insufficient. In such cases rather than the precise values of the probabilities, the interval-valued probabilities of the components can be considered. In order to find the interval-valued reliability of MSS, an interval UGF was built up. It can be seen from the results that this method is efficient when the state probabilities of components are inaccurate (or uncertain). Pan G. et al. [9] gave a method for the assessment of interval-valued reliability of MSS considering uncertainty. They defined the algorithm for IUGF approach and verified their method by taking examples. Kumar et al. [6] determined the interval-valued reliability of a 2-out-of-4 system consisting of two components which are configured in series. To evaluate the interval-valued reliability of the system, interval UGF approach has been used. A numerical example has also been illustrated. Meenakshi and Singh [8] evaluated the reliability and MTTF of a non-repairable MSS by using IUGF. They analyzed the system's reliability by including the uncertainties in the probabilities and the failure rates of the components of the considered system.

Further, in analyzing the SENs any of the researchers does not take into account the SEN whose SEs can be replaced when become faulty so that the whole network does not collapse.

Keeping these facts in view, in this chapter, we have considered an SEN, in which the SEs have some uncertainties associated with them and if any of the SEs of the considered SEN becomes faulty, then it can be replaced by a certain replacement rate. In this work we propose to analyze its reliability on the basis of three indices, namely, terminal reliability, broadcast reliability, and network reliability. The reliability evaluation has been done by using IUGF approach and the probabilities are obtained in intervals.

The SEN examined here is of size 8×8, i.e., has eight inputs and eight outputs. The RBDs for the terminal, broadcast, and network reliability of the SEN are presented in Figs. 11.1, 11.3, and 11.5, respectively. By using supplementary variable technique, the differential equations of various states governing the network's performance are obtained and by applying Laplace transform, the transition state probabilities of different components are computed. By using the probabilities obtained by the IUGF approach, the upper and lower bounds of the three different reliability parameters and the bounds of MTTF for each reliability of the considered network have been evaluated. A numerical example is also provided to give a practical explanation of the proposed model.

2. Assumptions

1. Initially the network is in good condition, i.e., all the nodes and links are operating properly.
2. The network considered is an 8×8 SEN in which each SE is of size 2×2.
3. All the network's components are either in working stage or in failed stage.
4. If the network fails completely then only will go for replacement and after replacement the network is as good as new.
5. The failure rates of different components are taken different while the replacement rates of all the components are supposed to be same.

3. Acronyms

BR	Broadcast reliability
IUGF	Interval-valued universal generating function
MTTF	Mean time to failure
NR	Network reliability
SEN	Shuffle-exchange network
SE	Switching element
TR	Terminal reliability

4. Notations

$\underset{par}{\otimes}$	Composition operator for parallel configuration
$\underset{ser}{\otimes}$	Composition operator for series configuration

x	Elapsed replacement time
λ_i	Failure rate of ith component
$u_i(z)$	IUGF of component i
$\underline{P}_i$	Lower bound of the probability of ith component, represented by a real number
η	Replacement rate for the entire network's component
$u(z)$	UGF of the network
$\overline{P}_i$	Upper bound of the probability of ith component, represented by a real number
$[R]$	Interval-valued reliability of the network

5. Definitions

(i) Terminal Reliability: The terminal reliability (TR) of a network is defined as the probability of existence of at least one fault-free path between a source—destination pair.

(ii) Broadcast Reliability: The broadcast reliability (BR) of a network is defined as the probability of successful communication between a single source node and all the destination nodes.

(iii) Network Reliability: The network reliability (NR) of a network is defined as the probability that all the source nodes are connected to all the destination nodes.

6. Terminal reliability of SEN

The RBD for TR of SEN is shown in Fig. 11.2 and the transition state diagram for the considered network is presented in Fig. 11.3.

(a) Formulation of mathematical model:

By applying supplementary variable technique to the assumed model, the difference-differential equations for lower bound probabilities of the components of the network in different states, governing the network's behavior obtained, are as follows:

$$\left(\frac{d}{dt}+\underline{\lambda}_1+\underline{\lambda}_2+\underline{\lambda}_3\right)\underline{P}_0(t)=\int_0^\infty \underline{P}_1(x,t)\eta(x)dx+\int_0^\infty \underline{P}_2(x,t)\eta(x)dx+\int_0^\infty \underline{P}_3(x,t)\eta(x)dx \quad (11.1)$$

$$\left\{\frac{\partial}{\partial x}+\frac{\partial}{\partial t}+\eta(x)\right\}\underline{P}_1(x,t)=0 \quad (11.2)$$

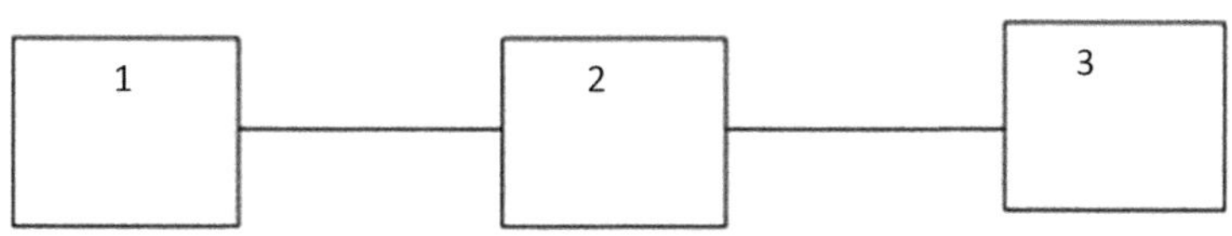

FIGURE 11.2

Terminal RBD for an 8 × 8 SEN.

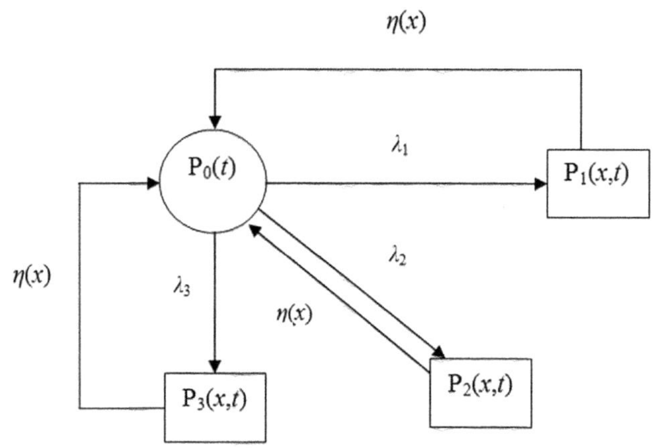

FIGURE 11.3

Transition state diagram for TR of SEN.

$$\left\{\frac{\partial}{\partial x}+\frac{\partial}{\partial t}+\eta(x)\right\}\underline{P}_2(x,t)=0 \tag{11.3}$$

$$\left\{\frac{\partial}{\partial x}+\frac{\partial}{\partial t}+\eta(x)\right\}\underline{P}_3(x,t)=0 \tag{11.4}$$

Boundary conditions:

$$\underline{P}_1(0,t) = \lambda_1\underline{P}_0(t) \tag{11.5}$$

$$\underline{P}_2(0,t) = \lambda_2\underline{P}_0(t) \tag{11.6}$$

$$\underline{P}_3(0,t) = \lambda_3\underline{P}_0(t) \tag{11.7}$$

Initial conditions:
$\underline{P}_0(t) = 0$ at $t=0$ and is zero at all other values of t.

(b) Solution of the model:

On taking the Laplace transform of Eqs. (11.1) to (11.4) along with the boundary conditions (11.5)–(11.7) and applying the initial conditions, we get:

$$\{s+\lambda_1+\lambda_2+\lambda_3\}\underline{P}_0(s)=1+\int_0^\infty \underline{P}_1(s,x)\eta(x)dx+\int_0^\infty \underline{P}_2(s,x)\eta(x)dx+\int_0^\infty \underline{P}_3(s,x)\eta(x)dx \tag{11.8}$$

$$\left\{s+\frac{\partial}{\partial x}+\eta(x)\right\}\underline{P}_1(s,x)=0 \tag{11.9}$$

$$\left\{s+\frac{\partial}{\partial x}+\eta(x)\right\}\underline{P}_2(s,x)=0 \tag{11.10}$$

$$\left\{s+\frac{\partial}{\partial x}+\eta(x)\right\}\underline{P}_3(s,x)=0 \tag{11.11}$$

Boundary conditions:

$$\underline{P}_1(0,s)=\underline{P}_0(s)\underline{\lambda}_1 \tag{11.12}$$

$$\underline{P}_2(0,s)=\underline{P}_0(s)\underline{\lambda}_2 \tag{11.13}$$

$$\underline{P}_3(0,s)=\underline{P}_0(s)\underline{\lambda}_3 \tag{11.14}$$

(c) Transition state probabilities:

On solving the equations from (11.8) to (11.11), the following transition state probabilities are obtained:

$$\underline{P}_0(s) = \frac{1}{s+\underline{A}} \tag{11.15}$$

$$\underline{P}_1(s) = \frac{\underline{\lambda}_1\underline{P}_0(s)(1-S(s))}{s} \tag{11.16}$$

$$\underline{P}_2(s) = \frac{\underline{\lambda}_2\underline{P}_0(s)(1-S(s))}{s} \tag{11.17}$$

$$\underline{P}_3(s) = \frac{\underline{\lambda}_3\underline{P}_0(s)(1-S(s))}{s} \tag{11.18}$$

where $\underline{A} = \underline{\lambda}_1 + \underline{\lambda}_2 + \underline{\lambda}_3$

Similarly, again by applying supplementary variable technique to the considered model the upper bounds of the probabilities for different states of the network's components are obtained as:

$$\left(\frac{d}{dt}+\overline{\lambda}_1+\overline{\lambda}_2+\overline{\lambda}_3\right)\overline{P}_0(t) = \int_0^\infty \overline{P}_1(x,t)\eta(x)dx + \int_0^\infty \overline{P}_2(x,t)\eta(x)dx + \int_0^\infty \overline{P}_3(x,t)\eta(x)dx \tag{11.19}$$

$$\left\{\frac{\partial}{\partial x}+\frac{\partial}{\partial t}+\eta(x)\right\}\overline{P}_1(x,t)=0 \tag{11.20}$$

$$\left\{\frac{\partial}{\partial x}+\frac{\partial}{\partial t}+\eta(x)\right\}\overline{P}_2(x,t)=0 \tag{11.21}$$

$$\left\{\frac{\partial}{\partial x}+\frac{\partial}{\partial t}+\eta(x)\right\}\overline{P}_3(x,t)=0 \tag{11.22}$$

Boundary conditions:

$$\overline{P}_1(0,t) = \overline{\lambda}_1\overline{P}_0(t) \tag{11.23}$$

$$\overline{P}_2(0,t) = \overline{\lambda}_2\overline{P}_0(t) \tag{11.24}$$

$$\overline{P}_3(0,t) = \overline{\lambda}_3\overline{P}_0(t) \tag{11.25}$$

Initial conditions:

At $t = 0$, $\overline{P}_0(t) = 1$ and is 0 at all other values of t.

Solving the equations from (11.19) to (11.22) and using the boundary conditions from (11.23) to (11.25), the upper bounds of the transition state probabilities are obtained as:

$$\overline{P}_0(s) = \frac{1}{(s+\overline{A})} \tag{11.26}$$

$$\overline{P}_1(s) = \frac{\overline{\lambda}_1 \overline{P}_0(s)(1-S(s))}{s} \tag{11.27}$$

$$\overline{P}_2(s) = \frac{\overline{\lambda}_2 \overline{P}_0(s)(1-S(s))}{s} \tag{11.28}$$

$$\overline{P}_3(s) = \frac{\overline{\lambda}_3 \overline{P}_0(s)(1-S(s))}{s} \tag{11.29}$$

where $\overline{A} = \overline{\lambda}_1 + \overline{\lambda}_2 + \overline{\lambda}_3$

(i) Interval-valued terminal reliability of the network:

The interval-valued TR of the network is given as:

$$R = \left[\underline{P}_0(t), \overline{P}_0(t)\right] \tag{11.30}$$

(ii) Mean time to failure:

The MTTF of the considered network can be determined by using the expression:

$$(\text{MTTF})_{\text{TR}} = \left\{\frac{1}{(s+\underline{\lambda}_1+\underline{\lambda}_2+\underline{\lambda}_3)}, \frac{1}{(s+\overline{\lambda}_1+\overline{\lambda}_2+\overline{\lambda}_3)}\right\} \tag{11.31}$$

7. Broadcast reliability of SEN

The RBD for computing the broadcast reliability of SEN is shown in Fig. 11.4 and the transition state diagram for the BR of SEN is shown in Fig. 11.5.

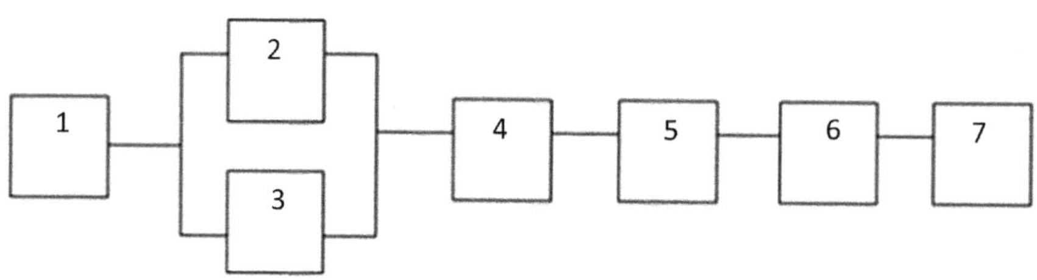

FIGURE 11.4

Broadcast RBD for SEN.

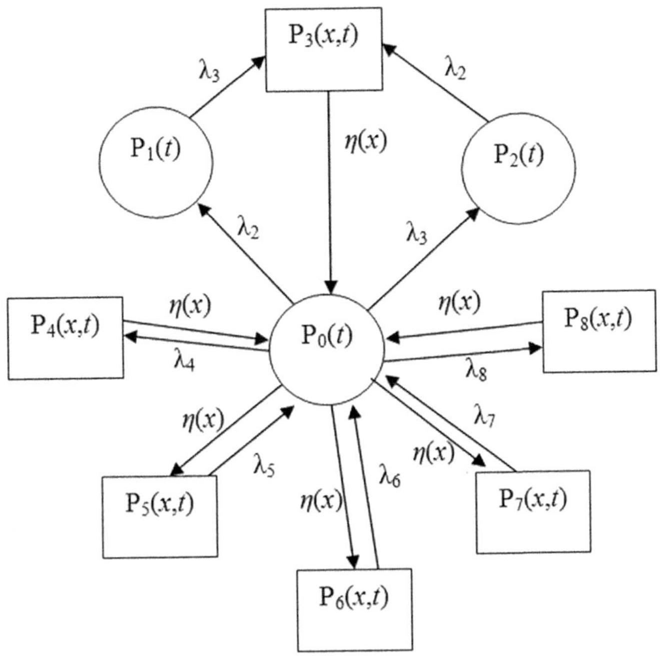

FIGURE 11.5

Transition state diagram for BR of SEN.

(a) Formulation of mathematical model:

By applying the supplementary variable technique to the considered model, the following set of equations for various states is obtained:

$$\left(\frac{d}{dt}+\lambda_1+\lambda_2+\lambda_3+\lambda_4+\lambda_5+\lambda_6+\lambda_7\right)P_0(t) = \int_0^\infty P_3(x,t)\eta(x)dx+\int_0^\infty P_4(x,t)\eta(x)dx + \int_0^\infty P_5(x,t)\eta(x)dx + \int_0^\infty P_6(x,t)\eta(x)dx + \int_0^\infty P_7(x,t)\eta(x)dx + \int_0^\infty P_8(x,t)\eta(x)dx \tag{11.32}$$

$$\left(\frac{d}{dt}+\lambda_3\right)P_1(t) = \lambda_2 P_0(t) \tag{11.33}$$

$$\left(\frac{d}{dt}+\lambda_2\right)\underline{P}_2(t) = \lambda_3\underline{P}_0(t) \tag{11.34}$$

$$\left\{\frac{\partial}{\partial x}+\frac{\partial}{\partial t}+\eta(x)\right\}\underline{P}_3(x,t)=0 \tag{11.35}$$

$$\left\{\frac{\partial}{\partial x}+\frac{\partial}{\partial t}+\eta(x)\right\}\underline{P}_4(x,t)=0 \tag{11.36}$$

$$\left\{\frac{\partial}{\partial x}+\frac{\partial}{\partial t}+\eta(x)\right\}\underline{P}_5(x,t)=0 \tag{11.37}$$

$$\left\{\frac{\partial}{\partial x}+\frac{\partial}{\partial t}+\eta(x)\right\}\underline{P}_6(x,t)=0 \tag{11.38}$$

$$\left\{\frac{\partial}{\partial x}+\frac{\partial}{\partial t}+\eta(x)\right\}\underline{P}_7(x,t)=0 \tag{11.39}$$

$$\left\{\frac{\partial}{\partial x}+\frac{\partial}{\partial t}+\eta(x)\right\}\underline{P}_8(x,t)=0 \tag{11.40}$$

Boundary conditions:

$$\underline{P}_3(0,t) = \lambda_3\underline{P}_1(t) + \lambda_2\underline{P}_2(t) \tag{11.41}$$

$$\underline{P}_4(0,t) = \lambda_1\underline{P}_0(t) \tag{11.42}$$

$$\underline{P}_5(0,t) = \lambda_4\underline{P}_0(t) \tag{11.43}$$

$$\underline{P}_6(0,t) = \lambda_5\underline{P}_0(t) \tag{11.44}$$

$$\underline{P}_6(0,t) = \lambda_5\underline{P}_0(t) \tag{11.45}$$

$$\underline{P}_8(0,t) = \lambda_7\underline{P}_0(t) \tag{11.46}$$

Initial conditions:

$\underline{P}_0(t) = 1$ at $t=0$ and is zero at all other values of t.

(b) Solution of the model:

The following equations are obtained by taking the Laplace transform of the equations from (11.31) to (11.45) and using the initial conditions.

$$\begin{aligned}\{s+\underline{B}\}\underline{P}_0(s) = 1 &+ \int_0^\infty \underline{P}_3(s,x)\eta(x)dx + \int_0^\infty \underline{P}_4(s,x)\eta(x)dx + \int_0^\infty \underline{P}_5(s,x)\eta(x)dx \\ &+ \int_0^\infty \underline{P}_6(s,x)\eta(x)dx + \int_0^\infty \underline{P}_7(s,x)\eta(x)dx + \int_0^\infty \underline{P}_8(s,x)\eta(x)dx\end{aligned} \tag{11.47}$$

$$\{s+\underline{\lambda}_3\}\underline{P}_1(s)=1+\underline{\lambda}_2\underline{P}_0(s) \tag{11.48}$$

$$\{s+\underline{\lambda}_2\}\underline{P}_2(s)=1+\underline{\lambda}_3\underline{P}_0(s) \tag{11.49}$$

$$\left\{s+\frac{\partial}{\partial x}+\eta(x)\right\}\underline{P}_3(s,x)=0 \tag{11.50}$$

$$\left\{s+\frac{\partial}{\partial x}+\eta(x)\right\}\underline{P}_4(s,x)=0 \tag{11.51}$$

$$\left\{s+\frac{\partial}{\partial x}+\eta(x)\right\}\underline{P}_5(s,x)=0 \tag{11.52}$$

$$\left\{s+\frac{\partial}{\partial x}+\eta(x)\right\}\underline{P}_6(s,x)=0 \tag{11.53}$$

$$\left\{s+\frac{\partial}{\partial x}+\eta(x)\right\}\underline{P}_7(s,x)=0 \tag{11.54}$$

$$\left\{s+\frac{\partial}{\partial x}+\eta(x)\right\}\underline{P}_8(s,x)=0 \tag{11.55}$$

Boundary conditions:

$$\underline{P}_3(0,s)=\underline{P}_1(s)\underline{\lambda}_3+\underline{P}_2(s)\underline{\lambda}_2 \tag{11.56}$$

$$\underline{P}_4(0,s)=\underline{P}_0(s)\underline{\lambda}_1 \tag{11.57}$$

$$\underline{P}_6(0,s)=\underline{P}_0(s)\underline{\lambda}_5 \tag{11.58}$$

$$\underline{P}_7(0,s)=\underline{P}_0(s)\underline{\lambda}_6 \tag{11.59}$$

$$\underline{P}_8(0,s)=\underline{P}_0(s)\underline{\lambda}_7 \tag{11.60}$$

Initial conditions:

$\underline{P}_0(t)=1$ at $t=0$ and is zero at all other values of t.

(c) Transition state probabilities:

On solving the equations from (11.46) to (11.54) and using the boundary conditions from (11.55) to (11.60), the following transition state probabilities are obtained:

$$\underline{P}_0(s)=\frac{1}{(s+\underline{B})} \tag{11.61}$$

$$\underline{P}_1(s)=\frac{\underline{\lambda}_2\underline{P}_0(s)}{s+\underline{\lambda}_3} \tag{11.62}$$

$$\underline{P}_2(s)=\frac{\underline{\lambda}_3\underline{P}_0(s)}{s+\underline{\lambda}_2} \tag{11.63}$$

$$\underline{P}_3(s) = \{\underline{\lambda}_3\underline{P}_1(s) + \underline{\lambda}_2\underline{P}_2(s)\}\left(\frac{1 - S(s)}{s}\right) \tag{11.64}$$

$$\underline{P}_4(s) = \{\underline{\lambda}_1\underline{P}_0(s)\}\left(\frac{1 - S(s)}{s}\right) \tag{11.65}$$

$$\underline{P}_5(s) = \{\underline{\lambda}_4\underline{P}_0(s)\}\left(\frac{1 - S(s)}{s}\right) \tag{11.66}$$

$$\underline{P}_6(s) = \{\underline{\lambda}_5\underline{P}_0(s)\}\left(\frac{1 - S(s)}{s}\right) \tag{11.67}$$

$$\underline{P}_7(s) = \{\underline{\lambda}_6\underline{P}_0(s)\}\left(\frac{1 - S(s)}{s}\right) \tag{11.68}$$

$$\underline{P}_8(s) = \{\underline{\lambda}_7\underline{P}_0(s)\}\left(\frac{1 - S(s)}{s}\right) \tag{11.69}$$

where $\underline{B} = \underline{\lambda}_1 + \underline{\lambda}_2 + \underline{\lambda}_3 + \underline{\lambda}_4 + \underline{\lambda}_5 + \underline{\lambda}_6 + \underline{\lambda}_7$

Similarly we can find the expressions for upper bounds of the transition state probabilities of the network's components by replacing $\underline{P}_i(s)$ by $\overline{P}_i(s)$ ($i = 0$ to 8) and $\underline{\lambda}_j$ by $\overline{\lambda}_j$ ($j = 1$ to 7) in Eqs. (11.61) to (11.69).

(i) Interval-valued broadcast reliability of the network:

The interval-valued BR of the 8 × 8 SEN is given as:

$$R = \left[\underline{P}_0(t) + \underline{P}_1(t) + \underline{P}_2(t),\ \overline{P}_0(t) + \overline{P}_1(t) + \overline{P}_2(t)\right] \tag{11.70}$$

(ii) Mean time to failure:

The MTTF of the network is given by Eq. (11.71) as:

$$(\text{MTTF})_{\text{BR}} = \left[\frac{1}{\underline{B}} + \frac{\underline{\lambda}_2}{\underline{\lambda}_3\underline{B}} + \frac{\underline{\lambda}_3}{\underline{\lambda}_2\underline{B}},\ \frac{1}{\overline{B}} + \frac{\overline{\lambda}_2}{\overline{\lambda}_3\overline{B}} + \frac{\overline{\lambda}_3}{\overline{\lambda}_2\overline{B}}\right] \tag{11.71}$$

8. Network reliability of SEN

The RBD for evaluating the NR of SEN is shown in Fig. 11.6 and the transition state diagram for the NR of the network under consideration is shown in Fig. 11.7.

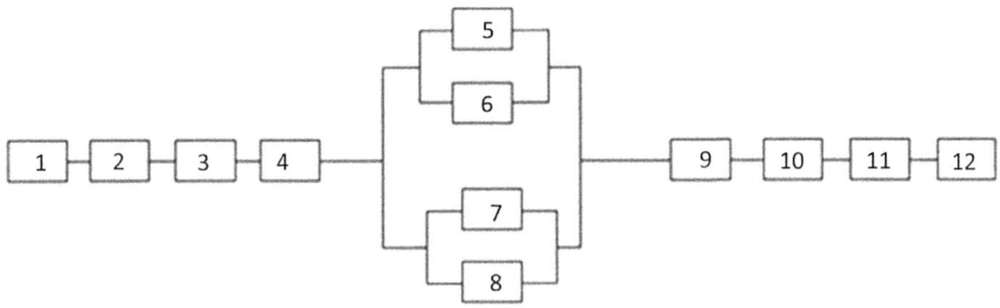

FIGURE 11.6

Network RBD for an 8 × 8 SEN.

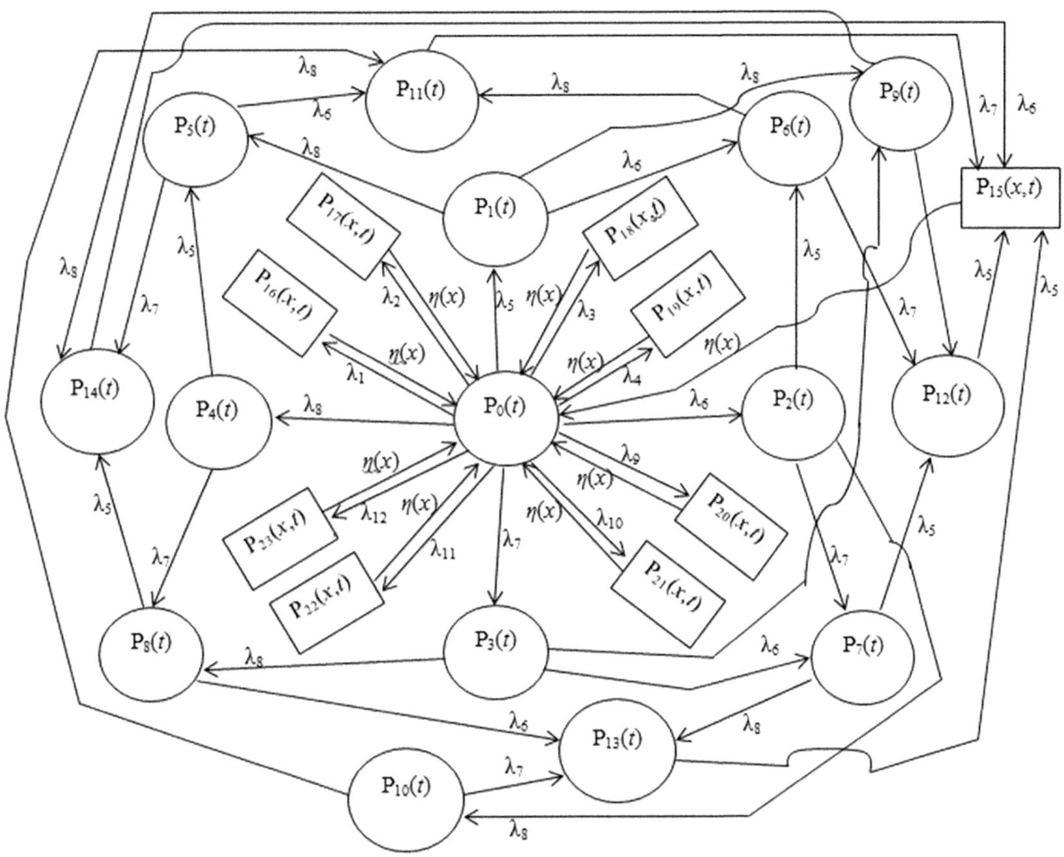

FIGURE 11.7

Transition state diagram.

(a) Formulation of mathematical model:

On applying supplementary variable technique to the model under consideration, the following set of difference-differential equations for lower bound probabilities of the components of the network in different states is obtained.

$$\begin{aligned}\left(\frac{d}{dt}+\underline{C}\right)\underline{P}_0(t) &= \int_0^\infty \underline{P}_{15}(x,t)\eta(x)dx+\int_0^\infty \underline{P}_{16}(x,t)\eta(x)dx+\int_0^\infty \underline{P}_{17}(x,t)\eta(x)dx \\ &+\int_0^\infty \underline{P}_{18}(x,t)\eta(x)dx+\int_0^\infty \underline{P}_{19}(x,t)\eta(x)dx+\int_0^\infty \underline{P}_{20}(x,t)\eta(x)dx \\ &+\int_0^\infty \underline{P}_{21}(x,t)\eta(x)dx+\int_0^\infty \underline{P}_{22}(x,t)\eta(x)dx+\int_0^\infty \underline{P}_{23}(x,t)\eta(x)dx\end{aligned} \tag{11.72}$$

$$\left(\frac{d}{dt}+\underline{\lambda}_6+\underline{\lambda}_7+\underline{\lambda}_8\right)\underline{P}_1(t)=\underline{\lambda}_5\underline{P}_0(t) \tag{11.73}$$

$$\left(\frac{d}{dt}+\underline{\lambda}_5+\underline{\lambda}_7+\underline{\lambda}_8\right)\underline{P}_2(t)=\underline{\lambda}_6\underline{P}_0(t) \tag{11.74}$$

$$\left(\frac{d}{dt}+\underline{\lambda}_5+\underline{\lambda}_6+\underline{\lambda}_8\right)\underline{P}_3(t)=\underline{\lambda}_7\underline{P}_0(t) \tag{11.75}$$

$$\left(\frac{d}{dt}+\underline{\lambda}_5+\underline{\lambda}_6+\underline{\lambda}_7\right)\underline{P}_4(t)=\underline{\lambda}_8\underline{P}_0(t) \tag{11.76}$$

$$\left(\frac{d}{dt}+\underline{\lambda}_6+\underline{\lambda}_7\right)\underline{P}_5(t)=\underline{\lambda}_8\underline{P}_1(t)+\underline{\lambda}_5\underline{P}_4(t) \tag{11.77}$$

$$\left(\frac{d}{dt}+\underline{\lambda}_8+\underline{\lambda}_7\right)\underline{P}_6(t)=\underline{\lambda}_6\underline{P}_1(t)+\underline{\lambda}_5\underline{P}_2(t) \tag{11.78}$$

$$\left(\frac{d}{dt}+\underline{\lambda}_8+\underline{\lambda}_5\right)\underline{P}_7(t)=\underline{\lambda}_6\underline{P}_3(t)+\underline{\lambda}_7\underline{P}_2(t) \tag{11.79}$$

$$\left(\frac{d}{dt}+\underline{\lambda}_6+\underline{\lambda}_5\right)\underline{P}_8(t)=\underline{\lambda}_8\underline{P}_3(t)+\underline{\lambda}_7\underline{P}_4(t) \tag{11.80}$$

$$\left(\frac{d}{dt}+\underline{\lambda}_6+\underline{\lambda}_8\right)\underline{P}_9(t)=\underline{\lambda}_5\underline{P}_3(t)+\underline{\lambda}_7\underline{P}_1(t) \tag{11.81}$$

$$\left(\frac{d}{dt}+\underline{\lambda}_7+\underline{\lambda}_5\right)\underline{P}_{10}(t)=\underline{\lambda}_8\underline{P}_2(t)+\underline{\lambda}_6\underline{P}_4(t) \tag{11.82}$$

$$\left(\frac{d}{dt}+\lambda_7\right)\underline{P}_{11}(t)=\underline{\lambda}_6\underline{P}_5(t)+\underline{\lambda}_8\underline{P}_6(t)+\underline{\lambda}_5\underline{P}_{10}(t) \tag{11.83}$$

$$\left(\frac{d}{dt}+\underline{\lambda}_8\right)\underline{P}_{12}(t)=\underline{\lambda}_5\underline{P}_7(t)+\underline{\lambda}_7\underline{P}_6(t)+\underline{\lambda}_6\underline{P}_9(t) \tag{11.84}$$

$$\left(\frac{d}{dt}+\underline{\lambda}_5\right)\underline{P}_{13}(t)=\underline{\lambda}_8\underline{P}_7(t)+\underline{\lambda}_7\underline{P}_{10}(t)+\underline{\lambda}_6\underline{P}_8(t) \tag{11.85}$$

$$\left(\frac{d}{dt}+\underline{\lambda}_6\right)\underline{P}_{14}(t)=\underline{\lambda}_7\underline{P}_5(t)+\underline{\lambda}_8\underline{P}_9(t)+\underline{\lambda}_5\underline{P}_8(t) \tag{11.86}$$

$$\left(\frac{\partial}{\partial x}+\frac{\partial}{\partial t}+\eta(x)\right)\underline{P}_{15}(x,t)=0 \tag{11.87}$$

$$\left(\frac{\partial}{\partial x}+\frac{\partial}{\partial t}+\eta(x)\right)\underline{P}_{16}(x,t)=0 \tag{11.88}$$

$$\left(\frac{\partial}{\partial x}+\frac{\partial}{\partial t}+\eta(x)\right)\underline{P}_{17}(x,t)=0 \tag{11.89}$$

$$\left(\frac{\partial}{\partial x}+\frac{\partial}{\partial t}+\eta(x)\right)\underline{P}_{18}(x,t)=0 \tag{11.90}$$

$$\left(\frac{\partial}{\partial x}+\frac{\partial}{\partial t}+\eta(x)\right)\underline{P}_{19}(x,t)=0 \tag{11.91}$$

$$\left(\frac{\partial}{\partial x}+\frac{\partial}{\partial t}+\eta(x)\right)\underline{P}_{20}(x,t)=0 \tag{11.92}$$

$$\left(\frac{\partial}{\partial x}+\frac{\partial}{\partial t}+\eta(x)\right)\underline{P}_{21}(x,t)=0 \tag{11.93}$$

$$\left(\frac{\partial}{\partial x}+\frac{\partial}{\partial t}+\eta(x)\right)\underline{P}_{22}(x,t)=0 \tag{11.94}$$

$$\left(\frac{\partial}{\partial x}+\frac{\partial}{\partial t}+\eta(x)\right)\underline{P}_{23}(x,t)=0 \tag{11.95}$$

Boundary conditions:

$$\underline{P}_{15}(0,t)=\underline{\lambda}_7\underline{P}_{11}(t)+\underline{\lambda}_8\underline{P}_{12}(t)+\underline{\lambda}_5\underline{P}_{13}(t)+\underline{\lambda}_6\underline{P}_{14}(t) \tag{11.96}$$

$$\underline{P}_{16}(0,t)=\underline{\lambda}_1\underline{P}_0(t) \tag{11.97}$$

$$\underline{P}_{17}(0,t)=\underline{\lambda}_2\underline{P}_0(t) \tag{11.98}$$

$$\underline{P}_{18}(0,t)=\underline{\lambda}_3\underline{P}_0(t) \tag{11.99}$$

$$\underline{P}_{19}(0,t)=\underline{\lambda}_4\underline{P}_0(t) \tag{11.100}$$

$$\underline{P}_{20}(0,t)=\lambda_9\underline{P}_0(t) \tag{11.101}$$

$$\underline{P}_{21}(0,t)=\lambda_{10}\underline{P}_0(t) \tag{11.102}$$

$$\underline{P}_{22}(0,t)=\lambda_{11}\underline{P}_0(t) \tag{11.103}$$

$$\underline{P}_{23}(0,t)=\lambda_{12}\underline{P}_0(t) \tag{11.104}$$

Initial conditions:
$\underline{P}_0(t) \;=\; 1$ at $t=0$ and is zero at all other values of t.

(b) Solution of the model:

Finding the Laplace transform of Eqs. (11.72) to (11.95) and the boundary conditions (11.96) to (11.104) and using the initial conditions we get:

$$\begin{aligned}\{s+\underline{C}\}\underline{P}_0(s)=1&+\int_0^\infty \underline{P}_{15}(s,x)\eta(x)dx+\int_0^\infty \underline{P}_{16}(s,x)\eta(x)dx+\int_0^\infty \underline{P}_{17}(s,x)\eta(x)dx\\&+\int_0^\infty \underline{P}_{18}(s,x)\eta(x)dx+\int_0^\infty \underline{P}_{19}(s,x)\eta(x)dx\\&+\int_0^\infty \underline{P}_{20}(s,x)\eta(x)dx+\int_0^\infty \underline{P}_{21}(s,x)\eta(x)dx+\int_0^\infty \underline{P}_{22}(s,x)\eta(x)dx+\int_0^\infty \underline{P}_{23}(s,x)\eta(x)dx\end{aligned} \tag{11.105}$$

$$\{s+\lambda_6+\lambda_7+\lambda_8\}\underline{P}_1(s)=\lambda_5\underline{P}_0(s) \tag{11.106}$$

$$\{s+\lambda_5+\lambda_7+\lambda_8\}\underline{P}_2(s)=\lambda_6\underline{P}_0(s) \tag{11.107}$$

$$\{s+\lambda_5+\lambda_6+\lambda_8\}\underline{P}_3(s)=\lambda_7\underline{P}_0(s) \tag{11.108}$$

$$\{s+\lambda_5+\lambda_6+\lambda_7\}\underline{P}_4(s)=\lambda_8\underline{P}_0(s) \tag{11.109}$$

$$\{s+\lambda_6+\lambda_7\}\underline{P}_5(s)=\lambda_8\underline{P}_1(s)+\lambda_5\underline{P}_4(s) \tag{11.110}$$

$$\{s+\lambda_8+\lambda_7\}\underline{P}_6(s)=\lambda_6\underline{P}_1(s)+\lambda_5\underline{P}_2(s) \tag{11.111}$$

$$\{s+\lambda_8+\lambda_5\}\underline{P}_7(s)=\lambda_6\underline{P}_3(s)+\lambda_7\underline{P}_2(s) \tag{11.112}$$

$$\{s+\lambda_6+\lambda_5\}\underline{P}_8(s)=\lambda_8\underline{P}_3(s)+\lambda_7\underline{P}_4(s) \tag{11.113}$$

$$\{s+\lambda_6+\lambda_8\}\underline{P}_9(s)=\lambda_5\underline{P}_3(s)+\lambda_7\underline{P}_1(s) \tag{11.114}$$

$$\{s+\lambda_5+\lambda_7\}\underline{P}_{10}(s)=\lambda_6\underline{P}_4(s)+\lambda_8\underline{P}_2(s) \tag{11.115}$$

$$\{s+\lambda_7\}\underline{P}_{11}(s)=\lambda_6\underline{P}_5(s)+\lambda_8\underline{P}_6(s)+\lambda_5\underline{P}_{10}(s) \tag{11.116}$$

$$\{s+\lambda_5\}\underline{P}_{12}(s)=\lambda_7\underline{P}_{10}(s)+\lambda_6\underline{P}_8(s)+\lambda_8\underline{P}_7(s) \tag{11.117}$$

$$\{s+\lambda_5\}\underline{P}_{13}(s)=\lambda_7\underline{P}_{10}(s)+\lambda_6\underline{P}_8(s)+\lambda_8\underline{P}_7(s) \tag{11.118}$$

$$\{s+\lambda_6\}\underline{P}_{14}(s) = \lambda_7\underline{P}_5(s) + \lambda_5\underline{P}_8(s) + \lambda_8\underline{P}_9(s) \tag{11.119}$$

$$\left\{s+\frac{\partial}{\partial x}+\eta(x)\right\}\underline{P}_{15}(s,x)=0 \tag{11.120}$$

$$\left\{s+\frac{\partial}{\partial x}+\eta(x)\right\}\underline{P}_{16}(s,x)=0 \tag{11.121}$$

$$\left\{s+\frac{\partial}{\partial x}+\eta(x)\right\}\underline{P}_{17}(s,x)=0 \tag{11.122}$$

$$\left\{s+\frac{\partial}{\partial x}+\eta(x)\right\}\underline{P}_{18}(s,x)=0 \tag{11.123}$$

$$\left\{s+\frac{\partial}{\partial x}+\eta(x)\right\}\underline{P}_{19}(s,x)=0 \tag{11.124}$$

$$\left\{s+\frac{\partial}{\partial x}+\eta(x)\right\}\underline{P}_{20}(s,x)=0 \tag{11.125}$$

$$\left\{s+\frac{\partial}{\partial x}+\eta(x)\right\}\underline{P}_{21}(s,x)=0 \tag{11.126}$$

$$\left\{s+\frac{\partial}{\partial x}+\eta(x)\right\}\underline{P}_{22}(s,x)=0 \tag{11.127}$$

$$\left\{s+\frac{\partial}{\partial x}+\eta(x)\right\}\underline{P}_{23}(s,x)=0 \tag{11.128}$$

Boundary conditions:

$$\underline{P}_{15}(0,s)=\underline{P}_{11}(s)\underline{\lambda}_7+\ \underline{P}_{12}(s)\underline{\lambda}_8+\underline{P}_{13}(s)\underline{\lambda}_5+\underline{P}_{14}(s)\underline{\lambda}_6 \tag{11.129}$$

$$\underline{P}_{16}(0,s)=\underline{P}_0(s)\underline{\lambda}_1 \tag{11.130}$$

$$\underline{P}_{17}(0,s)=\underline{P}_0(s)\underline{\lambda}_2 \tag{11.131}$$

$$\underline{P}_{18}(0,s)=\underline{P}_0(s)\underline{\lambda}_3 \tag{11.132}$$

$$\underline{P}_{19}(0,s)=\underline{P}_0(s)\underline{\lambda}_4 \tag{11.133}$$

$$\underline{P}_{20}(0,s)=\underline{P}_0(s)\underline{\lambda}_9 \tag{11.134}$$

$$\underline{P}_{21}(0,s)=\underline{P}_0(s)\underline{\lambda}_{10} \tag{11.135}$$

$$\underline{P}_{22}(0,s)=\underline{P}_0(s)\underline{\lambda}_{11} \tag{11.136}$$

$$\underline{P}_{23}(0,s)=\underline{P}_0(s)\underline{\lambda}_{12} \tag{11.137}$$

(c) Transition state probabilities:

On solving the equations from (11.105) to (11.128) by taking their Laplace transform and using the boundary conditions from (11.129) to (11.137), we get the following transition state probabilities:

$$\underline{P}_0(s) = \frac{1}{(s+\underline{C})} \tag{11.138}$$

$$\underline{P}_1(s) = \frac{\lambda_5}{(s+\lambda_6+\lambda_7+\lambda_8)(s+\underline{C})} \tag{11.139}$$

$$\underline{P}_2(s) = \frac{\lambda_6}{(s+\lambda_5+\lambda_7+\lambda_8)(s+\underline{C})} \tag{11.140}$$

$$\underline{P}_3(s) = \frac{\lambda_7}{(s+\lambda_5+\lambda_6+\lambda_8)(s+\underline{C})} \tag{11.141}$$

$$\underline{P}_4(s) = \frac{\lambda_8}{(s+\lambda_5+\lambda_6+\lambda_7)(s+\underline{C})} \tag{11.142}$$

$$\underline{P}_5(s) = \frac{\lambda_8\underline{P}_1(s)+\lambda_5\underline{P}_4(s)}{(s+\lambda_6+\lambda_7)} \tag{11.143}$$

$$\underline{P}_6(s) = \frac{\lambda_6\underline{P}_1(s)+\lambda_5\underline{P}_2(s)}{(s+\lambda_8+\lambda_7)} \tag{11.144}$$

$$\underline{P}_7(s) = \frac{\lambda_6\underline{P}_3(s)+\lambda_7\underline{P}_2(s)}{(s+\lambda_8+\lambda_5)} \tag{11.145}$$

$$\underline{P}_8(s) = \frac{\lambda_8\underline{P}_3(s)+\lambda_7\underline{P}_4(s)}{(s+\lambda_6+\lambda_5)} \tag{11.146}$$

$$\underline{P}_9(s) = \frac{\lambda_5\underline{P}_3(s)+\lambda_7\underline{P}_1(s)}{(s+\lambda_6+\lambda_8)} \tag{11.147}$$

$$\underline{P}_{10}(s) = \frac{\lambda_6\underline{P}_4(s)+\lambda_8\underline{P}_2(s)}{(s+\lambda_5+\lambda_7)} \tag{11.148}$$

$$\underline{P}_{11}(s) = \frac{\lambda_6\underline{P}_5(s)+\lambda_8\underline{P}_6(s)+\lambda_5\underline{P}_{10}(s)}{(s+\lambda_7)} \tag{11.149}$$

$$\underline{P}_{12}(s) = \frac{\lambda_5\underline{P}_7(s)+\lambda_7\underline{P}_6(s)+\lambda_6\underline{P}_9(s)}{(s+\lambda_8)} \tag{11.150}$$

$$\underline{P}_{13}(s) = \frac{\lambda_8\underline{P}_7(s)+\lambda_7\underline{P}_{10}(s)+\lambda_6\underline{P}_8(s)}{(s+\lambda_5)} \tag{11.151}$$

$$\underline{P}_{14}(s) = \frac{\underline{\lambda}_8 \underline{P}_9(s) + \underline{\lambda}_7 \underline{P}_5(s) + \underline{\lambda}_5 \underline{P}_8(s)}{(s + \underline{\lambda}_6)} \tag{11.152}$$

$$\underline{P}_{15}(s) = \{\underline{\lambda}_7 \underline{P}_{11}(s) + \underline{\lambda}_8 \underline{P}_{12}(s) + \underline{\lambda}_5 \underline{P}_{13}(s) + \underline{\lambda}_6 \underline{P}_{14}(s)\}\left(\frac{1 - S(s)}{s}\right) \tag{11.153}$$

$$\underline{P}_{16}(s) = \{\underline{\lambda}_1 \underline{P}_0(s)\}\left(\frac{1 - S(s)}{s}\right) \tag{11.154}$$

$$\underline{P}_{17}(s) = \{\underline{\lambda}_2 \underline{P}_0(s)\}\left(\frac{1 - S(s)}{s}\right) \tag{11.155}$$

$$\underline{P}_{18}(s) = \{\underline{\lambda}_3 \underline{P}_0(s)\}\left(\frac{1 - S(s)}{s}\right) \tag{11.156}$$

$$\underline{P}_{19}(s) = \{\underline{\lambda}_4 \underline{P}_0(s)\}\left(\frac{1 - S(s)}{s}\right) \tag{11.157}$$

$$\underline{P}_{20}(s) = \{\underline{\lambda}_9 \underline{P}_0(s)\}\left(\frac{1 - S(s)}{s}\right) \tag{11.158}$$

$$\underline{P}_{21}(s) = \{\underline{\lambda}_{10} \underline{P}_0(s)\}\left(\frac{1 - S(s)}{s}\right) \tag{11.159}$$

$$\underline{P}_{22}(s) = \{\underline{\lambda}_{11} \underline{P}_0(s)\}\left(\frac{1 - S(s)}{s}\right) \tag{11.160}$$

$$\underline{P}_{23}(s) = \{\underline{\lambda}_{12} \underline{P}_0(s)\}\left(\frac{1 - S(s)}{s}\right) \tag{11.161}$$

where $\underline{C} = \underline{\lambda}_1 + \underline{\lambda}_2 + \underline{\lambda}_3 + \underline{\lambda}_4 + \underline{\lambda}_5 + \underline{\lambda}_6 + \underline{\lambda}_7 + \underline{\lambda}_8 + \underline{\lambda}_9 + \underline{\lambda}_{10} + \underline{\lambda}_{11} + \underline{\lambda}_{12}$

In a similar manner, we can evaluate the expressions for upper bounds of the transition state probabilities of the network's components by replacing $\underline{P}_i$(s) by $\overline{P}_i$(s) ($i = 0$ to 23) and $\underline{\lambda}_j$ by $\overline{\lambda}_j$ ($j = 1$ to 12) in Eqs. (11.138) to (11.161).

(i) Interval-valued NR of the network

The interval-valued TR of the SEN+ is given:

$$R = \left[\underline{P}_0 + \underline{P}_1 + \underline{P}_2 + \underline{P}_3 + \underline{P}_4 + \underline{P}_5 + \underline{P}_6 + \underline{P}_7 + \underline{P}_8 + \underline{P}_9 + \underline{P}_{10} + \underline{P}_{11} + \underline{P}_{12} + \underline{P}_{13} + \underline{P}_{14}, \overline{P}_0 + \overline{P}_1 + \overline{P}_2 + \overline{P}_3 + \overline{P}_4 + \overline{P}_5 + \overline{P}_6 + \overline{P}_7 + \overline{P}_8 + \overline{P}_9 + \overline{P}_{10} + \overline{P}_{11} + \overline{P}_{12} + \overline{P}_{13} + \overline{P}_{14}\right] \tag{11.162}$$

(ii) Mean time to failure:

The lower bound of the MTTF for the NR of the considered network is given:

$$
\begin{aligned}
(\mathrm{MTTF})_{\mathrm{NR}} = {} & \frac{1}{\underline{C}} + \frac{\underline{\lambda}_5}{\underline{C}(\underline{\lambda}_6+\underline{\lambda}_7+\underline{\lambda}_8)} + \frac{\underline{\lambda}_6}{\underline{C}(\underline{\lambda}_5+\underline{\lambda}_7+\underline{\lambda}_8)} + \frac{\underline{\lambda}_7}{\underline{C}(\underline{\lambda}_6+\underline{\lambda}_5+\underline{\lambda}_8)} + \frac{\underline{\lambda}_8}{\underline{C}(\underline{\lambda}_6+\underline{\lambda}_7+\underline{\lambda}_5)} + \frac{\underline{\lambda}_5\underline{\lambda}_8}{\underline{C}(\underline{\lambda}_6+\underline{\lambda}_7)} \\
& \left\{\frac{1}{\underline{\lambda}_6+\underline{\lambda}_7+\underline{\lambda}_8} + \frac{1}{\underline{\lambda}_6+\underline{\lambda}_7+\underline{\lambda}_5}\right\} + \frac{\underline{\lambda}_5\underline{\lambda}_6}{\underline{C}(\underline{\lambda}_8+\underline{\lambda}_7)}\left\{\frac{1}{\underline{\lambda}_6+\underline{\lambda}_7+\underline{\lambda}_8} + \frac{1}{\underline{\lambda}_8+\underline{\lambda}_7+\underline{\lambda}_5}\right\} + \frac{\underline{\lambda}_6\underline{\lambda}_7}{\underline{C}(\underline{\lambda}_5+\underline{\lambda}_8)} \\
& \left\{\frac{1}{\underline{\lambda}_5+\underline{\lambda}_7+\underline{\lambda}_8} + \frac{1}{\underline{\lambda}_6+\underline{\lambda}_8+\underline{\lambda}_5}\right\} + \frac{\underline{\lambda}_7\underline{\lambda}_8}{\underline{C}(\underline{\lambda}_6+\underline{\lambda}_5)}\left\{\frac{1}{\underline{\lambda}_6+\underline{\lambda}_5+\underline{\lambda}_8} + \frac{1}{\underline{\lambda}_6+\underline{\lambda}_7+\underline{\lambda}_5}\right\} + \frac{\underline{\lambda}_5\underline{\lambda}_7}{\underline{C}(\underline{\lambda}_6+\underline{\lambda}_8)} \\
& \left\{\frac{1}{\underline{\lambda}_6+\underline{\lambda}_7+\underline{\lambda}_8} + \frac{1}{\underline{\lambda}_6+\underline{\lambda}_8+\underline{\lambda}_5}\right\} + \frac{\underline{\lambda}_6\underline{\lambda}_8}{\underline{C}(\underline{\lambda}_5+\underline{\lambda}_7)}\left\{\frac{1}{\underline{\lambda}_6+\underline{\lambda}_7+\underline{\lambda}_5} + \frac{1}{\underline{\lambda}_8+\underline{\lambda}_7+\underline{\lambda}_5}\right\} + \frac{\underline{\lambda}_5\underline{\lambda}_6\underline{\lambda}_8}{\underline{C}\underline{\lambda}_7} \\
& \left\{\frac{1}{(\underline{\lambda}_6+\underline{\lambda}_7)}\left\{\frac{1}{\underline{\lambda}_6+\underline{\lambda}_7+\underline{\lambda}_8} + \frac{1}{\underline{\lambda}_6+\underline{\lambda}_7+\underline{\lambda}_5}\right\} + \frac{1}{(\underline{\lambda}_8+\underline{\lambda}_7)}\left\{\frac{1}{\underline{\lambda}_6+\underline{\lambda}_7+\underline{\lambda}_8} + \frac{1}{\underline{\lambda}_8+\underline{\lambda}_7+\underline{\lambda}_5}\right\}\right. \\
& \left. + \frac{1}{(\underline{\lambda}_5+\underline{\lambda}_7)}\left\{\frac{1}{\underline{\lambda}_6+\underline{\lambda}_7+\underline{\lambda}_5} + \frac{1}{\underline{\lambda}_8+\underline{\lambda}_7+\underline{\lambda}_5}\right\}\right\} \\
& + \frac{\underline{\lambda}_5\underline{\lambda}_6\underline{\lambda}_7}{\underline{C}\underline{\lambda}_8}\left\{\frac{1}{(\underline{\lambda}_8+\underline{\lambda}_7)}\left\{\frac{1}{\underline{\lambda}_6+\underline{\lambda}_7+\underline{\lambda}_8} + \frac{1}{\underline{\lambda}_8+\underline{\lambda}_7+\underline{\lambda}_5}\right\} + \frac{1}{(\underline{\lambda}_8+\underline{\lambda}_5)}\left\{\frac{1}{\underline{\lambda}_5+\underline{\lambda}_7+\underline{\lambda}_8} + \frac{1}{\underline{\lambda}_8+\underline{\lambda}_6+\underline{\lambda}_5}\right\}\right. \\
& \left. + \frac{1}{(\underline{\lambda}_6+\underline{\lambda}_8)}\left\{\frac{1}{\underline{\lambda}_6+\underline{\lambda}_7+\underline{\lambda}_8} + \frac{1}{\underline{\lambda}_6+\underline{\lambda}_7+\underline{\lambda}_5}\right\}\right\} \\
& + \frac{\underline{\lambda}_6\underline{\lambda}_7\underline{\lambda}_8}{\underline{C}\underline{\lambda}_5}\left\{\frac{1}{(\underline{\lambda}_8+\underline{\lambda}_5)}\left\{\frac{1}{\underline{\lambda}_5+\underline{\lambda}_7+\underline{\lambda}_8} + \frac{1}{\underline{\lambda}_6+\underline{\lambda}_8+\underline{\lambda}_5}\right\} + \frac{1}{(\underline{\lambda}_5+\underline{\lambda}_6)}\left\{\frac{1}{\underline{\lambda}_6+\underline{\lambda}_5+\underline{\lambda}_8} + \frac{1}{\underline{\lambda}_6+\underline{\lambda}_7+\underline{\lambda}_5}\right\}\right. \\
& \left. + \frac{1}{(\underline{\lambda}_5+\underline{\lambda}_7)}\left\{\frac{1}{\underline{\lambda}_6+\underline{\lambda}_7+\underline{\lambda}_5} + \frac{1}{\underline{\lambda}_8+\underline{\lambda}_7+\underline{\lambda}_5}\right\}\right\} \\
& + \frac{\underline{\lambda}_5\underline{\lambda}_7\underline{\lambda}_8}{\underline{C}\underline{\lambda}_6}\left\{\frac{1}{(\underline{\lambda}_6+\underline{\lambda}_7)}\left\{\frac{1}{\underline{\lambda}_6+\underline{\lambda}_7+\underline{\lambda}_8} + \frac{1}{\underline{\lambda}_6+\underline{\lambda}_7+\underline{\lambda}_5}\right\} + \frac{1}{(\underline{\lambda}_5+\underline{\lambda}_6)}\left\{\frac{1}{\underline{\lambda}_6+\underline{\lambda}_5+\underline{\lambda}_8} + \frac{1}{\underline{\lambda}_6+\underline{\lambda}_7+\underline{\lambda}_5}\right\}\right. \\
& \left. + \frac{1}{(\underline{\lambda}_6+\underline{\lambda}_8)}\left\{\frac{1}{\underline{\lambda}_6+\underline{\lambda}_7+\underline{\lambda}_8} + \frac{1}{\underline{\lambda}_6+\underline{\lambda}_8+\underline{\lambda}_5}\right\}\right\}
\end{aligned}
\tag{11.163}
$$

We can evaluate the upper bound of the network by replacing $\underline{\lambda}_i$ by $\overline{\lambda}_i$ ($i = 1$ to 12) in Eq. (11.163).

9. Numerical illustration

9.1 Terminal reliability of the replaceable SEN under consideration by using the IUGF approach

Let the upper and lower bounds of the failure rates of the SEN be $\underline{\lambda}_1 = 0.15$, $\underline{\lambda}_2 = 0.25$, $\underline{\lambda}_3 = 0.35$, $\overline{\lambda}_1 = 0.1$, $\overline{\lambda}_2 = 0.2$, $\overline{\lambda}_3 = 0.3$. On substituting the assumed values of failures rates in Eqs. (11.15) to (11.18) and Eqs. (11.26) to (11.29) and taking their inverse Laplace transform, the expressions for various transition state probabilities are obtained. Table 11.1 shows the changes in the transition state probabilities of the operating state with respect to time.

From Eq. (11.29), we can determine the variation on the reliability of the replaceable SEN with time which is given in Table 11.2 and is shown in Fig. 11.8.

Table 11.1 Variations in the probabilities of operating states w. r. t. time.

T	$[\underline{P}_0(t), \overline{P}_0(t)]$
0	[1,1]
1	[0.472366552, 0.5488116]
2	[0.22313016, 0.3011942]
3	[0.105399224, 0.1652988]
4	[0.049787068, 0.0907179]
5	[0.023517745, 0.049787068]
6	[0.011109, 0.027324]
7	[0.005248, 0.014996]
8	[0.002479, 0.00823]
9	[0.001171, 0.004517]
10	[0.000553, 0.002479]

Table 11.2 Changes in the terminal reliability bounds with respect to time.

t	$[\underline{R}(t), \overline{R}(t)]$
0	[1,1]
1	[0.472366552, 0.5488116]
2	[0.22313016, 0.3011942]
3	[0.10539922, 0.1652988]
4	[0.049787068, 0.0907179]
5	[0.023517745, 0.049787068]
6	[0.011109, 0.027324]
7	[0.005248, 0.014996]
8	[0.002479, 0.00823]
9	[0.001171, 0.004517]
10	[0.000553, 0.002479]

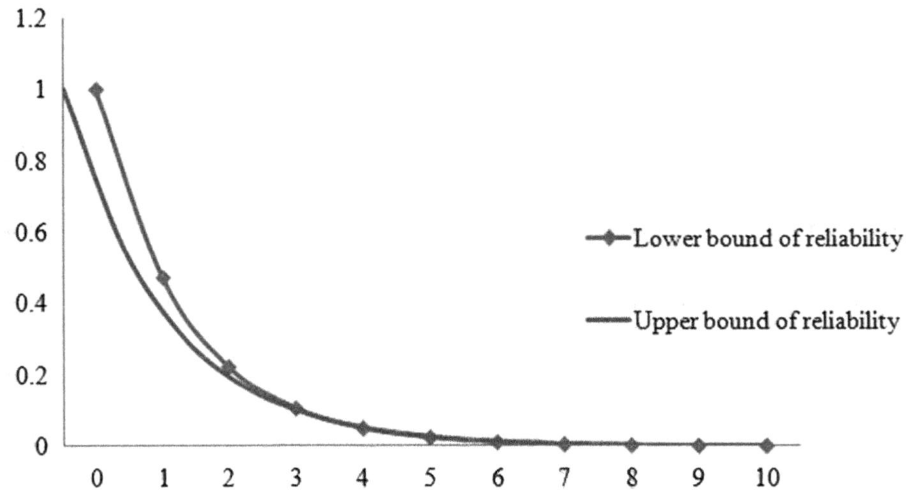

FIGURE 11.8

Change in TR of SEN with time.

9.2 MTTF of the replaceable SEN

Using Eq. (11.30), we can compute the bounds of the MTTF with respect to different parameters affecting it. Tables 11.3 and 11.4 show the changes in the MTTF w. r. t. different failure rates which are displayed in Figs. 11.9 and 11.10.

9.3 Broadcast reliability of the considered SEN by using the method of IUGF

Suppose the upper and lower bounds of the failure Eq. (11.70) of SEN be $\underline{\lambda}_1 = 0.09$, $\underline{\lambda}_2 = 0.11$, $\underline{\lambda}_3 = 0.13$, $\underline{\lambda}_4 = 0.15$, $\underline{\lambda}_5 = 0.17$, $\underline{\lambda}_6 = 0.19$, $\underline{\lambda}_7 = 0.21$, $\overline{\lambda}_1 = 0.01$, $\overline{\lambda}_2 = 0.03$, $\overline{\lambda}_3 = 0.05$, $\overline{\lambda}_4 = 0.07$,

Table 11.3 MTTF w. r. t. $\underline{\lambda}_1$, $\underline{\lambda}_2$, $\underline{\lambda}_3$.

t	MTTF w. r. t. $\underline{\lambda}_1$	MTTF w. r. t. $\underline{\lambda}_2$	MTTF w. r. t. $\underline{\lambda}_3$
0.1	[1.42857, 1.66667]	[1.66667, 1.66667]	[2, 1.66667]
0.2	[1.25, 1.66667]	[1.428571, 1.66667]	[1.66667, 1.66667]
0.3	[1.1111, 1.66667]	[1.25, 1.66667]	[1.42857, 1.66667]
0.4	[0.65, 1.66667]	[1.1111, 1.66667]	[1.25, 1.66667]
0.5	[0.90909, 1.66667]	[1, 1.66667]	[1.1111, 1.66667]
0.6	[0.83333, 1.66667]	[0.90909, 1.66667]	[1, 1.66667]
0.7	[0.76923, 1.66667]	[0.83333, 1.66667]	[0.90909, 1.66667]
0.8	[0.714285, 1.66667]	[0.76923, 1.66667]	[0.83333, 1.66667]
0.9	[0.66666, 1.66667]	[0.714285, 1.66667]	[0.7692307, 1.66667]
0.91	[0.66225, 1.66667]	[0.689655, 1.66667]	[0.7407407, 1.66667]

Table 11.4 MTTF w. r. t. $\overline{\lambda}_1$, $\overline{\lambda}_2$, $\overline{\lambda}_3$.

t	MTTF w. r. t. $\overline{\lambda}_1$	MTTF w. r. t. $\overline{\lambda}_2$	MTTF w. r. t. $\overline{\lambda}_3$
0.09	[1.3333, 1.6949]	[1.3333, 2.040816]	[1.3333, 2.5641025]
0.091	[1.3333, 1.69204]	[1.3333, 2.036659]	[1.3333, 2.5575447]
0.092	[1.3333, 1.689189]	[1.3333, 2.032520]	[1.3333, 2.5510204]
0.093	[1.3333, 1.686340]	[1.3333, 2.028397]	[1.3333, 2.5445292]
0.094	[1.3333, 1.683502]	[1.3333, 2.024291]	[1.3333, 2.538071066]
0.095	[1.3333, 1.680672]	[1.3333, 2.020202]	[1.3333, 2.531645]
0.096	[1.3333, 1.677852]	[1.3333, 2.016129]	[1.3333, 2.525252]
0.097	[1.3333, 1.675042]	[1.3333, 2.012072]	[1.3333, 2.5188916]
0.098	[1.3333, 1.6722408]	[1.3333, 2.008032]	[1.3333, 2.5125628]
0.099	[1.3333, 1.6722408]	[1.3333, 2.004008]	[1.3333, 2.5062656]

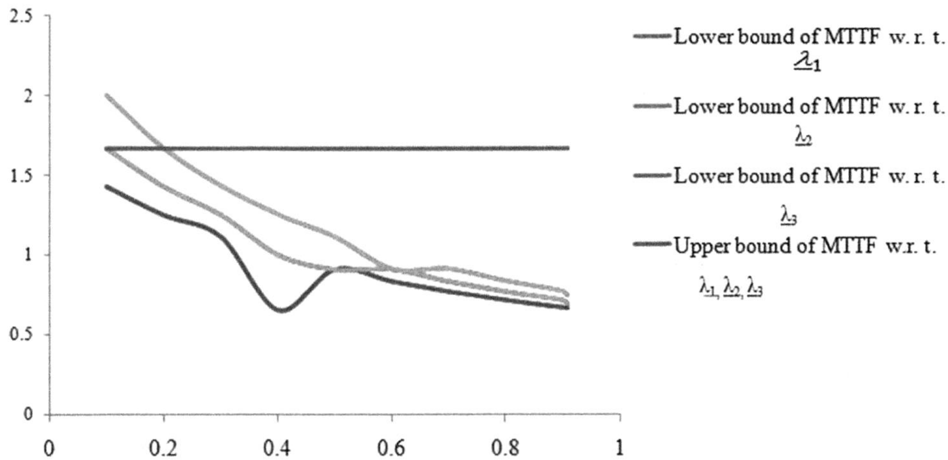

FIGURE 11.9

MTTF versus $\underline{\lambda}_1$, $\underline{\lambda}_2$, $\underline{\lambda}_3$.

$\overline{\lambda}_5 = 0.09$, $\overline{\lambda}_6 = 0.011$, $\overline{\lambda}_7 = 0.013$. Substituting the values of failures rates in Eqs. (11.61) to (11.69) and taking their inverse Laplace transform, we get the expressions for the transition state probabilities. Table 11.5 depicts the changes in the transition state probabilities of the operating states of SEN with time.

Using Eq. (2.40) and the transition state probabilities of the working states, we can evaluate the variation on the reliability of the considered replaceable SEN w. r. t. time which is given in Table 11.6 and is depicted in Fig. 11.11.

Using Eq. (11.70), the changes in the BR of the SEN with time are tabulated as shown in Table 11.6.

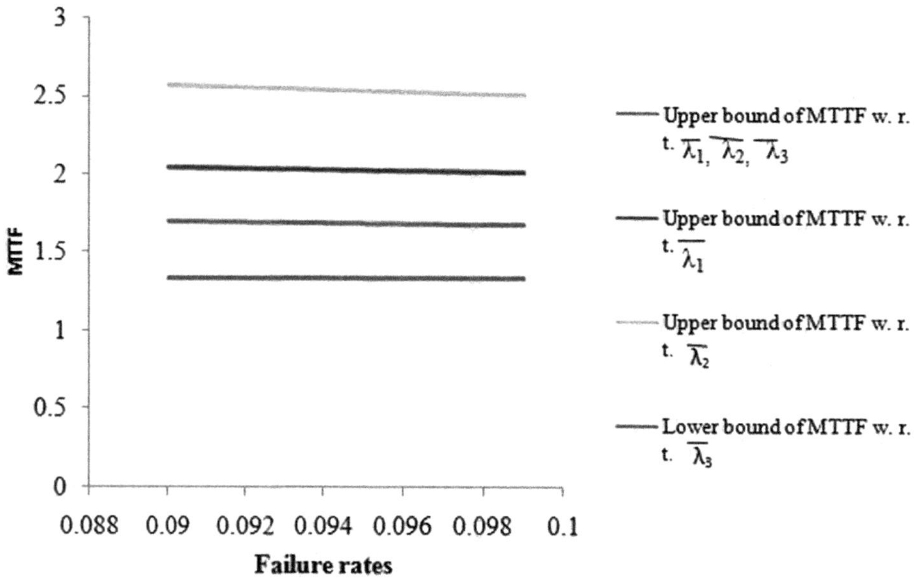

FIGURE 11.10

MTTF w. r. t. $\overline{\lambda}_1$, $\overline{\lambda}_2$, $\overline{\lambda}_3$.

Table 11.5 Variations in the probabilities of operating states w. r. t. time.

t	$[\underline{P}_0(t), \overline{P}_0(t)]$	$[\underline{P}_1(t), \overline{P}_1(t)]$	$[\underline{P}_2(t), \overline{P}_2(t)]$
0	[1,1]	[0,0]	[0,0]
1	[0.3499377, 0.612626394]	[0.023087, 0.063149]	[0.038893, 0.075496]
2	[0.122456, 0.375311098]	[0.036104, 0.077549]	[0.061571, 0.094051]
3	[0.042852, 0.229925485]	[0.043008, 0.075829]	[0.074348, 0.093499]
4	[0.014995, 0.14085842]	[0.046129, 0.069291]	[0.081094, 0.086995]
5	[0.005247, 0.0862935]	[0.047216, 0.061791]	[0.079065, 0.084175]
6	[0.001836, 0.0528657]	[0.046096, 0.05459]	[0.071225, 0.085044]
7	[0.0006425, 0.03238694]	[0.045839, 0.048051]	[0.063945, 0.084587]
8	[0.0002248, 0.011984109]	[0.042234, 0.04435]	[0.057332, 0.083346]
9	[0.00007868, 0.012155178]	[0.0371, 0.042646]	[0.051377, 0.081655]
10	[0.00002753, 0.0074466]	[0.03258, 0.040847]	[0.046032, 0.079714]

Table 11.6 Changes in the broadcast reliability bounds with time.

t	$[\underline{R}(t), \overline{R}(t)]$
0	[1, 1]
1	[0.4119177, 0.751271]
2	[0.220131, 0.54691109]
3	[0.160208, 0.3992534]
4	[0.142308, 0.29714442]
5	[0.131528, 0.2322595]
6	[0.119157, 0.192434]
7	[0.1104265, 0.1650249]
8	[0.0997908, 0.147536]
9	[0.08855568, 0.13648]
10	[0.07863953, 0.1280076]

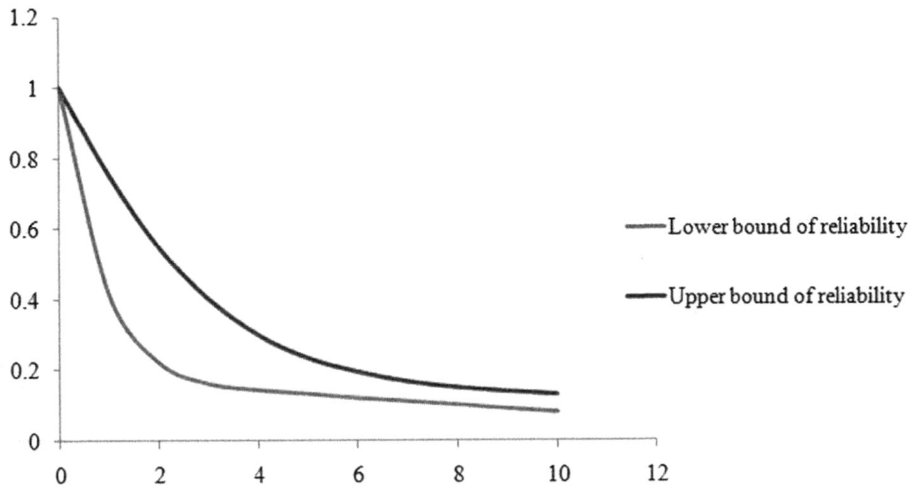

FIGURE 11.11

BR versus time.

9.4 MTTF of the considered replaceable SEN

By using Eq. (11.71), we can determine the bounds of MTTF with respect to the upper and lower bounds of failure rates which are given in Tables 11.7 to 11.12 and the changes in the bounds of MTTF with different failure rates are showed in Figs. 11.12 and 11.13.

Table 11.7 MTTF w. r. t. $\underline{\lambda}_1$, $\underline{\lambda}_2$, $\underline{\lambda}_3$.

t	MTTF w. r. t. $\underline{\lambda}_1$	MTTF w. r. t. $\underline{\lambda}_2$	MTTF w. r. t. $\underline{\lambda}_3$
0.1	[3.1541, 6.6767447]	[3.04539, 6.6767447]	[2.950089127, 6.6767447]
0.2	[2.61032, 6.6767447]	[3.03450, 6.6767447]	[2.8909428, 6.6767447]
0.3	[2.40315, 6.6767447]	[3.023968, 6.6767447]	[2.88001820, 6.6767447]
0.4	[2.33957, 6.6767447]	[3.079317, 6.6767447]	[2.7984891, 6.6767447]
0.5	[2.07395, 6.6767447]	[3.025697, 6.6767447]	[2.784826, 6.6767447]
0.6	[1.941007, 6.6767447]	[2.994386, 6.6767447]	[2.748489, 6.6767447]
0.7	[1.824079, 6.6767447]	[2.985162, 6.6767447]	[2.719327, 6.6767447]
0.8	[1.752232, 6.6767447]	[2.97804, 6.6767447]	[2.6909468, 6.6767447]
0.9	[1.627942, 6.6767447]	[2.967607, 6.6767447]	[2.68486, 6.6767447]
0.91	[1.619236, 6.6767447]	[2.95926, 6.6767447]	[2.65904876, 6.6767447]

Table 11.8 MTTF w. r. t. $\underline{\lambda}_4$, $\underline{\lambda}_5$, $\underline{\lambda}_6$.

t	MTTF w. r. t. $\underline{\lambda}_4$	MTTF w. r. t. $\underline{\lambda}_5$	MTTF w. r. t. $\underline{\lambda}_6$
0.1	[3.02797203, 6.6767447]	[3.08538967, 6.6767447]	[3.15437529, 6.6767447]
0.2	[3.00298486, 6.6767447]	[3.042962, 6.6767447]	[3.1202347, 6.6767447]
0.3	[2.989743, 6.6767447]	[2.984199, 6.6767447]	[3.0991243, 6.6767447]
0.4	[2.98481, 6.6767447]	[2.938494, 6.6767447]	[3.076883, 6.6767447]
0.5	[2.8746243, 6.6767447]	[2.908894, 6.6767447]	[3.00679418, 6.6767447]
0.6	[2.8599482, 6.6767447]	[1.998687, 6.6767447]	[2.943412, 6.6767447]
0.7	[2.8387624, 6.6767447]	[1.9385689, 6.6767447]	[2.8924664, 6.6767447]
0.8	[2.804849, 6.6767447]	[1.9052241, 6.6767447]	[2.808794, 6.6767447]
0.9	[2.7991889, 6.6767447]	[1.8294128, 6.6767447]	[2.799842, 6.6767447]
0.91	[2.796281, 6.6767447]	[1.8098967, 6.6767447]	[2.7844, 6.6767447]

Table 11.9 MTTF w. r. t. $\underline{\lambda}_7$.

t	MTTF
0.1	[3.2212246838, 6.6767447]
0.2	[3.21965327, 6.6767447]
0.3	[3.21199732, 6.6767447]
0.4	[3.2084612, 6.6767447]
0.5	[3.20071664, 6.6767447]
0.6	[3.1935338, 6.6767447]
0.7	[3.191664, 6.6767447]
0.8	[3.1873224, 6.6767447]
0.9	[3.184098, 6.6767447]
0.91	[3.1830056, 6.6767447]

Table 11.10 MTTF w. r. t. $\overline{\lambda}_1$, $\overline{\lambda}_2$, $\overline{\lambda}_3$.

	MTTF w. r. t. $\overline{\lambda}_1$	MTTF w. r. t. $\overline{\lambda}_2$	MTTF w. r. t. $\overline{\lambda}_3$
0.09	[3.11111111, 5.730994]	[3.11111111, 7.87878789]	[3.11111111, 6.101010101]
0.091	[3.11111111, 5.720957385]	[3.11111111, 7.918336956]	[3.11111111, 6.115155262]
0.092	[3.11111111, 5.710955711]	[3.11111111, 7.957886999]	[3.11111111, 6.129489603]
0.093	[3.11111111, 5.700988947]	[3.11111111, 7.997433355]	[3.11111111, 6.144004356]
0.094	[3.11111111, 5.6967532]	[3.11111111, 8.036971606]	[3.11111111, 6.159613889]
0.095	[3.11111111, 5.682439704]	[3.11111111, 8.07649755]	[3.11111111, 6.16985186]
0.096	[3.11111111, 5.669578326]	[3.11111111, 8.12623109]	[3.11111111, 6.17175276]
0.097	[3.11111111, 5.66146736]	[3.11111111, 8.166409123]	[3.11111111, 6.18647546]
0.098	[3.11111111, 5.651672434]	[3.11111111, 8.194962573]	[3.11111111, 5.18985845]
0.099	[3.11111111, 5.646891491]	[3.11111111, 8.210045294]	[3.11111111, 5.193768634]

Table 11.11 MTTF w. r. t. $\overline{\lambda}_4$, $\overline{\lambda}_5$, $\overline{\lambda}_6$.

	MTTF w. r. t. $\overline{\lambda}_4$	MTTF w. r. t. $\overline{\lambda}_5$	MTTF w. r. t. $\overline{\lambda}_6$
0.09	[3.11111111, 5.939393939]	[3.11111111, 6.6666667]	[3.11111111, 5.741066198]
0.091	[3.11111111, 5.92861464]	[3.11111111, 6.653088934]	[3.11111111, 5.730994152]
0.092	[3.11111111, 5.911480632]	[3.11111111, 6.639566396]	[3.11111111, 5.720957385]
0.093	[3.11111111, 5.907172996]	[3.11111111, 6.626098715]	[3.11111111, 5.710955711]
0.094	[3.11111111, 5.896510229]	[3.11111111, 6.61268556]	[3.11111111, 5.700988947]
0.095	[3.11111111, 5.88209438]	[3.11111111, 6.599326599]	[3.11111111, 5.691056911]
0.096	[3.11111111, 5.87496008]	[3.11111111, 6.587409274]	[3.11111111, 5.552231703]
0.097	[3.11111111, 5.864751646]	[3.11111111, 6.572769953]	[3.11111111, 5.671296296]
0.098	[3.11111111, 5.85127559]	[3.11111111, 6.55957162]	[3.11111111, 5.66146736]
0.099	[3.11111111, 5.843768634]	[3.11111111, 6.546426186]	[3.11111111, 5.65398205]

Table 11.12 MTTF w. r. t. $\overline{\lambda}_7$.

	MTTF
0.09	[3.11111111, 5.761316872]
0.091	[3.11111111, 5.750059036]
0.092	[3.11111111, 5.741066198]
0.093	[3.11111111, 5.730994152]
0.094	[3.11111111, 5.720957385]
0.095	[3.11111111, 5.714429847]
0.096	[3.11111111, 5.700988947]
0.097	[3.11111111, 5.691056911]
0.098	[3.11111111, 5.682984506]
0.099	[3.11111111, 5.671296296]

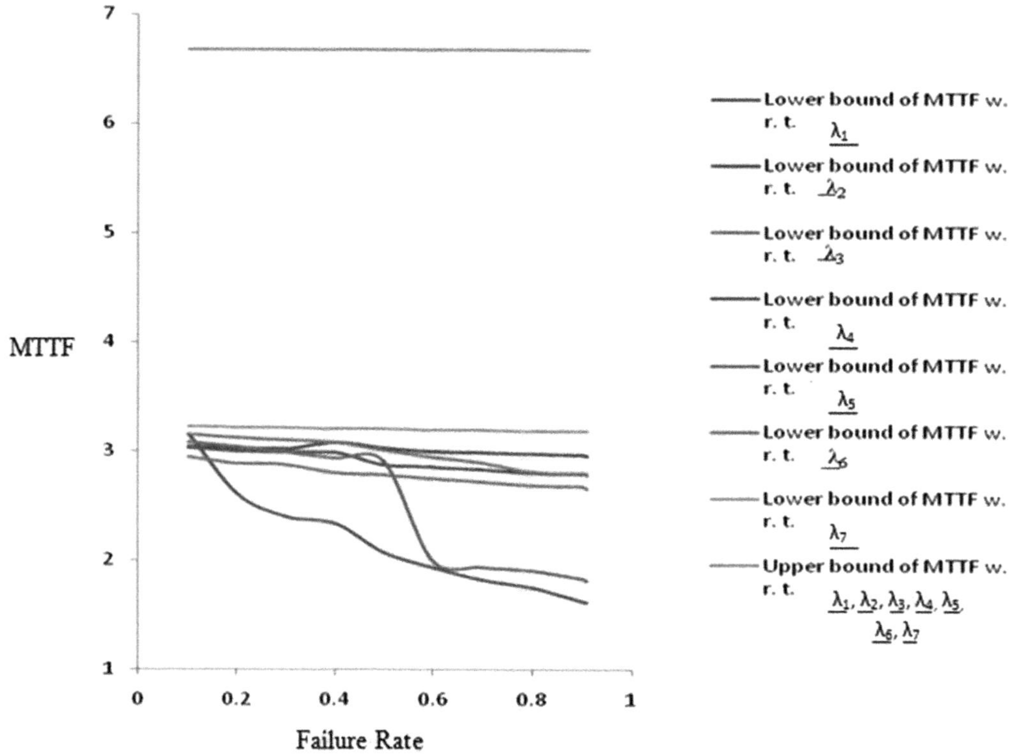

FIGURE 11.12

MTTF versus $\underline{\lambda}_1, \underline{\lambda}_2, \underline{\lambda}_3, \underline{\lambda}_4, \underline{\lambda}_5, \underline{\lambda}_6, \underline{\lambda}_7$.

9.5 Network reliability of the 8 × 8 replaceable SEN under consideration using IUGF approach

Let the upper and lower bounds of the failure rates of the proposed SEN be $\underline{\lambda}_1 = 0.03$, $\underline{\lambda}_2 = 0.08$, $\underline{\lambda}_3 = 0.1$, $\underline{\lambda}_4 = 0.15$, $\underline{\lambda}_5 = 0.19$, $\underline{\lambda}_6 = 0.22$, $\underline{\lambda}_7 = 0.225$, $\underline{\lambda}_8 = 0.24$, $\underline{\lambda}_9 = 0.26$, $\underline{\lambda}_{10} = 0.28$, $\underline{\lambda}_{11} = 0.288$, $\underline{\lambda}_{12} = 0.3$, $\overline{\lambda}_1 = 0.01$, $\overline{\lambda}_2 = 0.05$, $\overline{\lambda}_3 = 0.09$, $\overline{\lambda}_4 = 0.13$, $\overline{\lambda}_5 = 0.17$, $\overline{\lambda}_6 = 0.21$, $\overline{\lambda}_7 = 0.22$, $\overline{\lambda}_8 = 0.23$, $\overline{\lambda}_9 = 0.25$, $\overline{\lambda}_{10} = 0.27$, $\overline{\lambda}_{11} = 0.28$, $\overline{\lambda}_{12} = 0.29$. Substituting the values of failures rates in Eqs. (11.138)–(11.161) and taking their inverse Laplace transform, we get the expressions for various transition state probabilities. Tables 11.13 to 11.18 show the changes in the transition state probabilities of the working states of the considered SEN with respect to time.

Using Eq. (11.162), we can evaluate the variation on the reliability of the considered replaceable SEN w. r. t. time which is given in Table 11.19 and is depicted in Fig. 11.14.

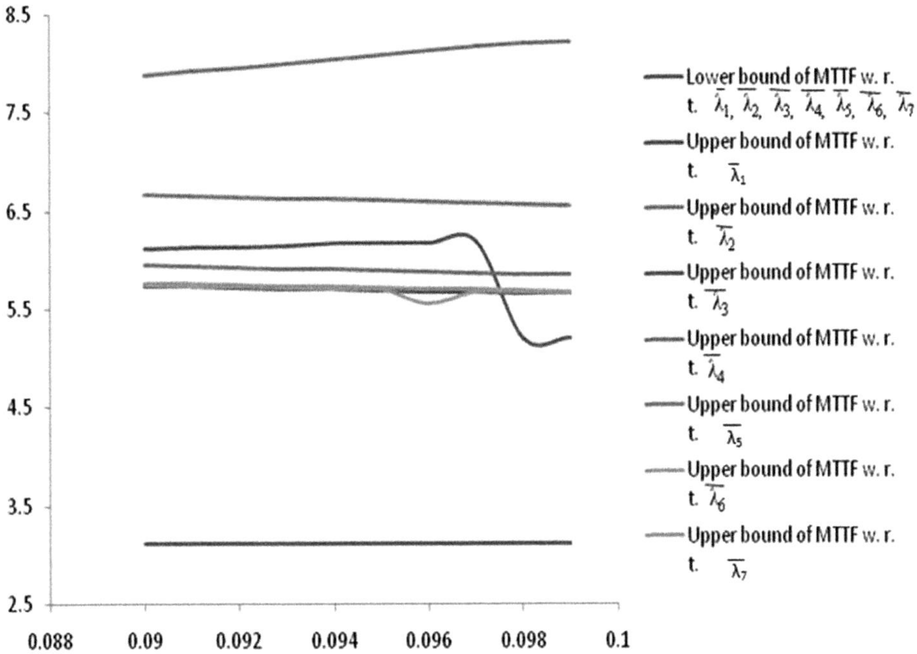

FIGURE 11.13

MTTF w. r. t. $\overline{\lambda}_1, \overline{\lambda}_2, \overline{\lambda}_3, \overline{\lambda}_4, \overline{\lambda}_5, \overline{\lambda}_6, \overline{\lambda}_7$.

Table 11.13 Changes in the probabilities of operating states of SEN w. r. t. time.

T	$[\underline{P}_0(t), \overline{P}_0(t)]$	$[\underline{P}_1(t), \overline{P}_1(t)]$
0	[1,1]	[0,0]
1	[0.094137387, 0.110803]	[0.044418945, 0.0648235]
2	[0.008861848, 0.012277]	[0.027769113, 0.028133671]
3	[0.0008342, 0.00136]	[0.0144095, 0.01509124]
4	[0.00007853, 0.000151]	[0.00730241, 0.0078609]
5	[0.000007392, 0.000016702]	[0.003684719, 0.004069675]
6	[0.000000696, 0.00000185]	[0.001857774, 0.002104165]
7	[0.0000000655, 0.000000205]	[0.000936518, 0.001087624]
8	[0.000000006167, 0.0000000227]	[0.000472093, 0.000562149]
9	[0.00000000058, 0.00000000251]	[0.000237978, 0.000290548]
10	[0.0000000000546, 0.000000000278]	[0.000119962, 0.00015017]

Table 11.14 Variations in the probabilities of operating states of SEN w. r. t. time.

t	$[\underline{P}_2(t), \overline{P}_2(t)]$	$[\underline{P}_3(t), \overline{P}_3(t)]$	$[\underline{P}_4(t), \overline{P}_4(t)]$
0	[0,0]	[0,0]	[0,0]
1	[0.0547816, 0.0567719]	[0.05620513, 0.059849369]	[0.060527514, 0.0629637]
2	[0.03361288, 0.036830661]	[0.034632655, 0.391507]	[0.0377358, 0.0415318]
3	[0.017945412, 0.020509858]	[0.018577912, 0.0220073]	[0.020553946, 0.023566162]
4	[0.0093673, 0.011110395]	0.009745409, 0.012039136]	[0.010942759, 0.013019038]
5	[0.004870073, 0.005985332]	[0.0050919, 0.0065505]	[0.00580371, 0.00715449]
6	[0.002530126, 0.00322072]	[0.00265865, 0.003560217]	[0.00307604, 0.00392752]
7	[0.001314292, 0.001732676]	[0.001387978, 0.00193456]	[0.001630145, 0.002155585]
8	[0.000682702, 0.000932095]	[0.000724592, 0.00105156]	[0.000863875, 0.001183023]
9	[0.000354625, 0.000501417]	[0.00037827, 0.000571148]	[0.000457799, 0.00064926]
10	[0.000354625, 0.000501417]	[0.000197474, 0.00031033]	[0.000242604, 0.000356321]

Table 11.15 Changes in the probabilities of operating states of SEN w. r. t. time.

T	$[\underline{P}_5(t), \overline{P}_5(t)]$	$[\underline{P}_6(t), \overline{P}_6(t)]$	$[\underline{P}_7(t), \overline{P}_7(t)]$
0	[0,0]	[0,0]	[0,0]
1	[0.014180225, 0.015616]	[0.012800842, 0.014152]	[0.017041057, 0.17097]
2	[0.0231898, 0.024536]	[0.02065343, 0.021934]	[0.027131, 0.028425]
3	[0.02324326, 0.023847]	[0.02324326, 0.023847]	[0.026655, 0.0291097]
4	[0.019552921, 0.019545]	[0.019545, 0.019552921]	[0.022092, 0.025045296]
5	[0.015096164, 0.014738]	[0.014738, 0.015096164]	[0.016852, 0.019790277]
6	[0.011091966, 0.010592]	[0.0010592, 0.0011091966]	[0.012255, 0.014890127]
7	[0.00789611, 0.007383]	[0.007383, 0.007896108]	[0.008645, 0.010859565]
8	[0.00550108, 0.00339]	[0.00504, 0.00550108]	[0.005974, 0.0077547381]
9	[0.003773949, 0.00339]	[0.00339, 0.003773949]	[0.004069, 0.005454769]
10	[0.00255985, 0.002256]	[0.002256, 0.00255985]	[0.002742, 0.003795334]

9.6 MTTF of the SEN under consideration

By using Eq. (11.163), one can evaluate the upper and lower bounds of MTTF by varying the bounds of failure rates which is presented in Tables 11.20 to 11.25 and the changes in the bounds of MTTF with different failure rates are displayed in Figs. 11.15 and 11.16, where $\underline{D} = \underline{\lambda}_1, \underline{\lambda}_2, \underline{\lambda}_3, \underline{\lambda}_4, \underline{\lambda}_5, \underline{\lambda}_6, \underline{\lambda}_7$ and $\overline{D} = \overline{\lambda}_1, \overline{\lambda}_2, \overline{\lambda}_3, \overline{\lambda}_4, \overline{\lambda}_5, \overline{\lambda}_6, \overline{\lambda}_7.$

Table 11.16 Variations in the probabilities of operating states of SEN w. r. t. time.

T	$[\underline{P}_8(t), \overline{P}_8(t)]$	$[\underline{P}_9(t), \overline{P}_9(t)]$	$[\underline{P}_{10}(t), \overline{P}_{10}(t)]$
0	[0,0]	[0,0]	[0,0]
1	[0.018866, 0.01887714]	[0.01348677, 0.014515]	[0.017917176, 0.018394]
2	[0.030348, 0.031913697]	[0.021907199, 0.022573]	[0.028489, 0.0300887]
3	[0.03026, 0.033166254]	[0.02179471, 0.021894]	[0.029298, 0.0310416]
4	[0.025479, 0.0289816]	[0.01619014, 0.017569]	[0.024571, 0.026916727]
5	[0.019754, 0.023272136]	[0.013929101, 0.013986]	[0.018973, 0.021442079]
6	[0.014607, 0.017801857]	[0.014607, 0.017801857]	[0.013972, 0.01626797]
7	[0.010482, 0.01320469]	[0.0074619, 0.008389]	[0.009984, 0.011966103]
8	[0,007371, 0.0095931]	[0,0041456, 0.004894]	[0,006991, 0.008619216]
9	[0.00511, 0.006867833]	[0.003362516, 0.004856]	[0.004825, 0.006117083]
10	[0.003506, 0.00486458]	[0.002260043, 0.002875]	[0.003296, 0.0042946431]

Table 11.17 Changes in the probabilities of operating states of SEN w. r. t. time.

T	$[\underline{P}_{11}(t), \overline{P}_{11}(t)]$	$[\underline{P}_{12}(t), \overline{P}_{12}(t)]$	$[\underline{P}_{13}(t), \overline{P}_{13}(t)]$
0	[0,0]	[0,0]	[0,0]
1	[0.003671448, 0.004256894]	[0.13766457, 0.00346929]	[0.004849701, 0.008122847]
2	[0.014398734, 0.016043895]	[0.100302535, 0.013428763]	[0.01948028, 0.038243624]
3	[0.025482219, 0.02748615]	[0.019127555, 0.023441357]	[0.0353916, 0.055730978]
4	[0.033285235, 0.034913454]	[0.030195137, 0.06578333]	[0.047551614, 0.063595279]
5	[0.037201905, 0.038063336]	[0.033279572, 0.055782812]	[0.054762233, 0.0649885]
6	[0.037894177, 0.037906925]	[0.03343099, 0.047380458]	[0.05756874, 0.062377196]
7	[0.035604155, 0.036335745]	[0.031618375, 0.039986228]	[0.057056122, 0.057525534]
8	[0,03212543, 0.0334165]	[0,028686443, 0.0334391]	[0,05162474, 0.054313357]
9	[0.0281681, 0.029821377]	[0.025260641, 0.027702867]	[0.045438296, 0.05023925]
10	[0.024184193, 0.026025473]	[0.021757738, 0.022754142]	[0.0394274, 0.045503708]

Table 11.18 Variations in the probabilities of operating states of SEN w. r. t. time.

T	$[\underline{P}_{14}(t), \overline{P}_{14}(t)]$
0	[0,0]
1	[0.003841521, 0.00436562]
2	[0.015056043, 0.016505295]
3	[0.026640307, 0.02837283]
4	[0.034812006, 0.036169455]
5	[0.0389485, 0.03958085]
6	[0.03957174, 0.039740104]
7	[0.037317141, 0.0381945]
8	[0,03381003, 0.03523007]
9	[0.029770542, 0.031552346]
10	[0.025428249, 0.027650956]

Table 11.19 Changes in the network reliability of SEN with time.

T	$[\underline{R}(t), \overline{R}(t)]$
0	[1,1]
1	[0.45698, 0.55439]
2	[0.42011, 0.4058109]
3	[0.389608, 0.3992534]
4	[0.3614208, 0.362971442]
5	[0.302322595, 0.3131528]
6	[0.27192434, 0.29157]
7	[0.261650249, 0.27110465]
8	[0.25147536, 0.26099708]
9	[0.2413648, 0.248855568]
10	[0.231280076, 0.2407863953]

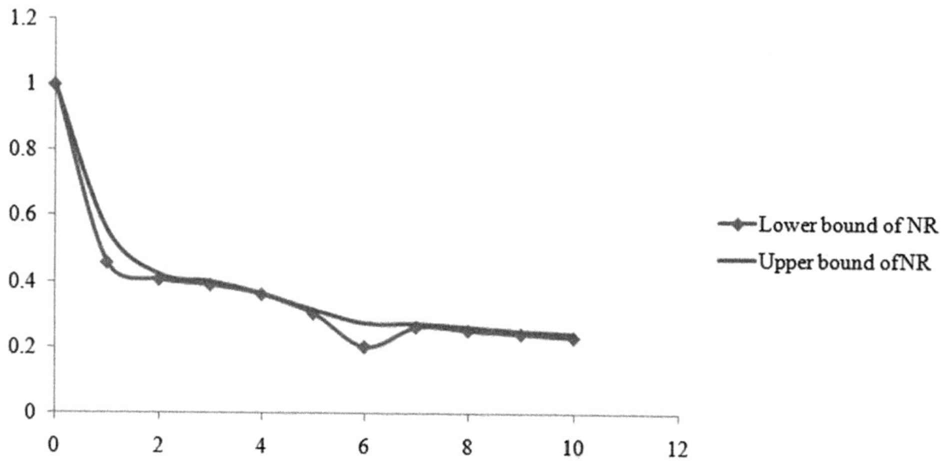

FIGURE 11.14

NR versus time.

10. Result and discussion

The proposed model reveals that how one can analyze the bounds of reliability of a replaceable SEN incorporating uncertainties by using the IUGF technique. The reliability bounds of SEN on the basis of three reliability indices, namely, terminal, broadcast, and network reliability, have been examined and it was concluded from Figs. 11.8, 11.9 and 11.10 that the lower as well as upper bounds of all the three reliability parameters decrease with increasing time, i.e., the performance of the network degrades with time. Also, it was found that the terminal reliability bounds of the SEN are the greatest followed by the bounds of broadcast reliability and the network reliability bounds are the least among all the three measures.

Table 11.20 MTTF w. r. t. $\underline{\lambda}_1$, $\underline{\lambda}_2$, $\underline{\lambda}_3$.

	MTTF w. r. t. $\underline{\lambda}_1$	MTTF w. r. t. $\underline{\lambda}_2$	MTTF w. r. t. $\underline{\lambda}_3$
0.1	[3.717309622, 5.1285649]	[3.78339513, 5.1285649]	[3.851872867, 5.1285649]
0.2	[3.561773655, 5.1285649]	[3.622399589, 5.1285649]	[3.685125123, 5.1285649]
0.3	[3.418730536, 5.1285649]	[3.474546545, 5.1285649]	[3.532215367, 5.1285649]
0.4	[3.286733218, 5.1285649]	[3.33828982, 5.1285649]	[3.291489655, 5.1285649]
0.5	[3.164549827, 5.1285649]	[3.28743096, 5.1285649]	[3.2615475, 5.1285649]
0.6	[3.051125102, 5.1285649]	[3.2006749, 5.1285649]	[3.14119521, 5.1285649]
0.7	[2.945549839, 5.1285649]	[3.19226708, 5.1285649]	[3.02940891, 5.1285649]
0.8	[2.847036467, 5.1285649]	[3.17868074, 5.1285649]	[2.92530551, 5.1285649]
0.9	[2.754899364, 5.1285649]	[3.18961125, 5.1285649]	[2.828119281, 5.1285649]
0.91	[2.746012592, 5.1285649]	[3.18102518, 5.1285649]	[2.737182969, 5.1285649]

Table 11.21 MTTF w. r. t. $\underline{\lambda}_4$, $\underline{\lambda}_5$, $\underline{\lambda}_6$.

	MTTF w. r. t. $\underline{\lambda}_4$	MTTF w. r. t. $\underline{\lambda}_5$	MTTF w. r. t. $\underline{\lambda}_6$
0.1	[3.922875, 5.1285649]	[4.0863218, 5.1285649]	[4.239564, 5.1285649]
0.2	[3.75006124, 5.1285649]	[4.00738701, 5.1285649]	[4.1757208, 5.1285649]
0.3	[3.591830816, 5.1285649]	[3.9015169, 5.1285649]	[4.10876915, 5.1285649]
0.4	[3.44641256, 5.1285649]	[3.8150989, 5.1285649]	[4.09431769, 5.1285649]
0.5	[3.31231091, 5.1285649]	[3.75872913, 5.1285649]	[3.9783104, 5.1285649]
0.6	[3.18825432, 5.1285649]	[3.685162318, 5.1285649]	[3.9153843, 5.1285649]
0.7	[3.073154886, 5.1285649]	[3.5487630301, 5.1285649]	[3.8087306, 5.1285649]
0.8	[2.96607632, 5.1285649]	[3.474998214, 5.1285649]	[3.774302, 5.1285649]
0.9	[2.866208, 5.1285649]	[3.33542873, 5.1285649]	[3.694094385, 5.1285649]
0.91	[2.8565902, 5.1285649]	[3.318920313, 5.1285649]	[3.69150718, 5.1285649]

Table 11.22 MTTF w. r. t. $\underline{\lambda}_7$

	MTTF w. r. t. $\underline{\lambda}_7$
0.1	[4.582672, 5.1285649]
0.2	[4.4568725, 5.1285649]
0.3	[4.341216, 5.1285649]
0.4	[4.31078, 5.1285649]
0.5	[4.25165488, 5.1285649]
0.6	[4.1607532, 5.1285649]
0.7	[4.0991589, 5.1285649]
0.8	[3.98362107, 5.1285649]
0.9	[3.879364, 5.1285649]
0.91	[3.8625622, 5.1285649]

Table 11.23 MTTF w. r. t. $\overline{\lambda}_1$, $\overline{\lambda}_2$, $\overline{\lambda}_3$.

	MTTF w. r. t. $\overline{\lambda}_1$	MTTF w. r. t. $\overline{\lambda}_2$	MTTF w. r. t. $\overline{\lambda}_3$
0.1	[3.8693814, 4.981010674]	[3.8693814, 5.085522018]	[3.8693814, 5.1285649]
0.2	[3.8693814, 4.784365957]	[3.8693814, 4.880708405]	[3.8693814, 4.920340629]
0.3	[3.8693814, 4.607908354]	[3.8693814, 4.691753376]	[3.8693814, 4.728364795]
0.4	[3.8693814, 4.439120502]	[3.8693814, 4.516883701]	[3.8693814, 4.550806973]
0.5	[3.8693814, 4.28226112]	[3.8693814, 4.354581017]	[3.8693814, 4.3610169]
0.6	[3.8693814, 4.136108863]	[3.8693814, 4.203537624]	[3.8693814, 4.23290219]
0.7	[3.8693814, 3.995647534]	[3.8693814, 4.062621177]	[3.8693814, 4.090043527]
0.8	[3.8693814, 3.8681133]	[3.8693814, 3.930846244]	[3.8693814, 3.956512886]
0.9	[3.8693814, 3.748468596]	[3.8693814, 3.807351232]	[3.8693814, 3.831425536]
0.91	[3.8693814, 3.71730962]	[3.8693814, 3.795427175]	[3.8693814, 3.819350448]

Table 11.24 MTTF w. r. t. $\overline{\lambda}_4$, $\overline{\lambda}_5$, $\overline{\lambda}_6$.

	MTTF w. r. t. $\overline{\lambda}_4$	MTTF w. r. t. $\overline{\lambda}_5$	MTTF w. r. t. $\overline{\lambda}_6$
0.1	[3.8693814, 5.239428867]	[3.8693814, 5.38762109]	[3.8693814, 5.4870522]
0.2	[3.8693814, 5.02229547]	[3.8693814, 5.217198215]	[3.8693814, 5.30181401]
0.3	[3.8693814, 4.822442885]	[3.8693814, 5.18529412]	[3.8693814, 5.2529919]
0.4	[3.8693814, 4.637887091]	[3.8693814, 5.09374592]	[3.8693814, 5.16620418]
0.5	[3.8693814, 4.46693659]	[3.8693814, 5.002738633]	[3.8693814, 5.06671564]
0.6	[3.8693814, 4.308140409]	[3.8693814, 4.9376402]	[3.8693814, 4.9896421]
0.7	[3.8693814, 4.160246814]	[3.8693814, 4.80441012]	[3.8693814, 4.83973746]
0.8	[3.8693814, 4.022170252]	[3.8693814, 4.729954308]	[3.8693814, 4.71078506]
0.9	[3.8693814, 3.892964655]	[3.8693814, 4.69425576]	[3.8693814, 4.62075346]
0.91	[3.8693814, 3.88049919]	[3.8693814, 4.66196539]	[3.8693814, 4.599755321]

Table 11.25 MTTF w. r. t. $\overline{\lambda}_7$.

	MTTF w. r. t. $\overline{\lambda}_7$
0.1	[3.8693814, 5.596132785]
0.2	[3.8693814, 5.4129532]
0.3	[3.8693814, 5.30076423]
0.4	[3.8693814, 5.21985243]
0.5	[3.8693814, 5.14637955]
0.6	[3.8693814, 5.029854187]
0.7	[3.8693814, 4.92324657]
0.8	[3.8693814, 4.843567145]
0.9	[3.8693814, 4.73145297]
0.91	[3.8693814, 4.710963245]

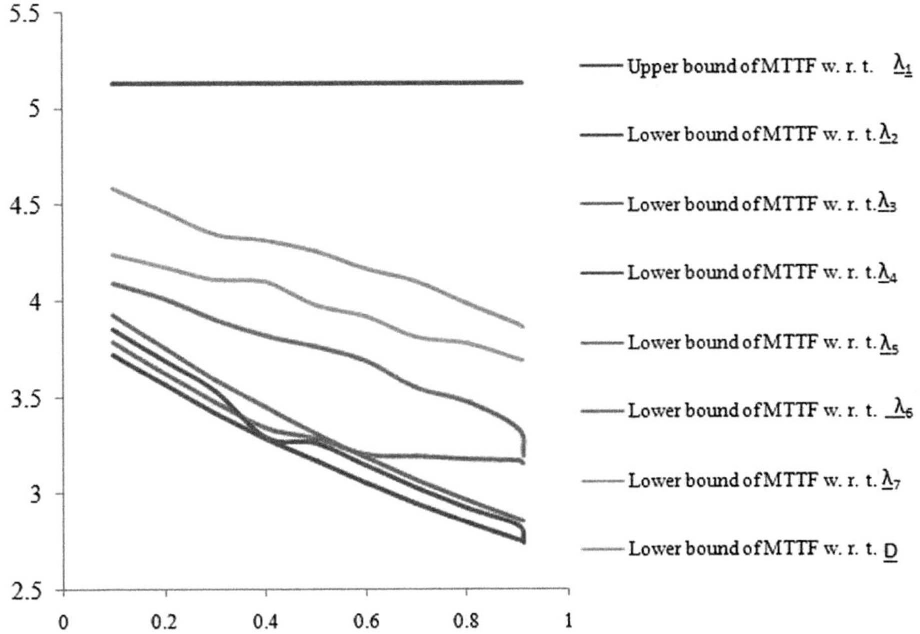

FIGURE 11.15

MTTF w. r. t. $\underline{\lambda}_1$, $\underline{\lambda}_2$, $\underline{\lambda}_3$, $\underline{\lambda}_4$, $\underline{\lambda}_5$, $\underline{\lambda}_6$, $\underline{\lambda}_7$.

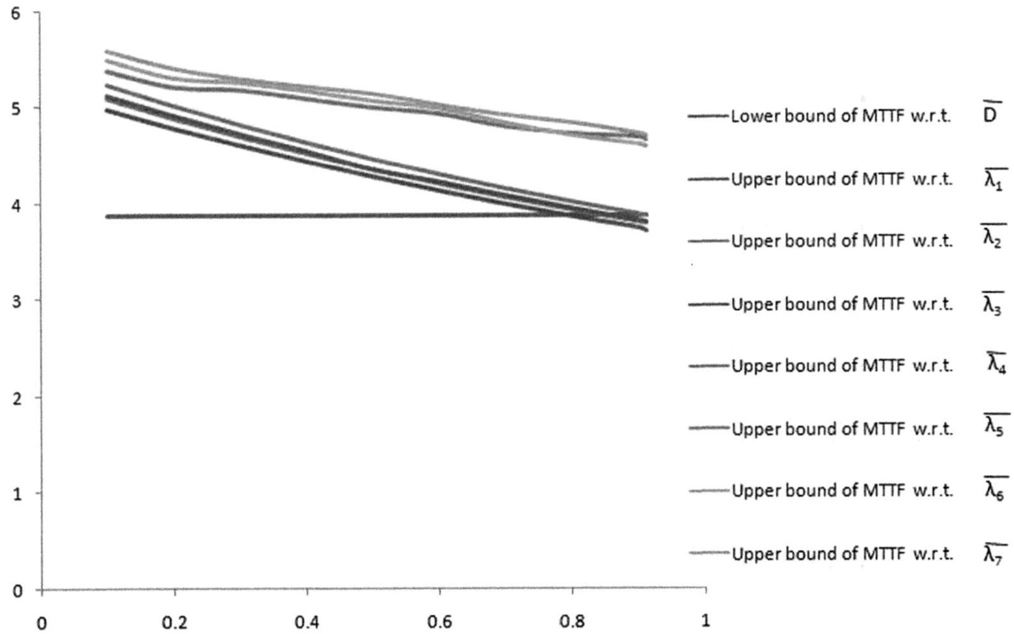

FIGURE 11.16

MTTF versus $\bar{\lambda}_1$, $\bar{\lambda}_2$, $\bar{\lambda}_3$, $\bar{\lambda}_4$, $\bar{\lambda}_5$, $\bar{\lambda}_6$, $\bar{\lambda}_7$.

Also, the bounds of MTTF of the considered network corresponding to each reliability parameter have been evaluated and the following conclusions are made:

(a) MTTF w. r. t. terminal reliability:
  - **(i)** It can be observed from Fig. 11.10 that as the values of $\underline{\lambda}_1$, $\underline{\lambda}_2$, and $\underline{\lambda}_3$ increases lower bound of the MTTF decreases while the upper bound of MTTF remains constant for all values of the mentioned failure rates.
  - **(ii)** Fig. 11.10 shows that with increasing values of $\overline{\lambda}_1$, $\overline{\lambda}_2$, and $\overline{\lambda}_3$ upper bound of MTTF of the network decreases slowly while the lower bound remains unchanged.

(b) MTTF w. r. t. broadcast reliability:
  - **(i)** By observing Fig. 11.12, it can be visualized that the lower bound of MTTF decreases with increasing $\underline{\lambda}_1$, $\underline{\lambda}_2$, $\underline{\lambda}_3$, $\underline{\lambda}_4$, $\underline{\lambda}_5$, $\underline{\lambda}_6$, $\underline{\lambda}_7$ while the upper bound of the MTTF remains constant with respect to all the parameters.
  - **(ii)** From Fig. 11.13, it can be concluded that the upper bound of the MTTF of the considered network decreases with increase in the values of $\overline{\lambda}_1$, $\overline{\lambda}_4$, $\overline{\lambda}_5$, $\overline{\lambda}_6$, $\overline{\lambda}_7$ and increases with increasing values of $\overline{\lambda}_2$, $\overline{\lambda}_3$. The lower bound of the MTTF remains unchanged for all the mentioned parameters.

(c) MTTF w. r. t. network reliability:
  - **(i)** From Fig. 11.15, it can be detected that on increasing the values of the failure rates $\underline{\lambda}_1$, $\underline{\lambda}_2$, $\underline{\lambda}_3$, $\underline{\lambda}_4$, $\underline{\lambda}_5$, $\underline{\lambda}_6$, $\underline{\lambda}_7$ lower bound of MTTF of the proposed SEN decreases whereas the upper bound of MTTF remains constant.
  - **(ii)** On examining Fig. 11.16 we can visualize that the upper bound of the MTTF decreases with increasing values of the parameters $\overline{\lambda}_1$, $\overline{\lambda}_2$, $\overline{\lambda}_3$, $\overline{\lambda}_4$, $\overline{\lambda}_5$, $\overline{\lambda}_6$, $\overline{\lambda}_7$ while its lower bound remains constant for all the failure rates.

It can also be visualized that the lower bound of the MTTF of the network is lowest with respect to $\underline{\lambda}_1$ which is corresponding to the TR and is highest with respect to $\underline{\lambda}_7$ which is corresponding to the NR. The lowest and the highest values of the MTTF are 0.6625 and 4.582672, respectively. Also, the upper bound of the MTTF is lowest with respect to the parameters $\underline{\lambda}_1$, $\underline{\lambda}_2$, $\underline{\lambda}_3$ with value 1.66667 and is highest with respect to $\overline{\lambda}_2$ with value 7.8787879. The lowest and highest values of the upper bound of the MTTF are obtained corresponding to the TR and BR of the SEN, respectively.

11. Conclusion

In this chapter, we have considered an SEN, in which the state probabilities of the SEs are uncertain and if any of the SEs of the considered SEN becomes faulty, then it can be replaced by a certain replacement rate. We analyzed its reliability on the basis of three indices, namely, terminal reliability, broadcast reliability, and network reliability. The reliability evaluation has been done by using IUGF approach and the probabilities are obtained in intervals.

The SEN examined is of size 8×8, i.e., has eight inputs and eight outputs. The RBD for the terminal, broadcast, and network reliability of the SEN has been presented. By using supplementary variable technique, the differential equations of various states governing the network's performance have been obtained and by applying Laplace transform, the transition state probabilities of different components are computed. By using the probabilities obtained by the IUGF approach, the upper and lower bounds of the three different reliability parameters and the bounds of MTTF for each reliability of the considered network have been evaluated. A numerical example was also provided to give a practical explanation of the proposed model.

References

[1] S. Bisht, Reliability indices and Signature Analysis of Complex Networks, Doctoral Thesis, GB Pant University of Agriculture and Technology, Pantnagar-263145 (Uttarakhand), 2018.

[2] S. Bisht, S.B. Singh, Signature reliability of binary state node in complex bridge networks using universal generating function, Int. J. Qual. Reliab. Manag. 36 (2) (2019) 186−201.

[3] F. Bistouni, M. Jahanshahi, Analyzing the reliability of shuffle-exchange networks using reliability block diagrams, Reliab. Eng. Syst. Saf. 132 (2014) 97−106.

[4] F. Bistouni, M. Jahanshahi, Determining the reliability importance of switching elements in the shuffle-exchange networks, Int. J. Parallel, Emergent Distributed Syst. 34 (4) (2019) 448−476.

[5] N.S. Fard, I. Gunawan, Terminal reliability improvement of shuffle-exchange network systems, Int. J. Reliab. Qual. Saf. Eng. 12 (01) (2005) 51−60.

[6] A. Kumar, S.B. Singh, M. Ram, Interval-valued reliability assessment of 2-out-of-4 system, in: 2016 International Conference on Emerging Trends in Communication Technologies (ETCT) (1-4). IEEE, 2016.

[7] C.Y. Li, X. Chen, X.S. Yi, J.Y. Tao, Interval-valued reliability analysis of multi-state systems, IEEE Trans. Reliab. 60 (1) (2011) 323−330.

[8] Meenakshi, S.B. Singh, Reliability analysis of multi-state complex system having two multi-state subsystems under uncertainty, J. Reliab. Stat. Stud. 10 (1) (2017) 161−177.

[9] G. Pan, C.X. Shang, Y.Y. Liang, J.Y. Cai, D.Y. Li, November). Analysis of interval-valued reliability of multi-state system in consideration of epistemic uncertainty, in: International Conference on P2P, Parallel, Grid, Cloud and Internet Computing, Springer, Cham, 2016, pp. 69−80.

[10] S. Rajkumar, N.K. Goyal, Review of multistage interconnection networks reliability and fault-tolerance, IETE Tech. Rev. 33 (3) (2016) 223−230.

[11] S. Sharma, K.S. Kahlon, P.K. Bansal, Reliability and path length analysis of irregular fault tolerant multistage interconnection network, Comput. Architect. News 37 (5) (2009) 16−23.

[12] N.A.M. Yunus, M. Othman, Shuffle exchange network in multistage interconnection network: a review and challenges, Int. J. Comput. Electr. Eng. 3 (5) (2011) 724.

[13] N.A.M. Yunus, M. Othman, Reliability evaluation for shuffle exchange interconnection network, Procedia Comput. Sci. 59 (2015) 162−170.

CHAPTER

12 Reliability, MTTF, and sensitivity evaluation of a computer network system connected in star topology

Kuldeep Nagiya[1], Akshay Kumar[2], Mangey Ram[3], Adarsh Anand[4]

[1]*Department of Mathematics, S.K.I.C. (U.P. Secondary Education Board, Prayagraj), Aligarh, Uttar Pradesh, India;* [2]*Department of Mathematics, Graphic Era Hill University, Dehradun, Uttarakhand, India;* [3]*Department of Mathematics; Computer Science & Engineering, Graphic Era Deemed to be University, Dehradun, Uttarakhand, India;* [4]*Department of Operational Research, University of Delhi, Delhi, India*

1. Introduction

The high dependable and high-class computing networking is developed by the rapid growing of a new digital economy [11]. The current technologies allow the adequate redundancy in fault-tolerant computer system network to ensure cost analysis a very low percentage of exhaustion of hardware. The system designer can be encouraged that fault and errors are minimized by the design of network, quality of computer systems, and hardware. This chapter is to evaluate the computer network performance for a specific time interval by using the star topology. In a star topology, all the cables are connected from the computer systems to a central location, called a hub. Each computer on a star network communicates with a central hub that resends the message either to all the computers in a star network or only to the destination computer. An active hub generates the electrical signal and sends it to all the computers connected to it. The main benefit of star topology is it reduces the chance of network failure by connecting all of the systems to a central node called hub. This type of networking can be seen easily in high-performance computer laboratories, offices, homes, etc.

A numbers of authors have done the significant work in reliability network modeling [14]. Mahmoud and Daoud [8] have given an analytical approach to calculate the reliability of mobile agent—based systems, and they investigated the reliability and availability of that system with the help of stochastic Petri Nets modeling. Levitin and Dai have done an extensive research work in the star topology. They analyzed the reliability and performance of star topology grid service with precedence constraints on subtask execution [7], service reliability performance and reliability of a star topology grid service with data dependency and two types of failure [5,6], and reliability and performance of tree-structured grid services [2]. They suggested a numerical algorithm for evaluating any subtask distribution. They also considered two types of failure: permanent failure and transient failure. They also proposed a virtual tree model of grid services. Fitzgerald et al. [3] discussed the reliability of the star graph architecture. They analyzed the node failure, link failure, and combined node and link failure. Mehmet-Ali et al. [9] studied on traffic analysis of a local area

The Handbook of Reliability, Maintenance, and System Safety through Mathematical Modeling.
https://doi.org/10.1016/B978-0-12-819582-6.00012-5

network with a star topology. They indicated a better performance having the modest expand in processing power. Guo et al. [4] studied the grid service reliability modeling and optimal task scheduling considering fault recovery. They presented a multiobjective task scheduling. Saltzer and Pogran [13] discussed a star-shaped ring network having high maintainability. They described a physical organization of a ring network. Raghavendra et al. [10] studied on the reliable loop topologies for large local computer networks. They proposed highly reliable and efficient double-loop network architecture. Abo-El-Fotoh et al. [1] considered on the reliability of wireless sensor networks. They defined a wireless sensor network reliability to calculate that considers the aggregate flow of sensor data into a sink node. Furthermore the reliability problems have been analyzed by a numbers of authors including Ram et al. [12] in various field and techniques. Hazra et al. [15] calculated the reliability of the Transformed-Transformer using family of distributions and studied various kinds of stochastic aging properties and various types of stochastic distributions. Sen et al. [16] determined the reliability of acceptance sampling plans and generalized hybrid censoring schemes with the help of Weibull distribution and evaluated specified producer's and consumer's from asymptotic normality of maximum likelihood parameters. Kumar and Singh [17] Kumar and Ram [18] evaluated the signature reliability, mean time to failure, expected cost, and expected value of sliding window system with the help of universal generating function and interval-valued techniques. Kaplesh et al. [19] discussed various techniques and methods for evaluating reliability measures of different kinds of system and gave a brief review in reliability and uses in engineering field. Li [20] computed the reliability and mean time between failure of the Dormant k-out-of-n Systems using Periodic Maintenance.

In this chapter, we consider three types of failure. The failure of an individual computer system unit, failure of hub, and the failure rate of cable connection. If one of the n computer systems failed then the system is partially failed because other computer systems work properly. Further if a cable connection is failed then the system is again partially failed. Further if the hub is failed then the system is completely failed because there is no data transfer possible among different computer systems. So, the complete system contains three types of states, i.e., good, partial, and failed states. The system configuration and transition state diagram are shown in Figs. 12.1 and 12.2, respectively.

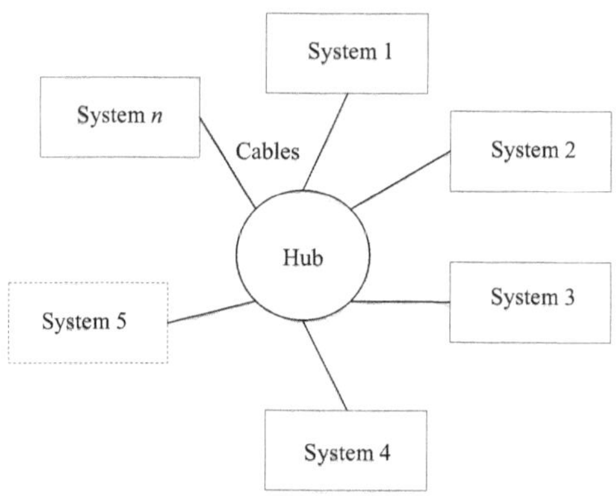

FIGURE 12.1

System configuration: a star topology.

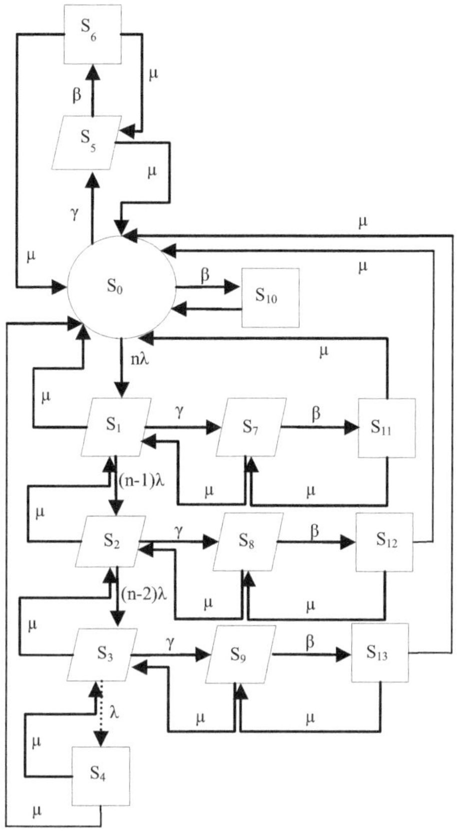

FIGURE 12.2

State transition diagram.

2. Assumptions and notations of the proposed model

(i) Assumptions

- **(i)** All elements are working at the initial stage.
- **(ii)** There are different states of the system for working, degraded, and failed conditions.
- **(iii)** All the repair rates and failure rates are supposed to be constant and vary with time.
- **(iv)** Repair space is available.
- **(v)** A repaired unit works considering a new one.

(ii) Notations

t	Scale time.
s	Laplace transform variable.
$P_0(t)$	Probability at time t when state is 0.
$P_i(t)$	Probability at time t in ith state; $i = 1, 2, 3, 5, 7, 8, 9$.

$P_i(x,t)$	Probability density function that the system is in ith state at time t and repair time of x, where $i = 4, 6, 10, 11, 12, 13$.
λ	Failure rate of individual computer system.
β	Failure rate of hub.
γ	Failure rate of cable.
μ	Repair rate from no state to yes state.

3. Formulation of the model

Let us take the possible state transitions, we can find the following set of differential equations:

$$\left[\frac{\partial}{\partial t} + n\lambda + \beta + \gamma\right] P_0(t) = \sum_i P_i(t)\mu + \sum_j \int_0^\infty P_j(x,t)\mu\, dx; i = 1,5;\ j = 4,6,10,11,12,13 \tag{12.1}$$

$$\left[\frac{\partial}{\partial t} + \beta + \mu\right] P_5(t) = \gamma P_0(t) + \int_0^\infty P_6(x,t)\mu\, dx \tag{12.2}$$

$$\left[\frac{\partial}{\partial t} + \frac{\partial}{\partial x} + 2\mu\right] P_6(x,t) = 0 \tag{12.3}$$

$$\left[\frac{\partial}{\partial t} + (n-1)\lambda + \gamma + \mu\right] P_1(t) = P_2(t)\mu + P_7(t)\mu + n\lambda P_0(t) \tag{12.4}$$

$$\left[\frac{\partial}{\partial t} + (n-2)\lambda + \gamma + \mu\right] P_2(t) = P_3(t)\mu + P_8(t)\mu + (n-1)\lambda P_1(t) \tag{12.5}$$

$$\left[\frac{\partial}{\partial t} + (n-3)\lambda + \gamma + \mu\right] P_3(t) = P_9(t)\mu + (n-2)\lambda P_2(t) \tag{12.6}$$

$$\left[\frac{\partial}{\partial t} + \frac{\partial}{\partial x} + \mu\right] P_4(x,t) = 0 \tag{12.7}$$

$$\left[\frac{\partial}{\partial t} + \frac{\partial}{\partial x} + \mu\right] P_{10}(x,t) = 0 \tag{12.8}$$

$$\left[\frac{\partial}{\partial t} + \beta + \mu\right] P_7(t) = \gamma P_1(t) + \int_0^\infty P_{11}(x,t)\mu\, dx \tag{12.9}$$

$$\left[\frac{\partial}{\partial t} + \frac{\partial}{\partial x} + 2\mu\right] P_{11}(x,t) = 0 \tag{12.10}$$

$$\left[\frac{\partial}{\partial t} + \beta + \mu\right] P_8(t) = \gamma P_2(t) + \int_0^\infty P_{12}(x,t)\mu\, dx \tag{12.11}$$

$$\left[\frac{\partial}{\partial t} + \frac{\partial}{\partial x} + 2\,\mu\right] P_{12}(x,t) = 0 \tag{12.12}$$

$$\left[\frac{\partial}{\partial t} + \beta + \mu\right] P_9(t) = \gamma P_3(t) + \int_0^\infty P_{13}(x,t)\mu\, dx \tag{12.13}$$

$$\left[\frac{\partial}{\partial t} + \frac{\partial}{\partial x} + 2\,\mu\right] P_{13}(x,t) = 0 \tag{12.14}$$

Boundary conditions are:

$$P_6(0,t) = \beta\, P_5(t) \tag{12.15}$$

$$P_4(0,t) = \lambda P_3(t) \tag{12.16}$$

$$P_{10}(0,t) = \beta\, P_0(t) \tag{12.17}$$

$$P_{11}(0,t) = \beta\, P_7(t) \tag{12.18}$$

$$P_{12}(0,t) = \beta\, P_8(t) \tag{12.19}$$

$$P_{13}(0,t) = \beta\, P_9(t) \tag{12.20}$$

Initial condition is $P_0(0) = 1$ and other probabilities are zero.
Taking Laplace transformation from Eqs. (12.1)–(12.20)

$$[s + n\lambda + \beta + \gamma]\overline{P}_0(s) = 1 + \sum_i \overline{P}_i(s)\mu + \sum_j \int_0^\infty \overline{P}_j(x,s)\mu\, dx;\ i = 1, 5;\ j = 4, 6, 10, 11, 12, 13 \tag{12.21}$$

$$[s + \beta + \mu]\overline{P}_5(s) = \gamma\overline{P}_0(s) + \int_0^\infty \overline{P}_6(x,s)\mu\, dx \tag{12.22}$$

$$\left[s + \frac{\partial}{\partial x} + 2\,\mu\right]\overline{P}_6(x,s) = 0 \tag{12.23}$$

$$[s + (n-1)\lambda + \gamma + \mu]\overline{P}_1(s) = \overline{P}_2(s)\,\mu + P_7(s)\,\mu + n\lambda\,\overline{P}_0(s) \tag{12.24}$$

$$[s + (n-2)\lambda + \gamma + \mu]\overline{P}_2(s) = \overline{P}_3(s)\,\mu + \overline{P}_8(s)\,\mu + (n-1)\lambda\,\overline{P}_1(s) \tag{12.25}$$

$$[s + (n-3)\lambda + \gamma + \mu]\overline{P}_3(s) = \overline{P}_9(s)\,\mu + (n-2)\lambda\,\overline{P}_2(s) \tag{12.26}$$

$$\left[s + \frac{\partial}{\partial x} + \mu\right]\overline{P}_4(x,s) = 0 \tag{12.27}$$

$$\left[s + \frac{\partial}{\partial x} + \mu\right]\overline{P}_{10}(x,s) = 0 \tag{12.28}$$

$$[s+\beta+\mu]\overline{P}_7(s)=\gamma\overline{P}_1(s)+\int_0^\infty \overline{P}_{11}(x,s)\mu\,dx \tag{12.29}$$

$$\left[s+\frac{\partial}{\partial x}+2\,\mu\right]\overline{P}_{11}(x,s)=0 \tag{12.30}$$

$$[s+\beta+\mu]\overline{P}_8(s)=\gamma\overline{P}_2(s)+\int_0^\infty \overline{P}_{12}(x,s)\mu\,dx \tag{12.31}$$

$$\left[s+\frac{\partial}{\partial x}+2\,\mu\right]\overline{P}_{12}(x,s)=0 \tag{12.32}$$

$$[s+\beta+\mu]\overline{P}_9(s)=\gamma\overline{P}_3(s)+\int_0^\infty \overline{P}_{13}(x,s)\mu\,dx \tag{12.33}$$

$$\left[s+\frac{\partial}{\partial x}+2\,\mu\right]\overline{P}_{13}(x,s)=0 \tag{12.34}$$

$$\overline{P}_6(0,s)=\beta\,\overline{P}_5(s) \tag{12.35}$$

$$\overline{P}_4(0,s)=\lambda\overline{P}_3(s) \tag{12.36}$$

$$\overline{P}_{10}(0,s)=\beta\,\overline{P}_0(s) \tag{12.37}$$

$$\overline{P}_{11}(0,s)=\beta\,\overline{P}_7(s) \tag{12.38}$$

$$\overline{P}_{12}(0,s)=\beta\,\overline{P}_8(s) \tag{12.39}$$

$$\overline{P}_{13}(0,s)=\beta\,\overline{P}_9(s) \tag{12.40}$$

From Eqs. 12.23, 12.27, 12.28, 12.30, 12.32, 12.34, we get

$$\overline{P}_i(0,s)=\overline{P}_i(0,s)\,\exp\left\{-sx-2\int_0^\infty \mu\,dx\right\};i=6,11,12,13 \tag{12.41}$$

$$\overline{P}_j(0,s)=\overline{P}_j(0,s)\,\exp\left\{-sx\quad\int_0^\infty \mu\,dx\right\};\ j=4,10 \tag{12.42}$$

After simplified these equations, we have

$$\overline{P}_0(s)=\frac{c_1c_2c_3-c_1(n-2)\lambda\,\mu-(n-1)\lambda\,\mu\,c_3}{\begin{array}{c}\left[\left(c+\beta\,\overline{T}_1(s)+\beta\overline{T}_2(s)\right)\left(c_1c_2c_3-c_1(n-2)\lambda\,\mu-(n-1)\lambda\,\mu\,c_3\right)\right.\\ -n\lambda\left(\mu+\beta A\overline{T}_1(s)\right)\left(c_2c_3-(n-2)\lambda\,\mu\right)-\beta A\overline{T}_1(s)n(n-1)\lambda^2c_3\\ \left.-n(n-1)(n-2)\lambda^3c_3\left(\beta A\overline{T}_1(s)+\lambda\overline{T}_2(s)\right)\right]\end{array}} \tag{12.43}$$

$$\overline{P}_1(s) = \frac{n\,\lambda[c_2c_3 - (n-2)\lambda\,\mu]\overline{P}_0(s)}{c_1c_2c_3 - c_1(n-1)\lambda\,\mu\,c_3} \tag{12.44}$$

$$\overline{P}_2(s) = \frac{(n-1)\lambda}{c_2 - \dfrac{(n-2)\,\lambda\mu}{c_3}} \times \frac{n\,\lambda[c_2c_3 - (n-2)\lambda\,\mu]\overline{P}_0(s)}{c_1c_2c_3 - c_1(n-1)\lambda\,\mu\,c_3} \tag{12.45}$$

$$\overline{P}_3(s) = \frac{(n-2)\lambda}{c_3} \times \frac{(n-1)\lambda}{c_2 - \dfrac{(n-2)\,\lambda\mu}{c_3}} \times \frac{n\,\lambda[c_2c_3 - (n-2)\lambda\,\mu]\overline{P}_0(s)}{c_1c_2c_3 - c_1(n-1)\lambda\,\mu\,c_3} \tag{12.46}$$

$$\overline{P}_4(s) = \lambda\overline{T}_2(s)\frac{(n-2)\lambda}{c_3} \times \frac{(n-1)\lambda}{c_2 - \dfrac{(n-2)\,\lambda\mu}{c_3}} \times \frac{n\,\lambda[c_2c_3 - (n-2)\lambda\,\mu]\overline{P}_0(s)}{c_1c_2c_3 - c_1(n-1)\lambda\,\mu\,c_3} \tag{12.47}$$

$$\overline{P}_5(s) = A\,\overline{P}_0(s) \tag{12.48}$$

$$\overline{P}_6(s) = \beta\overline{T}_1(s)A\,\overline{P}_0(s) \tag{12.49}$$

$$\overline{P}_7(s) = A\,\frac{n\,\lambda[c_2c_3 - (n-2)\lambda\,\mu]\overline{P}_0(s)}{c_1c_2c_3 - c_1(n-1)\lambda\,\mu\,c_3} \tag{12.50}$$

$$\overline{P}_8(s) = A\,\frac{(n-1)\lambda}{c_2 - \dfrac{(n-2)\,\lambda\mu}{c_3}} \times \frac{n\,\lambda[c_2c_3 - (n-2)\lambda\,\mu]\overline{P}_0(s)}{c_1c_2c_3 - c_1(n-1)\lambda\,\mu\,c_3} \tag{12.51}$$

$$\overline{P}_9(s) = A\,\frac{(n-2)\lambda}{c_3} \times \frac{(n-1)\lambda}{c_2 - \dfrac{(n-2)\,\lambda\mu}{c_3}} \times \frac{n\,\lambda[c_2c_3 - (n-2)\lambda\,\mu]\overline{P}_0(s)}{c_1c_2c_3 - c_1(n-1)\lambda\,\mu\,c_3} \tag{12.52}$$

$$\overline{P}_{10}(s) = \beta\overline{T}_2(s)\overline{P}_0(s) \tag{12.53}$$

$$\overline{P}_{11}(s) = \beta\overline{T}_1(s)\,A\,\frac{n\,\lambda[c_2c_3 - (n-2)\lambda\,\mu]\overline{P}_0(s)}{c_1c_2c_3 - c_1(n-1)\lambda\,\mu\,c_3} \tag{12.54}$$

$$\overline{P}_{12}(s) = \beta\overline{T}_1(s)\,A\,\frac{(n-1)\lambda}{c_2 - \dfrac{(n-2)\,\lambda\mu}{c_3}} \times \frac{n\,\lambda[c_2c_3 - (n-2)\lambda\,\mu]\overline{P}_0(s)}{c_1c_2c_3 - c_1(n-1)\lambda\,\mu\,c_3} \tag{12.55}$$

$$\overline{P}_{13}(s) = \beta\overline{T}_1(s)A\,\frac{(n-2)\lambda}{c_3} \times \frac{(n-1)\lambda}{c_2 - \dfrac{(n-2)\,\lambda\mu}{c_3}} \times \frac{n\,\lambda[c_2c_3 - (n-2)\lambda\,\mu]\overline{P}_0(s)}{c_1c_2c_3 - c_1(n-1)\lambda\,\mu\,c_3} \tag{12.56}$$

where,

$$c = s + n\lambda + \beta + \gamma + \mu,\ c_1 = s + (n-1)\lambda + \gamma + \mu - A,\ c_2 = s + (n-2)\lambda + \gamma + \mu - A$$

$$c_3 = s + (n-3)\lambda + \gamma + \mu - A,\ A = \frac{\gamma}{s + \beta + \mu - \beta \overline{T}_1(s)}$$

The Laplace transformations of the probabilities that the system is in working state and the failed state at any time are as follows:

$$\overline{P}_{up}(s) = \overline{P}_0(s) + \overline{P}_1(s) + \overline{P}_2(s) + \overline{P}_3(s) + \overline{P}_5(s) + \overline{P}_7(s) + \overline{P}_8(s) + \overline{P}_9(s)$$

$$= (1+A)\left\{1 + \frac{n\,\lambda[c_2c_3 - (n-2)\lambda\,\mu]}{c_1c_2c_3 - c_1(n-1)\lambda\,\mu\,c_3} + \frac{(n-1)\lambda}{c_2 - \dfrac{(n-2)\,\lambda\mu}{c_3}} \times \frac{n\,\lambda[c_2c_3 - (n-2)\lambda\,\mu]}{c_1c_2c_3 - c_1(n-1)\lambda\,\mu\,c_3}\right.$$

$$\left. + \frac{(n-2)\lambda}{c_3} \times \frac{(n-1)\lambda}{c_2 - \dfrac{(n-2)\,\lambda\mu}{c_3}} \times \frac{n\,\lambda[c_2c_3 - (n-2)\lambda\,\mu]}{c_1c_2c_3 - c_1(n-1)\lambda\,\mu\,c_3}\right\}\overline{P}_0(s) \tag{12.57}$$

$$\overline{P}_{down}(s) = \overline{P}_4(s) + \overline{P}_6(s) + \overline{P}_{10}(s) + \overline{P}_{11}(s) + \overline{P}_{12}(s) + \overline{P}_{13}(s) \tag{12.58}$$

4. Particular examples

4.1 Reliability

Taking the repair rate equal to zero and suppose the failure rate values of $\lambda = 0.002$, $\beta = 0.001$, $\gamma = 0.002$, $\mu = 0$, and $n = 100$ in the equations, the reliability of the system is given as

$$\begin{aligned} \mathrm{R(t)} = & -321.9488\, e^{0.0086769384\, t} + 396081.9488\, e^{-0.2076769384\, t} - 139268.4526\, e^{-0.209588267\, t} \\ & + 156.25747\, e^{0.008588267\, t} + 81102.01048\, e^{-0.203\, t} - 0.0099\, e^{-0.001\, t} - 337914.6761\, e^{-0.20576718\, t} \\ & + 165.8707\, e^{0.0087671898\, t} \end{aligned} \tag{12.59}$$

Putting $t = 0$ to 15, one can obtain reliability of the system (Table 12.1).

Table 12.1 Reliability as function of time.

Time	Reliability $R(t)$
0	1.00000
1	0.99501
2	0.99008
3	0.98423
4	0.97637
5	0.96552
6	0.95157
7	0.93384
8	0.91254
9	0.88798
10	0.86057
11	0.83083
12	0.79925
13	0.76654
14	0.73325
15	0.69994

4.2 Mean time to failure

Now we calculate the mean time to failure (MTTF) by taking the repair rate zero and letting the Laplace variable "s" approach zero in the equation

$$\begin{aligned}\mathrm{MTTF}=&\left(1-\frac{\gamma}{\beta}\right)\left\{\frac{1}{n\,\lambda+\beta+\gamma}+\frac{n\,\lambda}{(n\,\lambda+\beta+\gamma)\left((n-1)\lambda+\gamma-\frac{\gamma}{\beta}\right)}\right.\\&+\frac{(n-1)\,\lambda^2 n}{(n\,\lambda+\beta+\gamma)\left((n-1)\lambda+\gamma-\frac{\gamma}{\beta}\right)\left((n-2)\lambda+\gamma-\frac{\gamma}{\beta}\right)}\\&\left.+\frac{n(n-1)(n-2)\lambda^3}{(n\,\lambda+\beta+\gamma)\left((n-1)\lambda+\gamma-\frac{\gamma}{\beta}\right)\left((n-2)\lambda+\gamma-\frac{\gamma}{\beta}\right)\left((n-3)\lambda+\gamma-\frac{\gamma}{\beta}\right)}\right\}\end{aligned} \tag{12.60}$$

Setting $\lambda=0.002$, $\beta=0.1$, $\gamma=0.002$, and $n=100$ and varying λ, β, and γ, respectively, as 0.1, 0.2, 0.3, 0.4, 0.5, 0.6, 0.7, 0.8, 0.9 in Eq. (12.60), we find the variation of MTTF with respect to failure rates as shown in Fig. 12.4 (Table 12.2).

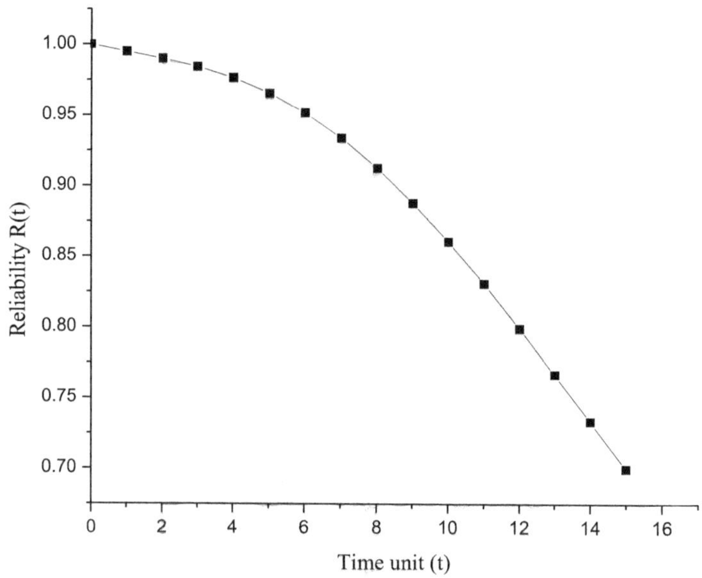

FIGURE 12.3

Reliability as function of time.

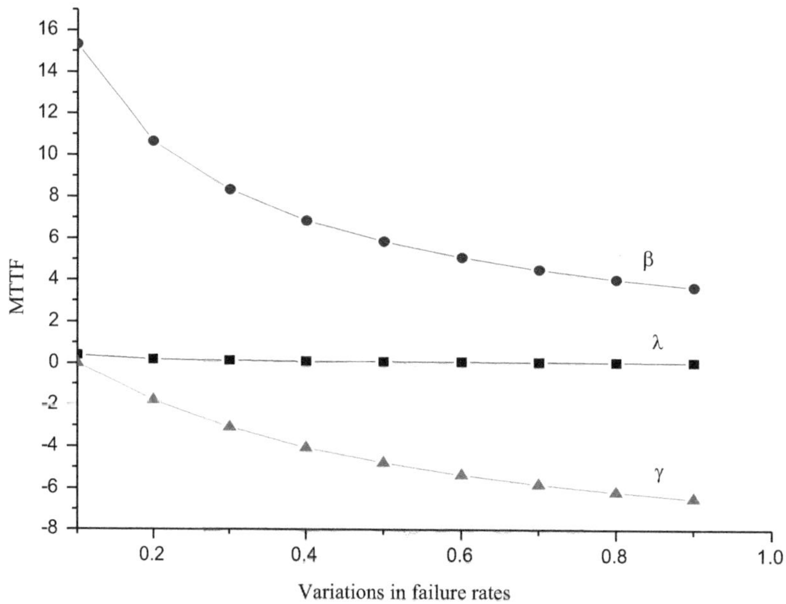

FIGURE 12.4

MTTF as function of failure rates.

Table 12.2 MTTF as function of failure rates.

Variations n λ, β, and γ	MTTF with respect to failure rates		
	λ	β	γ
0.1	0.395096	15.327818	0
0.2	0.198275	10.662152	−1.777372
0.3	0.132346	08.336940	−3.086301
0.4	0.099320	06.871797	−4.047573
0.5	0.079486	05.852564	−4.777756
0.6	0.066254	05.099615	−5.349783
0.7	0.056799	04.519657	−5.809517
0.8	0.049706	04.058814	−6.186864
0.9	0.044188	03.683622	−6.502055

4.3 Sensitivities

Reliability function is defined as the rate of change in output due to an input of the system known as the sensitivities. Some measures of failure rates are as follows.

4.3.1 Reliability sensitivity

Calculate the sensitivity of reliability by differentiating the equation of reliability with respect to failure rates λ, β, and γ, respectively, from the values of $\lambda = 0.002$, $\beta = 0.001$, $\gamma = 0.002$, and $n = 100$. We evaluate $\frac{\partial R(t)}{\partial \lambda}$, $\frac{\partial R(t)}{\partial \beta}$ & $\frac{\partial R(t)}{\partial \gamma}$ and taking $t = 0$ to 9, we obtain Table 12.3.

Table 12.3 Sensitivity of reliability as function of time.

Time	$\frac{\partial R(t)}{\partial \lambda}$	$\frac{\partial R(t)}{\partial \beta}$	$\frac{\partial R(t)}{\partial \gamma}$
0	−0.098519	−0.060000	0.198519
1	−0.203418	−0.960108	−1.696581
2	−1.008312	−1.640237	−3.931687
3	−4.793202	−2.230386	−4.906797
4	−12.518086	−2.680554	−6.261913
5	−25.932965	−3.090741	−6.377034
6	−44.927840	−3.354949	−6.512159
7	−68.932710	−3.599176	−5.467289
8	−97.197575	−3.797423	−3.932424
9	−128.532435	−3.907689	−1.787564
10	−161.857290	−3.979975	1.007290
11	−195.812140	−4.044280	4.512140
12	−229.166986	−4.081605	8.586986
13	−261.041826	−4.087949	13.053826
14	−290.696662	−4.096313	18.128662
15	−317.435493	−4.078696	23.459493

4.3.2 MTTF sensitivity

Estimate sensitivity analysis of MTTF by differentiating the equation of MTTF with respect to failure rates λ, β, and γ, respectively, from the values of $\lambda = 0.002$, $\beta = 0.001$, $\gamma = 0.002$, and $n = 100$. We find $\frac{\partial \text{MTTF}}{\partial \lambda}$, $\frac{\partial \text{MTTF}}{\partial \beta}$ and $\frac{\partial \text{MTTF}}{\partial \gamma}$.

Varying the failure rates, respectively, as 0.1, 0.2, 0.3, 0.4, 0.5, 0.6, 0.7, 0.8, 0.9 in the partial derivatives of MTTF, we get Fig. 12.6 (Table 12.4).

5. Conclusion

In this chapter, we analyze the reliability, MTTF, and sensitivity analysis of reliability as well as MTTF of networking system by using star topology also calculated the different reliability measures of networking system of hub, computer system, and cables. The value of reliability and MTTF is evaluated by assuming the value of different measures. Fig. 12.3 shows the reliability as a function of time. The graph shows that with the passage of time reliability decreases. Fig. 12.4 indicated the MTTF of the system and graph shows the variation of MTTF having failure rates. Fig. 12.5 discusses the sensitivity of reliability as a function of time also its graph defines the variation of partial derivative having different failure rates versus time increases. The graph in Fig. 12.6 shows the sensitivity of MTTF as function of failure rates and discusses the variation of failure rates in the partial derivatives of MTTF having failure rates.

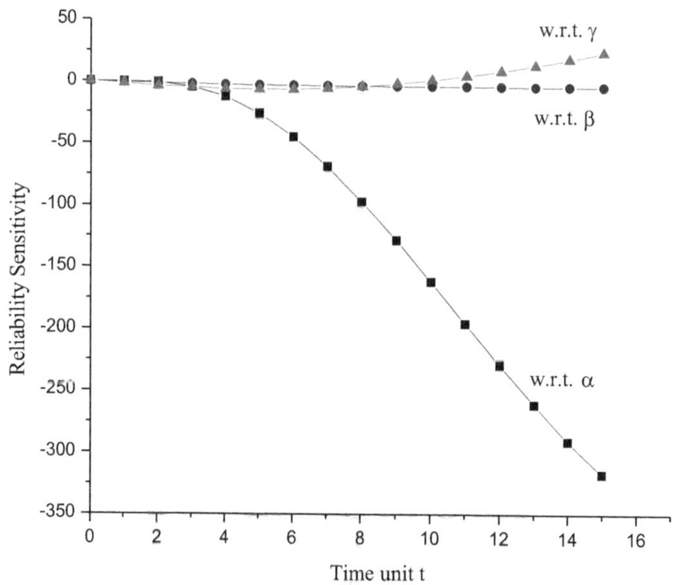

FIGURE 12.5

Sensitivity of reliability as function of time.

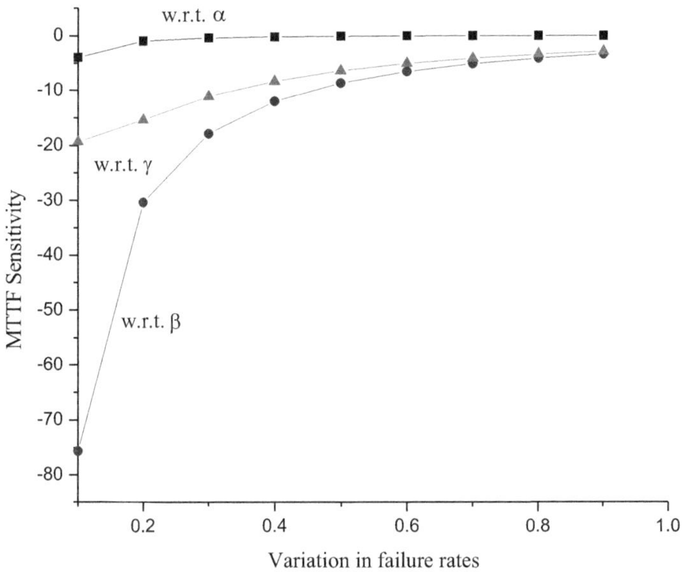

FIGURE 12.6

Sensitivity of MTTF as function of failure rates.

Table 12.4 Sensitivity of MTTF as function of failure rates.

Variations in λ, β, and γ	$\frac{\partial MTTF}{\partial \lambda}$	$\frac{\partial MTTF}{\partial \beta}$	$\frac{\partial MTTF}{\partial \gamma}$
0.1	−3.922049	−75.652037	−19.324567
0.2	−0.987720	−30.406676	−15.338941
0.3	−0.440066	−17.909871	−11.112387
0.4	−0.247842	−12.008185	−08.305376
0.5	−0.158736	−08.656931	−06.416814
0.6	−0.110288	−06.550858	−05.098373
0.7	−0.081056	−05.135250	−04.145157
0.8	−0.062075	−04.136020	−03.424935
0.9	−0.049057	−03.403663	−02.892099

References

[1] H.M. Abo-El-Fotoh, E.S. ElMallah, H.S. Hassanein, On the reliability of wireless sensor networks, in: Communications, 2006. ICC'06. IEEE International Conference on, vol. 8, IEEE, 2006, June, pp. 3455–3460.

[2] Y.S. Dai, G. Levitin, Reliability and performance of tree-structured grid services, Reliab. IEEE Trans. 55 (2) (2006) 337–349.

[3] K. Fitzgerald, S. Latifi, P.K. Srimani, Reliability modeling and assessment of the star-graph networks, Reliab. IEEE Trans. 51 (1) (2002) 49–57.
[4] S. Guo, H.Z. Huang, Z. Wang, M. Xie, Grid service reliability modeling and optimal task scheduling considering fault recovery, Reliab. IEEE Trans. 60 (1) (2011) 263–274.
[5] G. Levitin, Y.S. Dai, Performance and reliability of a star topology grid service with data dependency and two types of failure, IIE Trans. 39 (8) (2007) 783–794.
[6] G. Levitin, Y.S. Dai, Service reliability and performance in grid system with star topology, Reliab. Eng. Syst. Saf. 92 (1) (2007) 40–46.
[7] G. Levitin, Y.S. Dai, H. Ben-Haim, Reliability and performance of star topology grid service with precedence constraints on subtask execution, Reliab. IEEE Trans. 55 (3) (2006) 507–515.
[8] Q.H. Mahmoud, M. Daoud, An analytical approach to reliability estimation of mobile agent-based systems, J. Interconnect. Netw. 7 (02) (2006) 217–233.
[9] M.K. Mehmet-Ali, J.F. Hayes, A.K. Elhakeem, Traffic analysis of a local area network with a star topology, Commun. IEEE Trans. 36 (6) (1988) 703–712.
[10] C.S. Raghavendra, M. Gerla, A. Avizienis, Reliable loop topologies for large local computer networks, Comput. IEEE Transa. 100 (1) (1985) 46–55.
[11] M. Ram, On system reliability approaches: a brief survey, Int. J. Syst. Assur. Eng. Manag. 4 (2) (2013) 101–117.
[12] M. Ram, S.B. Singh, V.V. Singh, Stochastic analysis of a standby system with waiting repair strategy, IEEE Trans. Syst. Man Cybern. Syst. 43 (3) (2013) 698–707.
[13] J.H. Saltzer, K.T. Pogran, A star-shaped ring network with high maintainability, Comput. Network. 4 (5) (1980) 239–244, 1976.
[14] T. Takabatake, T. Nakamigawa, H. Ito, Connectivity of generalized hierarchical completely-connected networks, J. Interconnect. Netw. 9 (01n02) (2008) 127–139.
[15] N.K. Hazra, P. Kundu, A.K. Nanda, Some reliability properties of transformed-transformer family of distributions, Am. J. Math. Manag. Sci. (2018) 1–13.
[16] T. Sen, R. Bhattacharya, Y.M. Tripathi, B. Pradhan, Generalized hybrid censored reliability acceptance sampling plans for the weibull distribution, Am. J. Math. Manag. Sci. (2018) 1–20.
[17] A. Kumar, S.B. Singh, Signature of A-within-B-From-D/G sliding window system, Int. J. Math. Eng. Manag. Sci. 4 (1) (2019) 95–107.
[18] A. Kumar, M. Ram, Computation interval-valued reliability of sliding window system, Int. J. Math. Eng. Manag. Sci. 4 (1) (2019) 108–115.
[19] K.P. Amrutkar, K.K. Kamalja, An overview of various importance measures of reliability system, Int. J. Math. Eng. Manag. Sci. 2 (3) (2017) 150–171.
[20] L. Jamesh, Reliability calculation for dormant *k*-out-of-*n* systems with periodic maintenance, Int. J. Math. Eng. Manag. Sci. 1 (2) (2016) 68–76.

CHAPTER

Analysis of a system incorporating *k-out-of-n* structure with a warm standby redundancy: a reliability approach

13

Pardeep Kumar[1], **Amit Kumar**[1]

[1]*Department of Mathematics, Lovely Professional University, Phagwara, Punjab, India*

1. Introduction

Designing a good and reliable product keeping the cost under control is a very difficult task for reliability engineers. Every organization always make an attempt to launch a reliable product in the market by spending less money on it, so that the profit of the organization may increase by selling that product in the market. If this product performs well in the market, the reputation of the company grows and people just trust the brand name. On the contrary, if this product does not perform well as expected then organization has to bear warranty cost, test, and repair cost. It also has a bad effect on the reputation of the organization and they may lose future business. Hence, analysis of the various reliability indices is necessary to make the system more reliable. The main aim of the new system design is to reduce the down time of the system and increase its uptime. This can be done by increasing the reliability of the product or by reducing MTTR (mean time to repair) and increasing MTBF (mean time between failure). These two are very useful reliability metrics. This is possible with the proper maintenance strategies. For this, preventive maintenance or corrective maintenance or conditioning monitoring strategies can be used. Another approach to increase the system's reliability is to introduce redundancy in the system. It has been observed that redundancy is a better option for increasing the reliability of the product than increasing the reliability of a single product by spending money on it. There are many types of redundancies available such as parallel redundancy, *k-out-of-n*: F redundancy, *k-out-of-n*: G redundancy, warm redundancy, hot redundancy, and cold redundancy. A good and detailed explanation of all these redundancies can be found in Refs. [1–3]. Before we add redundancy in the system one needs to pay attention to cost, weight, space restriction, etc. No doubt when one more component is added in parallel for a component it increases the reliability of the system but this component should be chosen in such a way that it maximizes the system's reliability. Shen and Xie [4] investigated the effect of parallel redundancy upon system's reliability when components are added at different places in the system. In their investigation, they explained how to select a component for parallel redundancy which gives the maximum reliability of the system. Once the component in parallel redundancy has been selected it may have similar or different failure rates. Procedure for

The Handbook of Reliability, Maintenance, and System Safety through Mathematical Modeling.
https://doi.org/10.1016/B978-0-12-819582-6.00013-7

determining the reliability of the system with similar or different failure rates is quite different. Such an analysis is presented by Li [5]. In the literature, use of active redundancy and standby redundancy has been observed. In active redundancy, the parallel unit remains operational with the main unit. But in case of standby redundancy parallel unit only then becomes active when the main unit fails. Li [6] evaluated the reliability of the system for both active and standby redundancy. He determined that both redundancies improve the system's reliability and prolong system's uptime. System's reliability also depends upon many other factors such as common cause failure, hardware failure, software failure, temperature, and human error. All these factors have their direct influence on system's reliability and availability. Kumar and Ram [7] analyzed the performance of *2-out-of-3:* F incorporating human error and obtained the explicit expressions for reliability, availability, and mean time to failure (MTTF) of the system. Complex system's reliability incorporating common cause error and two types of repair facility is evaluated by Ram and Singh [8]. Ram and Manglik [9] investigated a system incorporating human error, common cause error, and catastrophic error. Various reliability characteristics such as availability, reliability, and MTTF of the system were evaluated.

In some real-life situation it has been observed that if out of n components k or more components fail then the whole system fails. This type of system is known as a *k-out-of-n*: F system. This system is widely used in industry and has many applications. This system was introduced initially by Birnbaum et al. [10] in 1961. When $k = 1$ then it becomes *1-out-of-n*: F system which is same as a series system and when $k = n$ then it becomes *n-out-of-n*: F system which is same as a parallel system. Very efficient algorithms have been developed by the researcher for the reliability evaluation of the *k-out-of-n* system. These algorithms were given by Refs. [11–13]. Koucky [14] developed a new method and gave an exact reliability formula and bounds for the *k-out-of-n* system when failures of the element are dependent and not identically distributed.

Markov modeling has been used by many researchers [5–7] for finding the various reliability indices. It is a very useful tool for describing the various possible states of the system. Markov model is considered a good model because the transition from one state to another state depends upon on present and future state and it does not depend on the past states. Many researchers used it in determining the various reliability indices. Applications of Markov modeling in sugar mill, communication system, thermal power plant, fertilizer plant, and paper plant can be found in Refs. [15–20].

In this chapter, we introduce warm standby redundancy, parallel redundancy, and *k-out-of-n*: F redundancy in the system and our aim is to determine the explicit expression of system's reliability and MTTF. Also, we perform sensitivity analysis on MTTF and reliability with the help of which we can find the most critical components of the system. The description of the system is given in the following section.

1.1 Model description

The system components are connected in a mixed configuration.

Subsystem A: This system has two components: one main component and another is warm standby. When the main component fails then standby component takes over.

Subsystem B: This subsystem is in a mixed configuration. Two units are working in parallel and the third component is connected with these components in a series configuration.

Subsystem C: This system is working in *2-out-of-3*: F configuration. When any two components of the system fail then the whole system fails (Fig. 13.1).

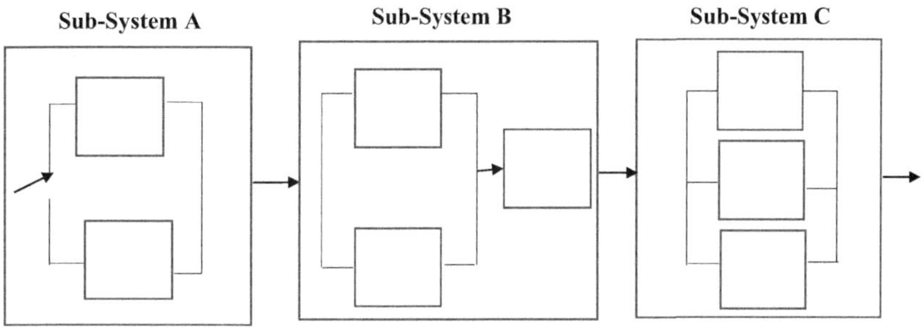

FIGURE 13.1

System configuration.

2. Nomenclature

See Table 13.1

Table 13.1 Nomenclature.

Notations	Description
t	Time variable
S	Laplace transform variable
$P_i(t)$	Probability of the system being in the degrade and fully working state
$\overline{P}_i(s)$	Laplace transformation of the probability of the system being in a degraded state and fully working state
$P_i(x,t)$	Probability of the system being in the failed state at instant t with an elapsed repair time x
$P_i(x,s)$	Laplace transformation of $P_i(x,\ t)$
$\lambda_i : i = 1,2,3$	The failure rate of components of subsystem A, subsystem B, and subsystem C which works with redundant components
$\mu_i : i = 1,2,3$	Repair rate of components of subsystem A, subsystem B, and subsystem C which works with redundant components
λ_{22}	The failure rate of a component of subsystem B which is in series connection with two parallel components
μ_{22}	Repair rate of a component of subsystem B which is in series connection with two parallel components
$\mu_{i,j} : i = 1,2,3$ $j = 1,2,3,22$	Simultaneous repair of two components of the system
$\mu_{i,j,k} : i = 1,2$ $j = 1,2,3$ $k = 2,3,22$	Simultaneous repair of three components of the system
$\mu_{i,j,k,l} : i = 1, j = 2,$ $k = 3, l = 1,2,3,22$	Simultaneous repair of four components of the system

3. State description

See Table 13.2

Table 13.2 Description of various states used.

State	Description
S_0	The system is in fully working condition and no component of the system has failed yet
S_1	The system is in the degraded state due to the failure of the main component of subsystem A
S_2	The system fails due to the failure of the both components of subsystem A
S_3	The system fails due to the failure of component of subsystem B which is in series connection with two parallel components
S_4	The system is in the degraded state due to the failure of one component of the parallel unit of subsystem B
S_5	The system fails due to the failure of both components of the parallel unit of subsystem B
S_6	The system fails due to the failure of the main component of subsystem A and failure of a component of subsystem B which is in series connection with two parallel components
S_7	The system fails due to the failure of one parallel component of subsystem B and failure of a component of subsystem B which is in series connection with two parallel components
S_8	The system is in the degraded state due to the failure of one component of the parallel unit of subsystem C
S_9	The system fails due to the failure of two components of the parallel unit of subsystem C
S_{10}	The system is in the degraded state due to the failure of the main component of subsystem A and one parallel component of subsystem B
S_{11}	The system fails due to the failure of two components of subsystem A and one parallel component of subsystem B
S_{12}	The system fails due to failure of the main component of subsystem A and two parallel components of subsystem B
S_{13}	The system fails due to failure of the main component of subsystem A and one parallel component of subsystem B and one component of subsystem B which is in series connection with two parallel components
S_{14}	The system is in the degraded state due to the failure of the main component of subsystem A and one parallel component of subsystem C
S_{15}	The system fails due to the failure of two components of subsystem A and one parallel component of subsystem C
S_{16}	The system fails due to failure of the main component of subsystem A and two parallel components of subsystem C
S_{17}	The system fails due to failure of main component of subsystem A and one parallel component of subsystem C and one component of subsystem B which is in series connection with two parallel components
S_{18}	The system is in the degraded state due to the failure of one parallel component of subsystem B and one parallel component of subsystem C
S_{19}	The system fails due to the failure of two components of subsystem B and one parallel component of subsystem C
S_{20}	The system fails due to the failure of one parallel component of subsystem B and two parallel components of subsystem C

Table 13.2 Description of various states used.—cont'd

State	Description
S_{21}	The system fails due to failure of one parallel component of subsystem B and one parallel component of subsystem C and one component of subsystem B which is in series connection with two parallel components
S_{22}	The system is in the degraded state due to the failure of the main component of subsystem A and one parallel component of subsystem B and failure of one parallel component of subsystem C
S_{23}	The system fails due to the failure of two components of subsystem A and one parallel component of subsystem B and one parallel component of subsystem C
S_{24}	The system fails due to failure of the main component of subsystem A and two parallel component of subsystem B and one parallel component of subsystem C
S_{25}	The system fails due to the failure of the main component of subsystem A and one parallel component of subsystem B and two parallel components of subsystem C
S_{26}	The system fails due to the failure of main component of subsystem A and one parallel component of subsystem B and one parallel component of subsystem C and one component of subsystem B which is in series connection with two parallel components
S_{27}	The system fails due to the failure of one component of subsystem C and failure of a component of subsystem B which is in series connection with two parallel components

4. Assumption of the system

- Initially, the system is in good working condition.
- Failure rate of the redundant component is the same as that of the main components.
- System repair and failure rates have been taken constant and follow exponential distribution.
- Repair facility is always available with the system.
- After complete failure the system components are repaired or replaced in such a way that after repair or replacement system works as good as a new one.
- Failure of the components of the system are independent of each other.
- When the main component fails the standby unit takes over the main unit without failure of its switching device.
- System continues to work in a degraded state.
- Failure rates of the main component and standby component are the same.

5. State transition diagram

The following Fig. 13.2 describes the various possible states of the considered system in which it can be present at any instant time t.

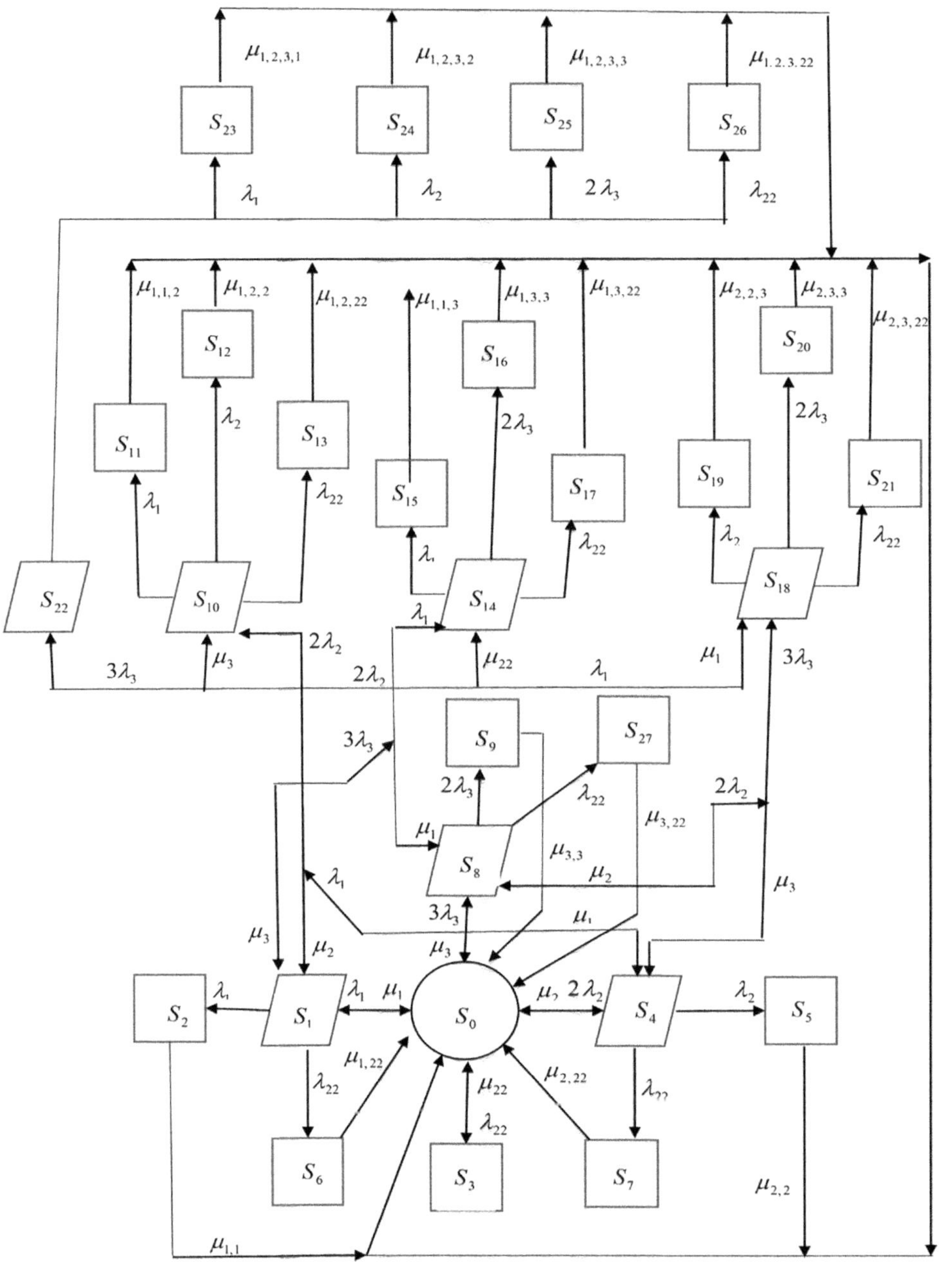

FIGURE 13.2

Transition state diagram.

6. Mathematical formulation and solution of the problem

From the above state transition diagram we develop the differential equations. Through Markov process intro-differential equations are developed in the time interval $(t, t + \Delta t)$, where Δt is very small time; letting $\Delta t \rightarrow 0$we obtain the following equations:

$$\left[\frac{\partial}{\partial t} + \lambda_1 + 2\lambda_2 + 3\lambda_3 + \lambda_{22}\right] P_0(t) = \mu_1 P_1(t) + \mu_2 P_4(t) + \mu_3 P_8(t) + \int_0^\infty \mu_{1,1} P_2(x,t)dx$$
$$+ \int_0^\infty \mu_{2,2} P_5(x,t)dx + \int_0^\infty \mu_{1,22} P_6(x,t)dx + \int_0^\infty \mu_{2,22} P_7(x,t)dx + \int_0^\infty \mu_{3,3} P_9(x,t)dx$$
$$+ \int_0^\infty \mu_{1,2,2} P_{12}(x,t)dx + \int_0^\infty \mu_{1,2,22} P_{13}(x,t)dx + \int_0^\infty \mu_{1,1,3} P_{15}(x,t)dx + \int_0^\infty \mu_{1,3,3} P_{16}(x,t)dx$$
$$+ \int_0^\infty \mu_{2,2,3} P_{19}(x,t)dx + \int_0^\infty \mu_{2,3,3} P_{20}(x,t)dx + \int_0^\infty \mu_{2,3,22} P_{21}(x,t)dx + \int_0^\infty \mu_{1,2,3,1} P_{23}(x,t)dx$$
$$+ \int_0^\infty \mu_{1,2,3,3} P_{25}(x,t)dx + \int_0^\infty \mu_{1,2,3,22} P_{26}(x,t)dx + \int_0^\infty \mu_{3,22} P_{27}(x,t)dx + \int_0^\infty \mu_{1,2,3,2} P_{24}(x,t)dx$$
$$+ \int_0^\infty \mu_{22} P_3(x,t)dx + \int_0^\infty \mu_{1,1,2} P_{11}(x,t)dx + \int_0^\infty \mu_{1,3,22} P_{17}(x,t)dx \tag{13.1}$$

$$\left[\frac{\partial}{\partial t} + \mu_1 + \lambda_1 + 2\lambda_2 + 3\lambda_3\right] P_1(t) = \lambda_1 P_0(t) + \mu_2 P_{10}(t) + \mu_3 P_{14}(t) \tag{13.2}$$

$$\left[\frac{\partial}{\partial t} + \mu_2 + \lambda_2 + 3\lambda_3 + \lambda_1\right] P_4(t) = 2\lambda_2 P_0(t) + \mu_1 P_{10}(t) + \mu_3 P_{18}(t) \tag{13.3}$$

$$\left[\frac{\partial}{\partial t} + \mu_1 + \mu_2 + \lambda_1 + \lambda_2 + \lambda_{22} + 3\lambda_3\right] P_{10}(t) = \mu_3 P_{22}(t) + \lambda_1 P_4(t) + 2\lambda_2 P_1(t) \tag{13.4}$$

$$\left[\frac{\partial}{\partial t} + \mu_1 + \mu_3 + \lambda_1 + 2\lambda_2 + \lambda_{22} + 2\lambda_3\right] P_{14}(t) = \mu_2 P_{22}(t) + \lambda_1 P_8(t) + 3\lambda_3 P_1(t) \tag{13.5}$$

$$\left[\frac{\partial}{\partial t} + \mu_3 + \lambda_1 + 2\lambda_2 + \lambda_{22} + 2\lambda_3\right] P_8(t) = \mu_1 P_{14}(t) + \mu_2 P_{18}(t) + 3\lambda_3 P_0(t) \tag{13.6}$$

$$\left[\frac{\partial}{\partial t} + \mu_2 + \mu_3 + \lambda_1 + \lambda_2 + \lambda_{22} + 2\lambda_3\right] P_{18}(t) = \mu_1 P_{22}(t) + 3\lambda_3 P_4(t) + 2\lambda_2 P_8(t) \tag{13.7}$$

$$\left[\frac{\partial}{\partial t}+\mu_1+\mu_2+\mu_3+\lambda_1+\lambda_2+\lambda_{22}+2\lambda_3\right]P_{22}(t)=3\lambda_3P_{10}(t)+2\lambda_2P_{14}(t)+\lambda_1P_{18}(t) \tag{13.8}$$

$$\left[\frac{\partial}{\partial x}+\frac{\partial}{\partial t}+\mu_{1,1}\right]P_2(x,t)=0 \tag{13.9}$$

$$\left[\frac{\partial}{\partial x}+\frac{\partial}{\partial t}+\mu_{22}\right]P_3(x,t)=0 \tag{13.10}$$

$$\left[\frac{\partial}{\partial x}+\frac{\partial}{\partial t}+\mu_{2,2}\right]P_5(x,t)=0 \tag{13.11}$$

$$\left[\frac{\partial}{\partial x}+\frac{\partial}{\partial t}+\mu_{1,22}\right]P_6(x,t)=0 \tag{13.12}$$

$$\left[\frac{\partial}{\partial x}+\frac{\partial}{\partial t}+\mu_{2,22}\right]P_7(x,t)=0 \tag{13.13}$$

$$\left[\frac{\partial}{\partial x}+\frac{\partial}{\partial t}+\mu_{3,3}\right]P_9(x,t)=0 \tag{13.14}$$

$$\left[\frac{\partial}{\partial x}+\frac{\partial}{\partial t}+\mu_{1,1,2}\right]P_{11}(x,t)=0 \tag{13.15}$$

$$\left[\frac{\partial}{\partial x}+\frac{\partial}{\partial t}+\mu_{1,2,2}\right]P_{12}(x,t)=0 \tag{13.16}$$

$$\left[\frac{\partial}{\partial x}+\frac{\partial}{\partial t}+\mu_{1,2,22}\right]P_{13}(x,t)=0 \tag{13.17}$$

$$\left[\frac{\partial}{\partial x}+\frac{\partial}{\partial t}+\mu_{1,1,3}\right]P_{15}(x,t)=0 \tag{13.18}$$

$$\left[\frac{\partial}{\partial x}+\frac{\partial}{\partial t}+\mu_{1,3,3}\right]P_{16}(x,t)=0 \tag{13.19}$$

$$\left[\frac{\partial}{\partial x}+\frac{\partial}{\partial t}+\mu_{1,3,22}\right]P_{17}(x,t)=0 \tag{13.20}$$

$$\left[\frac{\partial}{\partial x}+\frac{\partial}{\partial t}+\mu_{2,2,3}\right]P_{19}(x,t)=0 \tag{13.21}$$

$$\left[\frac{\partial}{\partial x}+\frac{\partial}{\partial t}+\mu_{2,3,3}\right]P_{20}(x,t)=0 \tag{13.22}$$

$$\left[\frac{\partial}{\partial x}+\frac{\partial}{\partial t}+\mu_{2,3,22}\right]P_{21}(x,t)=0 \tag{13.23}$$

$$\left[\frac{\partial}{\partial x}+\frac{\partial}{\partial t}+\mu_{1,2,3,1}\right]P_{23}(x,t)=0 \quad (13.24)$$

$$\left[\frac{\partial}{\partial x}+\frac{\partial}{\partial t}+\mu_{1,2,3,2}\right]P_{24}(x,t)=0 \quad (13.25)$$

$$\left[\frac{\partial}{\partial x}+\frac{\partial}{\partial t}+\mu_{1,2,3,3}\right]P_{25}(x,t)=0 \quad (13.26)$$

$$\left[\frac{\partial}{\partial x}+\frac{\partial}{\partial t}+\mu_{1,2,3,22}\right]P_{26}(x,t)=0 \quad (13.27)$$

$$\left[\frac{\partial}{\partial x}+\frac{\partial}{\partial t}+\mu_{3,22}\right]P_{27}(x,t)=0 \quad (13.28)$$

With initial conditions,

$$P_2(0,t)=\lambda_1 P_1(t) \quad (13.29)$$

$$P_3(0,t)=\lambda_{22} P_0(t) \quad (13.30)$$

$$P_5(0,t)=\lambda_2 P_4(t) \quad (13.31)$$

$$P_6(0,t)=\lambda_{22} P_1(t) \quad (13.32)$$

$$P_7(0,t)=\lambda_{22} P_4(t) \quad (13.33)$$

$$P_9(0,t)=2\lambda_3 P_8(t) \quad (13.34)$$

$$P_{11}(0,t)=\lambda_1 P_{10}(t) \quad (13.35)$$

$$P_{12}(0,t)=\lambda_2 P_{10}(t) \quad (13.36)$$

$$P_{13}(0,t)=\lambda_{22} P_{10}(t) \quad (13.37)$$

$$P_{15}(0,t)=\lambda_1 P_{14}(t) \quad (13.38)$$

$$P_{16}(0,t)=2\lambda_3 P_{14}(t) \quad (13.39)$$

$$P_{17}(0,t)=\lambda_{22} P_{14}(t) \quad (13.40)$$

$$P_{19}(0,t)=\lambda_2 P_{18}(t) \quad (13.41)$$

$$P_{20}(0,t)=2\lambda_3 P_{18}(t) \quad (13.42)$$

$$P_{21}(0,t)=\lambda_{22} P_{18}(t) \quad (13.43)$$

$$P_{23}(0,t)=\lambda_1 P_{22}(t) \quad (13.44)$$

$$P_{24}(0,t)=\lambda_2 P_{22}(t) \quad (13.45)$$

$$P_{25}(0,t)=2\lambda_3 P_{22}(t) \quad (13.46)$$

$$P_{26}(0,t)=\lambda_{22} P_{22}(t) \quad (13.47)$$

$$P_{27}(0,t)=\lambda_{22} P_8(t) \quad (13.48)$$

Taking Laplace transformation of Eqs. (13.1)–(13.48) we get the following equations:

$$[s+\lambda_1+2\lambda_2+3\lambda_3+\lambda_{22}]\overline{P}_0(s)=\mu_1\overline{P}_1(s)+\mu_2\overline{P}_4(s)+\mu_3\overline{P}_8(s)+\int_0^\infty \mu_{1,1}\overline{P}_2(x,s)dx$$
$$+\int_0^\infty \mu_{2,2}\overline{P}_5(x,s)dx+\int_0^\infty \mu_{1,22}\overline{P}_6(x,s)dx+\int_0^\infty \mu_{2,22}\overline{P}_7(x,s)dx+\int_0^\infty \mu_{3,3}\overline{P}_9(x,s)dx$$
$$+\int_0^\infty \mu_{1,2,2}\overline{P}_{12}(x,s)dx+\int_0^\infty \mu_{1,2,22}\overline{P}_{13}(x,s)dx+\int_0^\infty \mu_{1,1,3}\overline{P}_{15}(x,s)dx+\int_0^\infty \mu_{1,3,3}\overline{P}_{16}(x,s)dx$$
$$+\int_0^\infty \mu_{2,2,3}\overline{P}_{19}(x,s)dx+\int_0^\infty \mu_{2,3,3}\overline{P}_{20}(x,s)dx+\int_0^\infty \mu_{2,3,22}\overline{P}_{21}(x,s)dx+\int_0^\infty \mu_{1,2,3,1}\overline{P}_{23}(x,s)dx$$
$$+\int_0^\infty \mu_{1,2,3,3}\overline{P}_{25}(x,s)dx+\int_0^\infty \mu_{1,2,3,22}\overline{P}_{26}(x,s)dx+\int_0^\infty \mu_{3,22}\overline{P}_{27}(x,s)dx+\int_0^\infty \mu_{1,2,3,2}\overline{P}_{24}(x,s)dx$$
$$+\int_0^\infty \mu_{22}\overline{P}_3(x,s)dx+\int_0^\infty \mu_{1,1,2}\overline{P}_{11}(x,s)dx+\int_0^\infty \mu_{1,3,22}\overline{P}_{17}(x,s)dx \tag{13.49}$$

$$[s+\mu_1+\lambda_1+2\lambda_2+3\lambda_3]\overline{P}_1(s)=\lambda_1\overline{P}_0(s)+\mu_2\overline{P}_{10}(s)+\mu_3\overline{P}_{14}(s) \tag{13.50}$$

$$[s+\mu_2+\lambda_2+3\lambda_3+\lambda_1]\overline{P}_4(s)=2\lambda_2\overline{P}_0(s)+\mu_1\overline{P}_{10}(s)+\mu_3\overline{P}_{18}(s) \tag{13.51}$$

$$[s+\mu_1+\mu_2+\lambda_1+\lambda_2+\lambda_{22}+3\lambda_3]\overline{P}_{10}(s)=\mu_3\overline{P}_{22}(s)+\lambda_1\overline{P}_4(s)+2\lambda_2\overline{P}_1(s) \tag{13.52}$$

$$[s+\mu_1+\mu_3+\lambda_1+2\lambda_2+\lambda_{22}+2\lambda_3]\overline{P}_{14}(s)=\mu_2\overline{P}_{22}(s)+\lambda_1\overline{P}_8(s)+3\lambda_3\overline{P}_1(s) \tag{13.53}$$

$$[s+\mu_3+\lambda_1+2\lambda_2+\lambda_{22}+2\lambda_3]\overline{P}_8(s)=\mu_1\overline{P}_{14}(s)+\mu_2\overline{P}_{18}(s)+3\lambda_3\overline{P}_0(s) \tag{13.54}$$

$$[s+\mu_2+\mu_3+\lambda_1+\lambda_2+\lambda_{22}+2\lambda_3]\overline{P}_{18}(s)=\mu_1\overline{P}_{22}(s)+3\lambda_3\overline{P}_4(s)+2\lambda_2\overline{P}_8(s) \tag{13.55}$$

$$[s+\mu_1+\mu_2+\mu_3+\lambda_1+\lambda_2+\lambda_{22}+2\lambda_3]\overline{P}_{22}(s)=3\lambda_3\overline{P}_{10}(s)+2\lambda_2\overline{P}_{14}(s)+\lambda_1\overline{P}_{18}(s) \tag{13.56}$$

$$\left[\frac{\partial}{\partial x}+s+\mu_{1,1}\right]\overline{P}_2(x,s)=0 \tag{13.57}$$

$$\left[\frac{\partial}{\partial x}+s+\mu_{22}\right]\overline{P}_3(x,s)=0 \tag{13.58}$$

$$\left[\frac{\partial}{\partial x}+s+\mu_{2,2}\right]\overline{P}_5(x,s)=0 \tag{13.59}$$

$$\left[\frac{\partial}{\partial x}+s+\mu_{1,22}\right]\overline{P}_6(x,s)=0 \tag{13.60}$$

$$\left[\frac{\partial}{\partial x}+s+\mu_{2,22}\right]\overline{P}_7(x,s)=0 \tag{13.61}$$

$$\left[\frac{\partial}{\partial x}+s+\mu_{3,3}\right]\overline{P}_9(x,s)=0 \tag{13.62}$$

$$\left[\frac{\partial}{\partial x}+s+\mu_{1,1,2}\right]\overline{P}_{11}(x,s)=0 \tag{13.63}$$

$$\left[\frac{\partial}{\partial x}+s+\mu_{1,2,2}\right]\overline{P}_{12}(x,s)=0 \tag{13.64}$$

$$\left[\frac{\partial}{\partial x}+s+\mu_{1,2,22}\right]\overline{P}_{13}(x,s)=0 \tag{13.65}$$

$$\left[\frac{\partial}{\partial x}+s+\mu_{1,1,3}\right]\overline{P}_{15}(x,s)=0 \tag{13.66}$$

$$\left[\frac{\partial}{\partial x}+s+\mu_{1,3,3}\right]\overline{P}_{16}(x,s)=0 \tag{13.67}$$

$$\left[\frac{\partial}{\partial x}+s+\mu_{1,3,22}\right]\overline{P}_{17}(x,s)=0 \tag{13.68}$$

$$\left[\frac{\partial}{\partial x}+s+\mu_{2,2,3}\right]\overline{P}_{19}(x,s)=0 \tag{13.69}$$

$$\left[\frac{\partial}{\partial x}+s+\mu_{2,3,3}\right]\overline{P}_{20}(x,s)=0 \tag{13.70}$$

$$\left[\frac{\partial}{\partial x}+s+\mu_{2,3,22}\right]\overline{P}_{21}(x,s)=0 \tag{13.71}$$

$$\left[\frac{\partial}{\partial x}+s+\mu_{1,2,3,1}\right]\overline{P}_{23}(x,s)=0 \tag{13.72}$$

$$\left[\frac{\partial}{\partial x}+s+\mu_{1,2,3,2}\right]\overline{P}_{24}(x,s)=0 \tag{13.73}$$

$$\left[\frac{\partial}{\partial x}+s+\mu_{1,2,3,3}\right]\overline{P}_{25}(x,s)=0 \tag{13.74}$$

$$\left[\frac{\partial}{\partial x}+s+\mu_{1,2,3,22}\right]\overline{P}_{26}(x,s)=0 \tag{13.75}$$

$$\left[\frac{\partial}{\partial x}+s+\mu_{3,22}\right]\overline{P}_{27}(x,s)=0 \tag{13.76}$$

With initial conditions,

$$\overline{P}_2(0,s)=\lambda_1\overline{P}_1(s) \tag{13.77}$$

$$\overline{P}_3(0,s)=\lambda_{22}\overline{P}_0(s) \tag{13.78}$$

$$\overline{P}_5(0,s)=\lambda_2\overline{P}_4(s) \tag{13.79}$$

$$\overline{P}_6(0,s)=\lambda_{22}\overline{P}_1(s) \tag{13.80}$$

$$\overline{P}_7(0,s)=\lambda_{22}\overline{P}_4(s) \tag{13.81}$$

$$\overline{P}_9(0,s)=2\lambda_3\overline{P}_8(s) \tag{13.82}$$

$$\overline{P}_{11}(0,s)=\lambda_1\overline{P}_{10}(s) \tag{13.83}$$

$$\overline{P}_{12}(0,s)=\lambda_2\overline{P}_{10}(s) \tag{13.84}$$

$$\overline{P}_{13}(0,s)=\lambda_{22}\overline{P}_{10}(s) \tag{13.85}$$

$$\overline{P}_{15}(0,s)=\lambda_1\overline{P}_{14}(s) \tag{13.86}$$

$$\overline{P}_{16}(0,s)=2\lambda_3\overline{P}_{14}(s) \tag{13.87}$$

$$\overline{P}_{17}(0,s)=\lambda_{22}\overline{P}_{14}(s) \tag{13.88}$$

$$\overline{P}_{19}(0,s)=\lambda_2\overline{P}_{18}(s) \tag{13.89}$$

$$\overline{P}_{20}(0,s)=2\lambda_3\overline{P}_{18}(s) \tag{13.90}$$

$$\overline{P}_{21}(0,s)=\lambda_{22}P_{18}(s) \tag{13.91}$$

$$\overline{P}_{23}(0,s)=\lambda_1\overline{P}_{22}(s) \tag{13.92}$$

$$\overline{P}_{24}(0,s)=\lambda_2\overline{P}_{22}(s) \tag{13.93}$$

$$\overline{P}_{25}(0,s)=2\lambda_3\overline{P}_{22}(s) \tag{13.94}$$

$$\overline{P}_{26}(0,s)=\lambda_{22}\overline{P}_{22}(s) \tag{13.95}$$

$$\overline{P}_{27}(0,s)=\lambda_{22}\overline{P}_8(s) \tag{13.96}$$

On solving Eqs. (13.49)–(13.96), one can get the state transition probabilities. As there are many states of the system, the system is working in some states and in other states system is not working.

System's upstate is the probability that the system is in working/degraded state and system's downstate is the probability that the system is not working and it is the sum of all the probabilities of the failed states. These are given as following Eqs. (13.97) and (13.98), respectively.

$$P_{up}(t)=P_0(t)+P_1(t)+P_4(t)+P_8(t)+P_{10}(t)+P_{14}(t)+P_{18}(t)+P_{22}(t) \tag{13.97}$$

$$P_{down}(t)=P_2(t)+P_3(t)+P_5(t)+P_6(t)+P_7(t)+P_9(t)+P_{11}(t)+P_{12}(t)+P_{13}(t)+P_{15}(t)$$

$$+P_{16}(t)+P_{17}(t)+P_{19}(t)+P_{20}(t)+P_{21}(t)+P_{23}(t)+P_{24}(t)+P_{25}(t)+P_{26}(t)+P_{27}(t) \tag{13.98}$$

Now solving the above set of equations with initial and boundary condition, the various state probabilities of the considered system in terms of failures rates are obtained as follows.

$$\overline{P}_0(s)=\frac{1}{s+\lambda_1+2\lambda_2+3\lambda_3+\lambda_{22}} \tag{13.99}$$

$$\overline{P}_1(s)=\frac{\lambda_1}{H_1}\overline{P}_0(s) \tag{13.100}$$

$$\overline{P}_4(s) = \frac{2\lambda_2}{H_2}\overline{P}_0(s) \tag{13.101}$$

$$\overline{P}_8(s) = \frac{3\lambda_3}{H_3}\overline{P}_0(s) \tag{13.102}$$

$$\overline{P}_{10}(s) = \left[\frac{2\lambda_1\lambda_2}{H_2H_4} + \frac{2\lambda_1\lambda_2}{H_1H_4}\right]\overline{P}_0(s) \tag{13.103}$$

$$\overline{P}_{14}(s) = \left[\frac{3\lambda_1\lambda_3}{H_3H_5} + \frac{3\lambda_1\lambda_3}{H_5H_1}\right]\overline{P}_0(s) \tag{14.104}$$

$$\overline{P}_{18}(s) = \left[\frac{6\lambda_2\lambda_3}{H_6H_3} + \frac{6\lambda_2\lambda_3}{H_6H_2}\right]\overline{P}_0(s) \tag{13.105}$$

$$\overline{P}_{22}(s) = \left[\frac{6\lambda_1\lambda_2\lambda_3}{H_2H_4H_7} + \frac{6\lambda_1\lambda_2\lambda_3}{H_1H_4H_7} + \frac{6\lambda_1\lambda_2\lambda_3}{H_3H_5H_7} + \frac{6\lambda_1\lambda_2\lambda_3}{H_1H_5H_7} + \frac{6\lambda_1\lambda_2\lambda_3}{H_3H_6H_7} + \frac{6\lambda_1\lambda_2\lambda_3}{H_2H_6H_7}\right]\overline{P}_0(s) \tag{13.106}$$

where $H_1 = s + \lambda_1 + 2\lambda_2 + 3\lambda_3$

$$H_2 = s + \lambda_1 + \lambda_2 + 3\lambda_3$$

$$H_3 = s + \lambda_1 + 2\lambda_2 + 2\lambda_3 + \lambda_{22}$$

$$H_4 = s + \lambda_1 + \lambda_2 + 3\lambda_3 + \lambda_{22}$$

$$H_5 = s + \lambda_1 + 2\lambda_2 + 2\lambda_3 + \lambda_{22}$$

$$H_6 = s + \lambda_1 + \lambda_2 + 2\lambda_3 + \lambda_{22}$$

$$H_7 = s + \lambda_1 + \lambda_2 + 2\lambda_3 + \lambda_{22}$$

Substitute the values of these state probabilities (Eqs. (13.99)–(13.106)) in (13.97), and then use inverse Laplace transformation in the obtained equation. Authors obtain the explicit expression of the reliability of the system which is given in the following section.

7. Reliability indices

7.1 Reliability of the system

Reliability of a product or system is the probability that product/system performs its intended task for a specified period of time under random operating conditions. The expression for the reliability of the system is denoted by$R$ (t). The reliability of the present system is obtained as given in Eq. (13.107). (Taking various failure rates as $\lambda_1 = 0.03, \lambda_2 = 0.05, \lambda_3 = 0.07, \lambda_{22} = 0.02$)

$$R(t) = \left\{ \begin{array}{l} 5.500000000e^{(-.3400000000t)} + 2.8214228571e^{(-.3600000000t)} \\ +(6.450000000 + .2100000000t)e^{(-.2400000000t)} + 2.e^{(-.3100000000t)} + 3.e^{(-.3500000000t)} \\ \sinh(0.0100000000t) + 8.857142857e^{(-.3250000000t)}\sinh(0.0350000000t) \\ +(-0.0900000000t - 15.77142857)e^{(-.2900000000t)} \end{array} \right\} \tag{13.107}$$

In Eq. (13.107), on varying time t, we obtain the following Table 13.3 and Fig. 13.3.

Table 13.3 Reliability.

Time (t)	Reliability $R(t)$
0	1
1	0.96565
2	0.90921
3	0.83908
4	0.76183
5	0.68237
6	0.60424
7	0.52983
8	0.46064
9	0.39752
10	0.34080

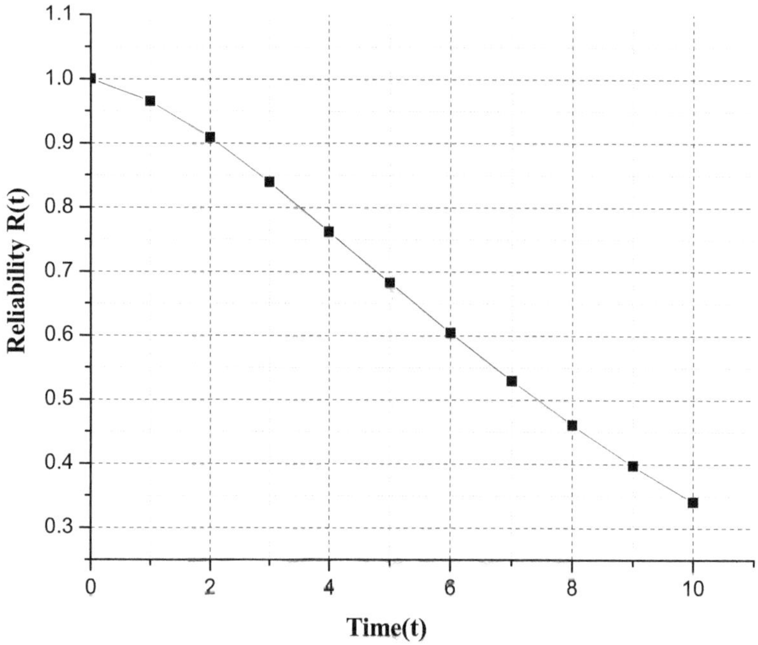

FIGURE 13.3

Reliability versus Time unit (t).

7.2 Mean time to failure

Mean time to system failure is a metric which is used for nonrepairable items. It is the expected time to the first failure of the system. Mathematically MTTF is expressed as it is given below:

$$\text{MTTF} = \int_0^\infty R(t)dt = \lim_{s \to 0} R(s) \tag{13.108}$$

The explicit expression for the system's MTTF is given below:

$$
\mathrm{MTTF}=\left\{\begin{array}{l}
\dfrac{1}{\lambda_1+2\lambda_2+3\lambda_3+\lambda_{22}}+\dfrac{\lambda_1}{(\lambda_1+2\lambda_2+3\lambda_3)(\lambda_1+2\lambda_2+3\lambda_3+\lambda_{22})}+\\[2ex]
\dfrac{2\lambda_2}{(\lambda_2+3\lambda_3+\lambda_1)(\lambda_1+2\lambda_2+3\lambda_3+\lambda_{22})}+\dfrac{3\lambda_3}{(2\lambda_2+\lambda_1+2\lambda_3+\lambda_{22})(\lambda_1+2\lambda_2+3\lambda_3+\lambda_{22})}\\[2ex]
+\dfrac{1}{\lambda_1+2\lambda_2+3\lambda_3+\lambda_{22}}\left(\begin{array}{c}\dfrac{2\lambda_1\lambda_2}{(\lambda_2+3\lambda_3+\lambda_1)(3\lambda_3+\lambda_1+\lambda_2+\lambda_{22})}+\\[2ex]\dfrac{2\lambda_1\lambda_2}{(\lambda_1+2\lambda_2+3\lambda_3)(3\lambda_3+\lambda_1+\lambda_2+\lambda_{22})}\end{array}\right)\\[2ex]
+\dfrac{\dfrac{3\lambda_1\lambda_3}{(2\lambda_2+\lambda_1+2\lambda_3+\lambda_{22})^2}+\dfrac{3\lambda_1\lambda_3}{(\lambda_1+2\lambda_2+3\lambda_3)(2\lambda_2+\lambda_1+2\lambda_3+\lambda_{22})}}{\lambda_1+2\lambda_2+3\lambda_3+\lambda_{22}}+\dfrac{1}{\lambda_1+2\lambda_2+3\lambda_3+\lambda_{22}}\\[2ex]
\left(\dfrac{6\lambda_2\lambda_3}{(\lambda_{22}+\lambda_1+\lambda_2+2\lambda_3)(2\lambda_2+\lambda_1+2\lambda_3+\lambda_{22})}+\dfrac{6\lambda_2\lambda_3}{(\lambda_{22}+\lambda_1+\lambda_2+2\lambda_3)(\lambda_2+3\lambda_3+\lambda_1)}\right)\\[2ex]
+\left(\dfrac{1}{\lambda_1+2\lambda_2+3\lambda_3+\lambda_{22}}\left(\begin{array}{l}
\dfrac{6\lambda_1\lambda_2\lambda_3}{(\lambda_{22}+\lambda_1+\lambda_2+2\lambda_3)(\lambda_2+3\lambda_3+\lambda_1)(3\lambda_3+\lambda_1+\lambda_2+\lambda_{22})}\\[2ex]
+\dfrac{6\lambda_1\lambda_2\lambda_3}{(\lambda_1+2\lambda_2+3\lambda_3)(3\lambda_3+\lambda_1+\lambda_2+\lambda_{22})(\lambda_{22}+\lambda_1+\lambda_2+2\lambda_3)}\\[2ex]
+\dfrac{6\lambda_1\lambda_2\lambda_3}{(2\lambda_2+\lambda_1+2\lambda_3+\lambda_{22})^2(\lambda_{22}+\lambda_1+\lambda_2+2\lambda_3)}+\\[2ex]
\dfrac{6\lambda_1\lambda_2\lambda_3}{(\lambda_1+2\lambda_2+3\lambda_3)(2\lambda_2+\lambda_1+2\lambda_3+\lambda_{22})(\lambda_{22}+\lambda_1+\lambda_2+2\lambda_3)}+\\[2ex]
\dfrac{6\lambda_1\lambda_2\lambda_3}{(2\lambda_2+\lambda_1+2\lambda_3+\lambda_{22})(\lambda_{22}+\lambda_1+\lambda_2+2\lambda_3)^2}+\\[2ex]
\dfrac{6\lambda_1\lambda_2\lambda_3}{(\lambda_2+3\lambda_3+\lambda_1)(\lambda_{22}+\lambda_1+\lambda_2+2\lambda_3)^2}
\end{array}\right)\right)
\end{array}\right\}
\tag{13.109}
$$

Now take particular values to the failure rate of components as $\lambda_1=0.03,\lambda_2=0.05,\lambda_3=0.07,$ $\lambda_{22}=0.02$in Eq. (13.109) and then vary failure rates, we obtain the following Table 13.4 and Fig. 13.4 for the MTTF of the considered system.

Table 13.4 Mean time to failure versus variation in failure rates.

Variation in failure rates	λ_1	λ_2	λ_3	λ_{22}
0.01	8.94672	9.73855	18.34031	9.25733
0.02	8.86545	9.56810	15.88974	8.74627
0.03	8.74627	9.32553	13.81623	8.28585
0.04	8.60072	9.04470	12.13551	7.86914
0.05	8.43732	8.74627	10.77552	7.49041
0.06	8.26238	8.44296	9.66471	7.14484
0.07	8.08059	8.14256	8.74627	6.82839
0.08	7.89541	7.84978	7.97739	6.53764
0.09	7.70941	7.56736	7.32615	6.26966

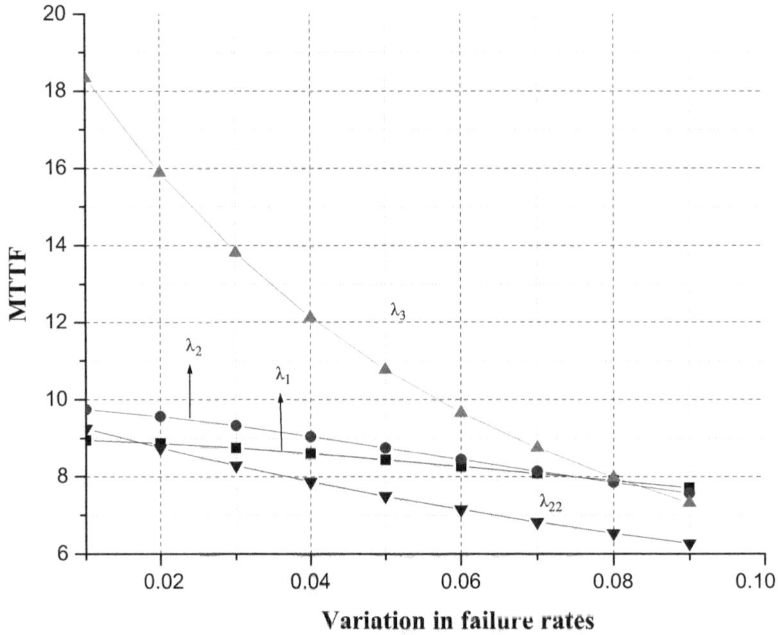

FIGURE 13.4

MTTF versus Variation in Failure rates.

7.3 Sensitivity analysis

Sensitivity analysis is a technique which helps to determine how the output of a function is affected by changing the input variables. Input variables are varied in a specified range. Another name of sensitivity analysis is what-if analysis. This analysis is very helpful in determining the most critical component of the system which affects the system's performance. Here we perform the sensitivity analysis on system's MTTF and system's reliability. It determines which failure rate affects the system's MTTF more and which failure affects the reliability of the system more. Table 13.5 and Fig. 13.5 give the sensitivity of system's MTTF and Table 13.6 and Fig. 13.6 give the sensitivity of system's reliability.

7.3.1 Sensitivity of MTTF

See Table 13.5.

Table 13.5 Sensitivity of MTTF.

Variation in failure rates	λ_1	λ_2	λ_3	λ_{22}
0.01	−5.74890	−11.81563	−255.09174	−53.90497
0.02	−10.24526	−21.36207	−228.26828	−48.44768
0.03	−13.40095	−26.60193	−186.70428	−43.75162
0.04	−15.56898	−29.22824	−150.74130	−39.68474
0.05	−17.00663	−30.25021	−122.46307	−36.14199
0.06	−17.90294	−30.28513	−100.64245	−33.03898
0.07	−18.39785	−29.71818	−83.75401	−30.30739
0.08	−18.59557	−28.79336	−70.54873	−27.89146
0.09	−18.57414	−27.66675	−60.09452	−25.74534

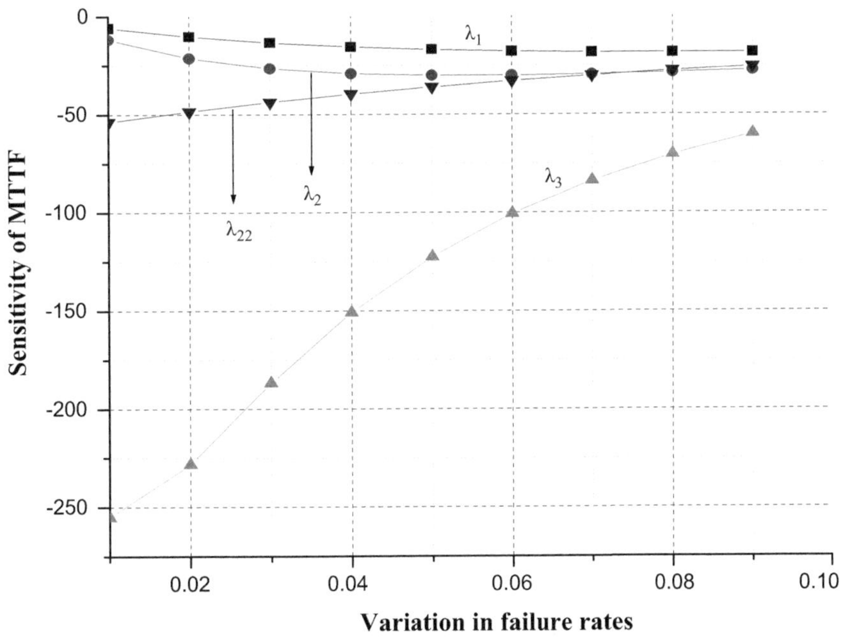

FIGURE 13.5

Sensitivity of system's MTTF versus Variation in Failure rates.

7.3.2 Sensitivity of reliability

See Table 13.6.

Table 13.6 Sensitivity of reliability versus Time unit (*t*).

Time(*t*)	λ_1	λ_2	λ_3	λ_{22}
1	−0.02106	−0.07534	−0.34675	−0.91433
2	−0.08301	−0.27390	−1.13921	−1.65654
3	−0.17632	−0.54623	−2.09584	−2.23037
4	−0.28720	−0.84473	−3.03450	−2.64608
5	−0.40210	−1.13163	−3.84806	−2.91914
6	−0.50998	−1.38110	−4.48318	−3.06835
7	−0.60303	−1.57833	−4.92329	−3.11396
8	−0.67661	−1.71742	−5.17525	−3.07619
9	−0.72880	−1.79899	−5.25960	−2.97411
10	−0.75972	−1.82794	−5.20359	−2.82490

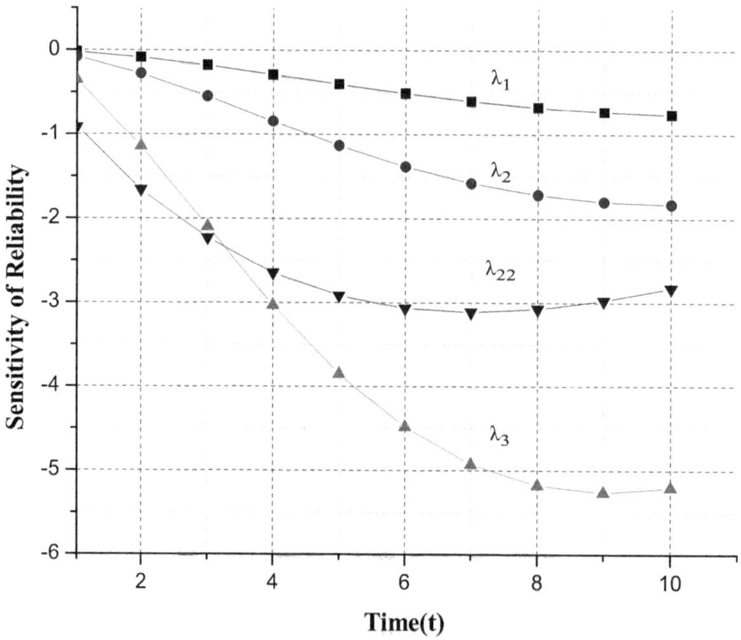

FIGURE 13.6

Sensitivity of System's reliability versus Time unit (*t*).

8. Result discussion

In the above section, after carefully examining the tables and graphs following results are obtained:

- ➢ From Table 13.3 and Fig. 13.3, we observe that the reliability of the system decreases as time increases. At a 10 unit of time the reliability of the considered system is observed as 0.34080.
- ➢ From Table 13.4 and Fig. 13.4, we observe that MTTF of subsystem C is more compared to components of subsystem A and subsystem B. But MTTF of subsystem C decreases very rapidly as compared to other components of the subsystem.
- ➢ From Table 13.5 and Fig. 13.5, we observe that system's MTTF is very sensitive with respect to the failure rate of the components of subsystem C.
- ➢ From Table 13.6 and Fig. 13.6, we observe that system's reliability is very sensitive for the failure rate of components of subsystem C.

9. Conclusion

In this chapter, we developed a model incorporating *2-out-of-3*: F and warm standby redundancy at the subsystem level in the system. With the help of Markov modeling, Chapman−Kolmogorov differential equations were developed and solved with the help of Laplace transformation. Explicit expressions of reliability and MTTF were obtained. Sensitivity analysis of MTTF and reliability were performed and critical components of the system were found. It was found after performing sensitivity analysis that failure rate of subsystem C affects the system's performance most. Hence the reliability of the conserved system will be increased if the more attention is given to the failure of this subsystem.

References

[1] L.S. Srinath, Reliability Engineering, 3rd edition, East-West Press Pvt. Ltd., New Delhi, India, 1994.

[2] E. Balagurusamy, Reliability Engineering, Tata McGraw-Hill Education, 1984.

[3] C.E. Ebeling, An Introduction to Reliability and Maintainability Engineering, Tata McGraw-Hill Education, 2004.

[4] K. Shen, M. Xie, On the increase of system reliability by parallel redundancy, IEEE Trans. Reliab. 39 (5) (1990) 607−611.

[5] J. Li, Reliability calculation of a parallel redundant system with different failure rate & repair rate using Markov modeling, J. Reliab. Stat. Stud. 9 (1) (2016) 1−10.

[6] J. Li, Reliability comparative evaluation of active redundancy vs. standby redundancy, Int. J.Math. Eng. Manag. Sci. 1 (3) (2016) 122−129.

[7] R. Mangey, A. Kumar, Performance of a structure consisting a 2-out-of-3:F substructure under human error, Arabian J. Sci. Eng. 39 (11) (2014) 8383−8394.

[8] M. Ram, S.B. Singh, Analysis of a complex system with common cause failure and two types of repair facilities with different distributions in failure, Int. J. Reliab. Saf. 4 (4) (2010) 381−392.

[9] M. Ram, M. Manglik, Stochastic behavior analysis of a Markov model under multi-state failures, Int. J. Syst. Assur. Eng. Manag. 5 (4) (2014) 686−699.

[10] Z.W. Birnbaum, J.D. Esary, S.C. Saunders, Multicomponent systems and structures and their reliability, Technometrics 3 (1) (1961) 55−77.

[11] A.M. Rushdi, Utilization of symmetric switching functions in the computation of k-out-of-n system reliability, Microelectron. Reliab. 26 (5) (1986) 973–987.
[12] R.E. Barlow, K.D. Heidtmann, Computing k-out-of-n system reliability, IEEE Trans. Reliab. 33 (4) (1984) 322–323.
[13] A.K. Sarje, On the reliability computation of a k-out-of-n system, Microelectron. Reliab. 33 (2) (1993) 267–269.
[14] M. Koucký, Exact reliability formula and bounds for general k-out-of-n systems, Reliab. Eng. Syst. Saf. 82 (2) (2003) 229–231.
[15] S.P. Sharma, Y. Vishwakarma, Application of Markov process in performance analysis of feeding system of sugar industry, J. Ind. Math. (2014) 1–9, 2014, Article ID 593176.
[16] A. Kumar, P. Kumar, Application of Markov process/mathematical modelling in analysing communication system reliability, Int. J. Qual. Reliab. Manag. (2019) 1–18.
[17] S. Gupta, P.C. Tewari, A.K. Sharma, Reliability and availability analysis of the ash handling unit of a steam thermal power plant, South Afr. J. Ind. Eng. 20 (1) (2009) 147–158.
[18] S. Gupta, P.C. Tewari, Markov approach for predictive modeling and performance evaluation of a thermal power plant, Int. J. Reliab. Qual. Saf. Eng. 17 (1) (2010) 41–55.
[19] A.K. Aggarwal, S. Kumar, V. Singh, T.K. Garg, Markov modeling and reliability analysis of urea synthesis system of a fertilizer plant, J. Ind. Eng. Int. 11 (1) (2015) 1–14.
[20] D. Kumar, Mathematical modeling and performance optimization for the digesting system of a paper plant, Int. J. Eng. 23 (3) (2010) 215–226.

Index

Note: 'Page numbers followed by "f" indicate figures and "t" indicate tables.'

T

U

V

W

Y

9780128195826